Heparin-Binding Proteins

Heparin-Binding
PROTEINS

H. Edward Conrad

Department of Biochemistry
School of Chemical Sciences
University of Illinois at Urbana–Campaign
Urbana, Illinois

Academic Press

San Diego   London   Boston   New York   Sydney   Tokyo   Toronto

Cover photograph: Figure 9.4. Model of the FGF-2/heparin complex.
Reprinted from [149], with permission.

This book is printed on acid-free paper. ∞

Academic Press
a division of Harcourt Brace & Company
525 B Street, Suite 1900, San Diego, California 92101-4495, USA
http://www.apnet.com

Academic Press Limited
24-28 Oval Road, London NW1 7DX, UK
http://www.hbuk.co.uk/ap/

Library of Congress Card Catalog Number: 97 074397

International Standard Book Number: 0-12-186060-4

Printed and bound in the United Kingdom
Transferred to Digital Printing, 2011

Table of Contents

Preface

Early in my career, I recognized that the methodology in carbohydrate structural analysis lagged well behind the methodologies in the protein and nucleic acid areas, and our early work was in the development of approaches for unraveling the structures of bacterial polysaccharides. Two types of monosaccharides that complicated structural studies of polysaccharides were the uronic acids and the amino sugars. Thus, I entered the field of heparin chemistry, and later biochemistry and cell biology, because heparin represented the ultimate challenge. If it had not been for our interest in the general problem of analysis of carbohydrate structures, we would have looked elsewhere, because, sad-to-recall, "mucopolysaccharides" seemed to me at the time to be rather uninteresting. The field was small and was populated with several centers of research, but was relatively quiet—a good area in which a young professor might seek tenure. Things have changed! Thanks to the persistence of the early workers,[1] there are many more centers of activity in this field now, among them the offspring of the early centers, and the grandchildren of these centers, all of whom earned their union cards during their initial research experience in the laboratories of their mentors. In addition, many, like myself, entered through the back door, without the proper prior training. Heparin, and more importantly heparan sulfate, are now a part of the center stage of research in biology. Almost weekly, a new biologist joins the field, probably reluctantly, as I did. The learning curve in this field is quite foreboding, and the literature is spread widely in the areas of chemistry, biochemistry, cell biology, and physiology. Clinical researchers represent perhaps half of all workers in the field. The neophyte, who by fortune (or misfortune) enters the heparinoid area, may quickly scan the literature and be tempted to turn back. However, the rewards that await the newcomers, as well as those who have been trying to fit the pieces of heparinoid structure and biology together for many years, promise to be great, because there is much virgin territory here. This book seeks to counter the temptation of newcomers to turn back, and indeed to lure them into the field. The emphasis here is on heparin and heparan sulfate and the proteins to which they bind. It also seeks to describe these carbohydrate/protein interactions in the context of their physiological settings. The array of heparin-binding proteins is broad and the author cannot pretend to be an expert in the many fields of biology where heparin-binding proteins play prominent roles. Thus, the presentations in these areas are derived primarily from the published literature, without any of the insight that comes from personal involvement. Hopefully,

---

[1] Excellent reviews of the history of heparin and heparan sulfate have been presented by Rodén [1] and by Marcum [2].

those who work in these areas will forgive the shortcomings that result from an outsider's view and, in fact, will offer suggestions for improving the discussions.

The chapters that follow are not intended to be comprehensive reviews of the status of these fields, although that aspect must be present in order to place the biological roles of heparan sulfate into proper perspectives. Several of the fields addressed here—especially hemostasis, growth factors, and lipoprotein metabolism—are moving so rapidly that they will certainly have advanced between the completion of this book and its publication. However, the concepts of heparan sulfate roles in cells and tissues have now clearly emerged, and they may be more enduring.

The premise of this book is that the learning curve can be eased, and the book is intended to smooth the path for the newcomer. Thus, the book is intended as a teaching monograph rather than a summary of all that is known. The now vast, and sometimes confusing, literature in this infant field of heparan sulfate cell biology cannot be totally encompassed in a single monograph, but the reader can be pointed to its locations and introduced to its themes. While the book is not intended primarily for those that are already converted, the latter may also find some value here.

During the course of this writing, several chapters were critiqued by Professors Ana Jonas and Bob Rosenberg. I offer my special thanks for their comments and encouragement. I would also like to express my thanks to Dr. Barbara Mulloy for many helpful comments on the three-dimensional structures of heparinoids.

H. Edward Conrad

Conventions, Abbreviations, and Terminology

Conventions

Upstream/Downstream Positions — The terms "upstream" and "downstream" are used to designate the relative positions of units in a polymer. In a polysaccharide the term, "upstream" will mean "on the nonreducing terminal side of the reference point." In the case of proteins, it will mean "on the C-terminal side of the reference point." The term "downstream" will mean "on the reducing terminal side of the reference point" in a carbohydrate polymer or "on the N-terminal side of the reference point" in a protein. See Figure 2.1.

Covalent/Non-covalent Complexes — For the description of various complexes between heparinoids and proteins, or between two proteins, the system that is adopted throughout this treatise is that *noncovalent complexes* will be designated by a slash between the two components of a complex (e.g., heparin/antithrombin), whereas *covalent complexes* will be designated by a dash between the two components of the complex (e.g., thrombin-antithrombin).

P1-P1' — The amino acids forming the peptde bond at the site of protease cleavage are designated P1-P1', with the surrounding region designated ...P3-P2-P1-P1'-P2'-P3'..., etc. The sequence shown proceeds from the N- to the C-terminal.

Abbreviations and Terminology

α_2M*	activated α_2-macroglobulin, generally containing an entrapped protease
α_2MR	α_2-macroglobulin receptor, same as LRP or LRP/α_2MR
α_2MRAP	α_2-macroglobulin receptor-associated protein
Aglycon	substituent attached to a monosaccharide at its downstream end
AMan	2,5-anhydro-D-mannose
AMan$_R$	aldehyde-reduced 2,5-anhydro-D-mannose
Amphiglycan	a heparan sulfate proteoglycan core protein
APC	activated protein C (same as Ca)
ApoA, apoB, apoC, apoE	apolipoproteins found in lipoproteins
AT	antithrombin (same as antithrombin III)
ATIII	antithrombin III (same as antithrombin)
β-VLDL	β-migrating VLDL, a mixture of abnormal cholesterol ester- and apoE-rich remnant-like particles that accumulate in animals fed diets high in saturated fat and cholesterol or in the plasma of patients with type III (apoE-defective) hyperlipoproteinemia. β-VLDL is used as a lipoprotein ligand in studies on the receptor role of LRP/α_2MR.
Betaglycan	a heparan sulfate proteoglycan core protein, also referred to as β-glycan or βetaglycan
bFGF	basic fibroblast growth factor
Ca	activated protein C (same as APC)
CD44E	a heparan sulfate proteoglycan core protein, also called Hermes antigen, Pgp-1, ECMR III, or epican
CE	cholesterol esters

Chondroitin sulfate	a glycosaminoglycan composed of repeating [βGlcA-1$\rightarrow$3-βGalNAc] disaccharides joined by 1$\rightarrow$4 linkages.
CS	chondroitin sulfate
CuZn-SOD	copper-zinc superoxide dismutase
Deaminative cleavage	cleavage of glycosidic linkages of hexosamines with nitrous acid; the amino group is converted to N_2 and the pyranose ring is contracted to a 2,5-anhydrosugar; see Figure 3.2
Dermatan sulfate	a glycosaminoglycan composed of repeating [HexA(either βGlcA or αIdoA)-1$\rightarrow$3-βGalNAc] disaccharides joined by 1$\rightarrow$4 linkages
Downstream	on the reducing terminal side of the reference point or, in the case of proteins, on the C-terminal side of the reference point
DS	dermatan sulfate
E. coli K5 polysaccharide	a polysaccharide composed of repeating [βGlcA $\rightarrow$ αGlcNAc] disaccharide units in which all of the monosaccharides are connected by 1$\rightarrow$4 linkages
EPI	extrinsic pathway inhibitor (same as LACI and TFPI)
Epican	a heparan sulfate proteoglycan core protein, also called CD44E, Hermes antigen, Pgp-1, or ECMR III
f.a.b.-m.s.	fast atom bombardment-mass spectrometry
FGF	fibroblast growth factor
FGFR	fibroblast growth factor receptor
GAG	glycosaminoglycan
Gal	D-galactose
GalN	D-galactosamine
GalNAc	N-acetyl-D-galactosamine

Glc   D-glucose

GlcA   D-glucuronic acid

GlcA-2-SO$_4$ D-glucuronic acid-2-sulfate

GlcN   D-glucosamine

GlcNAc   N-acetyl-D-glucosamine

GlcNR   D-glucosamine, substituted on the amino
                                   group with an R group that may be either a
                                   sulfate or an acyl (usually acetyl) group

GlcNS   N-sulfo-D-glucosamine

GlcNSO$_3$-3,6-di-SO$_4$ N-sulfo-D-glucosamine-3,6-disulfate

GlcNSO$_3$-3-SO$_4$ N-sulfo-D-glucosamine-3-sulfate

GlcNSO$_3$-6-SO$_4$ N-sulfo-D-glucosamine-6-sulfate

Glycosaminoglycan   a polysaccharide composed of repeating
                                   uronic acid→hexosamine disaccharide units

Glypican   a heparan sulfate proteoglycan core protein
                                   that has a GPI anchor at its C-terminal end

GPI anchor   a glycosylphosphatidylinositol unit attached
                                   to the C-terminal amino acid of some
                                   proteins

HA-heparin   high activity heparin; the fraction of heparin
                                   that binds to antithrombin affinity columns

HCII   heparin cofactor II

HDL   high density lipoprotein

Heparinoid   a generic term that includes heparin, heparan
                                   sulfate, and various fragments or derivatives
                                   of these structures

Heparitin sulfate   early name for heparan sulfate; still appears
                                   in the literature, especially in the naming of
                                   heparin lyases

HexA   hexuronic acid (i.e., GlcA, IdoA, GalA, etc.;
                                   UA is also used)

HL   hepatic lipase

HPLC   high pressure liquid chromatography

Hydrazinolysis	reaction of GlcNAc- or GalNAc-containing structures with hydrazine to remove the acetyl groups and form the corresponding GlcN- or GalN-containing structures
IDL	intermediate density lipoprotein
Ido	L-idose
IdoA	L-iduronic acid
IdoA-2-SO$_4$	L-iduronic acid-2-sulfate
KS	keratan sulfate
LACI	lipoprotein-associated coagulation inhibitor (same as EPI or TFPI)
LA-heparin	low activity heparin; the fraction of heparin that does not bind to antithrombin affinity columns
LDL	low density lipoprotein
LPL	lipoprotein lipase
LRP	low density lipoprotein receptor-related protein (same as α_2MR or LRP/α_2MR)
LRP/α_2MR	low density lipoprotein receptor-related protein (same as α_2MR or LRP)
Lyase	in the context used here, an enzyme that cleaves GlcNR→HexA linkages by a β-elimination reaction, converting the nonreducing terminal HexA to a $\Delta^{4,5}$-HexA
Mn-SOD	manganese superoxide dismutase
NLS	nuclear localization sequence (same as NTS)
NTS	nuclear translocation sequence (same as NLS)
PAGE	polyacrylamide gel electrophoresis
PAI	plasminogen activator inhibitor
PAI-I	plasminogen activator inhibitor-1
PAPS	3'-phosphoadenosine-5'-phosphosulfate
Perlecan	a heparan sulfate proteoglycan core protein

PI-PLC	phosphatidylinositol-specific phospholipase C
PL	pancreatic lipase
Ring contraction reaction	a side reaction in the deaminative cleavage of amino sugar glycosides, in which the pyranose ring of the amino sugar is converted to an aldehydopentose (furanose ring) without cleavage of the glycosidic bond, see Figure 3.4
Ryudocan	a heparan sulfate proteoglycan core protein
Scu-PA	single chain urokinase-type plasminogen activator
Serglycin	a heparin proteoglycan core protein
SOD	superoxide dismutase
Syndecans 1, 2, 3, and 4	heparan sulfate proteoglycan core proteins
TFPI	tissue factor pathway inhibitor (same as LACI or EPI)
t-PA	tissue plasminogen activator
UA	uronic acid (HexA is also used)
UDP	uridine-5'-diphosphate
UDPGlcA	uridine diphosphate-D-glucuronic acid
UDPGlcNAc	uridine diphosphate-N-acetyl-D-glucosamine
UDPHexNAc	uridine diphosphate-N-acetyl-D-hexosamine
u-PA	urokinase (a plasminogen activator)
Upstream	on the nonreducing terminal side of the reference point or, in the case of proteins, on the N-terminal side of the reference point

Chapter 1. Heparin vs Heparan Sulfate

I.   BACKGROUND

Heparin is a highly sulfated polysaccharide that is used clinically as an anticoagulant. It is obtainable as a relatively inexpensive reagent that has been available for use in a variety of research programs. Many heparin-binding proteins have been identified. The heparin-binding property has been used in the purification of these proteins by heparin affinity chromatography and has suggested further experiments that may define the effects of added heparin on the activities of the protein in question. Heparin also has a number of interesting effects on the growth and biological activities of cells in culture. It is fairly common for cell biologists to find that heparin, sometimes in combination with exogenous growth factors, has unique effects on the activities of cells. Although such results often represent quite interesting observations, their physiological significance, if any, is difficult to deduce. Scientists whose work leads them into the heparin field in this manner are usually faced with a rather steep learning curve if they seek to understand these heparin/protein or heparin/cell interactions. It is this difficult hurdle that this book addresses.

II.   HEPARINOIDS DEFINED

Heparin is found exclusively in mast cells [3] and is isolated for clinical uses from highly vascularized tissues, namely, hog mucosa and beef lung. It is a polydisperse mixture of structurally similar polymers made up of repeating units of disaccharides containing a uronic acid residue—either $\underline{D}$-glucuronic acid or

1

L-iduronic acid—and D-glucosamine, which is either N-sulfated or N-acetylated. The disaccharides may be further O-sulfated at C6 and/or C3 of the D-glucosamine and at C2 of the uronic acid residue. Heparin is an intracellular product found in the secretory granules of mast cells in complexation with basic proteases. It is released from mast cells only when the cells degranulate in response to some external signal. Thus, despite of the potent anticoagulant activity of exogenous heparin, it is doubtful that endogenous heparin plays a physiological role in the inhibition of blood coagulation. If this is true, and if heparin occurrence is confined to a single cell type, then it is reasonable to ask whether there is any physiological significance to the heparin-binding properties of the many other heparin-binding proteins produced by quite diverse cell types and found in a variety of tissues. Perhaps heparin binding by proteins is simply a coincidental, nonphysiological interaction.

Before accepting such a conclusion, note should be taken of a second sulfated polysaccharide, heparan sulfate (originally called "heparitin sulfate"), which occurs in most animal cells. Heparan sulfate is structurally similar to heparin. It is less highly sulfated than heparin but contains all of the structural motifs found in heparin. Although there are a number of reasons for distinguishing heparan sulfate from heparin,[1] it is often convenient, because of the structural similarities between these two polymers, to refer to them together as "heparinoids." Heparan sulfate has recently become available as a reagent, but it is more expensive than heparin. There are only a few reports of studies of the binding of heparan sulfate to heparin-binding proteins or the effects of exogenous heparan sulfate on the activities of cells. Heparan sulfates from different tissues may have significant differences in structure, and presumably biological activity, so that failure to obtain any interesting biological activities with heparan sulfate from a single source may not be reflective of the potential activities of heparan sulfates from other sources. Heparan sulfate occurs as a proteoglycan on the surface of most animal cells, i.e., in contrast to heparin, it is normally *secreted* from cells. As a major component of the extracellular matrix, it is well positioned to play a prominent role in the physiology of the cell. For example, heparan sulfate on the surface of vascular endothelial cells plays a role in the prevention of the clotting of blood even though, compared to heparin, it has very low anticoagulant activity. However, there are only a few other cases in which a physiological role for heparan sulfate been defined.

---

[1] See Chapter 2 for a more extensive discussion of the differences between heparin and heparan sulfate.

III. THE UBIQUITY OF HEPARINOIDS IN ANIMAL CELLS

Nader and Dietrich have summarized their extensive studies on the broad occurrence of heparin and heparan sulfate in vertebrate and invertebrate species [3]. They have developed a rigorous set of criteria that are used to distinguish heparin from heparan sulfate, and, using these criteria, demonstrate that heparin occurs in many vertebrate and invertebrate species and that heparan sulfate is in fact ubiquitous in animal species. Heparinoids are found in mollusks, annelids, and arthropods, including arachnids, crustaceans, and insects, and in organisms as ancient as coelenterates [4]. The significant conclusions from the very extensive work by this laboratory are that these heparinoids, especially heparan sulfate, are found in virtually every animal species that has been examined, suggesting that these polymers play a significant role in the physiology of animal tissues [5-7]. In fact, it has been reported that congenital deficiency in heparan sulfate results in severe clinical problems, and eventually death [8].

IV. HEPARIN-BINDING PROTEINS

More than 100 heparin-binding proteins have been identified, and this number is increasing. Many of these are listed in the Appendix. These proteins fall into quite diverse groups, including the proteins of the circulatory system involved in coagulation and fibrinolysis, many growth factors, several proteins of the extracellular matrix, a number of proteins involved in lipid metabolism, and others that are not easily included in any such group. Some of these proteins have been extensively characterized, but in only a few cases have both the amino acid and the monosaccharide sequences that mediate the interaction between a heparin-binding protein and heparin been defined.

As we shall see, there are amino acid sequence motifs that suggest that a heparinoid sequence will bind to a protein. These generally are represented by the presentation of certain distributions of basic amino acids, especially lysines and arginines, along a relatively short amino acid sequence in a protein. Undoubtedly, there are many proteins that contain such sequences and therefore will bind to heparin or heparan sulfate, even though, *in vivo*, they never encounter heparins or heparan sulfates. When such proteins are recovered from cell or tissue extracts, they will be found to bind to heparin, even though their binding has no physiological significance. As we have noted, the fact that mast cell heparin binds to antithrombin and markedly alters its activity does not necessarily signify that these two macromolecules come into contact *in vivo*. However, since heparan sulfate is found on the surface of virtually all types of animal cells, any protein that occurs extracellularly will have the opportunity to interact with heparan sulfate, and if these extracellular proteins can find access to cells that present heparan sulfate on their surfaces, it seems very probable that there will be an interaction between the heparan sulfate and the heparin-binding proteins. Whether such interactions influence the activities of the protein or the cell is a question that must be asked for each interaction. Nevertheless, it is clear that (a) the extracellular occurrence of a heparin-binding protein and (b) the demonstration that the protein does, in fact, interact with the heparan sulfate species that are found on the surfaces of cells suggests that the particular protein-heparan sulfate interaction is of some physiological significance (it is of interest to note that our list of heparin-binding proteins contains very few proteins that are not secreted). As a corollary, it would seem that proteins that are not secreted to the cell surface or the extracellular matrix have much less chance to participate in physiologically significant interactions with heparin or heparan sulfate, and a finding that these proteins bind to heparin may, at first glance, appear to be gratuitous. Once again, however, note must be taken of the fact that cell surface heparan sulfate proteoglycan is rapidly internalized by all cells and is largely catabolized in the lysosomes, and so, depending on the trafficking of the internalized heparan sulfate, there may be opportunities for an intracellular protein to undergo physiologically significant interactions with the heparan sulfate. The most obvious examples of proteins that are normally not secreted but that still interact with heparan sulfate are the enzymes involved in the biosynthesis and catabolism of heparinoids that are found in the Golgi and in the lysosomes, respectively. There are other examples as well, including, of course, the proteases that bind to heparin in the mast cells.

V.   HEPARIN BINDING TO ANTITHROMBIN: THE PROTOTYPE FOR HEPARIN-BINDING PROTEINS

Although we will return to a detailed discussion of the heparin/antithrombin interaction in Chapter 7, there will be a number of earlier occasions to refer to this, the best characterized example of a heparin-binding protein. Through the efforts of Lindahl and associates, and Rosenberg and associates, and Choay and associates, the structure of a pentasaccharide segment, isolated from fragmented heparin by affinity chromatography on heparin-Sepharose, has been established as:

$$GlcNAc\text{-}6\text{-}SO_4 \rightarrow GlcA \rightarrow GlcNSO_3\text{-}3,6\text{-}di\text{-}SO_4 \rightarrow IdoA\text{-}2\text{-}SO_4 \rightarrow GlcNSO_3\text{-}6\text{-}SO_4$$

This sequence occurs in only about one-third of the chains in a commercial heparin; i.e., it is a rare sequence. Its most distinguishing feature is the unusual

3-O-SO$_4$ on the internal glucosamine residue, which is absolutely essential for binding to antithrombin.

VI. THEME

The purpose of this treatise is severalfold.

- The structural bases for heparin-protein interactions will be described. This will involve a discussion of the structures of heparinoids and the amino acid sequences that will bind heparinoids.

- A summary of the widely dispersed literature on a number of prototypic heparin-binding proteins will be presented. The discussion will be confined to proteins from heparinoid-producing species, namely the eukaryotes.

- An attempt will be made to address the biological significance of the heparin-binding capacity of each protein and to place the heparinoid-protein interactions into a true physiological perspective.

- A major premise throughout is that it is heparan sulfate and not heparin that is the major physiological player in the interactions of heparin-binding proteins with heparinoids. Indeed, as noted above, heparan sulfate has already been established as the biologically significant heparinoid that controls blood coagulation by its interactions with antithrombin.

- An attempt will be made to define the kind of information that must be obtained in order to determine the physiological role of the proteins and to present experimental approaches for obtaining such information.

There is voluminous literature in the heparinoid field. The presentation here will restrict itself to those aspects of the literature that pertain to the theme described. For a comprehensive view of the field, the reader will be referred to reviews and to the primary literature.

Chapter 2. Structures of Heparinoids

I. THE COMPOSITION OF HEPARINOIDS

Although many details remain to be worked out, the basic structural features of heparinoids have been established. In contrast to other biopolymers, for which research on structures and biosynthetic pathways developed as separate disciplines, an understanding of heparinoid structures has evolved as a result of the integration of data covering (a) structural analyses, (b) pathways of heparinoid biosynthesis, and (c) the specificities of the enzymes involved in the biosynthesis. Thus, the present discussion of structure will be directed toward an understanding of the sequences that are found in heparin and heparan sulfate and how these structures are formed and modified *in vivo*. Extensive reviews on the

7

structures of heparinoids [9-15] and the biosynthetic pathways by which they are formed [16-20] can be consulted for additional details.

Heparins and heparan sulfates are mixtures of polydisperse, structurally similar, linear chains that cannot be separated into pure compounds. The present discussion will treat heparin and heparan sulfate together (i.e., all heparinoids). We will describe first the disaccharide building blocks and the sequences in which these disaccharides are found in heparins and heparan sulfates, and then the differences between heparin and heparan sulfate. With this as background, we will ask how the current information about biosynthesis of heparinoids contributes to what we may expect to find in primary heparinoid structures. We will address the sizes and shapes of these polymers, and finally, since heparinoids occur as proteoglycans, we will describe the current state of knowledge of the core proteins to which heparinoids are attached. Heparinoids are polymers of at least 18 different disaccharides, which represent their monomeric units. These are shown in Table 2.1. The disaccharides are made up of a hexuronic acid (HexA) residue and a D-glucosamine (GlcN) residue which are linked to each other and to the other disaccharides by 1→4 linkages. The uronic acid may be either β-D-glucuronic acid (GlcA) or α-L-iduronic acid (IdoA). Both of these uronic acids occur in heparinoids as underivatized monosaccharides or as 2-O-sulfated residues. Thus, each disaccharide monomer has one of four possible structures in the uronic acid position (GlcA, GlcA-2-SO4, IdoA, or IdoA-2-SO4).

Table 2.1. Heparin and Heparan Sulfate Disaccharides [1]

GlcA → GlcNAc	IdoA → GlcNAc
GlcA → GlcNAc-6-SO4	IdoA → GlcNAc-6-SO4
GlcA → GlcNSO3	IdoA → GlcNSO3
GlcA → GlcNSO3-6-SO4	IdoA → GlcNSO3-6-SO4
GlcA → GlcNSO3-3-SO4	IdoA → GlcNSO3-3-SO4
GlcA → GlcNSO3-3,6-di-SO4	IdoA → GlcNSO3-3,6-di-SO4
GlcA-2-SO4 → GlcNAc	IdoA-2-SO4 → GlcNAc
[GlcA-2-SO4 → GlcNAc-6-SO4]	IdoA-2-SO4 → GlcNAc-6-SO4
GlcA-2-SO4 → GlcNSO3	IdoA-2-SO4 → GlcNSO3
GlcA-2-SO4 → GlcNSO3-6-SO4	IdoA-2-SO4 → GlcNSO3-6-SO4
[GlcA-2-SO4 → GlcNSO3-3-SO4]	[IdoA-2-SO4 → GlcNSO3-3-SO4]
[GlcA-2-SO4 → GlcNSO3-3,6-di-SO4]	[IdoA-2-SO4 → GlcNSO3-3,6-di-SO4]

[1] Disaccharides in brackets have not yet been observed in heparinoids.

The α-$\underline{D}$-glucosamine residue may be either N-sulfated (GlcNSO$_3$) or N-acetylated (GlcNAc).[1] The GlcNSO$_3$ residues also may be O-sulfated at C3, at C6, at both C3 and C6, or they may not be O-sulfated at all. The GlcNAc residues may be O-sulfated at C6 or may be unsulfated, but GlcNAc residues that are sulfated at C3, or at both C3 and C6, have not been reported. Thus, each disaccharide monomer has one of six possible structures in the glucosamine position (GlcNSO$_3$, GlcNSO$_3$-6-SO$_4$, GlcNSO$_3$-3-SO$_4$, GlcNSO$_3$-3,6-di-SO$_4$, GlcNAc, or GlcNAc-6-SO$_4$).

Since uronic acid and glucosamine residues occur in alternating positions in heparinoids, the disaccharide monomers of heparinoids may be viewed either as uronic acid→glucosamine units or as glucosamine→uronic acid units. In the present discussion we will consider the HexA→GlcN disaccharides as the monomers of heparinoids, and later will discuss the sequences in which these disaccharides occur in heparinoids.

Since there are four possible uronic acid residues and six possible glucosamine residues in the heparinoid monomers, the potential number of different [uronic acid→glucosamine] disaccharides is 24 (Table 2.1). If these 24 disaccharides were combined into all possible tetrasaccharide sequences, the number of tetrasaccharides would be 576 (24^2). And if these combinations were carried one step further, there would be 13,824 possible hexasaccharides (24^3). An average heparin chain contains 15-20 disaccharides; thus, the potential for structural diversity in heparinoids is enormous. For heparinoid scientists who seek to identify these structures, Nature has been kind. Five of the 24 potential disaccharides have not (yet) been found in heparinoids; thus, for the moment at least, the actual number of disaccharides found in heparinoids is only 19 (mercifully, this reduces the number of possible tetrasaccharides to 324, and the number of hexasaccharides to 5832!).

In discussing the sequences in which these disaccharides occur in heparinoids, a few words about the representation of these sequences are in order. Just as linear sequences of the monomeric units in proteins and nucleic acids are customarily drawn with the N-terminal amino acid or the 5' nucleotide on the left and the C-terminal amino acid or the 3' nucleotide on the right, respectively, linear sequences of carbohydrate monomers are customarily drawn with the nonreducing terminal monosaccharide on the left and the reducing terminal monosaccharide on the right. This convention will be followed here. Also, for ease of describing the positions of one disaccharide, or group of disaccharides, in a sequence with respect to some reference disaccharide in the same chain, the nomenclature used in describing nucleic acid structures will be

---

[1] The N-sulfate group on all GlcNSO$_3$ residues is very labile under acidic conditions and may be hydrolyzed readily. However, there is now evidence that some N-unsulfated GlcN residues occur naturally in heparan sulfates [21].

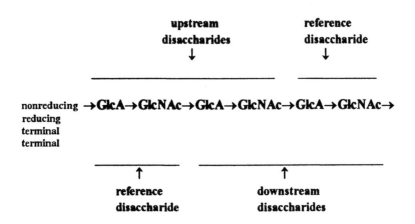

Figure 2.1. Structural Nomenclature

adopted, as shown in Figure 2.1. When two or more disaccharides are linked together, those disaccharides attached to the *nonreducing* end of the reference disaccharide(s) will be referred to as being *upstream* from the reference point, whereas those attached to the *reducing* end of the reference disaccharide(s) will be referred to as being *downstream*. For example, as illustrated in the sequence of three disaccharides in Figure 2.1, the middle and nonreducing terminal disaccharides will be considered upstream from the reducing terminal disaccharide, and the middle and reducing terminal disaccharides will be considered downstream from the nonreducing terminal disaccharide.[2] Arrows between monosaccharide residues will originate from the C1 of an upstream monosaccharide and point toward an attachment site at C4 of the downstream monosaccharide; i.e., the arrows will always point in the downstream direction.

II.   HEPARIN AND HEPARAN SULFATE STRUCTURES

A.   Disaccharides of Heparinoids

The literature contains a number of reports of the disaccharide compositions of heparins and heparan sulfates [see, for example, references 22,

---

[2] It is much easier to say "upstream" than to say "attached to the nonreducing terminal," or to say "downstream" than to say "attached to the reducing terminal."

23, 24]. Table 2.2 shows typical disaccharide compositions that are obtained
when heparinoids are cleaved to disaccharides by treatment with nitrous acid.
The analytical procedures used for these analyses are described in Chapter 3.
Here the yield of disaccharides is close to 100%, but the methods used for these
analyses do not give the percentage of each disaccharide that was N-sulfated in
the original polymer or the percentage that was N-acetylated. The results are
expressed as moles of each disaccharide per 100 moles of $IdoA-2-SO_4 \rightarrow AMan-6-SO_4$. The latter is the most abundant disaccharide in heparins, but it is much
lower in heparan sulfates. Data show that the disaccharide units of heparinoids
are $GlcA \rightarrow AMan$, $IdoA \rightarrow AMan$, $GlcA \rightarrow AMan-6-SO_4$, $IdoA-2-SO_4 \rightarrow AMan$,
and $IdoA-2-SO_4 \rightarrow AMan-6-SO_4$. In heparins, the most prominent disaccharide is
$IdoA-2-SO_4 \rightarrow AMan-6-SO_4$, whereas $GlcA \rightarrow AMan$ and $IdoA \rightarrow AMan$ are
relatively minor components. In heparan sulfates, however, the unsulfated and
monosulfated disaccharides are major components, whereas the $IdoA-2-SO_4 \rightarrow AMan-6-SO_4$ is relatively low. Several minor disaccharide units are also found
in heparinoids. These include $GlcA \rightarrow AMan-3,6-diSO_4$, which is derived in part
from the pentasaccharide segment that binds to antithrombin, and $GlcA-2-SO_4 \rightarrow AMan$ and $GlcA-2-SO_4 \rightarrow AMan-6-SO_4$. The yields of the latter are always low
and are not given in Table 2.2. It is of particular interest that none of the
heparan sulfates reported here contain detectable amounts of $GlcA \rightarrow AMan-3,6-diSO_4$, although this disaccharide is found in vascular endothelial cell heparan
sulfate (Chapter 7).

Heparinoids can be compared by disaccharide content, but it is also
helpful to note significant differences among heparinoids in terms of the ratios
of total IdoA to total GlcA. These ratios are high for beef lung heparin and lower
for hog mucosa and beef mucosa heparins. A comparison of the heparins with
the heparan sulfates shows clear distinctions in the IdoA/GlcA ratios, which are
much lower for heparan sulfates. In addition, there are 3-4-fold differences in
these ratios for heparan sulfates from different sources.

B. Higher Oligosaccharides of Heparinoids

Table 2.3 lists the sequences of the tetra- and higher oligosaccharides that
have been reported in heparinoids. Oligosaccharides are listed in order of size.
Within each size class, the oligosaccharides formed by nitrous acid cleavage are
listed first, followed by those formed by heparin lyase I cleavage (see Chapter
3). Also, within each group, the oligosaccharides are listed in the order of
increasing numbers of sulfate groups per oligosaccharide.

Table 2.2. Disaccharide Composition of Heparinoids [1]

Heparinoid [1]	Tissue	Source	Disaccharide							IdoA/GlcA
			1 GlcA→AMan	2 IdoA→AMan	3 GlcA→AMan-6-SO4	4 IdoA→AMan-6-SO4	5 GlcA→AMan-3,6-SO4	6 IdoA-2-SO4→AMan	7 IdoA-2-SO4→AMan-6-SO4	
			moles disaccharide per 100 moles IdoA-2-SO4→AMan-6-SO4							
Heparin	Beef lung	Upjohn	5	0	7.2	4.1	3.9	10.3	100	10.5
Heparin	Beef lung	SPL	5	0.6	6.7	5.5	4.2	9.0	100	10.8
Heparin	Hog mucosa	SPL			18.5	11.3	5.7	21.6	100	5.5
Heparin	Hog mucosa	Ming Han			18.6	13.0	7.4	14.6	100	4.9
Heparin	Beef mucosa	Opocrin			25.8	6.7	5.1	70.7	100	5.7
Heparin	Beef mucosa	Syntex			22.1	11.2	6.6	42.7	100	5.4
Heparan sulfate	Pancreas	Opocrin	949	93	101	426	0	212	100	0.5
Heparan sulfate	Spleen	Opocrin	1224	2316	106	776	0	197	100	2.0
Heparan sulfate	Beef Kidney	Sigma	300	406	113	638	0	60	100	1.5

[1] Data from [23]. Samples were analyzed by nitrous acid cleavage to disaccharides after de-N-acetylation by hydrazinolysis (Chapter 3). Anhydromannose residues in the disaccharides were derived from GlcNSO3 or GlcNAc residues in the heparinoids. All samples were analyzed by Dr. Patrick N. Shaklee, Scientific Protein Laboratories (SPL).

Table 2.3. Higher Oligosaccharides Identified in Heparinoids

No.	Reference	SO₃'s	Oligosaccharide
			TRISACCHARIDES
1	[27]	1	IdoA→GlcNAc-6-SO₄→GlcA
2	[27]	1	IdoA-2-SO₄→GlcNAc→GlcA
			TETRASACCHARIDES
3	[28, 29]	0	GlcA→GlcNAc→GlcA→AMan
4	[28]	0	IdoA→GlcNAc→GlcA→AMan
5	[28, 29]	1	GlcA→GlcNAc-6-SO₄→GlcA→AMan
6	[27-29]	1	IdoA→GlcNAc-6-SO₄→GlcA→AMan
7	[27]	1	IdoA→GlcNAc→GlcA→AMan-6-SO₄
8	[27, 28]	1	IdoA-2-SO₄→GlcNAc→GlcA→AMan
9	[29]	2	GlcA→GlcNAc→IdoA-2-SO₄→AMan-6-SO₄
10	[29]	2	GlcA→GlcNAc-6-SO₄→GlcA→AMan-6-SO₄
11	[28, 29]	2	IdoA→GlcNAc-6-SO₄ →GlcA→AMan-6-SO₄
12	[28]	2	IdoA→GlcNAc-6-SO₄→GlcA→AMan-3-SO₄
13	[27, 28]	2	IdoA-2-SO₄→GlcNAc→GlcA→AMan-6-SO₄
14	[27]	2	IdoA-2-SO₄→GlcNAc-6-SO₄→GlcA→AMan
15	[28]	3	IdoA→GlcNAc-6-SO₄→GlcA→AMan-3,6-di-SO₄
16	[28]	3	IdoA-2-SO₄→GlcNAc-6-SO₄→GlcA→AMan-6-SO₄
17	[30]	3	Δ⁴,⁵-UA→GlcNAc-6-SO₄→GlcA→GlcNSO₃-6-SO₄
18	[31]	3	Δ⁴,⁵-UA →GlcNAc-6-SO₄→GlcA→GlcNSO₃-3-SO₄
19	[30]	3	Δ⁴,⁵-UA-2-SO₄→GlcNAc→GlcA→GlcNSO₃-6-SO₄

No.	Oligosaccharide	SO's	Reference
20	$\Delta^{4,5}$-UA-2-SO$_4$→GlcNAc→IdoA →GlcNSO$_3$-6-SO$_4$	3	[30]
21	$\Delta^{4,5}$-UA→GlcNSO$_3$-6-SO$_4$→GlcA→GlcNSO$_3$-6-SO$_4$	4	[32]
22	$\Delta^{4,5}$-UA→GlcNAc-6-SO$_4$→GlcA→GlcNSO$_3$-6-SO$_4$	4	[30]
23	$\Delta^{4,5}$-UA→GlcNAc-6-SO$_4$→GlcA→GlcNSO$_3$-3,6-di-SO$_4$	4	[32]
24	$\Delta^{4,5}$-UA-2-SO$_4$→GlcNAc-6-SO$_4$→GlcA→GlcNSO$_3$-6-SO$_4$	4	[30, 32]
25	$\Delta^{4,5}$-UA-2-SO$_4$→GlcNSO$_3$→IdoA-2-SO$_4$→GlcNSO$_3$	4	[32]
26	$\Delta^{4,5}$-UA-2-SO$_4$→GlcNSO$_3$-6-SO$_4$→IdoA→GlcNAc-6-SO$_4$	4	[30, 32]
27	$\Delta^{4,5}$-UA-2-SO$_4$→GlcNSO$_3$→GlcA→GlcNSO$_3$-6-SO$_4$	4	[31]
28	$\Delta^{4,5}$-UA →GlcNSO$_3$-6-SO$_4$→GlcA→GlcNSO$_3$-3,6-diSO$_4$	5	[32]
29	$\Delta^{4,5}$-UA-2-SO$_4$→GlcNSO$_3$→IdoA-2-SO$_4$→GlcNSO$_3$-6-SO$_4$	5	[32]
30	$\Delta^{4,5}$-UA-2-SO$_4$→GlcNSO$_3$-6-SO$_4$→GlcA-2-SO$_4$→GlcNSO$_3$	5	[30, 32, 33]
31	$\Delta^{4,5}$-UA-2-SO$_4$→GlcNSO$_3$-6-SO$_4$→GlcA→GlcNSO$_3$-6-SO$_4$	5	[30]
32	$\Delta^{4,5}$-UA-2-SO$_4$→GlcNSO$_3$-6-SO$_4$→IdoA→GlcNSO$_3$-6-SO$_4$	5	[30, 32, 33]
33	$\Delta^{4,5}$-UA-2-SO$_4$→GlcNSO$_3$-6-SO$_4$→GlcA-2-SO$_4$→GlcNSO$_3$-6-SO$_4$	6	[32]
34	$\Delta^{4,5}$-UA-2-SO$_4$→GlcNSO$_3$-6-SO$_4$→IdoA-2-SO$_4$→GlcNSO$_3$-6-SO$_4$	6	[30, 32, 33]
35	$\Delta^{4,5}$-UA-2-SO$_4$→GlcNSO$_3$-6-SO$_4$→GlcA→GlcNSO$_3$-3,6-di-SO$_4$	6	[34]

HEXASACCHARIDES

No.	Oligosaccharide	SO's	Reference
36	IdoA→GlcNSO$_3$-6-SO$_4$→GlcA→GlcNSO$_3$-3,6-diSO$_4$→IdoA-2-SO$_4$→GlcNSO$_3$-6-SO$_4$	8	[35]
37	$\Delta^{4,5}$-UA-2-SO$_4$→GlcNSO$_3$-6-SO$_4$→IdoA→GlcNAc-6-SO$_4$→GlcA→GlcNSO$_3$-6-SO$_4$	6	[30, 36]
38	$\Delta^{4,5}$-UA-2-SO$_4$→GlcNSO$_3$-6-SO$_4$→IdoA→GlcNAc-6-SO$_4$→GlcA→GlcNSO$_3$-6-SO$_4$	6	[36]
39	$\Delta^{4,5}$-UA-2-SO$_4$→GlcNAc-6-SO$_4$→GlcA→GlcNSO$_3$-3,6-diSO$_4$→IdoA→GlcNSO$_3$-3-SO$_4$	7	[37]
40	$\Delta^{4,5}$-UA-2-SO$_4$→GlcNSO$_3$-6-SO$_4$→GlcA→GlcNAc-6-SO$_4$→GlcA→GlcNSO$_3$-6-SO$_4$	7	[36]
41	$\Delta^{4,5}$-UA-2-SO$_4$→GlcNSO$_3$-6-SO$_4$→IdoA→GlcNAc-6-SO$_4$→GlcA→GlcNSO$_3$-3,6-di-SO$_4$	7	[30, 36]
42	$\Delta^{4,5}$-UA-2-SO$_4$→GlcNSO$_3$-6-SO$_4$→IdoA-2-SO$_4$→GlcNAc-6-SO$_4$→GlcA→GlcNSO$_3$-6-SO$_4$	7	[36]
43	$\Delta^{4,5}$-UA-2-SO$_4$→GlcNSO$_3$-6-SO$_4$→IdoA-2-SO$_4$→GlcNSO$_3$-6-SO$_4$→IdoA →GlcNAc-6-SO$_4$	7	[33]
44	$\Delta^{4,5}$-UA-2-SO$_4$→GlcNSO$_3$-6-SO$_4$→IdoA-2-SO$_4$→GlcNSO$_3$-6-SO$_4$→IdoA-2-SO$_4$→GlcNSO$_3$	8	[33]

14

45	[33]	8	$\Delta^{4,5}$-UA-2-SO$_4$→GlcNSO$_3$-6-SO$_4$→IdoA-2-SO$_4$→GlcNSO$_3$-6-SO$_4$→GlcA→GlcNSO$_3$-6-SO$_4$
46	[30, 33]	9	$\Delta^{4,5}$-UA-2-SO$_4$→GlcNSO$_3$-6-SO$_4$→IdoA-2-SO$_4$→GlcNSO$_3$-6-SO$_4$→IdoA-2-SO$_4$→GlcNSO$_3$-6-SO$_4$

OCTASACCHARIDES

47	[38]	9	IdoA-2-SO$_4$→GlcNSO$_3$-6-SO$_4$→IdoA →GlcNAc-6-SO$_4$→GlcA→GlcNSO$_3$-6-SO$_4$→IdoA-2-SO$_4$→AMan-6-SO$_4$
48	[35]	9	IdoA→GlcNSO$_3$-6-SO$_4$→GlcA→GlcNSO$_3$-3,6-diSO$_4$→IdoA-2-SO$_4$→GlcNSO$_3$-6-SO$_4$→IdoA-2-SO$_4$→AMan
49	[38, 39]	9	IdoA→GlcNAc-6-SO$_4$→GlcA→GlcNSO$_3$-6-SO$_4$→IdoA-2-SO$_4$→GlcNSO$_3$-6-SO$_4$→IdoA-2-SO$_4$→AMan-6-SO$_4$
50	[40]	7	$\Delta^{4,5}$-UA-2-SO$_4$→GlcNSO$_3$→IdoA→GlcNAc→GlcA→GlcNSO$_3$-3-SO$_4$→IdoA-2-SO$_4$→GlcNSO$_3$
51	[41]	10	$\Delta^{4,5}$-UA →GlcNAc-6-SO$_4$→GlcA→GlcNSO$_3$-3,6-diSO$_4$→IdoA-2-SO$_4$→GlcNSO$_3$-6-SO$_4$→IdoA-2-SO$_4$→GlcNSO$_3$-6-SO$_4$
52	[37]	10	$\Delta^{4,5}$-UA-2-SO$_4$→GlcNSO$_3$-6-SO$_4$→IdoA →GlcNAc-6-SO$_4$→GlcA→GlcNSO$_3$-3,6-diSO$_4$→IdoA-2-SO$_4$→GlcNSO$_3$-6-SO$_4$
53	[30]	12	$\Delta^{4,5}$-UA-2-SO$_4$→GlcNSO$_3$-6-SO$_4$→IdoA-2-SO$_4$→GlcNSO$_3$-6-SO$_4$→IdoA-2-SO$_4$→GlcNSO$_3$-6-SO$_4$→IdoA-2-SO$_4$→GlcNSO$_3$-6-SO$_4$

DECASACCHARIDES

54	[37]	12	$\Delta^{4,5}$-UA-2-SO$_4$→GlcNSO$_3$-6-SO$_4$→IdoA→GlcNAc-6-SO$_4$→GlcA →GlcNSO$_3$-3,6-diSO$_4$→IdoA-2-SO$_4$→GlcNSO$_3$-6-SO$_4$→ IdoA-2-SO$_4$→GlcNSO$_3$-6-SO$_4$
55	[35]	14	IdoA-2-SO$_4$→GlcNSO$_3$-6-SO$_4$→IdoA→GlcNSO$_3$-6-SO$_4$→GlcA →GlcNSO$_3$-3,6-diSO$_4$→IdoA-2-SO$_4$→GlcNSO$_3$-6-SO$_4$→ IdoA-2-SO$_4$→GlcNSO$_3$-6-SO$_4$
56	[33]	14	$\Delta^{4,5}$-UA-2-SO$_4$→GlcNSO$_3$-6-SO$_4$→IdoA-2-SO$_4$→GlcNSO$_3$-6-SO$_4$→IdoA-2-SO$_4$→GlcNSO$_3$-6-SO$_4$→GlcA→GlcNSO$_3$-6-SO$_4$→ IdoA-2-SO$_4$→GlcNSO$_3$-6-SO$_4$
57	[33]	15	$\Delta^{4,5}$-UA-2-SO$_4$→GlcNSO$_3$-6-SO$_4$→IdoA-2-SO$_4$→GlcNSO$_3$-6-SO$_4$→IdoA-2-SO$_4$→GlcNSO$_3$-6-SO$_4$→IdoA-2-SO$_4$→ GlcNSO$_3$-6-SO$_4$→IdoA-2-SO$_4$→GlcNSO$_3$-6-SO$_4$

Note that the largest structures that have been characterized are decasaccharides. The relative amounts of each of these oligosaccharide sequences are difficult to quantify; thus, no comparisons of the amounts of these oligosaccharides from different heparinoids have been reported. A number of these oligosaccharides contain the relatively rare 3-O-sulfated GlcN residues, some of which are derived from the antithrombin-binding pentasaccharide (numbers 12, 15, 18, 22, 27, 35, 36, 38, 39, 41, 47-52, 54, and 55). Two of the oligosaccharides, 29 and 33, contain the equally rare 2-O-sulfated GlcA residues. Examination of the structures of oligosaccharides 29 and 33 show that these GlcA-2-SO_4 residues occur in highly sulfated sequences of the heparin.

As will be described below, the specificities of the enzymes that synthesize heparinoids suggest that immediately downstream from every GlcNAc there must be a GlcA residue. This, in fact, is found to be the case in all of the GlcNAc-containing oligosaccharides except 9 and 20, where the *unexpected* IdoA residues are underlined. Whether the report of these GlcNAc→IdoA sequences indicates broader specificities of the enzymes or an incorrect assignment of these structures is uncertain.

C.   Differences in Heparan Sulfates from Different Sources

Only a few heparinoids from unusual sources have been characterized in detail, but some of these are of particular interest. For example, Hovingh and Linker [25] have described a "heparan sulfate" isolated from lobsters that has GlcA→GlcNSO₃-6-SO_4 as the major repeating unit and a very low content of IdoA-2-SO_4 residues. Also, Pejler *et al.* [26] have characterized clam heparins with M_r 22,500 chains containing up to three antithrombin-binding sites per chain. This study showed that 25-30% of the total disaccharide units are present in GlcNSO₃→GlcA→GlcNSO₃-3-SO_4 and GlcNSO₃→GlcA→GlcNSO₃-3,6-di-SO_4 sequences, the content of which increases in chains selected for their affinity for antithrombin. An unusual sequence, GlcNSO₃→IdoA→GlcNSO₃-3,6-di-SO_4, was found in the clam heparins and represented 3-4% of the heparinoid. Further information on the variability of heparinoid structures comes from the extensive studies of Nader and Dietrich and their collaborators, who have identified and partially characterized heparins and heparan sulfates from a variety of animal species [3].

D. The Difference between Heparin and Heparan Sulfate

1. Introduction

Heparin and heparan sulfate have many common structural features. We have seen that heparins have higher IdoA/GlcA ratios than found in heparan sulfates and contain higher percentages of the more highly sulfated disaccharide species. However, the distinction between these two polymers is much more complex than this simple difference. Several criteria have been proposed to distinguish heparin from heparan sulfate. Lane and Lindahl [42] suggest that the term "heparin" be restricted to a heparinoid in which more than 80% of the GlcN residues are N-sulfated and the number of O-sulfates is greater than the number of N-sulfates; all other related heparinoids would be referred to as heparan sulfate. Gallagher and Walker [43] have stated that all heparan sulfates have about 50% of their GlcN residues N-sulfated and have ratios of O-sulfates to N- sulfates of 1 or less. According to Nader and Dietrich [3] heparins and heparan sulfates are distinguished by a combination of their physical and biological properties. Thus, all heparins are precipitated from tissue extracts by 2 $\underline{M}$ potassium acetate, pH 5.7, at 4°C; they show mobilities in agarose gels similar to those of commercial heparins; they have an average molecular weight of 10,700 to 68,000; they form multiple bands on isoelectric focusing in polyacrylamide gels [44]; the molar ratios of GlcN/HexA/SO$_4$ are compatible with those of commercial heparins; and they are digested by heparin lyase I (Chapter 3). Heparan sulfates, however, are soluble in 2 $\underline{M}$ potassium acetate; do not form multiple bands on isoelectrofocusing; have SO$_4$/hexosamine ratios of 0.8—1.8; and are minimally degraded by heparin lyase I (cleavage of <5% of the linkages between disaccharides). According to these workers, heparins are composed largely of IdoA-2-SO$_4$→GlcNSO$_3$-6-SO$_4$ and GlcA→GlcNSO$_3$-6-SO$_4$, whereas heparan sulfates contain only a few percent of the IdoA-2-SO$_4$→GlcNSO$_3$-6-SO$_4$ disaccharide, but contain a much more complex mixture of less highly sulfated disaccharides than heparin.

Since the distinctions between heparin and heparan sulfate are important for all that follows, and since these distinctions are based on a rather complex experimental background, this background is described here.

2. Heparin

Heparin appeared first on the scene. It has been known from the beginning [1, 2, 45] primarily for its anticoagulant activity, and it is used extensively in the clinical setting as an anticoagulant drug. Apparently, it is found only in mast cells and is isolated from tissues rich in these cells. The usual sources from which heparins are prepared (hog mucosa, beef lung) contain heparinoid chains, which have a range of anticoagulant activities; they also contain other glycosaminoglycans. In the manufacture of heparin, the goal is to extract the heparinoids from these mast cell-rich tissues and to remove proteins and fractionate the polysaccharides to obtain the heparinoid fraction that has the highest anticoagulant activity (150-180 IU/mg [3]) consistent with reasonable recoveries, while minimizing the amounts of other glycosaminoglycans. The heparinoids with the highest anticoagulant activities are recovered in the fractions that are most highly sulfated, but fractions with lower anticoagulant activities are also obtained and are set aside as "heparin by-products." The products with high anticoagulant activities are called "heparins." As we have seen, heparins have high IdoA/GlcA ratios, contain >2 SO_4 groups/disaccharide, and contain 7-9 times as many $GlcNSO_3$ residues as GlcNAc residues. Some of their structural features are indicated by the disaccharide contents and by the oligosaccharide fragments that have been prepared from them. The rules that control the relative amounts of the disaccharide components of these polymers and their positions in heparin sequences are derived from the biosynthesis/enzyme specificity studies described in Section III and are still "seen through a glass darkly." The GlcNAc residues in heparins occur in isolated disaccharides, positioned between $GlcNSO_3$-containing disaccharides. 3-O-Sulfated GlcN residues are critical for anticoagulant activity. Thus, heparin is enriched in these disaccharide units and in the pentasaccharide sequences that contain them (Chapter 7).

3.   Heparin by-products

The lower anticoagulant activity fractions obtained in the manufacture of heparin are referred to as "heparin by-products" and, in the early literature, were called "heparin monosulfate" [46], and later "heparitin sulfate" [47]. These by-products contain heparinoids that are less highly sulfated and that have lower IdoA/GlcA ratios than heparin; they also contain chondroitin sulfate and dermatan sulfate, which can be removed from the heparinoid fraction. More detailed studies of the structures of the heparinoids in these heparin by-products have demonstrated that they are composed of blocks of relatively unsulfated

---

[3] IU, international units, defined by the USP assay for anticoagulant activity; the values used here are arbitrarily chosen to suggest a range of values observed for commercial heparins.

disaccharide units, containing mostly GlcA and GlcNAc, interspersed with blocks of sulfated disaccharides [9, 29, 48-50].

Of further interest with regard to the structures of these heparin by-products, several groups have found that fractionation of these mixtures gives an extremely broad array of structures, ranging from fractions that have low IdoA/GlcA ratios and very high GlcNAc/GlcNSO$_3$ ratios to fractions that have high IdoA/GlcA ratios and low GlcNAc/ GlcNSO$_3$ ratios [47, 50-53]. As shown in Table 2.4, this range represents a *continuum* of structures and anticoagulant activities [53].

4.   Heparan sulfate

Heparitin sulfate was eventually given the name "heparan sulfate." From the previous discussion, it would seem that "heparan sulfate" must contain a continuum of structures ranging from fractions with low IdoA/GlcA ratios, low SO$_4$ content, and low anticoagulant activity to fractions with high IdoA /GlcA ratios, high SO$_4$ content, and high anticoagulant activity. And any reasonable scientist might ask, "where does heparan sulfate stop and heparin begin?"

As this field has evolved, it has become clear that proteoglycans containing heparin-like chains with relatively low sulfate and iduronic acid (heparan sulfate proteoglycans) are produced and secreted by *all* animal cells, and that the heparinoid chains from these proteoglycans are structurally similar to the heparin by-products obtained from tissues rich in mast cells. These non-mast cell heparan sulfate proteoglycans are believed to be distinct from heparin by-products. Structures of the heparan sulfate isolated from tissues that do not contain mast cells have been examined in some detail. They have relatively low IdoA /GlcA ratios, approximately equal amounts of GlcNSO$_3$ and GlcNAc, and approximately 1 SO$_4$/disaccharide unit, significantly below the SO$_4$ content of a clinical heparin. The structural features of the heparinoid chains from heparan sulfate proteoglycans vary significantly, depending on the cells that synthesize them [54] (see also Table 2.2). However, using structural approaches similar to those used earlier in characterizing heparin by-products, heparan sulfates isolated from tissues [55-57], or recovered from cell cultures [43, 50, 54, 58-61] all exhibit the types of block structures that have been found in heparin by-products, with variations in the relative lengths of the sulfated and the unsulfated blocks [54].

Most of the discussions about the distinctions between heparin and heparan sulfate ignore two facts: (a) "heparin" is not the total heparinoid pool obtained from mast cell-rich tissues, but is only that fraction of the heparinoids from these tissue extracts that has the highest anticoagulant activity, and (b)

Table 2.4. Continuum of Structures in Heparin By-Products [1]

Fraction	Disaccharides						AC Act.[2]	Ratio
	1	2	3	4	5	6		Col. 6/ Col. 5
	IdoA→ AM	GlcA →AM	GlcA→ AM-6-SO₄	IdoA→ AM-6-SO₄	IdoA-2- SO₄→AM	IdoA-2-SO₄→ AM-6-SO₄		
				mmoles/mol AMan				
LAH-1	5	78	72	35	127	425	19	3.3
LAH-2	3	79	71	30	104	587	45	5.6
LAH-3	10	69	65	26	94	571	60	6.1
LAH-4	12	68	65	37	93	578	56	6.2
LAH-5	8	60	88	56	88	773	75	8.8
LAH-6	10	79	79	58	65	711	85	10.9
LAH-7	2	86	94	129	80	868	104	10.9
LAH-8	13	83	70	62	44	669	96	15.2
LAH-9	6	89	93	80	70	918	130	13.1
LAH-10	8	99	79	68	51	918	125	18.0

[1] Data from reference [53]. A heparin by-product, LAH (low activity heparin, 76 i.u./mg), was separated by anion-exchange chromatography to obtain fractions with increasing degrees of sulfation, from fraction 1 to fraction 10. All fractions were analyzed for the disaccharides released by low pH nitrous acid (samples were not de-N-acetylated prior to nitrous acid cleavage); therefore, disaccharides were not obtained from regions composed of blocks of GlcNAc-containing disaccharides, and yields of total disaccharides increased as the percentage of N-sulfated GlcN increased with increasing total sulfation—see Chapter 3). AM, anhydromannose. See Table 2.1 for other abbreviations.

[2] Anticoagulant activity: units/mg

"heparan sulfate" generally is the *total* heparinoid pool that is extracted from tissues that contain few, if any, mast cells. Thus, "heparin" is a fractionated product whereas "heparan sulfate" is not. The fact that heparan sulfates contain a range of structures is only beginning to emerge. For example, the fractionation of *heparan sulfate* from bovine liver has yielded three fractions with SO_4/disaccharide ratios of 0.8, 1.2, and 2.1 [62]. The most highly sulfated of these fractions fits many of the criteria generally used to describe a heparin. Similar studies on heparan sulfates from other tissues may show that there is a continuum of structures in most heparan sulfates that mimic those of the heparin by-products shown in Table 2.4. If this turns out to be the case, then our ability to distinguish heparin and heparan sulfate on *structural* grounds will once again be called into question. Thus, although it is common to describe the high anticoagulant fraction obtained in the manufacture of commercial heparin as being identical to the product produced by mast cells, in fact, the lower activity heparin by-products are also recovered from tissues rich in mast cells. One does not know whether the "heparin by-products" are derived solely (or even partially) from mast cells or from the non-mast cells in the tissues taken for heparin extraction.[4] However, there is no question that *mast cell-rich tissues* yield such a continuum of structures that it is impossible to draw a line that would place heparin on one side and heparan sulfate on the other. Thus, the "heparin" that occurs in mast cells per se *may* contain significant levels of low anticoagulant "heparin by-products".[5] Since the apparent role of the heparinoids in mast cells does not appear to depend on their anticoagulant activity, heparinoids with low anticoagulant activity may serve the physiological needs of mast cells just as well as the high anticoagulant activity heparinoids.

Another facet of heparin biology that is critical to this discussion is that heparin chains, during their biosynthesis, are attached to a *unique* core protein, serglycin, that is found only in mast cells and in other hematopoietic cells as well (Section VI.B.1., p. 55). This proteoglycan form of heparin, referred to as "macromolecular heparin" [63-65], is not secreted by mast cells; instead, the heparin chains are cleaved from the core protein within the cells by an endoglucuronidase and stored in complexes with basic proteases within the

---

[4] One feature of the heparin by-products that is distinct from most heparan sulfates is that even those heparin by-products with low IdoA/ GlcA ratios have quite significant anticoagulant activities. Perhaps this is a signal that these heparin by-products are actually derived from mast cells.

[5] Mastocytoma cell microsomal fractions synthesize a typical heparin *in vitro*, but there is also a significant portion of the final product that is less highly matured than the portion that has the heparin-like properties. However, this observation does not really indicate what mast cells produce *in vivo*. The mechanisms that regulate heparinoid synthesis in whole cells are lost in broken cell preparations.

secretory granules in the cells [66]. In contrast, the heparan sulfate proteoglycans are clearly distinguished from macromolecular heparins by virtue of their distinctive core proteins (Section VI, p. 52) and by the fact that intact heparan sulfate proteoglycans are *secreted continuously* by most animal cells (Chapter 5).

In fact, what has happened is that our *concept* of heparan sulfate has changed. Originally, heparan sulfate (heparin monosulfate or heparitin sulfate) referred to those structures found in heparin by-products. Now, almost by unwritten consent, heparan sulfate has become a term that refers to the heparan sulfate chains in heparan sulfate proteoglycans obtained from tissues that do not contain mast cells and that appear on the surfaces of cells. Thus, in the future, we may all come to a consensus that it is the *metabolism* and the *physiological* roles of heparin and heparan sulfate that are their most distinguishing features. In fact, Kjellén and Lindahl [13] have suggested that the term "heparin" be reserved for those glycosaminoglycans that are synthesized by mast cells and attached to serglycin, and that all other structurally related glycosaminoglycans should be named "heparan sulfate."

III. BIOSYNTHESIS OF HEPARINOIDS: WHAT IT TEACHES ABOUT STRUCTURES

A. The Biosynthetic Pathway

The present understanding of the extremely complex pathway of heparin biosynthesis is due almost exclusively to the work of Lindahl and co-workers in Uppsala and to collaborators in other laboratories, primarily those of Rodén and Feingold. Detailed descriptions of the synthesis of heparinoids have been given in several reviews [18-20]. The reader is referred to these reviews and to the primary literature for a more comprehensive description of heparin biosynthesis. The present discussion will be confined to those aspects of heparinoid metabolism that control the structural complexities of the heparinoids that occur in Nature. Unfortunately, there are many gaps in our knowledge of both the structure and the biosynthesis of heparinoids, and this ignorance makes a complex topic even more difficult.

Connective tissue mast cells synthesize heparin. Most of the studies on the biosynthesis of heparinoids have been carried out by incubating microsomal preparations from mastocytoma cells with radiolabeled metabolic precursors of heparin, namely the sugar donors, UDPGlcA and UDPGlcNAc, and the sulfate donor, 3'-phosphoadenosine-5'-phosphosulfate (PAPS). Such cell extracts catalyze the synthesis of a typical heparin when all of the precursors are present.[6] In order to segregate the pre-sulfation steps from the post-sulfation steps in the reaction sequence, incubations have been carried out with the monosaccharide donors but without PAPS, thus precluding any sulfation. In such incubations, a polymeric heparin precursor that lacks sulfate substituents accumulates. When 3'phosphoadenosine-5'-phosphosulfate is added, the precursor is converted to a heparin-like product.

An outline of the major steps in heparinoid biosynthesis is shown in Figure 2.2. Heparinoids are synthesized as polysaccharide chains attached to a serine residue on a core protein via a βGlcA-1,3→βGal-1,3→βGal-1,4→βXyl-1 →Ser linkage region [67-69]. The core proteins of heparin and of heparan sulfates from different sources serve as acceptors, or primers, for the polysaccharide synthesis. If core protein synthesis is blocked or the linkage region cannot be synthesized (as, for example, in mutants lacking the glycosyl transferases for attaching the linkage region monosaccharides [70]), the heparinoid chains cannot be synthesized [16]. The main polysaccharide chains, which are added to the linkage region and which will become the protein-binding portions of these heparinoid structures, are initially simple polymers of repeating βGlcA-1,4-αGlcNAc disaccharide units [see references in 71, 72] which are joined by 1,4 linkages to each other and to the linkage region GlcA. Interestingly, the enzyme that attaches the first GlcNAc to the linkage region is a different transferase from the one that is involved in the elongation of the main chain [73, 74]. The initial polymer structures become much more complex as a result of a series of "maturation reactions" that take place after formation of the (GlcA → GlcNAc)$_n$ polymer. The first maturation step, i.e., the de-N-acetylation of some of the GlcNAc residues (Figure 2.2, reaction 1), is followed by N-sulfation of these residues (reaction 2) [72, 75-81]. Subsequent reactions occur in sequence and involve C5 epimerization of some of the D-glucuronic

---

[6] The actual products contain a heparin-like product as a portion of the final mixture of products (perhaps 50-60% of the total product).

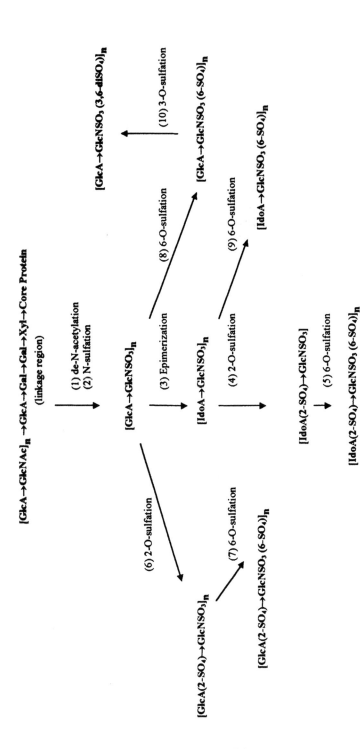

Figure 2.2. Outline of the pathway of biosynthesis of heparin.

24

acid residues to form L-iduronic acid residues [7] (reaction 3) [77, 82-89], O-sulfation of some of the uronic acid residues at C2 (reactions 4 and 6) [8] [77, 84, 90, 91], O-sulfation of some of the GlcNAc and GlcNSO$_3$ residues at C6 (reactions 5, 7, 8, and 9) [92], and finally O-sulfation of a few of the GlcNSO$_3$ and GlcNSO$_3$-6-SO$_4$ residues at C3 (reaction 10) [89, 90, 93-95]. This process yields proteoglycans with M$_r$=750,000-1,000,000, containing heparin chains with M$_r$= ~ 60,000-100,000 [65, 96]. An additional step in the "biosynthesis" (of heparin in mast cells at least) is the cleavage of the glycosaminoglycan chains at some of the GlcA residues by an endo-β-glucuronidase to yield shortened chains (average molecular weight ~ 12 kDa), most of which are no longer attached to the core protein [63-65, 97, 98]. Several aspects of this pathway are important for an understanding of how the complexities of the final structures are controlled.

First, the de-N-acetylation reaction (reaction 1) occurs randomly at some of the GlcNAc residues along the repeating disaccharide chain [75, 99]. When the incubation is carried out in the absence of PAPS *until maximal conversion has occurred*, ~50% of the GlcNAc residues in the polymer are de-N-acetylated [9]. Thus, in the very first maturation step, a mixture of polymer chains is formed (as measured by the sequence of GlcA→GlcNAc and GlcA→GlcN disaccharide units upstream from the linkage region). When PAPS, the SO$_4$ donor, is present (or added), all of the GlcN residues are converted to GlcNSO$_3$ residues (reaction 2) [75, 99].

Second, following the initial de-N-acetylation/N-sulfation reactions (the "first pass"), the subsequent steps, including additional de-N-acetylation/N-sulfation reactions, occur on, or adjacent to, the GlcNSO$_3$-containing disaccharide units, and these reactions continue in either the upstream or the downstream direction, or both, until no further de-N-acetylation/N-sulfation occurs.[10] Thus, the GlcNSO$_3$-containing disaccharides serve as foci for further de-N-acetylation/N-sulfation, epimerization, and O-sulfation [86, 100]. The

---

[7] As a result of a quirk of carbohydrate nomenclature, the D-GlcA to L-IdoA conversion, which results from enzymatic epimerization at C5 of the GlcA, results in a change in the description of the anomeric configuration from β- (for the GlcA) to α- (for the IdoA) even though there is no change in the actual configuration at C1. This is because it is the C5 configuration that serves as a reference point of naming the anomeric configuration.

[8] 2-O-Sulfation occurs primarily on the IdoA units (reaction 4); a very small proportion of the GlcA residues become 2-O-sulfated (reaction 6).

[9] It may be that a lower percentage of the GlcNAc residues are deacetylated *in vivo* in heparin biosynthesis. In heparan sulfate biosynthesis, it is clear that only a few of the GlcNAc's are de-N-acetylated in the first pass.

[10] The mechanisms that control the degree of de-N-acetylation/N-sulfation are unknown.

[unsulfated disaccs]$_{n \text{ is large}}$→[sulfated disaccs]$_{n \text{ is small}}$→[unsulfated disaccs]$_{n \text{ is large}}$

↓
↓
↓

[unsulfated disaccs]$_{n \text{ is small}}$→[sulfated disaccs]$_{n \text{ is large}}$→[unsulfated disaccs]$_{n \text{ is small}}$

Figure 2.3. Structural changes during biosynthetic maturation of heparinoids.

result is that the most highly sulfated disaccharide units in each chain occur in blocks that contain the original GlcNSO$_3$'s, while the unsulfated disaccharide units occur in blocks between the sulfated blocks. As illustrated in Figure 2.3, the relative lengths of the unsulfated blocks and the highly sulfated blocks depend on the extent of further de-N-acetylation/N-sulfation upstream and/or downstream from each of the original GlcNSO$_3$ residues. In heparin biosynthesis the reactions occur to such an extent that only single GlcNAc-containing disaccharides remain in the unsulfated "blocks." In heparan sulfate, the unsulfated blocks contain multiple GlcA→GlcNAc disaccharide units.

Third, the maturation reactions occur at some but not all of the potential disaccharide precursor units, i.e., maturation is incomplete. This is due, in part, to the substrate specificities of the maturation enzymes (below), but probably in larger part to the poorly understood mechanisms that regulate the maturation reactions. In other words, the specificities of the maturation enzymes would permit much more extensive maturation than actually occurs. Thus, the final polymer is composed of a mixture of chains that contain varying amounts of each of the disaccharide units (Table 2.2), with blocks of highly sulfated disaccharide units separated by intervening blocks of unsulfated disaccharides.

Fourth, since most of the sulfation occurs following the conversion of GlcA to IdoA [82, 86], the most highly sulfated heparinoid chains also have the highest IdoA/GlcA ratios.

Fifth, most of the studies on the biosynthesis of heparinoids have been carried out with extracts of mouse mastocytoma cells, cells that presumably synthesize only heparin *in vivo*. However, there are reasons to believe that the basic features of the pathway for heparin and heparan sulfate synthesis are identical. For example, everything that is known about the structures of heparan sulfates is consistent with our understanding of heparin synthesis based on the mastocytoma cell work. Also, Riesenfeld *et al.* [101] have shown that microsomal extracts from rat hepatoma cells, which produce a typical heparan sulfate in whole cell cultures, produce a product that is very similar to that obtained with the mast cell microsomes. The final product made by the hepatoma cell extracts is a typical, highly sulfated heparin-like product, significantly different from the heparan sulfate produced by the whole cells. Apparently, the degree of maturation is somehow regulated in whole cells, and these regulatory mechanisms are lost in cell extracts.

B.   Substrate Specificities of the Heparinoid Biosynthetic Enzymes

1.   Introduction

The specificities of the enzymes involved in the biosynthesis of heparin chains limit the number of possible combinations of disaccharides. At least 13 enzyme activities are required for synthesis of the carbohydrate portions of heparinoid proteoglycans.[11] Since we are interested only in the structures that mediate the binding of heparin to heparin-binding proteins, namely the structures of the main heparinoid chains and, more specifically, the structures formed by the maturation enzymes that act after the formation of the [GlcA$\rightarrow$ GlcNAc]$_n$ polymer, we will consider here the maturation enzymes only. These are all membrane-bound (Golgi) enzymes whose combined activities have been studied by incubation of mouse mastocytoma microsomal fractions with the glycosyl donors (UDPGlcA and UDPGlcNAc) in the absence or presence of the sulfate donor, and whose individual activities have been studied with detergent-treated mouse mastocytoma microsomal fractions or with purified or cloned enzymes.

For the specificity studies, oligosaccharide substrates of known structure have been prepared by fragmentation of heparin or the *E. coli* K5 polysaccharide [72, 102]. The latter is a polymer of repeating β-GlcA 1,4-α-GlcNAc disaccharide units, identical to the initial [β-GlcA$\rightarrow\alpha$-GlcNAc]$_n$ product formed in heparin biosynthesis (but lacking a linkage region and a core protein). Procedures used for the preparation and characterization of oligosaccharide substrates will be described in Chapter 3. For now, let it be said that a valid test for substrate specificity, as it operates *in vivo*, may require tri-, tetra-, penta-, or higher heparinoid oligosaccharide acceptors and that the isolation and structural characterization of the desired acceptor substrates is extremely difficult (in some

---

[11] For synthesis of the linkage region, xylosyl transferase, galactosyl transferases I and II, and glucuronosyl transferase I; for the synthesis of the initial polymer, glucuronosyl transferase II and N-acetylglucosamine transferase; and for the maturation steps, the de-N-acetylase, the N-sulfotransferase, the uronic acid epimerase, the uronic acid 2-O-sulfotransferases, the GlcN 6- O-sulfotransferase(s), and the GlcN 3-O-sulfotransferase(s).

cases bordering on the impossible!). As a consequence, only a limited number of substrates have been tested. In the present discussion, we can consider only those results of the specificity studies with the more readily available substrates that have been reported in the literature. This discussion then leads to suggestions of how these specificities control the final structures of heparinoids.

2.   The de-N-acetylase/N-sulfotransferase

The *de-N-acetylase* initiates the maturation process by removing acetyl groups from GlcNAc residues at random sites along the $[GlcA{\rightarrow}GlcNAc]_n$ chain, before any N- or O-sulfation takes place. The enzyme can recognize any GlcNAc in the $[GlcA{\rightarrow}GlcNAc]_n$ polymer. Interestingly, when microsomal preparations are incubated with UDPGlcA and UDPGlcNAc in the absence of PAPS (to preclude the sulfation steps), the de-N-acetylated product that accumulates contains approximately equal numbers or GlcNAc and GlcN residues [75, 99]. This suggests that the positively charged GlcN residues located close to otherwise susceptible GlcNAc residues prevent further de-N-acetylation, a possibility that has not yet been tested with oligosaccharide substrates. Although few, if any, N-unsubstituted glucosamine residues are found in heparin, there is now good evidence that a few N-unsubstituted GlcN residues occur in heparan sulfates [103]. Thus, in most cases the N-sulfotransferase can transfer SO_4 from PAPS to the GlcN residues that are formed by the de-N-acetylase. Of course, if PAPS is not available, some N-unsubstituted GlcN residues may accumulate.

It has been found that both the deacetylase and the N-sulfotransferase activities reside in a single protein [80, 81, 104-106], suggesting that these steps occur in concert *in vivo*. In fact, it may be appropriate to think of the de-N-acetylation and the N-sulfation as though they were a single reaction *in vivo*.[12] Following the initial round of de-N-acetylation/N-sulfation ("first pass"), the de-N-acetylase catalyzes further de-N-acetylation of some of the remaining GlcNAc residues, which is followed by N-sulfation of the resulting GlcN residues. Thus, GlcNAc residues that were not converted in the first pass become substrates for further N-sulfation [77]. These subsequent de-N-acetylation/N-sulfation reactions may be considered as a part of the "further maturation." With synthetic substrates, it has been shown that a GlcNAc residue can be deacetylated by the enzyme when there is a $GlcNSO_3$ on either the

---

[12] Perhaps the reason the N-unsubstituted GlcN residues are found in heparan sulfates is that the heparan sulfate-producing cells cannot produce enough PAPS to complete the N-sulfation reaction (see Section III.D., p. 34)

immediate upstream or the immediate downstream disaccharide, and that such adjacent GlcNSO$_3$ residues markedly increase the rate of deacetylation of the GlcNAc residues [72, 77]. Thus, the GlcNAc-containing disaccharides around the initial GlcNSO$_3$ residues become *preferred substrates* for the further de-N-sulfation/N-acetylation. What is not clear is whether the enzyme can act if *both* adjacent disaccharides contain GlcNSO$_3$'s, i.e., can GlcNAc residues in GlcNSO$_3 \rightarrow$GlcA$\rightarrow$GlcNAc$\rightarrow$GlcA$\rightarrow$GlcNSO$_3$ sequences be converted by the enzyme? The fact that isolated GlcNAc residues persist in heparins suggests the possibility that GlcNSO$_3$-containing disaccharides on both sides of a GlcA$\rightarrow$ GlcNAc disaccharide may preclude the final de-N-acetylation, but such substrates have not been tested. Several forms of this enzyme have been cloned [107-110], and it is as yet uncertain whether the different forms exhibit different substrate specificities [reviewed in reference 20].

3.   The C2 epimerase

The *C2 epimerase* catalyzes the interconversion of unsulfated β-$\underline{D}$-glucuronic and α-$\underline{L}$-iduronic acid residues.[7] The enzyme has been purified from bovine liver [111]. In isolation, this reaction is reversible [83-85, 87, 111] and reaches an equilibration with a ratio of GlcA/IdoA of approximately 4:1 (as expected from the relative thermodynamic stabilities of the GlcA and the IdoA rings). The reaction is pulled in the direction of IdoA by the 2-O-sulfotransferase that converts IdoA to IdoA-2-SO$_4$ and (presumably) in the direction of the GlcA by the 2-O-sulfotransferase that converts GlcA to GlcA-2-SO$_4$. 2-O-Sulfation of iduronic acid prevents further action of the epimerase.[13] There is an absolute requirement for N-sulfation before the epimerase can act; i.e., the epimerase will not attack the [GlcA$\rightarrow$GlcNAc]$_n$ polymer [82, 85]. The enzyme will use

$$GlcNSO_3 \rightarrow HexA \rightarrow GlcNSO_3$$

sequences as substrates. Also, the enzyme acts on substrates with an upstream GlcNSO$_3$ and a downstream GlcNAc, *but does not attack the opposite sequence* [89], i.e., the enzyme acts on

$$GlcNSO_3 \rightarrow GlcA \rightarrow GlcNAc,$$

but *not* on

$$GlcNAc \rightarrow GlcA \rightarrow GlcNSO_3.$$

---

[13] Although it is likely that sulfation of the GlcA blocks action of the epimerase, GlcA-2-O-sulfate-containing substrates have not been available to test with the epimerase.

Thus,

$$GlcNSO_3 \rightarrow IdoA \rightarrow GlcNAc$$

sequences can occur in heparinoids, but

$$GlcNAc \rightarrow IdoA \rightarrow GlcNSO_3$$

sequences *cannot be formed*. In effect, the upstream GlcNAc blocks the epimerization of its immediate downstream GlcA. We have already noted that, with only two exceptions, the GlcNAc-containing tetrasaccharides that have been identified to date (Table 2.3) contain a GlcA downstream from the GlcNAc, as predicted by these enzyme specificity studies. This suggests that the GlcNAc→IdoA sequences in Table 2.3 may be incorrectly identified.

4.   The iduronosyl-2-O-sulfotransferase

The *iduronosyl-2-O-sulfotransferase* catalyzes the O-sulfation of some of the IdoA residues that are formed by the epimerase. It has been shown that the IdoA in an IdoA→GlcNSO$_3$ disaccharide can be sulfated by the enzyme [92]. Such disaccharide acceptors must occur in a sequence with an upstream GlcNSO$_3$ (e.g., GlcNSO$_3$→IdoA→GlcNSO$_3$), since, as described above, the epimerase cannot form GlcNAc→IdoA→GlcNSO$_3$ sequences. The 2-O-sulfotransferase does not act on the IdoA→GlcNSO$_3$-6-SO$_4$ units; apparently the 6-O-sulfation at the downstream GlcNSO$_3$ blocks the action of the 2-O-sulfotransferase [92]. The enzyme has been purified to homogeneity from mouse mastocytoma cell extracts [112] and from Chinese hamster cell culture medium [113]. The mastocytoma enzyme apparently has both 2-O-sulfotransferase and 6-O-sulfotransferase activity, whereas the CHO cell enzyme has only the 2-O-sulfotransferase activity.

5.   The glucuronosyl 2-O-sulfotransferase

Heparinoids contain small amounts of O-sulfated glucuronosyl residues in GlcA-2-SO$_4$→GlcNSO$_3$ and GlcA-2-SO$_4$→GlcNSO$_3$-6-SO$_4$ disaccharide units, which appear to be located in the highly sulfated regions of these polymers (Table 2.3) [32]. Kusche and Lindahl [91] have shown that the pentasaccharide

$$GlcNSO_3 \rightarrow GlcA \rightarrow GlcNSO_3 \rightarrow GlcA \rightarrow GlcNSO_3$$

serves as a sulfate acceptor for the *glucuronosyl-O-sulfotransferase* and that there is a very active glucuronosyl 2-O-sulfotransferase in mouse mastocytoma microsomal fractions. Note that this substrate is the initial maturation product, formed by the de-N-acetylase/N-sulfotransferase. Since GlcA-2-SO$_4$ residues can occur either upstream or downstream from a GlcNSO$_3$-6-SO$_4$ residue (Table 2.3, structures 29 and 33), it would appear that the 2-O-sulfation of GlcA residues does not prevent 6-O-sulfation on adjacent GlcN residues. Alternatively, the 2-O-sulfation may occur after the adjacent GlcNSO$_3$-6-SO$_4$ residues are formed. We will return to this point later.

The demonstration that a mutant CHO cell line that lacks IdoA 2-O-sulfotransferase fails to form GlcA-2-SO$_4$ residues as well suggests that the glucuronosyl 2-O-sulfotransferase is the same enzyme as the IdoA 2-O-sulfotransferase [114].

6.   The GlcN-6-O-sulfotransferase

The *GlcN 6-O-sulfotransferase* will sulfate GlcNAc residues in heparinoids, provided there is a GlcNSO$_3$ residue on either the upstream or the downstream disaccharide [92]. It will also sulfate a GlcNSO$_3$ residue with an upstream IdoA or GlcA (i.e., *any* GlcNSO$_3$), but GlcNAc residues in the [GlcNAc→GlcA]$_n$ block sequences in heparan sulfates cannot be sulfated. It is of particular interest that both the GlcNAc and the GlcNSO$_3$ residues that are found at the *boundaries* between the sulfated blocks and the unsulfated blocks can be sulfated by the enzyme, even though GlcNAc residues *within* the blocks cannot. The enzyme has been purified to apparent homogeneity from mouse mastocytoma cell extracts [112] and from Chinese hamster cell culture medium [115]. The mastocytoma enzyme apparently has both 2-O-sulfotransferase and 6-O-sulfotransferase activity, whereas the CHO cell enzyme has only the 6-O-sulfotransferase activity

7.   The 3-O-sulfotransferase

A late reaction in the biosynthetic pathway is the 3-O-sulfation of the GlcNSO$_3$ residue that yields the GlcA→GlcNSO$_3$ (3,6-diSO$_4$) disaccharide unit shown in the upper right-hand corner of Figure 2.2 [89, 90, 93-95, 116]. The enzyme that catalyzes this reaction requires the specific and rare pentasaccharide sequence below as an acceptor:

$$\downarrow$$
GlcNR-6-SO$_4$→GlcA→GlcNSO$_3$-6-SO$_4$→IdoA-2-SO$_4$ → GlcNSO$_3$-6-SO$_4$,

where R is either acetyl or sulfate, and the $\downarrow$ indicates the site of 3-O-sulfation. The GlcNSO$_3$ residue that becomes sulfated does not necessarily have to be 6-O-sulfated (see structures 12, 18, 38, and 50 in Table 2.3), but the corresponding acceptor without this 6-O-sulfate has not been tested as a substrate for the 3-O-sulfotransferase. These 3-O-sulfated GlcN residues occur primarily in heparinoid sequences that are rich in IdoA and sulfate. The 3-O-sulfation is required for the anticoagulant activity of heparinoids, and the pentasaccharide sequence formed by the 3-O-sulfotransferase is the minimum structure required for tight binding to antithrombin (Chapter 7). Thus, the *GlcNSO$_3$ -3-O-sulfotransferase* that catalyzes this late step in heparinoid polymer maturation converts the nonanticoagulant polymer into an anticoagulant. The enzyme has recently been highly purified from LTA cells and it has been suggested that there may be multiple species of the 3-O-sulfotransferase with different specificities [117].

C.   Implications of Biosynthesis Results for Heparinoid Structures

The structural features of heparinoids that are suggested by the available information on the biosynthesis pathway and the specificity of the maturation enzymes are summarized as follows:

*   The random nature of the initial de-N-acetylation reaction would be expected to introduce structural heterogeneity at an early stage of the biosynthesis. This, in fact, happens.

*   Because the de-N-acetylation reactions that take place after the first pass occur much more rapidly at GlcNAc residues adjacent to GlcNSO$_3$-containing disaccharides, and because epimerization can occur only at uronic acid residues adjacent to GlcNSO$_3$ residues, the initial GlcNSO$_3$ residues represent the foci for further maturation reactions. These are the specificities that apparently give rise to block structures in heparan sulfate in which there are sequences of sulfated disaccharides separated by sequences of nonsulfated disaccharides.

*   The epimerase can use either

$$\text{GlcNSO}_3 \rightarrow \text{GlcA} \rightarrow \text{GlcNSO}_3$$

    or

$$\text{GlcNSO}_3 \rightarrow \text{GlcA} \rightarrow \text{GlcNAc}$$

    sequences as substrates, but cannot act on

$$\text{GlcNAc} \rightarrow \text{GlcA} \rightarrow \text{GlcNSO}_3$$

sequences. Therefore,

$$GlcNAc \rightarrow IdoA \rightarrow GlcNSO_3$$

sequences *cannot occur* in heparinoids. This conclusion is consistent with most of the studies of the GlcNAc-containing sequences in heparin, which show that all GlcNAc residues have a downstream GlcA.

- Both the

$$IdoA \rightarrow GlcNSO_3\text{-}6\text{-}SO_4$$

and the

$$IdoA\text{-}2\text{-}SO_4 \rightarrow GlcNSO_3$$

disaccharides that occur in heparinoids are formed from the same precursor disaccharide, namely

$$IdoA \rightarrow GlcNSO_3.$$

If the 6-O-sulfation of this precursor occurs first, the 2-O-sulfotransferase cannot act, and

$$IdoA \rightarrow GlcNSO_3\text{-}6\text{-}SO_4$$

becomes an end product disaccharide in the biosynthesis. If the 2-O-sulfation occurs first, the 6-O-sulfotransferase *can* act; if it does,

$$IdoA\text{-}2\text{-}SO_4 \rightarrow GlcNSO_3\text{-}6\text{-}SO_4$$

may become an end-product disaccharide structure in the biosynthesis. If, however, some regulatory mechanism prevents the 6-O-sulfation following the 2-O-sulfation of the IdoA,

$$IdoA\text{-}2\text{-}SO_4 \rightarrow GlcNSO_3$$

may become an end product. This disaccharide occurs in significant amounts in heparinoids (see Tables 2.2, and 2.3).

- The specificity of the GlcN-6-O-sulfotransferase requires that, in order to become 6-O-sulfated, GlcNAc residues must occur adjacent to GlcNSO$_3$-containing disaccharides. This explains why GlcNAc residues within extended $[GlcA \rightarrow GlcNAc]_n$ blocks of heparinoids are not sulfated.

- The maturation enzymes acting in concert fail to convert the initial

$$[GlcA \rightarrow GlcNAc]_n$$

sequences into the fully matured products that are suggested by their known specificities. Instead, incomplete maturation is obtained.

Many of the remaining problems in the understanding of heparin biosynthesis have been outlined by Kjellén and Lindahl [13].

D.   Branching Pathways Lead to Structural Diversity

Figure 2.4 shows a hypothetical reaction sequence that would be expected when all of the known enzyme specificities are taken into account and all of the enzymes catalyze complete transformation of all of their potential disaccharide substrates. The initial reactant (I) is the product of the polymerization of the GlcA and the GlcNAc from their respective UDP sugar donors. For the subsequent reactions in Figure 2.4, two assumptions are made: (a) 2-O-sulfation takes place prior to 6-O-sulfation, thus allowing the O-sulfation of both the IdoA residues and the $GlcNSO_3$ residues, and (b) the de-N-acetylation/N-sulfation reactions cannot occur at isolated GlcNAc-containing disaccharides when these are flanked by $GlcNSO_3$-containing disaccharides (the latter assumption is predicated on the fact that heparins always contain a number of isolated GlcNAc-containing disaccharides). The enzyme specificities are invoked for the epimerase and the 6-Osulfotransferase, i.e., GlcA cannot be epimerized in the following sequence,

$$GlcNAc \rightarrow GlcA \rightarrow GlcNSO_3,$$

and GlcNAc residues can be sulfated if they are adjacent to a $GlcNSO_3$-containing disaccharide. If de-N-acetylation/N-sulfation is initiated at several points along the chain (at the center and the two extreme disaccharide units in our example, yielding structure II), the first four reactions generate a sequence containing several GlcNAc residues that are flanked by $GlcNSO_3$-containing disaccharides and that have GlcA residues in their immediate downstream positions. As a consequence of such a reaction sequence, product V contains a number of

$$GlcNAc-6-SO_4 \rightarrow GlcA \rightarrow GlcNSO_3-6-SO_4 \rightarrow IdoA-2-SO_4 \rightarrow GlcNSO_3-6-SO_4$$

sequences. These represent the pentasaccharide acceptor sequence required for the 3-O-sulfotransferase reaction. Thus, the final reaction in the sequence, reaction 5, generates antithrombin-binding sites, e.g., D-E-F-G-H and P-Q-R-S-T (in Figure 2.4, the latter sequence does not extend to the S and T residues), at several positions in the chain. Clearly, such a reaction sequence, which is fully compatible with all of the enzyme specificities, does not generate a typical heparin. It makes a product that would have a much greater anticoagulant activity than typically found in heparin. Furthermore, product VI, the final product, contains only 3 of the 19 disaccharide building blocks that are actually found in heparinoids. What is the cause of the much greater complexity than predicted by these reasonable assumptions? There are several possibilities.

First, as noted earlier, the GlcA 2-O-sulfotransferase acts on the sequence

$$GlcNSO_3 \rightarrow GlcA \rightarrow GlcNSO_3 \rightarrow GlcA \rightarrow GlcNSO_3,$$

which occurs prominently in intermediate II of Figure 2.4. Such a sequence is a substrate for both the epimerase (reaction 2) and the GlcA 2-O-sulfotransferase (Figure 2.2, reaction 6). It is likely that the GlcA 2-O-sulfotransferase competes with the epimerase for these disaccharide substrates. However, the epimerase competes so effectively that the amount of $GlcA\text{-}2\text{-}SO_4$ that is formed is very small, exactly as is found in a typical heparin. Furthermore, if the GlcA 2-O-sulfotransferase captures some of intermediate II, then a small amount of an additional minor disaccharide,

$$GlcA\text{-}2\text{-}SO_4 \rightarrow GlcNSO_3\text{-}6\text{-}SO_4,$$

can be formed also (Figure 2.2, reaction 7). Thus, the diversion of a small amount of intermediate II by the GlcA 2-O-sulfotransferase would account for the synthesis of two more of the disaccharides found in heparin.

A second reason for the higher complexity of heparinoid sequences is that

$$\text{the } IdoA \rightarrow GlcNSO_3$$

units in intermediate III are potential substrates for both the IdoA 2-O-sulfotransferase and the 6-O-sulfotransferase. Thus, a competition between these enzymes for the disaccharide acceptors may occur. As was noted, if the 2-O-sulfotransferase acts on the acceptor disaccharide first, generating

$$IdoA\text{-}2\text{-}SO_4 \rightarrow GlcNSO_3,$$

then this disaccharide can serve as an acceptor for the 6-O-sulfotransferase. If, however, the 6-O-sulfotransferase acts on the acceptor disaccharide first, generating

$$IdoA \rightarrow GlcNSO_3\text{-}6\text{-}SO_4,$$

then this disaccharide cannot serve as an acceptor for the 2-O-sulfotransferase. The latter reaction sequence is shown in Figure 2.5 where product IV accumulates. Product IV contains three disaccharide units that are not found in the final product in Figure 2.4, but does not yield a typical heparinoid. Product IV is not an effective substrate for the 3-O-sulfotransferase. Consequently, an anticoagulantly active heparin could not be formed. The actual presence in heparinoids of these additional disaccharides resulting from the competition

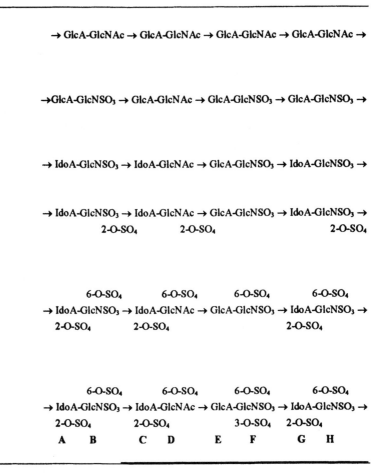

→ GlcA-GlcNAc → GlcA-GlcNAc → GlcA-GlcNAc → GlcA-GlcNAc →

→GlcA-GlcNSO₃ → GlcA-GlcNAc → GlcA-GlcNSO₃ → GlcA-GlcNSO₃ →

→ IdoA-GlcNSO₃ → IdoA-GlcNAc → GlcA-GlcNSO₃ → IdoA-GlcNSO₃ →

→ IdoA-GlcNSO₃ → IdoA-GlcNAc → GlcA-GlcNSO₃ → IdoA-GlcNSO₃ →
     2-O-SO₄   2-O-SO₄   2-O-SO₄

     6-O-SO₄   6-O-SO₄   6-O-SO₄   6-O-SO₄
→ IdoA-GlcNSO₃ → IdoA-GlcNAc → GlcA-GlcNSO₃ → IdoA-GlcNSO₃ →
     2-O-SO₄   2-O-SO₄   2-O-SO₄

     6-O-SO₄   6-O-SO₄   6-O-SO₄   6-O-SO₄
→ IdoA-GlcNSO₃ → IdoA-GlcNAc → GlcA-GlcNSO₃ → IdoA-GlcNSO₃ →
     2-O-SO₄   2-O-SO₄   3-O-SO₄   2-O-SO₄
     A   B   C   D   E   F   G   H

Figure 2.4. Sequential synthesis of heparin: 2-O-sulfation precedes 6-O-GlcNSO₃-containing disaccharides occur on both sides of the GlcNAc.

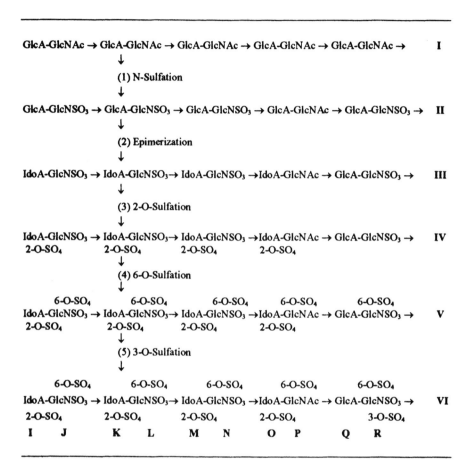

GlcA-GlcNAc → GlcA-GlcNAc → GlcA-GlcNAc → GlcA-GlcNAc → GlcA-GlcNAc →   I
                ↓
                (1) N-Sulfation
                ↓
GlcA-GlcNSO₃ → GlcA-GlcNSO₃ → GlcA-GlcNSO₃ → GlcA-GlcNAc → GlcA-GlcNSO₃ →   II
                ↓
                (2) Epimerization
                ↓
IdoA-GlcNSO₃ → IdoA-GlcNSO₃→ IdoA-GlcNSO₃ →IdoA-GlcNAc → GlcA-GlcNSO₃ →   III
                ↓
                (3) 2-O-Sulfation
                ↓
IdoA-GlcNSO₃ → IdoA-GlcNSO₃ → IdoA-GlcNSO₃ →IdoA-GlcNAc → GlcA-GlcNSO₃ →   IV
2-O-SO₄   2-O-SO₄   2-O-SO₄   2-O-SO₄
                ↓
                (4) 6-O-Sulfation
                ↓
    6-O-SO₄   6-O-SO₄   6-O-SO₄   6-O-SO₄   6-O-SO₄
IdoA-GlcNSO₃ → IdoA-GlcNSO₃ → IdoA-GlcNSO₃ →IdoA-GlcNAc → GlcA-GlcNSO₃ →   V
2-O-SO₄   2-O-SO₄   2-O-SO₄   2-O-SO₄
                ↓
                (5) 3-O-Sulfation
                ↓
    6-O-SO₄   6-O-SO₄   6-O-SO₄   6-O-SO₄   6-O-SO₄
IdoA-GlcNSO₃ → IdoA-GlcNSO₃→ IdoA-GlcNSO₃ →IdoA-GlcNAc → GlcA-GlcNSO₃ →   VI
2-O-SO₄   2-O-SO₄   2-O-SO₄   2-O-SO₄   3-O-SO₄
I   J   K   L   M   N   O   P   Q   R

sulfation. It is assumed here that no further de-N-acetylation occurs if

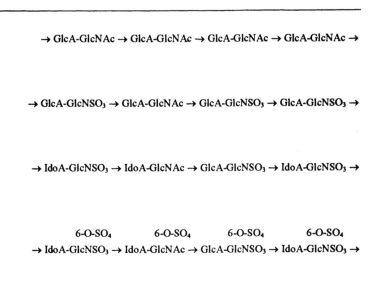

→ GlcA-GlcNAc → GlcA-GlcNAc → GlcA-GlcNAc → GlcA-GlcNAc →

→ GlcA-GlcNSO$_3$ → GlcA-GlcNAc → GlcA-GlcNSO$_3$ → GlcA-GlcNSO$_3$ →

→ IdoA-GlcNSO$_3$ → IdoA-GlcNAc → GlcA-GlcNSO$_3$ → IdoA-GlcNSO$_3$ →

    6-O-SO$_4$ 6-O-SO$_4$ 6-O-SO$_4$ 6-O-SO$_4$
→ IdoA-GlcNSO$_3$ → IdoA-GlcNAc → GlcA-GlcNSO$_3$ → IdoA-GlcNSO$_3$ →

Figure 2.5. Sequential synthesis of heparin: 6-O-sulfation prevents 2-O-sulfation. disaccharides occur on both sides of the GlcNAc. NR = no reaction.

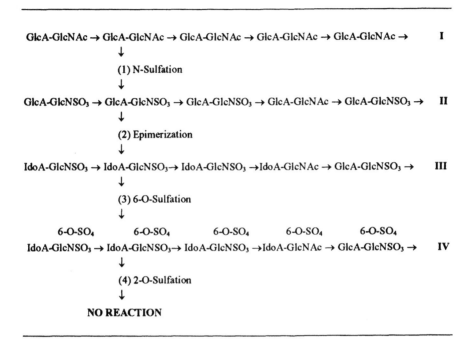

GlcA-GlcNAc → GlcA-GlcNAc → GlcA-GlcNAc → GlcA-GlcNAc → GlcA-GlcNAc →   I
                    ↓
            (1) N-Sulfation
                    ↓
GlcA-GlcNSO₃ → GlcA-GlcNSO₃ → GlcA-GlcNSO₃ → GlcA-GlcNAc → GlcA-GlcNSO₃ →   II
                    ↓
            (2) Epimerization
                    ↓
IdoA-GlcNSO₃ → IdoA-GlcNSO₃→ IdoA-GlcNSO₃ →IdoA-GlcNAc → GlcA-GlcNSO₃ →   III
                    ↓
            (3) 6-O-Sulfation
                    ↓
     6-O-SO₄   6-O-SO₄   6-O-SO₄   6-O-SO₄   6-O-SO₄
IdoA-GlcNSO₃ → IdoA-GlcNSO₃→ IdoA-GlcNSO₃ →IdoA-GlcNAc → GlcA-GlcNSO₃ →   IV
                    ↓
            (4) 2-O-Sulfation
                    ↓
            NO REACTION

It is assumed that no further de-N-acetylation occurs if GlcNSO₃-containing between enzymes at the branch points, plus the disaccharides in the final products of Figures 2.4 and 2.5, is consistent with the idea that there is a true competition between the GlcA-2-O-sulfotransferase and the epimerase and between the 2-O- and 6-O-sulfotransferases. In the latter case, the high levels of

$$IdoA-2-SO_4 \rightarrow GlcNSO_3-6-SO_4$$

that are found in heparins (Table 2.4) suggests that the 2-O-sulfotransferase must fare quite well in this competition.

E.   Progressive Maturation Steps in the Biosynthesis of Heparinoids

There is clearly a required order in which the maturation enzymes must act, e.g., the epimerase cannot act on the $[GlcA \rightarrow GlcNAc]_n$ sequences and the 6-O-sulfotransferase cannot act on GlcNAc residues unless they occur next to a GlcNSO$_3$-containing disaccharide. Thus, there is an ordered reaction sequence. What is less clear is whether the whole maturing heparinoid chain must be completely converted by one maturation enzyme before the next can act, as depicted in Figures 2.4 and 2.5. In this context, the two cases above in which a single disaccharide unit, or sequence of disaccharides, can serve as a substrate for two different maturation enzymes illustrates how such competition can affect the nature of the disaccharides formed and the complexities of the sequences in which they occur in the final product. However, they still do not account for the actual structural complexity of heparinoids. Figures 2.6 and 2.7 show modified schemes for the maturation of the initial $[GlcA \rightarrow GlcNAc]_n$ polymer. In both cases, the starting material is depicted as the product of the first pass of the de-N-acetylation/N-sulfation reaction that yields a chain with widely spaced GlcNSO$_3$ residues—in this case, one in the middle of the chain and one each at the two extremes. Figure 2.6 shows a stepwise progression of subsequent de-N-acetylation/N-sulfation reactions that would occur when the epimerase is not present. This sequence starts in the middle and progresses in both the upstream and the downstream directions. For the sake of simplicity, progressive steps from the two extreme GlcNSO$_3$-containing disaccharides toward the middle are not shown, although such progressions would be consistent with the proposed biosynthetic scheme. Again it is assumed that the de-N-acetylase cannot act on a GlcNAc between two GlcNSO$_3$-containing disaccharides. As a result, the progression stops when a single isolated GlcA$\rightarrow$GlcNAc disaccharide remains.

Figure 2.7 shows the same type of progressive synthesis that would occur when the epimerase is also present. This series of reactions begins with the same partially N-sulfated "first pass" product found in Figure 2.6. This intermediate is then subjected to alternating, stepwise epimerase and deacetylase/N-sulfotransferase reactions, invoking once again the specificity of the epimerase as the reaction sequence progresses. Interestingly, this reaction sequence suggests that the deacetylase may act on

$$IdoA \rightarrow GlcNAc,$$

but this substrate has not been tested.

Although not shown in Figure 2.7, it is obvious that all of the O-sulfotransferases would participate in such a progressive maturation reaction sequence as their specific sulfate acceptor di- or oligosaccharide structures are generated by the reactions shown. Furthermore, the competition of the pairs of enzymes that act on the

$$GlcA \rightarrow GlcNSO_3$$

and the

$$IdoA \rightarrow GlcNSO_3$$

disaccharides (Section III.D., p. 34) can also contribute to the generation of complex structures.

F.   Unsulfated Blocks as Intermediates in Heparin Biosynthesis

The progressive reaction sequences shown in Figures 2.6 and 2.7 yield intermediates that contain blocks of $GlcA \rightarrow GlcNAc$ disaccharides and blocks that are rich in IdoA and $GlcNSO_3$. Addition of the O-sulfotransferase reactions to these sequences would show the appearance of $IdoA-2-SO_4$ and $GlcNSO_3-6-SO_4$, as well as 3-O-sulfated GlcN residues in the blocks that are rich in IdoA and $GlcNSO_3$. Thus, the progressive reaction sequences would generate heparan sulfate-like products as intermediates in the heparin biosynthesis. Furthermore, any regulatory mechanism that would arrest the de-N-acetylase/N-sulfotransferase reaction early would yield a heparan sulfate-like product. One simple mechanism that might control the progression would be a limitation of the amount of 3'-phosphoadenosine-5'-phosphosulfate, the sulfate donor, perhaps by regulation of its synthesis or its transport into the Golgi where these reactions take place. A shortage of PAPS would prevent the N-sulfation but would not stop the conversion of GlcNAc residues to N-unsubstituted GlcNs. It would also limit the amount of O-sulfation in the regions rich in IdoA and $GlcNSO_3$. This is precisely what is found in heparan sulfate; namely, blocks of $GlcA \rightarrow GlcNAc$, reduced sulfation in the IdoA-rich blocks, and some N-unsubstituted GlcN residues at the junctions between the blocks.

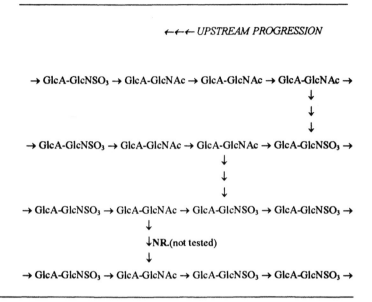

← ← ← *UPSTREAM PROGRESSION*

Figure 2.6. Proposed progression of de-N-acetylation/N-sulfation during heparinoid bio
residues in I are formed in the "first pass" of de-N-sulfation/N-acetylation (not shown).
lustrative purposes, the progressions here proceed only from the central GlcNSO$_3$
GlcNSO$_3$-containing disaccharides occur on both sides of the GlcNAc. NR = no reaction

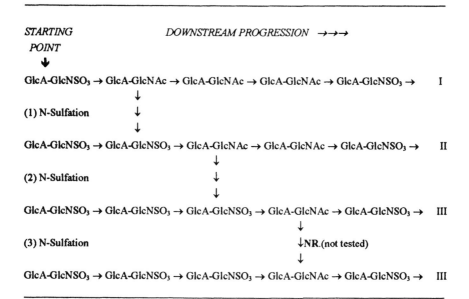

STARTING   *DOWNSTREAM PROGRESSION* →→→
  POINT
    ↓

GlcA-GlcNSO₃ → GlcA-GlcNAc → GlcA-GlcNAc → GlcA-GlcNAc → GlcA-GlcNSO₃ →   I
                     ↓
(1) N-Sulfation   ↓
                     ↓
GlcA-GlcNSO₃ → GlcA-GlcNSO₃ → GlcA-GlcNAc → GlcA-GlcNAc → GlcA-GlcNSO₃ →   II
                                  ↓
(2) N-Sulfation   ↓
                                  ↓
GlcA-GlcNSO₃ → GlcA-GlcNSO₃ → GlcA-GlcNSO₃ → GlcA-GlcNAc → GlcA-GlcNSO₃ →   III
                                                  ↓
(3) N-Sulfation   ↓NR.(not tested)
                                                  ↓
GlcA-GlcNSO₃ → GlcA-GlcNSO₃ → GlcA-GlcNSO₃ → GlcA-GlcNAc → GlcA-GlcNSO₃ →   III

synthesis (without epimerase action). It is assumed that the central and end GlcNSO₃
Although progression can occur in both directions from each GlcNSO₃ residue, for il
formed in the first pass. It is also assumed that no further de-N-acetylation occurs if

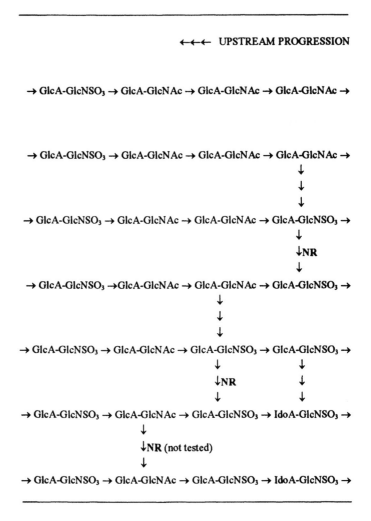

Figure 2.7. Proposed progression of de-N-acetylation/sulfation reactions

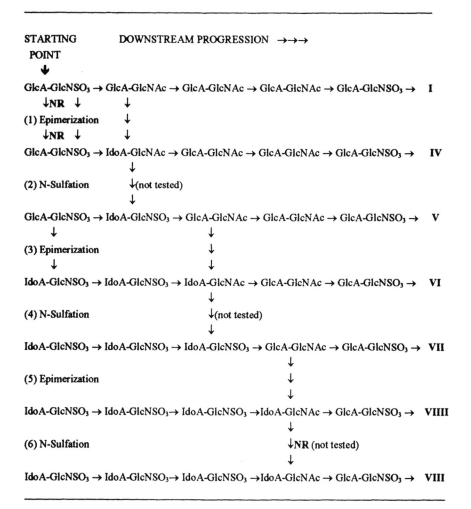

STARTING   DOWNSTREAM PROGRESSION →→→
  POINT
  ↓

GlcA-GlcNSO₃ → GlcA-GlcNAc → GlcA-GlcNAc → GlcA-GlcNAc → GlcA-GlcNSO₃ →   I
  ↓NR ↓   ↓
(1) Epimerization   ↓
  ↓NR ↓   ↓
GlcA-GlcNSO₃ → IdoA-GlcNAc → GlcA-GlcNAc → GlcA-GlcNAc → GlcA-GlcNSO₃ →   IV
                        ↓
(2) N-Sulfation   ↓(not tested)
                        ↓
GlcA-GlcNSO₃ → IdoA-GlcNSO₃ → GlcA-GlcNAc → GlcA-GlcNAc → GlcA-GlcNSO₃ →   V
  ↓   ↓
(3) Epimerization   ↓
  ↓   ↓
IdoA-GlcNSO₃ → IdoA-GlcNSO₃ → IdoA-GlcNAc → GlcA-GlcNAc → GlcA-GlcNSO₃ →   VI
                         ↓
(4) N-Sulfation   ↓(not tested)
                         ↓
IdoA-GlcNSO₃ → IdoA-GlcNSO₃ → IdoA-GlcNSO₃ → GlcA-GlcNAc → GlcA-GlcNSO₃ → VII
                         ↓
(5) Epimerization   ↓
                         ↓
IdoA-GlcNSO₃ → IdoA-GlcNSO₃→ IdoA-GlcNSO₃ →IdoA-GlcNAc → GlcA-GlcNSO₃ → VIIII
                         ↓
(6) N-Sulfation   ↓NR (not tested)
                         ↓
IdoA-GlcNSO₃ → IdoA-GlcNSO₃→ IdoA-GlcNSO₃ →IdoA-GlcNAc → GlcA-GlcNSO₃ →   VIII

during heparinoid biosynthesis—with epimerase action. NR = no reaction.

The one observation that must be reconciled with the concept of heparan sulfate-like structures among the heparin biosynthesis intermediates is the finding that, in the first pass of de-N-acetylation/N-sulfation in microsomal preparations, approximately 50% of the GlcNAc's are converted to $GlcNSO_3$'s. Analyses of these initial biosynthetic products show that most of the GlcA→ $GlcNSO_3$ disaccharides are adjacent to each other or separated by only one or two GlcA→GlcNAc disaccharide units [14] [99], leaving very short blocks of unsulfated disaccharides. While it is easy to imagine that further de-N-acetylation/N-sulfation reactions can reduce the number of GlcNAc-containing disaccharides to the level of 10-15% of the total GlcN residues that are found in heparin, the fact that some GlcNAc residues remain between two $GlcNSO_3$-containing disaccharides while others in the same structural environment become N-sulfated leads to two suggestions. The first, which we have already mentioned, is that the deacetylase cannot act on a GlcNAc residue that is flanked by $GlcNSO_3$-containing disaccharides. The second is that, *in vivo*, in contrast to the *in vitro* results, only 10-15% of the GlcNAc's become N-sulfated in the first pass.The progressive biosynthesis pathway in Figure 2.7, with the additional provisions for the O-sulfation and for the branches in the pathway described above, (a) would be consistent with the biosynthesis studies, (b) would ultimately yield a product having all of the properties of heparin, and (c) would account for unsulfated block structures as intermediates in heparin biosynthesis. Such blocks of unsulfated disaccharides are found in heparin by-products, and Cifonelli and Dorfman [47] suggested long ago that these heparin by-products are intermediates in heparin biosynthesis. Additional support for the idea that block structures are (or can be) intermediates in heparin biosynthesis comes from the observations of Jacobsson and Lindahl [92] that exogenous heparan sulfate can serve as a substrate for de-N-acetylation/N-sulfation in mouse mastocytoma extracts. The possibility of unsulfated block structures in the biosynthesis of heparin also adds weight to the suggestion that the pathways of heparin and heparan sulfate synthesis are essentially the same. As mentioned in Section III.A., p. 22, that microsomal extracts from mastocytoma cells, which make a typical heparin, and from hepatoma cells, which make typical heparan sulfate, both yield similar heparin-like products when incubated with UDPGlcA, UDPGlclNAc, and PAPS *in vitro*. Here, the regulatory mechanisms that operate in intact hepatoma cells are lost *in vitro*. The suggestion earlier that the availability of PAPS could serve as a simple regulatory mechanism in whole cells would explain these results, as PAPS is provided in excess in microsomal incubations, but may be less available in intact cells.

---

[14] There are, however, significant amounts of the products from the microsomal syntheses that do show at least 4-5 GlcNAc- containing disaccharides in blocks.

There are other explanations for the incomplete maturation of some of the disaccharide units in heparinoids. These include the possibility that the maturation enzymes are compartmentalized in the Golgi and that the passage of the maturing heparinoid chain through some of these compartments occurs so rapidly that some of the maturation reactions are simply incomplete. It is also possible that such compartments in the Golgi, if they exist, present localized environments (pH, metal ion concentrations, etc.) that regulate the specificities of the maturation enzymes.

IV. THE SIZE AND VARIABILITY OF HEPARINOIDS

We have already mentioned that the biosynthesis of heparin yields macromolecular heparin, a proteoglycan composed of the serglycin core protein and multiple chains of heparin, having a molecular weight of 750,000-1,000,000. The heparin chains have molecular weights of ~60,000-100,000 [96]. As the "biosynthesis" continues, the heparin chains of the glycosaminoglycan

are cleaved by an endo-β-glucuronidase to yield shortened chains with average molecular weights of ~12,000 [63-65, 97, 98]. However, a typical heparin displays molecular weights ranging from ~5000 to ~25,000; i.e., the heparinoids, like all polysaccharides, are polydisperse.

Given the polydispersity and range of structures in all heparinoid preparations, it is interesting to consider what constitutes a "pure heparin" or a "pure heparan sulfate". All that one can say is that a pure heparinoid is a preparation that contains only oligosaccharides or polysaccharides composed solely of the disaccharide units that we have described. Such a glycosaminoglycan preparation would be free of protein and other glycosaminoglycans, most notably dermatan sulfate and chondroitin sulfate. It may be noted that even heparin preparations that are pure by these criteria can be separated into several well-resolved fractions by electrophoresis on cellulose acetate strips or agarose gels in barium acetate buffers. Typically such a treatment yields two peaks, referred to as the "fast-moving" and the "slow-moving" components. In fact, under some electrophoretic conditions, it is possible to resolve intermediate components [118]. This phenomenon appears to be related to the molecular weight species and the degree of sulfation of the components of the heparin mixtures. The higher the average molecular weight and the degree of sulfation, the higher the percentage of slow-moving component [119, 120]. It has been suggested that the slow-moving fractions complex more strongly to the Ba^+ ions of the electrophoresis buffer [121].

V.   THE SHAPE OF HEPARINOIDS

A.   Introduction

The discussion to this point has dealt with the linear sequences that are found in heparinoids. An understanding of the binding of heparinoids to proteins ultimately must address the interactions of three-dimensional heparinoid structures with three-dimensional proteins. In this area, the tertiary structures and the flexibilities are understood much more clearly for proteins than for heparinoids. The knowledge of heparinoid shapes and dynamics is based largely on NMR and X-ray diffraction studies and is described in several reviews [11, 12, 122-126]. In general terms, the X-ray diffraction analyses of ordered fibers in heparin films have shown that the crystalline portions of these heparin films form twofold helices with a disaccharide axial periodicity of 0.82 to 0.87 nm., depending on the humidity and the salt form, whereas heparan sulfate has a disaccharide axial periodicity of 0.93 [127-130]. Models based on these results suggest that the helix is stabilized by H-bonding between the ring oxygen of the uronic acid and C3 hydroxyl of the adjacent GlcN [130]. More recent NMR data yield structures that are consistent with those found in the crystallography studies [131, 132]. Thus, it is generally true that these are quite linear structures, and the question now becomes, how much flexibility is there in these chains? Two factors control flexibility: (a) the shapes taken by the individual monosaccharides; i.e., the ring conformations, and (b) the bending that can occur at each glycosidic linkage. We will discuss these separately. The influences of the overall flexibility of heparinoid chains on heparinoid-protein interactions will be discussed in Chapter 6.

B.   Ring Conformations

The monosaccharides of heparinoids are pyranose ring structures that take chair (C) or skew boat (S) conformations. Other ring conformations are possible, but have not been reported for heparinoid monosaccharide residues. A scheme for designating the ring conformations of monosaccharides has been discussed by Stoddart [133]. According to this system, the ring carbons are numbered 1 to 5, as usual, and the ring oxygen is designated by O. A reference plane is then chosen that contains four of these atoms and that has the lowest numbered ring carbon displaced from this plane. The ring conformation is then described by indicating whether the ring is a chair, C, or a skew boat, S (sometimes designated as a twist boat, T). Then those atoms that lie *above* the reference

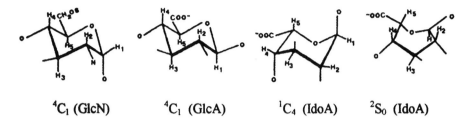

$^4C_1$ (GlcN) $^4C_1$ (GlcA) $^1C_4$ (IdoA) $^2S_0$ (IdoA)

Figure 2.8. Ring conformations of heparin monosaccharides.

plane are indicated as a *superscript* to the left of the C or S, and those atoms that lie *below* the reference plane are indicated as a *subscript* to the right of the C or S; e.g., $^1C_4$ indicates a chair form in which carbons 2, 3, and 5, and the ring oxygen form the plane, and carbons 1 and 4 lie above and below the plane, respectively. The ring conformations for the monosaccharides found in heparinoids are illustrated in Figure 2.8.

Based on X-ray crystallography and NMR studies (see Chapter 3), it has been shown that the *GlcN* and the *GlcA* residues of heparinoids form rather rigid and stable $^4C_1$ rings. All of the nonhydrogen ring substituents of the GlcN and the GlcA residues are positioned equatorially except for the anomeric OH of the GlcN, which is α-linked. When the monosaccharides are linked in oligo- or polysaccharides, the ring conformations of these monosaccharides do not seem to be altered by the immediately adjacent residues linked to C4 (upstream) or C1 (downstream).

The IdoA residues equilibrate between the $^1C_4$ and the $^2S_0$ conformations, which are essentially equi-energetic [see 133, 134]. The equilibrium between the $^1C_4$ and the $^2S_0$ conformations is extremely rapid—up to 10^6/sec—under physiological conditions [124], indicating that the energy barrier between these two forms is relatively low. Calculations based on NMR vicinal coupling constants have shown that the $^4C_1$ conformation of IdoA is of similar potential energy to the $^1C_4$ and $^2S_0$ conformations [124, 135], suggesting that the $^4C_1$ ring form may also participate in the equilibrium. In fact, there is evidence for participation of the $^4C_1$ form of IdoA when it occurs at the nonreducing terminal of heparin di- and oligosaccharides [136, 137], and presumably when the IdoA is free in solution (not reported). However, for IdoA residues in internal positions of heparinoid chains, the $^4C_1$ conformation does not occur and the IdoA residues equilibrate between the $^1C_4$ and the $^2S_0$ conformations.[15] The positioning of the nonhydrogen ring substituents in IdoA depends on the ring

---

[15] However, the likelihood that IdoA may take the $^4C_1$ conformation when heparinoids are bound to proteins seems high.

conformation. Thus, when IdoA is in the $^1C_4$ conformation, the bulky carboxyl group is equatorial, whereas all other substituents, including the anomeric OH, which is α-linked, are in axial positions. For the $^2S_0$ conformation, the nonhydrogen ring substituents exhibit minimal steric interactions, and substituents at C4 might be expected to cause little distortion of the ring conformation (Figure 2.9). The $^1C_4 \leftrightarrow {}^2S_0$ equilibrium of IdoA lies toward the $^1C_4$ conformation (60:40) when it occurs as in internal residue in a heparinoid chain, but the equilibrium shifts toward the $^2S_0$ conformation (40:60) when the IdoA is a nonreducing terminal residue [138]. Furthermore, the sulfation patterns of neighboring residues may affect the conformational equilibrium [132, 139].

C. Rotations about the Glycosidic Linkages

A major consideration in the three-dimensional structures assumed by carbohydrate polymers is the rotation that can occur about the two bonds that form the glycosidic linkage in heparinoids, the C1-O and the O-C4 bonds. For most polysaccharides, the rotations about these bonds are the only possible conformationally adaptable aspects of the structure [140]. For heparinoids, rotations about these bonds also represent flexible portions of the polymer, but in the case of IdoA-containing sequences, variation in ring conformations also must be taken into account. These rotations are expressed by the dihedral angles, ϕ and ψ, where $\phi = H_1$—C1—O_4—C4 and $\psi = $ C1—O_4—C4—H_4. Calculations of the degrees of rotation of the ϕ and ψ angles for each value of the other are expressed as contour plots, such as those shown in Figure 2.9 [132]. The contour lines in such plots connect ϕ,ψ coordinates of equal potential energies and the outer contour depicts the limit of ϕ,ψ rotations that are possible before steric interactions block further rotation. The column of plots on the left of Figure 2.9, labeled column 1, is for heparinoids in which the GlcN residues are N-sulfated, whereas the column on the right, labeled column 2, represents heparinoids in which the GlcN residues are N-acetylated.[16] The labels *between* the two columns distinguish the A-I, or GlcN→IdoA, linkages [(a) and (b)], and the I-A, or IdoA→GlcN, linkages [(c) and (d)]. As also noted by the labels between the columns, the (a) and (c) plots represent the IdoA-containing disaccharides in which the IdoA residues are in the $^1C_4$ conformations, whereas the (b) and (d) plots represent the IdoA-containing disaccharides in which the

---

[16] As we have discussed, the GlcNAc→IdoA linkage does not appear in heparinoids because of the specificity of the epimerase that converts GlcA to IdoA, but the results are shown here to illustrate several points.

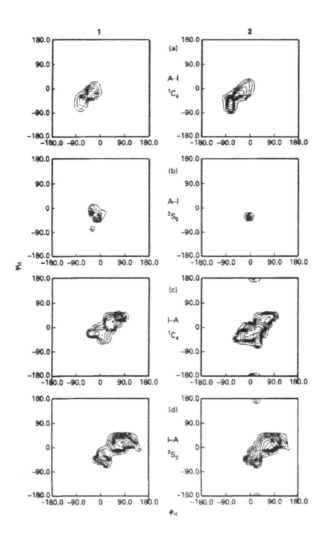

Figure 2.9. Contour plots showing permissible rotations about glycosidic bonds. See text for further discussion. Reprinted from [132], with permission.

IdoA residues are in the $^2S_0$ conformations. From a study of these diagrams, several things are clear:

- There is little difference in the degree of rotation possible when GlcNAc- and GlcNSO$_3$-containing disaccharides are compared.

- There is a higher degree of rotation about the IdoA→GlcN bonds than about the GlcN→IdoA bonds, where rotation is quite limited.

- In all cases, there are differences in the possible degrees of rotation as a result of switching the IdoA ring from the $^1C_4$ to the $^2S_0$ conformation.

- Since the $^1C_4$ and $^2S_0$ conformers are in equilibrium in IdoA-rich segments of heparin and heparan sulfate, the IdoA→GlcN bonds can rotate over the entire range of ϕ,ψ angles depicted in *both* the $^1C_4$ and the $^2S_0$ contour plot

Although not shown, it has been observed that the possible degrees of rotation that can occur at GlcA-containing disaccharides, where the GlcA is in the $^4C_1$ conformation, are, in fact, greater than those shown here for the IdoA-containing disaccharides.[17]

We might take note of several things depicted in these plots. First, the allowed rotations about a *glycosidic* linkage are much more limited than the rotations that can occur at *peptide* linkages, i.e., the heparinoid chains are quite linear, with limited bending possible. Although the degree of bending of the chain is not easily deduced from these diagrams, it has been estimated by Mulloy that, in order to bend 180^0 and go back in the opposite direction, a heparinoid sequence must be more than 20 disaccharides long, i.e., the bend at each glycosidic bond is less than $4\text{-}5^{0}$.[17] Since greater degrees of rotational freedom are observed for the GlcA-containing disaccharides than for the IdoA-containing disaccharides, fewer GlcA-containing disaccharides may be required for a similar 180^0 bend.

VI.   PROTEOGLYCAN FORMS OF HEPARINOIDS

A.   Introduction

Although heparin and heparan sulfate are synthesized as proteoglycans, they occur in cells and tissues as both proteoglycans and free heparinoid chains. This is a result of their normal metabolism after the biosynthesis of the proteoglycans is completed (Chapter 5). The heparan sulfate proteoglycans may be associated with the cell membranes via transmembrane core proteins or glycosyl phosphatidylinositol-anchored core proteins. They may be shed from the cell surface into the extracellular matrix or internalized by the cell and metabolized further. The proteoglycan nature of the heparinoids has been

---

[17] Personal communication from Dr. Barbara Mulloy.

described in a number of reviews which may be consulted for more detailed discussions of the proteoglycan structures and for reference to the extensive primary literature [10, 13, 14, 141-148]. The proteoglycan structures will be reviewed here briefly, and the metabolism of heparan sulfate proteoglycans will be discussed in Chapter 5.

Both heparin and heparan sulfate, as well as other glycosaminoglycans (chondroitin sulfate and dermatan sulfate), are attached to core proteins through a βGlcA-1,3-βGal-1,3-βGal-1,4-βXyl-1→Ser linkage region that couples the repeating disaccharide portions of the glycosaminoglycans to their core proteins [59, 61, 68]. Fransson *et al.* [149] have reported that the linkage region may be phosphorylated on the C2 of the xylose residue in the linkage region of bovine lung heparan sulfate proteoglycan. The glycosaminoglycan chains are attached to the GlcA residue of the linkage region through the first downstream amino sugar of their respective repeating disaccharide sequences denoted by the "↓" in the following types of structures:

for heparinoids,

$$\downarrow$$
$$(\text{HexA} \rightarrow \text{GlcNAc})_n \rightarrow \text{GlcA} \rightarrow \alpha \text{GlcNAc} \rightarrow \text{GlcA} \rightarrow \text{Gal} \rightarrow \text{Gal} \rightarrow \text{Xyl} \rightarrow \text{Ser} ,$$

and for chondroitin sulfate and dermatan sulfate,

$$\downarrow$$
$$(\text{HexA} \rightarrow \text{GalNAc})n \rightarrow \text{GlcA} \rightarrow \beta \text{GalNAc} \rightarrow \text{GlcA} \rightarrow \text{Gal} \rightarrow \text{Gal} \rightarrow \text{Xyl} \rightarrow \text{Ser}.$$

The amino acid recognition sites for attachment of glycosaminoglycans to core proteins are Ser-Gly sequences. Although it is not totally clear how the chain attachment of the different glycosaminoglycan types is controlled, there are distinct groups of core proteins that act as acceptors for heparins, for heparan sulfates, for chondroitin sulfates, or for dermatan sulfates,[18] A recent review describes both chondroitin sulfate/dermatan sulfate- and heparinoid-containing proteoglycans [148]. Our discussion here will be limited to the well-characterized heparinoid-bearing core proteins. Some light has been shed on the question of specific sequences that control chondroitin sulfate and heparinoid addition to the core protein by the recent work of Esko *et al.* [150, 151], who have shown that the Ser-Gly acceptor sequences for heparinoid chains occur in

---

[18] It appears that a single set of enzymes synthesizes the linkage region for both heparinoids and chondroitin sulfates/dermatan sulfates, and that the unique biosynthetic step that leads to the synthesis of the heparinoid vs the chondroitin sulfate/dermatan sulfate groups is the addition of the first amino sugar, shown by the arrows (see Chapter 5).

clusters, whereas chondroitin sulfate and dermatan sulfate acceptor sequences tend to be isolated. Both are flanked by acidic amino acid residues and an adjacent tryptophan.

Although there is a relatively small number of types of core proteins that carry heparan sulfate, these are found in a large variety of cells, and expression of the individual core proteins may be highly specific for different cell types and stages of development (Chapter 5). Thus, the control of the amount of heparan sulfate proteoglycan on cell surfaces is subject to regulatory mechanisms that control the expression of the core proteins. It has been suggested that the structure of these core proteins is the principal determinant of the functional characteristics of a proteoglycan [152], but a growing body of evidence seems to support the conclusion that the carbohydrate portion of the proteoglycans mediates most of the proteoglycan functions on cell surfaces [see, for example, 153]. Interestingly, although expression of the core proteins in cells is necessary in order to form cell surface heparan sulfate proteoglycans, the fine structures of the glycosaminoglycans that are attached to a given core protein, and thus their abilities to bind proteins, appear to be controlled by the heparan sulfate chain maturation reactions that take place in the individual cells and not by the structures of the core proteins [see for example, 154, 155-157].

Heparinoid core proteins are developmentally regulated [144, 158-167]. They vary in their molecular weight from ~13,000 to ~400,000. The number of heparinoid chains attached to a core protein may vary from 2-3 to 12 (or higher—see serglycin in Section IV.B.1., below), with molecular weights of the heparinoid chains ranging from ~15,000 to ~100,000 [13, 70, 142, 168]. In addition to the glycosaminoglycan chains, the core proteins usually have N- and/or O-linked oligosaccharides attached to arginine and serine residues, respectively, as a result of their processing via the secretory pathways of cells [13, 169, 170]. Finally, the various proteoglycans and their core proteins have been given names that describe some physical, chemical, or biological (or philosophical!) property of the structure in question. However, these names are often used in the literature interchangeably to indicate *both* the proteoglycan *and* the core protein. Since the heparan sulfate chains on a given core protein have structures that depend on the biosynthetic machinery of the cell and since it is the core protein that distinguishes one type of heparan sulfate proteoglycan from another, we will try to avoid confusion in the present discussion by using these names to refer only to the core proteins.

B. Core Proteins

1. Serglycin

Mast cells and other hematopoietic cells (basophils, natural killer cells, cytotoxic T lymphocytes, eosinophils, macrophages, and platelets) contain a unique core protein called serglycin [171-174]. All of these cells store a family of proteoglycans in complexes with various basic proteases in their secretory granules. In connective tissue mast cells, serglycin carries heparin; in natural killer cells it carries chondroitin sulfate; and in mucosal mast cells it carries a mixture of chondroitin sulfate E and heparin [175, 176]. Serglycin was originally identified as a core protein that contained predominantly Ser and Gly residues [65, 177, 178]. Core proteins from various rodent and human cells [179, 180] and from rat yolk sac tumor cells [181] have been described. The core proteins have been cloned and characterized in detail [180-184] and the mechanisms regulating gene expression have been studied [185, 186]. The core is synthesized as a precursor protein of 16.7-18.6 kDa that has an extensive sequence of consecutive Ser-Gly repeats. Glycosaminoglycan chains are attached to up to two-thirds of these serines. The unique Ser-Gly repeat structure and the high degree of glycosaminoglycan substitution render these proteoglycans resistant to proteases.

2. The syndecan family

a. Introduction

The syndecans are a family of transmembrane core proteins [144, 187, 188]. The core proteins are small, cysteine-free type 1 glycoproteins that have an extracellular domain to which the glycosaminoglycans are attached, a transmembrane domain, and a relatively short (<35 amino acids) cytoplasmic domain (Figure 2.10). There is a high degree of sequence identity in the transmembrane and cytoplasmic domains, leading to the grouping of the syndecans into a family of four. David [146] refers to the family as "syndecan-like intercalated proteoglycans" (SLIPS). The extracellular domains are highly divergent; the main conserved features are the glycosaminoglycan attachment sites and a putative proteolytic cleavage site (dipeptide sequence containing basic amino acids) near the transmembrane domain. In the cytoplasmic domains, there is a strict conservation of four tyrosines, one at the junction of the membrane-spanning domain and the cytoplasmic domain that may be a signal for internalization [189]. One of the internal tyrosines fits a consensus sequence

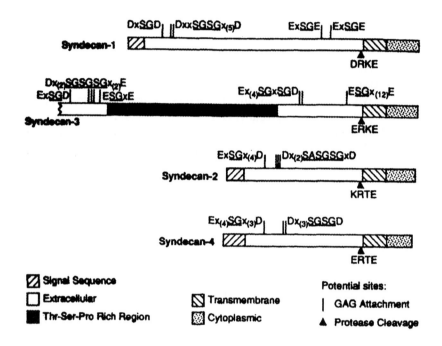

Figure 2.10. Domain structures of syndecans. The positions and amino acid sequences at the glycosaminoglycan attachment sites are shown. Reprinted from [144], with permission from the Annual Review of Cell Biology, © 1992, by Annual Reviews Inc.

for phosphorylation. An ancestral gene of the syndecans has been cloned from *Drosophila* [190].

b. Syndecan-1

Syndecan-1 is the original member of the syndecan family and is usually referred to simply as syndecan. It has been cloned from mouse and human mammary epithelial cells [191, 192], rat microvascular endothelia [193], rat vascular smooth muscle cells [165], and hamster kidney cells [194]. It is expressed in rat liver cells, including hepatocytes and Kupffer cells, but at only low levels in Ito and endothelial cells [195]. Syndecan-1 has a common dipeptide of basic amino acids at a site close to the membrane which is a putative protease-cleavage site. The ectodomain has five potential glycosaminoglycan attachment sites (see Figure 2.10). Syndecan-1 forms a hybrid proteoglycan containing both heparan sulfate and chondroitin sulfate [196]. The ratio of heparan sulfate to chondroitin sulfate varies among tissues, and heparan sulfate and chondroitin sulfate chains are attached at specific glycosylation sites [197]. Syndecan-1 polarizes to the basolateral surface of cells and colocalizes at basal cell surfaces with F-actin. Thus, syndecan-1 appears to link the extracellular matrix to the cytoskeleton and could be involved in the transduction of external signals (from growth factors, for example) through the plasma membrane to the cytoplasm. Syndecan-1 is the major proteoglycan product in epithelial cells but is hardly detectable in mesenchymal cells. The gene structure of the mouse syndecan-1 gene has been described [198, 199].

c.   Syndecan-2

Syndecan-2, originally called *fibroglycan* because of its presence in fibroblastic cells, is expressed at high levels in lung and skin fibroblasts and in some cell lines of mesodermal and neuroectodermal origin, but is virtually absent in most epithelial cell lines. It has been cloned from human fetal lung fibroblasts [200] and rat liver [201]. It has four potential heparan sulfate attachment sites.

d.   Syndecan-3

Syndecan-3 has been cloned from chick limb buds [166] and from a neonatal rat Schwann cell line (*N-syndecan*) [202]. It has three to five potential glycosaminoglycan attachment sites. It is mainly associated with nervous tissue and developing mesenchyme [202].

e.   Syndecan-4

Syndecan-4 includes two members: *amphiglycan* and *ryudocan*. An additional member of this group has been cloned from chick embryos [203]. Syndecan-4 is the most widely expressed member of the syndecan family [see, for example, 204]. Amphiglycan is a core protein cloned from human lung fibroblasts [205], named amphiglycan from the Greek words amphy, "around," and "both sides" and amphoo, "both," referring to its transmembrane structure and its presence in both epithelial and fibroblastic cells. The protein contains 198 amino acids with four potential heparan sulfate attachment sites and all of the structural features of the syndecans (above).

Ryudocan is a core protein cloned from rat and human vascular endothelial cells [193, 206, 207]. The rat ryudocan has three potential glycosaminoglycan attachment sites; the human core protein has four. The mRNA for ryudocan is found in lung, liver, skeletal muscle, and kidney. It is widely expressed in humans.

3. The glypican family

Glypicans are glycosylphosphatidylinositol (GPI)-anchored core proteins. GPI-anchored proteins occur widely in eukaryotic cells [208, 209]. GPI-anchors consist of a common core oligosaccharide,

Protein-CONH→ethanolamine-PO_4-6→Man-1→2αMan-1→6-αMan-1→4αGlcN-1→ myo-inositol-1-PO_4-diacylglycerol,

that joins, through an amide linkage, the C-terminal amino acid of the core protein to the phosphatidylinositol which is embedded in the outer leaflet of the cell membrane. Species-specific carbohydrate (galactose or galactosamine-containing oligosaccharides) or ethanolamine-PO_4 side chains may be attached to these common core structures. The inositol-PO_4 is linked to a lipid moiety containing glycerol with attached fatty acyl or fatty alkyl ether chains, or a sphingosine derivative. An additional fatty acid may sometimes be found in ester linkage to the C2 or C3 hydroxyl groups of the inositol ring. A unique feature of all anchors is the N-unsubstituted GlcN residue. Thus, the GPI anchors can be cleaved with nitrous acid (Chapter 3). Another common way to recognize GPI-anchored proteins is by their release from cell surfaces by treatment with phosphatidylinositol-specific phospholipase C, which cleaves the GPI anchor between the phosphate and the diacylglycerol of the phosphatidylinositol. This releases the core protein with the anchor linked to the C-terminal amino acid and terminated by an inositol phosphate. Alternatively, one may release the protein with phospholipase D, which cleaves between the inositol and the phosphate, leaving the anchor with an unphosphorylated inositol attached to the protein.[19] The GPI anchor localizes the proteins in the lipid bilayer of the membrane through the embedded lipoidal structure attached to the inositol.

Glypican from human lung fibroblasts is a cysteine-rich protein of 558 amino acids with a Ser-Gly repeat near its C-terminal and one additional glycosylation site near the N-terminal. The C-terminal structure is similar to that of other GPI-anchored proteins. Most of the glycosaminoglycan is concentrated next to the plasma membrane. Glypican is expressed in many different cell lines—both of fibroblastic and of epithelial origin. It is highly conserved, with 90% sequence identity in rat, mouse, and human. GPI-anchored heparan sulfate proteoglycans have been cloned from human lung fibroblasts (the original

---

[19] The phospholipase C may be prevented from action by some of the structural features of some of the GPI-anchors (e.g., the fatty acylation of the inositol moiety); [see 210, 211-213]. Phospholipase D, however, will cleave GPI anchors that contain fatty acylated inositol residues [discussed in 214].

glypican [215, 216], brain (*cerebroglycan*) [100, 162, 163, 217, 218], rat brain [219, 220], mouse kidney and brain (*K-glypican*) [163], rat intestine (*OCI-5*) [158, 221], hepatoma cells [222], ovarian granulosa cells [223], human hepatocellular carcinoma cells [224], and osteogenic cells [225], suggesting that glypican is a core protein for a new family of heparan sulfate proteoglycans. This family is referred to by David [146] as "glypican-related integral membrane heparan sulfate proteoglycans" (GRIPS).

4. Perlecan

Perlecan, so-named because of its "beads-on-a-string" appearance in the electron microscope [226], is the core protein of basement membrane heparan sulfate proteoglycans. The proteoglycan produced by the Engelbreth Holm Swarm tumor is readily available and has stood as a prototype for basement heparan sulfate proteoglycans [see 227]. However, it is also found in cartilage matrix; i.e., it is not restricted to basement membranes [228]. It has been cloned by several laboratories [226, 229-231], and the structure of the human gene has been described [232]. The core protein has a molecular weight of 396,000-467,000, depending on the source, with three potential sites for heparan sulfate attachment, and with a domain structure that includes regions of homologies with the LDL receptor, the laminin A chain, immunoglobulin-like repeats, and epidermal growth factor. The proteoglycan appears to span the whole thickness of the basement membrane. When the mouse perlecan is expressed in CHO cells, both heparan sulfate and chondroitin sulfate/dermatan sulfate become attached to the protein [233]. The recent literature has been reviewed by Iozzo and colleagues [148, 234, 235].

5. CD44E

A recent membrane protein to be identified as a potential HSPG core protein is a member of the CD44 family of transmembrane glycoproteins, which includes Hermes antigen, Pgp-1, and ECMR III. These proteins exhibit exceptional heterogeneity that occurs as a result of alternative splicing and variable glycosylation. CD44 exists as a number of variants, most of which occur without attached glycosaminoglycan. One form, CD44E (epidermal), has been identified in keratinocytes [236, 237] and given the name "*epican*" (epidermal intercellular proteoglycan). Consequently, CD44 has been referred to as a "part-time" or "amateur" core protein [13, 146].

6.   Betaglycan

Betaglycan (β-glycan, βetaglycan) is another "part-time" core protein for heparan sulfate proteoglycans. It may carry heparan sulfate or chondroitin sulfate chains or no glycosaminoglycan at all. It contains two cysteine-rich regions and has no tyrosines in cytoplasmic domain. It is the type III receptor for TGFβ [100, 238-240].

7.   Others

Agrin [241] and testican [242] have also been identified as core proteins for heparan sulfate.

ADDITIONAL REFERENCES

Bame, K. J., and Robson, K. (1997) Heparanases produce distinct populations of heparan sulfate glycosaminoglycans in Chinese hamster ovary cells, *J. Biol. Chem.* **272**, 2245-2251.

Carey, D. J., Conner, K., Asundi, V. K., O'Mahony, D. J., Stahl, R. C., Showalter, L., Cizmeci-Smith, G., Hartman, J., and Rothblum, L. I. (1997) cDNA cloning, genomic organization, and *in vivo* expression of rat N-syndecan, *J. Biol. Chem.* **272**, 2873-2879.

Edgren, G., Havsmark, B., Jönsson, M., and Fransson, L.-Å. (1997) Glypican (heparan sulfate proteoglycan) is palmitoylated, deglycanated and reglycanated during recycling in skin fibroblasts, *Glycobiology* **7**, 103-112.

Hileman, R. E., Smith, A. E., Toida, T., and Linhardt, R. J. (1997) Preparation and structure of heparin lyase-derived heparan sulfate oligosaccharides, *Glycobiology* **7**, 231-239.

Iozzo, R. V., Pillarisetti, J., Sharma, B., Murdoch, A. D., Danielson, K. G., Uitto, J., and Mauviel, A. (1997) Structural and functional characterization of the human perlecan gene promoter. Transcriptional activation by transforming growth factor-β via a nuclear factor 1-binding element, *J. Biol. Chem.* **272**, 5219-5228.

Kobayashi, M., Habuchi, H., Yoneda, M., Habuchi, O., and Kimata, K. (1997) Molecular cloning and expression of Chinese hamster ovary cell heparan-sulfate 2-sulfotransferase, *J. Biol. Chem.* **272**, 13980-13985.

Litwack, E. D., Stipp, C. S., Kumbasar, A., and Lander, A. D. (1994) Neuronal expression of glypican, a cell-surface glycosylphosphatidylinositol-anchored heparan sulfate proteoglycan, in the adult rat nervous system, *J. Neurosci.* **14**, 3713-3724.

McFall, A. J., and Rapraeger, A. C. (1997) Identification of an adhesion site within the syndecan-4 protein domain, *J. Biol. Chem.* **272**, 12901-12904.

Mikhailov, D., Mayo, K. H., Vlahov, I. R., Toida, T., Pervin, A., and Linhardt, R. J. (1996) NMR solution conformation of heparin-derived tetrasaccharide, *Biochem. J.* **318**, 93-102.

Song, H. H., Shi, W., and Filmus, J. (1997) OCI-5/rat glypican-3 binds to fibroblast growth factor-2 but not to insulin-like growth factor-2, *J. Biol. Chem.* **272**, 7574-7577.

Chapter 3. Determination of Heparinoid Structures

I. INTRODUCTION

Heparinoids, by their very nature, can be obtained only as mixtures of closely related polymers. Such mixtures are polydisperse. Commercial heparins exhibit molecular weight ranging from 7000-35,000, whereas heparan sulfate chains may show a similar broad range of molecular weights that are generally two- to three-fold higher than those of heparins [see, for example, 243]. In addition to their molecular weight ranges, the chains in all heparinoid preparations show a range of structural properties. For example, these mixtures may contain polymer chains that exhibit significant differences in their IdoA/GlcA ratios and degrees of sulfation, expressed as the SO_4/disaccharide ratio.[1] The SO_4/disaccharide ratio is easily calculated from the disaccharide composition of the heparinoid, obtained by methods described below. The average ratios are usually <1 for heparan sulfates and >2 for heparins, but each heparinoid preparation contains a range of structures that that may have quite different SO_4/disaccharide ratios. Although heparinoids may be fractionated according to size, further separated according to their SO_4/disaccharide ratios, and finally selected by affinity chromatography on matrices containing covalently bound heparin-binding proteins, the final subfractions obtained are still mixtures. One cannot sequence a polymer that is not pure, and so cannot obtain disaccharide sequences for polymeric heparinoids, even though the methodology currently available could be adapted to sequencing a pure heparinoid. Thus, the only useful structural data that can be obtained for polymeric heparinoids are the relative amounts of mono-, di-, or higher oligosaccharides that are recovered when oligomeric or polymeric heparinoids are fragmented by specific cleavage procedures. Even the fragments obtained by cleaving heparinoids are complex mixtures and must be extensively purified prior to characterization. Methods for separation of mono- and disaccharides are available, but the fractionation of larger fragments is a difficult task, even for tetrasaccharides [244]. The larger the oligosaccharide fragments, the more difficult the separation (remember that there are 324 possible tetrasaccharides and 5832 possible hexasaccharides and that these structures have closely related, overlapping physical properties—see Chapter 2).

Some structural features of heparinoids are so well established that they are assumed to be present in all heparinoid fragments. These include (a) the structures of the glycosidic linkages, (b) the pyranose ring structures of all of the monosaccharide residues, and (c) the repeating uronic acid-glucosamine disaccharide nature of the polymers. Furthermore, heparinoid fragments formed by nitrous acid cleavage always have anhydromannose residues at their reducing terminals, whereas fragments formed by heparin lyase cleavage always have $\Delta^{4,5}$-HexA residues at their nonreducing terminals. The determination of the further aspects of the structures of heparinoids has required the development of heparinoid-specific methodology to supplement both the more classical methodology and the physical methodology that has been applied to other carbohydrate polymers. This chapter describes the approaches that have been used to bring us to our present state of knowledge of heparinoid structures and that will serve as the bases for the further structure studies that must be performed in examination of structure-activity relationships.

---

[1] The SO_4/COOH ratio is also used to express degrees of sulfation. Since each disaccharide contains one COOH group, each COOH represents a single disaccharide unit.

II.   ACID HYDROLYSIS AND MONOSACCHARIDE ANALYSIS

Typically, one determines the monosaccharide composition of a polysaccharide by hydrolysis to yield mixtures of monosaccharides and then separation of the monosaccharides by chromatography and quantification of each separated component. The monosaccharides formed by hydrolysis of heparinoids are GlcN, GlcA, and IdoA. The total uronic acid content is equal to the total GlcN. There may also be trace amounts of Gal and Xyl from linkage region sequences that remain in the purified heparinoid. If other monosaccharides are present, these represent contaminants from other polysaccharides (e.g., GalN from chondroitin SO_4 and dermatan SO_4) or glycoproteins (including the typical monosaccharides of N- and O-linked oligosaccharides and GPI anchors). When a pure heparinoid is analyzed for its monosaccharide composition, the only useful information that can be obtained is the relative amount of GlcA and IdoA. This information can also be obtained from the disaccharide analysis methods described below. Thus, if a thorough characterization of a heparinoid is undertaken, analysis of the monosaccharide composition is redundant. However, some interesting features of the heparinoid structures and chemistry emerge from a discussion of methods for monosaccharide analysis.

Classical colorimetric methods have been used for analysis of the monosaccharide content of heparinoids. Generally, such assays use concentrated H_2SO_4 in the presence of phenolic reagents to obtain hydrolysis of the polymer and color formation in the same reaction mixtures. GlcN can be assayed using the Elson-Morgan method [245] or by the indole method, which measures anhydromannose formed by treatment of amino sugars with nitrous acid [246]. Uronic acids can be measured by carbazole and orcinol assays [247]. Standard curves that relate color yield to the amount of each monosaccharide are required for these assays, but, in most carbohydrate reactions, different monosaccharides give different molar color yields. For example, although GlcA and IdoA can both be measured using the carbazole assay, they give different color yields [discussed in 248]. Consequently, the use of a GlcA standard curve for measuring color yield per uronic acid does not give an accurate measure of the amount of uronic acid present when there are significant levels of IdoA in the sample being analyzed.[2] These colorimetric methods are useful only for obtaining some general sense of monosaccharide composition and for following carbohydrate elution profiles during chromatographic separations. It is a sufficient critique of these methods to note that all published analyses of

---

[2] Iduronic acid is not readily available to use in preparing a standard curve (but see footnote 3).

heparinoids using these colorimetric assays have failed to yield 1:1 ratios of uronic acids to glucosamine. Therefore, results of the uronic acid or the GlcN analyses by these procedures are inaccurate (or our concept of heparinoid structure is in need of revision!).

A more accurate way to obtain the monosaccharide composition of a polysaccharide is to convert the polymer to monosaccharides and to analyze the mixture of monosaccharides for the individual components. Polysaccharides that are made up of neutral sugars are converted completely to monosaccharides by hydrolysis in 1 $\underline{N}$ mineral acid at 100 °C for 4-6 hr. Under these hydrolysis conditions, the N- and O-sulfate and the N-acetyl substituents of heparinoids are all released as free SO_4 and acetate, respectively, but the glycosidic linkages of both the uronic acids and amino sugars found in heparinoids are quite resistant to acid and are incompletely hydrolyzed.

The typical pathway for acid hydrolysis for GlcNAc residues in heparinoids is shown in Figure 3.1. As long as the acetyl substituents remain attached to the amino groups of GlcN, the glycosides are hydrolyzed rapidly (reaction a). However, the substituents attached to the amino groups are also released under the hydrolysis conditions, thus converting the GlcNAc residues in the polymer to N-unsubstituted GlcN residues (reactions b and c). For GlcN residues that are N-sulfated, the N-sulfate groups are extremely labile, and are completely released in only a few minutes under the above hydrolysis conditions, but the N-acetyl groups are released over a 2- to 3-hr period. The glycosidic bonds of the resulting GlcN residues are extremely resistant to acid (reaction d), apparently because the positively charged amino group prevents the protonation of the glycosidic oxygen, a necessary step in the acid hydrolysis of

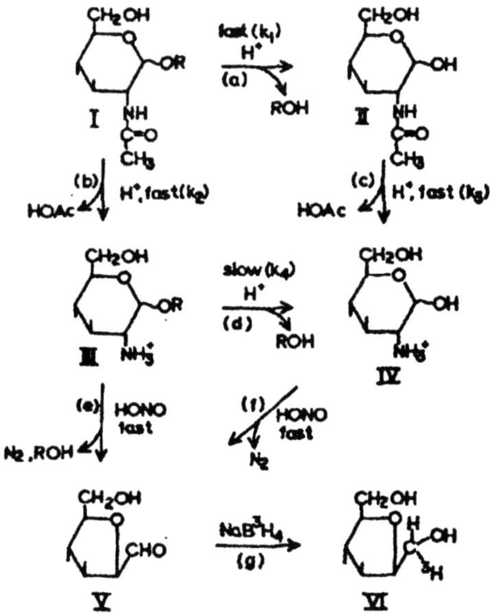

Figure 3.1. Hydrolysis of GlcNAc glycosides. Reprinted from [249] with permission, © 1970 by the American Chemical Society.,

glycosides. In fact, α-linked GlcN glycosides are more resistant to acid than β-linked GlcN glycosides [249]. Table 3.1 shows the rate constants for acid hydrolysis of the N-acetyl group and compares the rates of acid hydrolysis of amino sugar glycosides, before and after the loss of N-acetyl groups, with hydrolysis rates for other glycosides. As can be seen, the GlcN glycosides are very resistant to acid. For practical purposes, one can consider that it is impossible to obtain complete hydrolysis of the GlcN glycosides in heparinoids after the substituents are lost from the amino groups. An alternate way to cleave the glycosidic bonds of GlcN rapidly and (almost) completely is by an elimination reaction with nitrous acid (reactions e and f). In this reaction the glucosamine residue is converted to a 2,5-anhydro-$\underline{D}$-mannose residue (AMan). Free GlcN also reacts with nitrous acid and is converted to AMan. The nitrous acid cleavage of GlcN bonds is described in more detail in Section III.B.1., p. 69.

Uronic acid glycosides present an additional problem in the acid hydrolysis. It has been known for many years that the glycosidic bonds of $\underline{D}$-glucuronic acid and $\underline{D}$-galacturonic acid are hydrolyzed very slowly. Indeed, the glycosidic bonds of the $\underline{D}$-glucuronic acid residues in heparinoids exhibit the same extreme acid resistance noted for GlcA in other polysaccharides [253]. After $\underline{L}$-iduronic acid was found to be a major component of heparinoids (and dermatan SO$_4$), it was assumed that the glycosidic bonds of this uronic acid were also very acid resistant. Surprisingly,

Table 3.1. Rate Constants for Acid Hydrolysis of Amino Sugar and Uronic Acid Glycosides[1]

Reaction	Rate constant $(k \times 10^4 \text{ sec}^{-1})$	Ref.
AMINO SUGARS		
αMeGlcNAc → αMeGlcN + HOAc	1.44	[249]
βMeGlcNAc → βMeGlcN + HOAc	--[2]	[249]
αMeGlcNAc → MeOH + GlcNAc	1.85	[249]
αMeGlcN → MeOH + GlcN	0.0021	[249]
βMeGlcNAc → MeOH + GlcNAc	29.9	[249]
βMeGlcN → MeOH + GlcN	0.0043	[249]
GlcNAc → GlcN + HOAc	2.10	[249]
GlcNSO$_3$ → GlcN + SO$_4{}^{2-}$	--[3]	[249]
URONIC ACIDS		
βGlcA-1,4-AMan$_R$ → GlcA + AMan$_R$	0.15	[250]
αGlcA-1,3-Man → GlcA + AMan$_R$	0.11	[250]
αIdoA-1,4-AMan$_R$ → IdoA + AMan$_R$	2.14	[251]
αGlcA-1,3-Fuc → GlcA + Fucose	0.04	[252]
NEUTRAL SUGARS		
βGlc-1,4-GlcA → Glc + GlcA	0.5	[252]
βGlc-1,4-Glc → 2 Glc	1.3	[252]
αGlc-1,3-Glc → 2 Glc	2.85	[251]

[1] Hydrolysis conditions: 0.5 M H$_2$SO$_4$, 100 °C
[2] Too rapid relative to glycoside bond hydrolysis to measure
[3] Too rapid to measure

however, L-iduronic acid glycosides are no more resistant to acid cleavage than neutral sugars, as shown in Table 3.1, and are completely hydrolyzed under the above conditions.

The events that occur when heparinoids are hydrolyzed in 1 $\underline{N}$ acid at 100 °C for 4- to 6-hr are dictated by the above observations. The sequence of events, shown in Figure 3.1, is as follows:

- The N-sulfate groups, when they are present, are lost almost immediately. Since 80-90% of the GlcN residues in heparins are N-sulfated, most of the GlcN glycosides in the heparins become resistant to further hydrolysis very early.

- The GlcNAc residues lose their acetyl groups over 2- to 3-hr, and then their glycosides become acid resistant. Prior to the loss of the N-acetyl groups, the GlcNAc glycosidic bonds continue to be hydrolyzed at a rate similar to that of the IdoA glycosidic bonds.

- Virtually all of the IdoA bonds are hydrolyzed over a 4- to 6-hr period of hydrolysis, but most of the GlcA and the GlcN bonds remain intact [250].

After hydrolysis for 4- to 6-h, there is some free GlcN present, but most of the GlcN is present in GlcN→HexA combinations, which occur as GlcN→GlcA and GlcN→IdoA disaccharides or as higher oligosaccharides composed primarily of GlcN→GlcA disaccharide units (since most of the GlcA→GlcN disaccharides occur in blocks; see Chapter 2). If the hydrolysis mixture is now treated with nitrous acid (pH 4, see Section III.B., p. 69), all of the free GlcN is converted to free AMan and all of the remaining GlcN bonds are cleaved. The GlcN→IdoA disaccharides yield equal amounts of anhydromannose and IdoA, whereas the GlcN→GlcA-containing *oligosaccharides* yield primarily GlcA→ AMan disaccharides. Thus, the hydrolysis/nitrous acid reaction sequence yields a mixture containing most of the IdoA as free iduronic acid,[3] free AMan, a small amount of free GlcA, and the GlcA→AMan disaccharide, which contains most of the GlcA.

The individual components in the final depolymerization mixture can be analyzed in a variety of ways. In one approach, the mixture is reduced with NaB^3H_4 to convert the free or combined anhydromannose to [^3H]anhydro-mannitol and the other products to their respective ^3H-labeled glycitols (reaction g, Figure 3.1). All of the ^3H-labeled products now have the same molar specific radioactivity. The mixtures are then separated by high performance liquid chromatography or paper chromatography and the amount of each labeled

---

[3] Since heparin is relatively inexpensive and contains a high proportion of L-IdoA, the hydrolysis/nitrous acid cleavage sequence offers a way to release large amounts of the otherwise unavailable L-IdoA to use as a standard. Methods for separating it from the mixture are available [250].

component is quantified by scintillation counting [250]. The ^{3}H cpm in each peak are summed and the relative amounts of each monosaccharide component are calculated as shown in Table 3.2.

This protocol, in contrast to the colorimetric assays for uronic acids and GlcN, yields total uronic acid and total glucosamine in equimolar amounts. Identical results for the analytical procedure described here are obtained whether the acid hydrolysis is carried out for 4, 6, or 8 hr before adding the nitrous acid reagent [250]. This is a reflection of the fact that all N- and O-sulfate groups and all N-acetyl groups have been completely removed after 4 hr of hydrolysis. From the 4th to the 8th hr, there is some further hydrolysis, which reduces the yields of disaccharides and increases the yield of monosaccharides in the assay mixture, but this does not change the final analytical results. Following acid hydrolysis and nitrous acid cleavage of unlabeled glycosaminoglycans, it would be equally feasible to convert the products to chromophoric derivatives and to analyze these by gas chromatography or high performance liquid chromatography (see Section III.E.1., p. 96). Such analytical schemes would avoid the use of radioisotopes.

The above hydrolysis/deamination reaction sequence gives complete

Table 3.2. Calculation of Equivalents of Each
Monosaccharide from cpm [1,2]

Amount of monosaccharide:	is equal to:	Sum of total ^{3}H cpm in:
GlcA$_R$	=	GlcA$_R$ + βGlcA 1→4 AMan$_R$
IdoA$_R$	=	IdoA$_R$ + αIdoA 1→4 AMan$_R$
AMan$_R$	=	AManR + -βGlcA 1→4 AMan$_R$+ α-IdoA 1→4 AMan$_R$

[1] The amounts of monosaccharides in any heparinoid can be calculated by measuring the number of ^{3}H cpm in the monosaccharides and disaccharides formed on acid hydrolysis. The relative yields of each monosaccharide can be calculated from the ^{3}H cpm alone. For the molar amounts of each monosaccharide, the specific radioactivity of the NaB^{3}H$_4$ used to reduce the hydrolysate must be used to convert cpm into moles.

[2] The subscript "R" indicates the aldehyde-reduced form of each mono- or disaccharide.

cleavage of the IdoA and GlcN glycosidic bonds, but only partial cleavage of the GlcA bonds. An alternate approach to monosaccharide analysis that results in complete conversion of heparinoids to monosaccharides involves first converting the polymeric heparinoid to its carboxyl-reduced derivative. This can be accomplished easily and stoichiometrically by reacting the heparinoid with a water-soluble carbodiimide to esterify all of the uronic carboxyl groups. The esterified uronic acids can then be reduced to their corresponding neutral monosaccharide residues by reaction with NaBH₄ [253]. This reaction sequence can be carried out in aqueous solution without depolymerization of the heparinoid (see Chapter 4). In the carboxyl reduction reaction, the IdoA and GlcA residues are converted to idose and glucose residues, respectively. The glycosidic linkages of these neutral monosaccharide residues are much more labile to acid than their corresponding uronic acids and are completely hydrolyzed by the typical hydrolysis conditions described earlier. Thus, in a 4-hr hydrolysis period, all of the glycosidic bonds of the idose and glucose residues are cleaved, and treatment with nitrous acid then cleaves all remaining GlcN glycosides, to yield a mixture containing only anhydromannose, glucose, and idose. Interestingly, approximately 90% of the idose is recovered, not as the normal reducing sugar, but as its 1,6-anhydro derivative, L-idosan [253, 254]. The mixture of monosaccharides can now be analyzed by a variety of procedures. The detection method chosen for such an analysis must be one that does not rely on the free reducing group of the monosaccharides, since idosan is a nonreducing sugar.

III. CLEAVAGE OF HEPARIN WITH RETENTION OF SULFATE SUBSTITUENTS

A. Introduction

Since each disaccharide type that occurs in heparinoids is uniquely distinguished by its uronic acid component and its sulfation pattern, most of the methodologies that have been developed for analysis of heparinoids are designed to preserve these structural elements. Only two cleavage methods meet this criterion: (a) cleavage of heparinoids with nitrous acid and (b) cleavage of heparinoids with microbial lyases. Both nitrous acid and the enzymes cause elimination reactions that cleave bonds between an upstream GlcN residue and a downstream HexA residue and yield an upstream fragment that has the GlcN, or a modified form of GlcN, at its reducing terminal, and a downstream fragment that has the HexA, or a modified form of the HexA, at its nonreducing terminal. Both cleavage methods yield fragments that contain one of the terminal residues in an altered form of the original monosaccharide. However, in both approaches, it is possible to identify these altered residues with one of the original monosaccharide structures.

B.   Cleavage with Nitrous Acid

1.   The cleavage reaction

Nitrous acid was used in heparin structure studies by some of the earliest workers in the field [255, 256]. Glycosidic bonds of N-unsubstituted GlcNs are cleaved when treated with nitrous acid (Figure 3.1, reaction e). This reaction occurs rapidly at room temperature and at pH's above 4, but the rate of the reaction falls off sharply at pH's below 4 [248]. If the amino group of the GlcN is acetylated, as found in most amino sugar-containing carbohydrates, it is not cleaved by nitrous acid under any conditions.

Glycosidic bonds of *N-sulfated* GlcN residues react with nitrous acid at pH 1.5 and room temperature, causing a similar deamination reaction that cleaves heparinoids between each N-sulfated GlcN and its downstream uronic acid residue. This reaction, illustrated in Figure 3.2. occurs at maximal rates at pH 1.5 and below, but very slowly at pH's above 3-3.5 [248]. Since heparinoids are the only structures found in Nature that have N-sulfated amino groups, the pH 1.5 nitrous acid reaction cleaves heparinoids but does not cleave proteins, nucleic acids,[4] or closely related carbohydrates (e.g., other glycosaminoglycans). Each cleavage occurs by an attack of nitrous acid on the GlcNSO$_3$ residue that results in the loss of the N-SO$_3$ as free SO$_4^{2-}$ and the amino group as nitrogen gas [5] (deaminative cleavage). The GlcNSO$_3$ residues at the site of cleavage are converted to 2,5-anhydro-<u>D</u>-mannose reducing terminal residues in the upstream fragments. The structure of the uronic acid at the site of cleavage is not altered. This is not an hydrolysis reaction, as it is often mistakenly described in the literature. Nitrous acid is consumed in the reaction (in contrast, H$^+$ is a catalyst in an hydrolysis reaction). Therefore the extent of bond cleavage can be controlled by limiting the levels of nitrous acid. The reaction occurs rapidly at room temperature and almost stoichiometrically (but

---

[4] Nitrous acid reacts with NH$_2$ groups of proteins and nucleic acids, but these reactions do not result in polymer cleavage.

[5] For the nitrogen gas that is formed, one of the nitrogens is derived from the GlcN and the other from the nitrous acid.

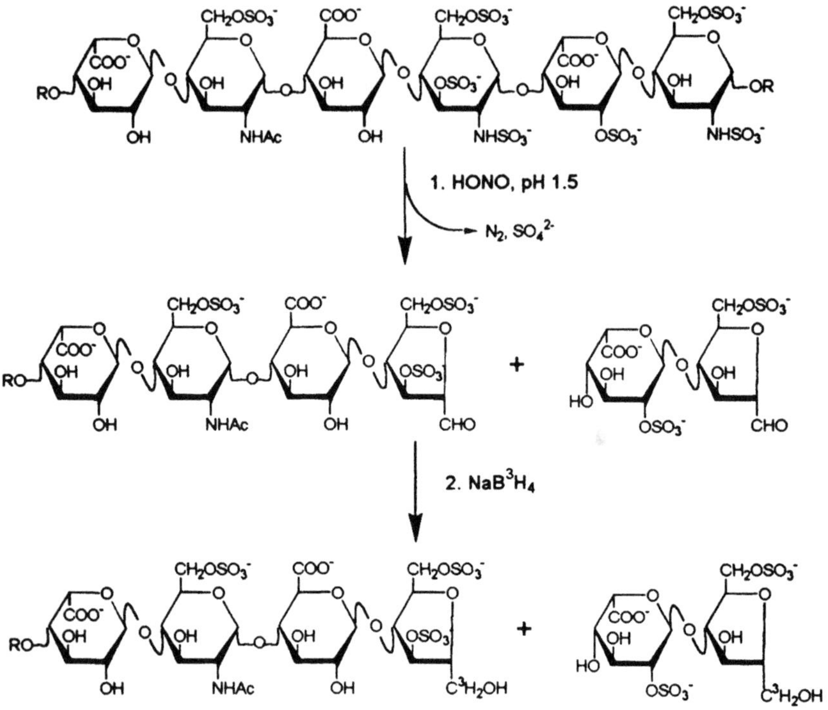

Figure 3.2. Nitrous acid cleavage of glycosides of GlcNSO$_3$.

see Section III.B.3., p. 72). Figure 3.2 also shows that the GlcNAc residues do
not react with nitrous acid. When heparin is treated with nitrous acid at pH 1.5,
the products are primarily disaccharides, with smaller yields of tetrasaccharides.
This is due to the fact that (a) 80-90% of the glucosamine residues in heparin are
GlcNSO$_3$ residues and are susceptible to nitrous acid, and (b) the nitrous acid-
resistant GlcNAc residues occur in disaccharides that have GlcNSO$_3$-containing
disaccharides at their immediate upstream and downstream positions. Thus, after
nitrous acid treatment, these GlcNAc's are recovered in tetrasaccharides. For
heparan sulfate, only 40-60% of the glucosamines are N-sulfated. Because these
GlcNSO$_3$'s occur in blocks, treatment with nitrous acid at pH 1.5 converts
heparan sulfate to a mixture containing smaller yields of disaccharides, identical
in kind, but not in proportion, to those obtained from heparin (see Chapter 2)

and larger yields of higher oligosaccharides that contain the blocks of unsulfated GlcNAc-containing disaccharides.[6]

2. Hydrazinolysis

Because the glycosidic bonds of the GlcNAc residues are not cleaved by nitrous acid, one does not obtain complete conversion of the heparinoids to disaccharides by cleavage with nitrous acid at pH 1.5. However, N-acetylated amino sugars can be de-N-acetylated with hydrazine (Figure 3.3, reaction 1) to convert the GlcNAc residues in these polymers to GlcN residues [23, 257].[7] The glycosidic bonds of the resulting N-unsubstituted amino sugars can then be cleaved at room temperature with nitrous acid at pH 4 (Figure 3.3, reaction 2). The N-sulfate groups in heparin and heparan sulfate are stable under the hydrazinolysis conditions [23, 257], and the glycosidic bonds of the GlcNSO$_3$ residues are not cleaved by nitrous acid at pH 4. Thus, following de-N-acetylation by hydrazinolysis, all heparinoids can be converted completely to their constituent disaccharides by treatment with nitrous acid at pH 4 and then at pH 1.5. Alternatively, both cleavages will occur at pH 3, albeit at somewhat slower rates [23]. The resulting disaccharides contain a nonreducing terminal uronic acid and a reducing terminal 2,5-anhydro-D-mannose residue. For quantitative analysis of the disaccharide mixture, the AMan residues can be reduced stoichiometrically with NaB^{3}H$_4$ to [^{3}H]anhydro-D-mannitols, thus labeling the disaccharides so that they can be assayed qualitatively and/or quantitatively (see Figure 3.2, reaction 2). Note that the same procedure may be used to analyze metabolically labeled glycosaminoglycans (with the use of unlabeled NaBH$_4$).

It should be noted also that all other glycosaminoglycans (chondroitin sulfate, dermatan sulfate, keratan sulfate, hyaluronic acid) contain N-acetylated amino sugars at every second residue. When such glycosaminoglycans are present as contaminants in the heparinoid sample, they will be completely de-N-acetylated by hydrazinolysis and will then be converted quantitatively to *disaccharides* by nitrous acid at pH 4. There are several implications of this fact. First, all fragments *larger* than disaccharides must be derived from heparinoids (assuming that only glycosaminoglycans were present in the original sample).

---

[6] These oligosaccharides contain small amounts of SO$_4$, presumably at their reducing and nonreducing terminal disaccharides. Unfortunately, the reducing terminal and nonreducing terminal disaccharide units have not been characterized. Their identification would represent one step toward the characterization of the structures at the junctions between the highly sulfated and the unsulfated blocks (Chapter 2).

[7] See Chapter 4 for a more extensive discussion of the hydrazinolysis reaction.

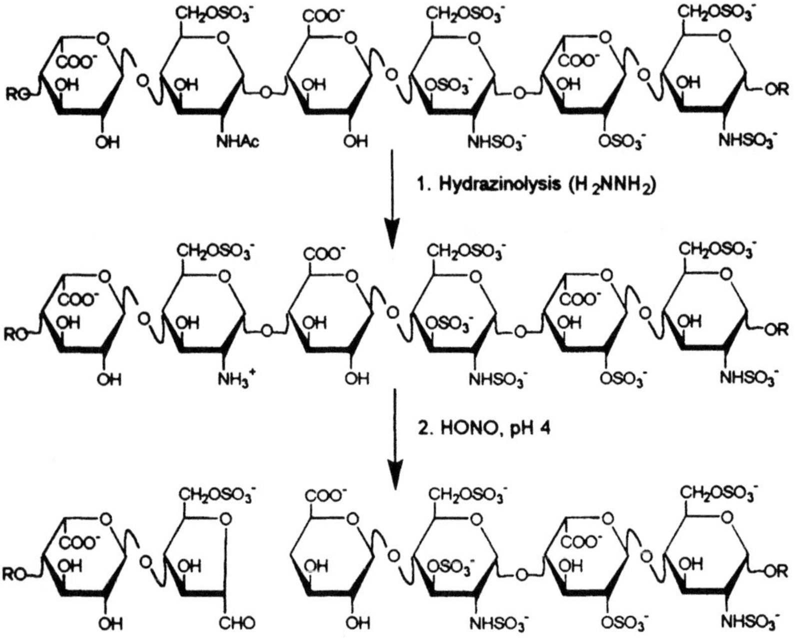

Figure 3.3. Nitrous acid cleavage of glycosidic bonds at de-N-acetylated GlcNAc residues.

Second, in the analysis of the heparinoid disaccharide fractions formed by such cleavages, one must be alert to the possibility that some disaccharides may be obtained from these contaminating glycosaminoglycans; these disaccharides can be distinguished from heparinoid-derived disaccharides by their chromatographic behaviors [258]. Finally, if the sample contains significant amounts of glycoproteins, both the O-linked and the N-linked oligosaccharides attached to the protein(s) may yield fragments as a result of cleavage at their GlcNAc or GalNAc residues (hydrazinolysis/nitrous acid treatment may also modify the structures of sialic acids).

3. Ring contraction

The nitrous acid reaction is virtually stoichiometric for β-linked amino sugars [249]. Consequently, nitrous acid treatment of de-N-acetylated glycosaminoglycans that contain β-linked amino sugars (chondroitin SO_4, dermatan SO_4, hyaluronic acid, and keratan SO_4) gives stoichiometric yields of disaccharides. However, a side reaction, the "ring contraction reaction," occurs to the extent of 10-15% for α-linked amino sugars, such as those found in heparin and heparan sulfate [248, 249]. This occurs for both N-unsubstituted and N-sulfated GlcN residues. This reaction is shown in Figure 3.4 for a segment of a heparinoid in which all of the GlcN residues are N-sulfated. The abnormal ring contraction side-reaction (reaction 3) converts the GlcN to a pentose with an exocyclic aldehyde group but does not cleave the glycosidic bond. The ring contraction reaction proceeds through the same initial steps as the normal cleavage reaction, with deamination of the N-sulfate groups which yields free SO_4^{2-} and nitrogen gas and forms a carbonium ion at C2 (reaction 1). In the normal bond cleavage reaction, the electron migration to the carbonium ion is from the C1-ring oxygen bond (reaction 2), but in the ring contraction reaction, the electrons of the C3-C4 bond migrate to the carbonium ion (reaction 3). The former migration leads to bond cleavage whereas the latter converts the GlcN residue to an aldehydopentose without bond cleavage. The result is that 10-15%

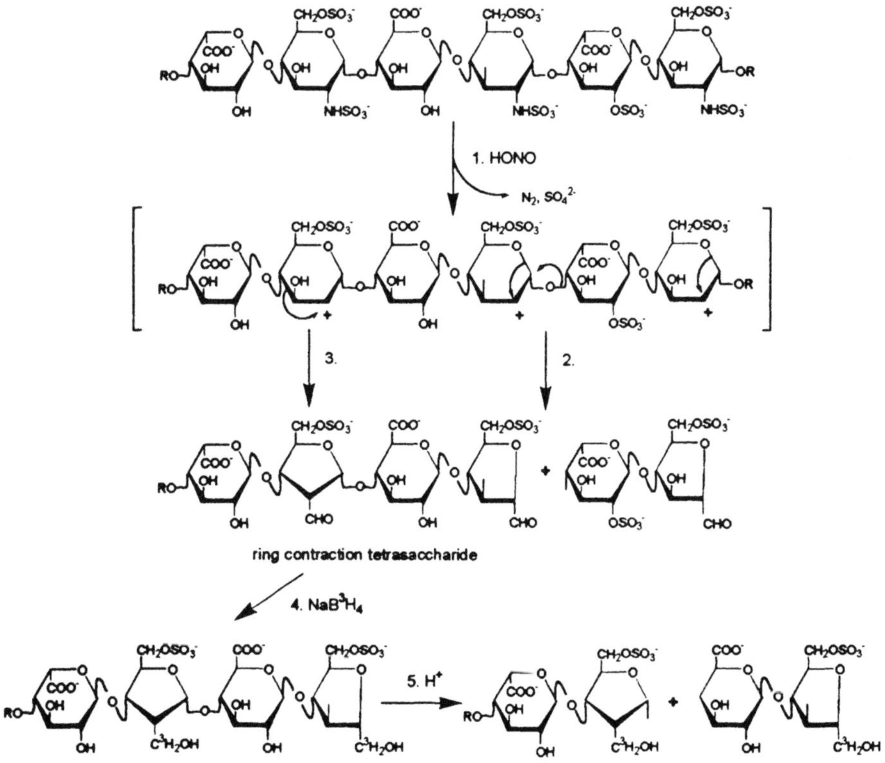

Figure 3.4. Ring contraction in the nitrous acid reaction. See text for description.

of the $GlcNSO_3$-containing disaccharides are converted into "ring contraction tetrasaccharides" instead of the expected disaccharides. Thus, when heparinoids are treated with nitrous acid, both ring contraction tetrasaccharides and the *bona fide* tetrasaccharides (those that contain internal GlcNAc residues—see Figure 3.2) are formed. If the mixture of deamination products is reduced with NaB^3H_4 and separated by HPLC using both a UV detector (set at 195 nm, for detection of N-acetyl groups) and a radioactivity detector, only the *bona fide* tetrasaccharides are observed in the UV, but both the *bona fide* and the ring contraction tetrasaccharides are observed by the radioactivity detector. In fact, the ring contraction tetrasaccharides incorporate two gram atoms of ^3H per mole (reaction 4), while the *bona fide* tetrasaccharides incorporate only one (Figure 3.2, reaction 2). To complicate matters further, prior to NaB^3H_4 reduction, both the anhydromannose and the aldehydopentose can undergo isomerization under mildly alkaline conditions, so that these tetrasaccharides may form mixtures of four isomers.

In addition to their failure to absorb in the UV, the ring contraction tetrasaccharides can be distinguished from the GlcNAc-containing disaccharides by their marked acid lability (reaction 5). They are completely converted to their two disaccharide components by mild acid hydrolysis (pH 2, 80°C, 120 min) [244]. Under such hydrolysis conditions, neither the glycosidic bonds nor the sulfate ester bonds of *bona fide* di- and tetrasaccharides show measurable hydrolysis. Mild acid hydrolysis of each NaB^3H_4-reduced ring contraction tetrasaccharide yields one normal nitrous acid disaccharide (with an $[^3H]AMan_R$ reducing terminal) and the new ring contracted disaccharide that is ^3H labeled in its exocyclic hydroxymethyl group.

No method for controlling the extent of the ring contraction reaction has been found. Thus, heparin and heparan sulfate yield significant amounts of the ring contraction tetrasaccharides [28]. However, there seems to be little structural selectivity for the ring contraction, as the ring contraction tetrasaccharides contain disaccharide units in the same proportions as found in the original glycosaminoglycan. Thus, even for heparin and heparan sulfate, the normal disaccharides formed during nitrous acid cleavage are obtained in the same proportions that are present in the original glycosaminoglycan. Consequently, when yields are expressed in *percentage of total disaccharides recovered*, the presence of ring contraction tetrasaccharides does not interfere with the quantitation; the yields of di- and monosaccharides are truly representative of their proportions in the starting glycosaminoglycan. Some of these heparin ring contraction tetrasaccharides have been identified and their high performance liquid chromatography migration has been recorded [28, 244].

4. HPLC profiling of nitrous acid cleavage products

The mixture of ^{3}H-labeled oligosaccharides formed by cleavage of heparinoids with nitrous acid, either with or without prior hydrazinolysis, can be resolved by high performance liquid chromatography on either a strong anion exchange (SAX) column [28, 259] or a reversed phase ion pairing column [23, 117, 260]. The highest resolution is obtained with the latter column system, for which the effluent peaks can be monitored with a radioactivity flow detector coupled to a computer for data analysis. When heparin is cleaved without hydrazinolysis, a mixture of di- and tetrasaccharides is obtained. Under optimized conditions, the disaccharides and tetrasaccharides can be resolved in a single analysis, as illustrated in Figure 3.5, which shows three panels of a single elution profile, each with a progressively higher amplification of the data. At the bottom is a reagent blank elution profile at the highest amplification, which shows that the sample peaks observed at the highest amplification level in the heparin fragment profiles are indeed derived from heparin. Figure 3.5 illustrates the relative amounts of the disaccharides, all of which are observed at the lowest amplification level, and the tetrasaccharides, which are observed only at the intermediate and highest amplification levels. Many of the tetrasaccharides have been structurally characterized and are included in the structures given in Table 2.3 in Chapter 2. These results illustrate the complexity of the mixture of tetrasaccharide sequences that contain the GlcNAc residues, as well as their relatively small contribution to the total heparin structure.

C. Cleavage with Heparin Lyases

1. Introduction

Enzymatic degradation has been used extensively in the characterization of heparinoids. Although the literature in this area spans many years and is quite complex, the wide spread use and the easy availability of the bacterial lyases for heparinoid characterization make an understanding of the activities of these enzymes essential for appreciating the applications and limitations of these structural tools.

Heparinoids are converted to small oligosaccharides by treatment with lyases that cleave heparinoids between glucosamine residues and their downstream uronic acids, as illustrated in Figure 3.6. The upstream fragment formed by the cleavage is an oligosaccharide that contains the glucosamine at

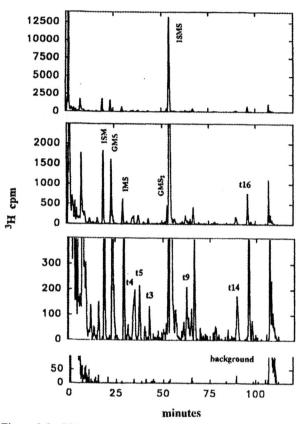

Figure 3.5. HPLC analysis of oligosaccharides formed by pH
1.5 nitrous acid cleavage of heparin. Abbreviations are: ISMS,
IdoA-2-SO$_4$→ AMan$_R$-6-SO$_4$; ISM, IdoA-2-SO$_4$→ AMan$_R$;
GMS, GlcA→ AMan$_R$-6-SO$_4$; IMS, IdoA→ AMan$_R$-6-SO$_4$;
GMS$_2$; GlcA→ AMan$_R$-3,6-(SO$_4$)$_2$; t3, t4, t5, etc. are
tetrasaccharides identified in [28]. Unpublished work of Y.
Guo.

the site of cleavage as its reducing terminal. The downstream fragment has a Δ
^{4,5}-uronic acid at its nonreducing terminal. All N- and O-sulfates are retained in
the products. As shown in the bracketed reactions in Figure 3.6, the lyase-
catalyzed reaction may be considered as a deprotonation at the C5 position of
the uronic acid converting C5 to a carbanion. The pair of electrons from the
carbanion then migrate to form the Δ^{4,5} double bond and to break the glycosidic

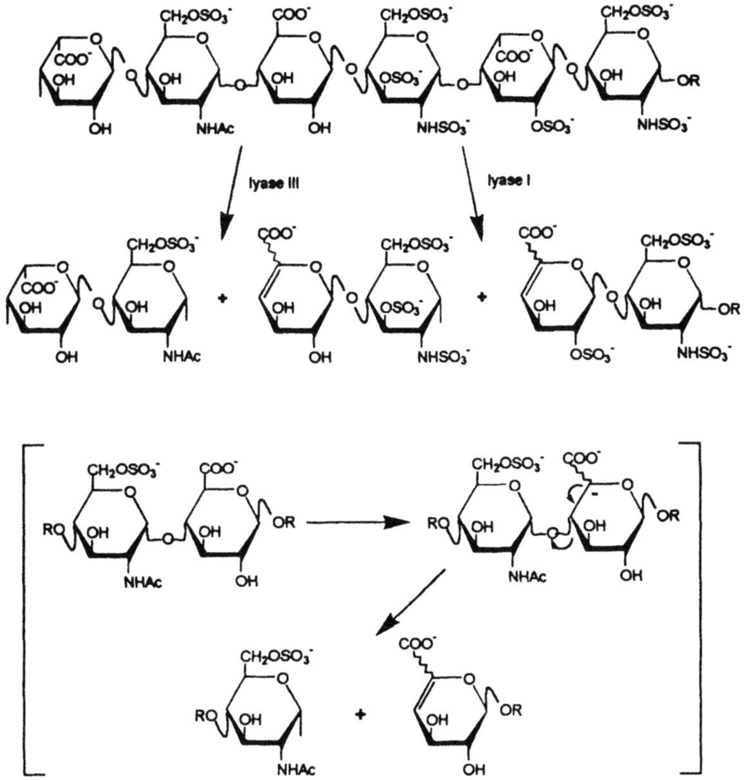

Figure 3.6. Cleavage of heparinoids by heparin lyases. See text for
description.

bond with the elimination of the upstream GlcN residue. Several general reviews
of the actions of these enzymes have appeared [261-263].

The lyases that have been most extensively studied are the enzymes
purified from *Flavobacterium heparinum*. This organism was originally isolated
from soil by enrichment for growth on heparin as a sole carbon source [264].
These enzymes are induced by heparin and heparin degradation products, and by
heparan sulfate, N-acetylglucosamine, and maltose. *F. heparinum* produces
three heparin-cleaving lyases, heparin lyases I, II, and III [265], which, in
combination, convert heparinoids to disaccharides. Since the polymer cleavage
reactions are β-elimination reactions, the enzymes are referred to as lyases rather
than hydrolases. In addition to the lyases, *F. heparinum* produces a
glycuronidase, which converts the disaccharides to monosaccharides, and
several sulfatases, which cleave the sulfate groups from the mono- or

disaccharides, which are then further catabolized [266-270]. Heparin lyase I from *F. heparinum* has been cloned and expressed in *E. coli*, and its mechanism of action has been studied [271-274]. Lyases II and III have also been cloned and expressed [275].

2.   Nomenclature of lyases

Unfortunately, the nomenclature of these enzymes has evolved slowly and is not uniform. Thus, one must be aware that several different names are used for each of the enzymes.

The names are based on the relative rates and extents of cleavage of heparin and heparan sulfate, as shown in Table 3.3, compiled by Linhardt [262]. Here, the relative rates of cleavage of heparin by lyase I are normalized to 100 with hog mucosa heparin as a substrate, whereas the rates of cleavage of heparan sulfate by lyases II and III are normalized to 100 with bovine kidney heparan sulfate as a substrate. Table 3.3 also presents an estimate of the fraction of the total disaccharide linkages in the substrates that are cleaved by exhaustive treatment with each enzyme. The results show that lyase I is very effective on heparin, cleaving the polymer rapidly at 58% (hog mucosa) or 76% (beef lung heparin) of the total linkages between disaccharides, but is very much less active on heparan sulfate. Lyase III shows just the opposite type of specificity for the hog mucosa heparin and bovine kidney heparan sulfate. Heparan sulfate is cleaved rapidly and almost completely, whereas heparin is minimally cleaved. Finally, for lyase II, the relative rates of cleavage for heparin and heparan sulfate are similar.

To avoid some of the confusion concerning the enzyme nomenclature, it is important to remember that an early name for heparan sulfate was heparitin sulfate (Chapter 2). Thus, enzymes that cleave heparan sulfate have been referred to as "heparitinases," whereas enzymes that cleave heparin have been referred to as "heparinases." Since all of the heparin lyases cleave both heparins and heparan sulfates, they all have been referred to both as heparinases and as heparitinases. The names given to these enzymes were numbered in order of decreasing effectiveness on their substrates. Thus, when the enzymes are named as heparinases, lyases I, II, and III were named heparinases I, II, and III, respectively. However, when the enzymes were named as heparitinases, the names heparitinase I, II, and III were given to lyases III, II, and I, respectively. Thus, two sets of names were coined for these three enzymes. Later, in order to avoid the implication that these enzymes are hydrolases, the enzymes were given the systematic names of lyases. Unfortunately, we now have three sets of names for these enzymes, plus several other names that are also sometimes used.

Table 3.3. Activities of Heparin Lyases on Heparin
and Heparan Sulfate [1,2]

Substrate	Enzyme		
	Lyase I	Lyase II	Lyase III
Heparin			
Relative rate	100	58	<1
% of bonds cleaved	58 (76)[3]	85	6
Heparan sulfate			
Relative rate	13	100	100
% of bonds cleaved	19	39	94

[1] Data from [262]

[2] Data represent the relative rates of cleavage and the degree of cleavage for each of the heparin lyases. The heparin is hog mucosa heparin. The lyase I rate for heparan sulfate is compared to the rate for the heparin, which is taken as 100. The heparan sulfate is bovine kidney heparan sulfate. The lyase II and III rates for heparin are compared to the rate for the heparan sulfate, which is taken as 100.

[3] Value in () is that obtained for beef lung heparin.

The three lyases have been purified to apparent homogeneity [276]. They are available through Sigma Chemical Co., St. Louis, Missouri., USA, Seikagaku Fine Chemicals, Tokyo, Japan, and Grampian Enzymes, Aberdeenshire, Scotland. The names that are used are:

- heparin lyase I (EC 4.2.2.7) was the first of these enzymes to be widely available and is the most extensively studied and utilized. It is commonly called heparinase (offered under this name by Seikagaku), but is also called heparinase I (Sigma) and heparitinase III. This enzyme is an endolytic enzyme with M_r = 42.50 kDa [271].

- heparin lyase II (no EC number) is also called heparinase II (Sigma) or heparitinase II (Seikagaku). This enzyme, M_r = 84.5 kDa, is apparently an endolytic enzyme, but its action pattern has not been studied in detail [275].

- heparin lyase III (EC 4.2.2.8) is also called heparinase III (Sigma), heparitinase, heparitinase I (Seikagaku), and heparan sulfate lyase. This enzyme, M_r = 73.1 kDa, is apparently an endolytic enzyme, but it action pattern has not been studied in detail [275].

3.   Assay of lyases

Each lyase-catalyzed cleavage generates a $\Delta^{4,5}$-HexA residue at the nonreducing terminal of the downstream fragment, which can be monitored by its UV absorbency. Molar extinction coefficients at 233 nm of all of the $\Delta^{4,5}$-HexA's are similar, ranging from 5063 to 5657 cm^2/mmole [277, 278]. The ΔA_{233}, the most convenient and most generally used assay for these enzymes, does not require identification of the oligosaccharide fragments, since each cleavage generates a $\Delta^{4,5}$-HexA. The rate of formation of the *individual* oligosaccharides can also be obtained by following the ΔA_{233} of each $\Delta^{4,5}$-oligosaccharide after separation from the mixture by high performance liquid chromatography, paper chromatography, or paper electrophoresis. The disaccharides could also be detected following radiolabeling of the disaccharides by reduction with NaB^3H_4 to incorporate 1 g atom of ^3H per mole of disaccharide (see Section IV.D., p. 103). The label recovered in the separated disaccharides can be used for qualitative and quantitative assays of each separated disaccharide.

4.   Specificity of lyases

a.   Introduction

Several problems are encountered in the determination of the substrate specificities of these enzymes. First, in contrast to the chemical reactions with nitrous acid, for which cleavage depends only on the substituents on the amino group of the glucosamine residues, the heparin lyases exhibit specificities that are very dependent on the positions of the N- and O-sulfates. Furthermore, polymeric heparinoids contain multiple sites of cleavage for each lyase. Thus, determination of the specific structures that are attacked by each enzyme requires the isolation of oligosaccharides with desired structures. Ideally, one would like to be able to measure the *kinetics* of cleavage of individual bonds in heparin oligosaccharides and to show how the upstream and downstream disaccharide structures, as well as the sizes of the substrates, affect the rates. We have already discussed the difficulties in the isolation of pure heparinoid oligosaccharides (Chapter 2). It is an impossible task, given our limited abilities to obtain pure oligosaccharides from heparin, to determine how the rates of cleavage of specific glycosidic linkages are affected by (a) all of the desired structural permutations and (b) molecular size.

A second problem is that the formation of the $\Delta^{4,5}$-double bond during the cleavage destroys the C5 asymmetry of the uronic acid that distinguishes GlcA from IdoA; i.e., GlcA and IdoA yield the same unsaturated uronic acid in the products (see Figure 3.6). Thus, the structure of the original uronic acid residue at the site of cleavage cannot be deduced from the structure of the product, but must be determined from a complete knowledge of the structures of the substrates and the specificities of the lyases. As shown later, the $\Delta^{4,5}$-HexA's formed by any of the three lyases may be derived from either GlcA or IdoA residues.

It has become clear that the disaccharide structures on both sides of the cleavage site are important in assigning the lyase specificities. Most attempts to define the specificities of these enzymes have been based on the structures of the products formed when *polymeric* heparin or heparan sulfate is totally depolymerized with one of these enzymes. Such studies have given general insights into the specificities of the enzymes, but have not provided the full structural features of the disaccharide units at the cleavage sites. One of the definitive studies on the specificities of the lyases has been reported by Linhardt and associates [279], who have prepared a number of pure oligosaccharide substrates from partial lyase I digests of heparin. The structures of the oligosaccharides were determined and the substrates were tested with all three lyases. Table 3.4 gives the structures of all of the oligosaccharides tested in these specificity studies for lyase I (Table 3.5), lyase II (Table 3.6), and lyase III (Table 3.7). Table 3.4 also includes the structures of all of the cleavage products that are formed by the lyases in these studies, as well as some of the structures that are described in the oligosaccharide mapping studies (Section III.C.6., p. 91). Each oligosaccharide is denoted by a number. These oligosaccharides have been described in a number of publications. However, the numbering systems used to identify them differs from one publication to another. *In the substrate specificity description here, the numbers that are used in Table 3.4 to identify each oligosaccharide will be used for the same structures throughout the discussion of Tables 3.5, 3.6, 3.7, and 3.9 and in Figures 3.7 and 3.8.* In Tables 3.5, 3.6, and 3.7, the substrates are presented in the order of decreasing rates of cleavage; thus, the orders of the structures, and their numbers, are not the same in each table.[8]

---

[8] Since these oligosaccharides have been given different numbers in different publications, it is difficult to follow these structures through these several papers. Also, it should be noted that, in the identification of oligosaccharide 2 in the initial work [280], it was observed that there was one GlcNSO$_3$ and one GlcNSO$_3$-6-SO$_4$ in the tetrasaccharide, but the positions of these two residues were mis-assigned. The correct structure was reported in 1992 [281], but all papers from this lab between 1986 and 1992

In these tables, the products formed from each of the pure substrates are given beneath each substrate. All substrates were assayed at a concentration of 10 μg/50 μl, equivalent to 167 μM for a typical tetrasaccharide with a molecular weight of 1200. Although the K_m values are not known for these substrates, it is likely that the substrate concentrations represent saturating levels. Due to the problems in obtaining large amounts of substrates, it has not been possible to measure the actual *kinetics* of cleavage of the substrates. Instead, the products, and any remaining substrate, were determined after a 24 h incubation period. Activities are given in terms of the products formed. Also, the fraction of each

Table 3.4. Oligosaccharides Used in the Study of Specificities
of Heparin Lyases

---

1. $\Delta^{4,5}$-UA-2-SO$_4$ → GlcNSO$_3$-6-SO$_4$

2. $\Delta^{4,5}$-UA-2-SO$_4$ → GlcNSO$_3$-6-SO$_4$ → IdoA-2-SO$_4$ → GlcNSO$_3$

3. $\Delta^{4,5}$-UA-2-SO$_4$ → GlcNSO$_3$-6-SO$_4$ → GlcA → GlcNSO$_3$-6-SO$_4$

4. $\Delta^{4,5}$-UA-2-SO$_4$ → GlcNSO$_3$-6-SO$_4$ → IdoA-2-SO$_4$ → GlcNSO$_3$-6-SO$_4$

5. $\Delta^{4,5}$-UA-2-SO$_4$ → GlcNSO$_3$-6-SO$_4$ → IdoA → GlcNAc-6-SO$_4$ →GlcA →
   GlcNSO$_3$-3,6-di-SO$_4$

6. $\Delta^{4,5}$-UA → GlcNAc

7. $\Delta^{4,5}$-UA → GlcNSO$_3$

8. $\Delta^{4,5}$-UA → GlcNAc-6-SO$_4$

9. $\Delta^{4,5}$-UA-2-SO$_4$ → GlcNSO$_3$

10. $\Delta^{4,5}$-UA → GlcNSO$_3$-6-SO$_4$

11. $\Delta^{4,5}$-UA → GlcNSO$_3$-3,6-di-SO$_4$)

12. $\Delta^{4,5}$-UA-2-SO$_4$ → GlcNSO$_3$-6-SO$_4$ → IdoA → GlcNAc-6-SO$_4$)

13. $\Delta^{4,5}$-UA-2-SO$_4$ → GlcNSO$_3$-6-SO$_4$ → IdoA-2-SO$_4$ → GlcNSO$_3$-6-SO$_4$ →
    IdoA-2-SO$_4$ → GlcNSO$_3$-6-SO$_4$

14. $\Delta^{4,5}$-UA-2-SO$_4$ → GlcNSO$_3$-6-SO$_4$ → IdoA-2-SO$_4$ → GlcNSO$_3$-6-SO$_4$ →
    IdoA-2-SO$_4$ → GlcNSO$_3$-6-SO$_4$ → IdoA-2-SO$_4$ → GlcNSO$_3$-6-SO$_4$

15. $\Delta^{4,5}$-UA-2-SO$_4$ → GlcNSO$_3$-6-SO$_4$ → IdoA → GlcNAc-6-SO$_4$ →GlcA → GlcNSO$_3$-6-SO$_4$

16. $\Delta^{4,5}$-UA-2-SO$_4$ → GlcNSO$_3$ → IdoA-2-SO$_4$ →GlcNSO$_3$

17. $\Delta^{4,5}$-UA-2-SO$_4$ → GlcNSO$_3$-6-SO$_4$ → IdoA → GlcNSO$_3$-6-SO$_4$

18. $\Delta^{4,5}$-UA-2-SO$_4$ → GlcNSO$_3$ → GlcA → GlcNSO$_3$-6-SO$_4$

19. $\Delta^{4,5}$-UA → GlcNAc-6-SO$_4$ → GlcA → GlcNSO$_3$-6-SO$_4$

20. $\Delta^{4,5}$-UA → GlcNAc-6-SO$_4$ → GlcA → GlcNSO$_3$-3,6-di-SO$_4$

Structures from reference [279].

---

contained the incorrect structure. The corrected structure is used in all references to this tetrasaccharide in this book.

substrate converted to the products in the 24 hr incubation is given in parentheses in these Tables.

Another definitive study, by Yamada *et al.* [32], has provided additional information concerning the specificities of these enzymes, especially with regard to the cleavages that occur at GlcA-2-SO$_4$ residues. These data are presented in Table 3.8, where the substrates are designated using the same Roman numerals used by the authors of this study. All of the substrates that were tested in these latter studies were tetrasaccharides; thus, the cleavage products are simply the two component disaccharides.[9] No statements concerning the degrees of cleavages of these substrates are given. The results of both studies will be discussed together below. Notably, in neither study were sequences containing multiple GlcA→GlcNAc disaccharides tested as substrates.Specificity of heparin lyase I

At the outset, note should be taken that all of the oligosaccharide substrates and products listed in Tables 3.4 and 3.8 were formed by the cleavage of heparin with heparin lyase I. Thus, the structures at both the reducing terminals and the nonreducing terminals of these oligosaccharides give information concerning the specificity of heparin lyase I. We will examine these structures first, and then turn to a discussion of the further cleavage of these oligosaccharides by lyases I, II, and III.

We will examine first the *reducing terminal disaccharide units* of these substrates. With a single exception (compound VII in Table 3.8), the glucosamine residues are N-sulfated. Note also that, with only a few exceptions, all of these GlcN residues are 6-O-sulfated. Oligosaccharide 5 has a GlcNSO$_3$-3,6-di-SO$_4$ residue at its reducing terminal, indicating that heparin lyase I can cleave within an apparent antithrombin-binding site (see Chapter 7).

Examination of the *nonreducing terminal disaccharide units* of oligosaccharide substrates shows that, with only a few exceptions, the $\Delta^{4,5}$-HexA is 2-O-sulfated. Also, the GlcN residue of the nonreducing terminal disaccharide is both N- and O-sulfated, except for several products. Thus, lyase I cleavage takes place within highly sulfated regions of heparinoids.

Turning now to the tests of the further cleavage of these substrates by lyase I (Table 3.5), it is clear that heparin lyase I exhibits the same exacting specificity that is implied by the residues at the end groups of its products. It gives facile cleavage of the

$$GlcNSO_3\text{-}6\text{-}SO_4 \rightarrow IdoA\text{-}2\text{-}SO_4$$

linkages in

---

[9] Six of the substrates listed in Table 3.8 are different from any of those tested in Tables 3.5-3.7. Substrates IV, VI, VII, VIII, and X in Table 3.8 are the same as substrates 18, 16, 12, 3, and 2, respectively, in Table 3.4.

Table 3.5. Substrate Specificity of Heparin Lyase I [1,2]

Cleaved Substrates:

$\downarrow$

4. $\Delta^{4,5}$-UA-2-SO$_4$ → GlcNSO$_3$-6-SO$_4$ → IdoA-2-SO$_4$ → GlcNSO$_3$-6-SO$_4$
   Products: 1 + 1 (100)

                $\downarrow$ $\downarrow$

13. $\Delta^{4,5}$-UA-2-SO$_4$ → GlcNSO$_3$-6-SO$_4$ → IdoA-2-SO$_4$ → GlcNSO$_3$-6-SO$_4$ → IdoA-2-SO$_4$ →
    GlcNSO$_3$-6-SO$_4$
    Products: 1 + 1 + 1 (100)

                $\downarrow$ $\downarrow$

14. $\Delta^{4,5}$-UA-2-SO$_4$ → GlcNSO$_3$-6-SO$_4$ → IdoA-2-SO$_4$ → GlcNSO$_3$-6-SO$_4$ → IdoA-2-SO$_4$ →
            $\downarrow$
    GlcNSO$_3$-6-SO$_4$ → IdoA-2-SO$_4$ → GlcNSO$_3$-6-SO$_4$
    Products: 1 + 1 + 1 + 1 (100)

                    $\downarrow$

15. $\Delta^{4,5}$-UA-2-SO$_4$ → GlcNSO$_3$-6-SO$_4$ → IdoA → GlcNAc-6-SO$_4$ →GlcA → GlcNSO$_3$-6-SO$_4$
    Products: 1 + 19 (80)

                $\downarrow$

16. $\Delta^{4,5}$-UA-2-SO$_4$ → GlcNSO$_3$ → IdoA-2-SO$_4$ →GlcNSO$_3$
    Products: 9+9 (30)

                $\downarrow$

17. $\Delta^{4,5}$-UA-2-SO$_4$ → GlcNSO$_3$-6-SO$_4$ → IdoA → GlcNSO$_3$-6-SO$_4$
    Products: 1 +10 (20)

                $\downarrow$

2. $\Delta^{4,5}$-UA-2-SO$_4$ → GlcNSO$_3$-6-SO$_4$ → IdoA-2-SO$_4$ → GlcNSO$_3$
   Products: 1+9 (20)

Uncleaved Substrates:
3. $\Delta^{4,5}$-UA-2-SO$_4$ → GlcNSO$_3$-6-SO$_4$ → GlcA → GlcNSO$_3$-6-SO$_4$
5. $\Delta^{4,5}$-UA-2-SO$_4$ → GlcNSO$_3$-6-SO$_4$ → IdoA → GlcNAc-6-SO$_4$ →GlcA →
   GlcNSO$_3$-3,6-di-SO$_4$
18. $\Delta^{4,5}$-UA-2-SO$_4$ → GlcNSO$_3$ → GlcA → GlcNSO$_3$-6-SO$_4$

---

[1] Data taken from [279].
[2] Numbers of the substrates and products are listed in Table 3.4. Arrows indicate the positions of cleavage. The products formed from each of the substrates are given beneath each substrate. All substrates were assayed at a concentration of 10 µg/ 50 µl (not at the same molar concentrations). The products, and any remaining substrate, were determined after a 24-hr incubation period. Activities are given in terms of the products formed and the fraction of each substrate converted to the products (in parentheses) in the 24-hr incubation. The substrates are presented in the order of decreasing rates of cleavage.

Table 3.6. Substrate Specificity of Heparin Lyase II [1,2]

Cleaved Substrates:

$\downarrow$

2. $\Delta^{4,5}$-UA-2-SO$_4$ → GlcNSO$_3$-6-SO$_4$ → IdoA-2-SO$_4$ → GlcNSO$_3$
   Products: 1 + 9 (100)

$\downarrow$

3. $\Delta^{4,5}$-UA-2-SO$_4$ → GlcNSO$_3$-6-SO$_4$ → GlcA → GlcNSO$_3$-6-SO$_4$
   Products: 1 + 10 (100)

$\downarrow$

4. $\Delta^{4,5}$-UA-2-SO$_4$ → GlcNSO$_3$-6-SO$_4$ → IdoA-2-SO$_4$ → GlcNSO$_3$-6-SO$_4$
   Products: 1 + 1 (100)

$\downarrow$ $\downarrow$

13. $\Delta^{4,5}$-UA-2-SO$_4$ → GlcNSO$_3$-6-SO$_4$ → IdoA-2-SO$_4$ → GlcNSO$_3$-6-SO$_4$ →
    IdoA-2-SO$_4$ → GlcNSO$_3$-6-SO$_4$
    Products: 1 + 1 + 1 (100)

$\downarrow$

16. $\Delta^{4,5}$-UA-2-SO$_4$ → GlcNSO$_3$ → IdoA-2-SO$_4$ → GlcNSO$_3$
    Products: 9 + 9 (100)

$\downarrow$

17. $\Delta^{4,5}$-UA-2-SO$_4$ → GlcNSO$_3$-6-SO$_4$ → IdoA → GlcNSO$_3$-6-SO$_4$
    Products: 1 + 10 (100)

$\downarrow$

18. $\Delta^{4,5}$-UA-2-SO$_4$ → GlcNSO$_3$ → GlcA → GlcNSO$_3$-6-SO$_4$
    Products: 9 + 10 (95)

$\downarrow$ $\downarrow$

15. $\Delta^{4,5}$-UA-2-SO$_4$ → GlcNSO$_3$-6-SO$_4$ → IdoA → GlcNAc-6-SO$_4$ → GlcA → GlcNSO$_3$-6-SO$_4$
    Products: 1 + 8 + 10 (90)

$\downarrow$

5.[3] $\Delta^{4,5}$-UA-2-SO$_4$ → GlcNSO$_3$-6-SO$_4$ → IdoA → GlcNAc-6-SO$_4$ → GlcA → GlcNSO$_3$-3,6-di-SO$_4$
    Products: 1 + 20 (80)

---

[1] Data taken from [279].

[2] The numbers of the substrates and products are listed in Table 3.4. The products formed from each of the pure substrates are given beneath each substrate. All substrates were assayed at a concentration of 10 µg/ 50 µl (not at the same molar concentrations). The products, and any remaining substrate, were determined after a 24-hr incubation period. Activities are given in terms of the products formed and the fraction of each substrate converted to the products (in parentheses) in the 24-hr incubation. The substrates are presented in the order of decreasing rates of cleavage.

[3] According to an earlier report, oligosaccharide 5 is cleaved by lyase II at both internal uronic acid units [282].

Table 3.7. Substrate Specificity of Heparin Lyase III [1, 2]

Cleaved Substrates:

↓

18. $\Delta^{4,5}$-UA-2-SO$_4$ → GlcNSO$_3$ → GlcA → GlcNSO$_3$-6-SO$_4$
   Products: 9 + 10 (100)

↓   ↓

15. $\Delta^{4,5}$-UA-2-SO$_4$ → GlcNSO$_3$-6-SO$_4$ → IdoA → GlcNAc-6-SO$_4$ →GlcA → GlcNSO$_3$-6-SO$_4$
   Products: 1 + 8 + 10 (75)

↓

17. $\Delta^{4,5}$-UA-2-SO$_4$ → GlcNSO$_3$-6-SO$_4$ → IdoA → GlcNSO$_3$-6-SO$_4$
   Products: 1 + 10 (70)

↓

3. $\Delta^{4,5}$-UA-2-SO$_4$ → GlcNSO$_3$-6-SO$_4$ → GlcA → GlcNSO$_3$-6-SO$_4$
   Products: 1 + 10 (50)

Uncleaved Substrates: [3]

2. $\Delta^{4,5}$-UA-2-SO$_4$ → GlcNSO$_3$-6-SO$_4$ → IdoA-2-SO$_4$ → GlcNSO$_3$

4. $\Delta^{4,5}$-UA-2-SO$_4$ → GlcNSO$_3$-6-SO$_4$ → IdoA-2-SO$_4$ → GlcNSO$_3$-6-SO$_4$

5.[3]. $\Delta^{4,5}$-UA-2-SO$_4$ → GlcNSO$_3$-6-SO$_4$ → IdoA → GlcNAc-6-SO$_4$ →GlcA → GlcNSO$_3$-3,6-di-SO$_4$

$\Delta^{4,5}$-UA-2-SO$_4$ → GlcNSO$_3$-6-SO$_4$ → IdoA-2-SO$_4$ → GlcNSO$_3$-6-SO$_4$ → IdoA-2-SO$_4$ →
   GlcNSO$_3$-6-SO$_4$

16. $\Delta^{4,5}$-UA-2-SO$_4$ → GlcNSO$_3$ → IdoA-2-SO$_4$ →GlcNSO$_3$

---

[1] Data taken from [279].

[2] The numbers of the substrates and products are listed in Table 3.4. The products formed from each of the pure substrates are given beneath each substrate. All substrates were assayed at a concentration of 10 µg/50 µl (not at the same molar concentrations). The products, and any remaining substrate, were determined after a 24-hr incubation period. Activities are given in terms of the products formed and the fraction of each substrate converted to the products (in parentheses) in the 24-hr incubation. The substrates are presented in the order of decreasing rates of cleavage.

[3] According to an earlier report, substrate 5 was cleaved by heparin lyase III [283].

Table 3.8. Heparin Lyase Specificities

Lyase I Substrates

XI $\Delta^{4,5}$-UA-2-SO$_4$→GlcNSO$_3$→IdoA-2-SO$_4$→GlcNSO$_3$-6-SO$_4$

XII $\Delta^{4,5}$-UA-2-SO$_4$→GlcNSO$_3$-6-SO$_4$→GlcA-2-SO$_4$→GlcNSO$_3$-6-SO$_4$

XIII $\Delta^{4,5}$-UA-2-SO$_4$→GlcNSO$_3$-6-SO$_4$→IdoA-2-SO$_4$→GlcNSO$_3$-6-SO$_4$

Lyase II Substrates

II $\Delta^{4,5}$-UA→GlcNSO$_3$-6-SO$_4$→GlcA→GlcNSO$_3$-6-SO$_4$

IV $\Delta^{4,5}$-UA-2-SO$_4$→GlcNSO$_3$→GlcA→GlcNSO$_3$-6-SO$_4$

V $\Delta^{4,5}$-UA-2-SO$_4$→GlcNAc-6-SO$_4$→GlcA→GlcNSO$_3$-6-SO$_4$

VI $\Delta^{4,5}$-UA-2-SO$_4$→GlcNSO$_3$→IdoA-2-SO$_4$→GlcNSO$_3$

VII $\Delta^{4,5}$-UA-2-SO$_4$→GlcNSO$_3$-6-SO$_4$→IdoA→GlcNAc-6-SO$_4$

VIII $\Delta^{4,5}$-UA-2-SO$_4$→GlcNSO$_3$-6-SO$_4$→GlcA→GlcNSO$_3$-6-SO$_4$

IX $\Delta^{4,5}$-UA-2-SO$_4$→GlcNSO$_3$-6-SO$_4$→GlcA-2-SO$_4$→GlcNSO$_3$

X $\Delta^{4,5}$-UA-2-SO$_4$→GlcNSO$_3$-6-SO$_4$→IdoA-2-SO$_4$→GlcNSO$_3$

XI $\Delta^{4,5}$-UA-2-SO$_4$→GlcNSO$_3$→IdoA-2-SO$_4$→GlcNSO$_3$-6-SO$_4$

XII $\Delta^{4,5}$-UA-2-SO$_4$→GlcNSO$_3$-6-SO$_4$→GlcA-2-SO$_4$→GlcNSO$_3$-6-SO$_4$

XIII $\Delta^{4,5}$-UA-2-SO$_4$→GlcNSO$_3$-6-SO$_4$→IdoA-2-SO$_4$→GlcNSO$_3$-6-SO$_4$

Lyase III Substrates

II $\Delta^{4,5}$-UA→GlcNSO$_3$-6-SO$_4$→GlcA→GlcNSO$_3$-6-SO$_4$

IV $\Delta^{4,5}$-UA-2-SO$_4$→GlcNSO$_3$→GlcA→GlcNSO$_3$-6-SO$_4$

V $\Delta^{4,5}$-UA-2-SO$_4$→GlcNAc-6-SO$_4$→GlcA→GlcNSO$_3$-6-SO$_4$

VII $\Delta^{4,5}$-UA-2-SO$_4$→GlcNSO$_3$-6-SO$_4$→IdoA→GlcNAc-6-SO$_4$

VIII $\Delta^{4,5}$-UA-2-SO$_4$→GlcNSO$_3$-6-SO$_4$→GlcA→GlcNSO$_3$-6-SO$_4$

Not cleaved by Lyase I

II $\Delta^{4,5}$-UA→GlcNSO$_3$-6-SO$_4$→GlcA→GlcNSO$_3$-6-SO$_4$

IV $\Delta^{4,5}$-UA-2-SO$_4$→GlcNSO$_3$→GlcA→GlcNSO$_3$-6-SO$_4$

V $\Delta^{4,5}$-UA-2-SO$_4$→GlcNAc-6-SO$_4$→GlcA→GlcNSO$_3$-6-SO$_4$

VI $\Delta^{4,5}$-UA-2-SO$_4$→GlcNSO$_3$→IdoA-2-SO$_4$→GlcNSO$_3$

VII $\Delta^{4,5}$-UA-2-SO$_4$→GlcNSO$_3$-6-SO$_4$→IdoA→GlcNAc-6-SO$_4$

VIII $\Delta^{4,5}$-UA-2-SO$_4$→GlcNSO$_3$-6-SO$_4$→GlcA→GlcNSO$_3$-6-SO$_4$

IX $\Delta^{4,5}$-UA-2-SO$_4$→GlcNSO$_3$-6-SO$_4$→GlcA-2-SO$_4$→GlcNSO$_3$

X $\Delta^{4,5}$-UA-2-SO$_4$→GlcNSO$_3$-6-SO$_4$→IdoA-2-SO$_4$→GlcNSO$_3$

Table 3.8. Heparin Lyase Specificities, continued.

Not cleaved by Lyase III

VI	$\Delta^{4,5}$-UA-2-SO$_4$→GlcNSO$_3$→IdoA-2-SO$_4$→GlcNSO$_3$
IX	$\Delta^{4,5}$-UA-2-SO$_4$→GlcNSO$_3$-6-SO$_4$→GlcA-2-SO$_4$→GlcNSO$_3$
X	$\Delta^{4,5}$-UA-2-SO$_4$→GlcNSO$_3$-6-SO$_4$→IdoA-2-SO$_4$→GlcNSO$_3$
XI	$\Delta^{4,5}$-UA-2-SO$_4$→GlcNSO$_3$→IdoA-2-SO$_4$→GlcNSO$_3$-6-SO$_4$
XII	$\Delta^{4,5}$-UA-2-SO$_4$→GlcNSO$_3$-6-SO$_4$→GlcA-2-SO$_4$→GlcNSO$_3$-6-SO$_4$
XIII	$\Delta^{4,5}$-UA-2-SO$_4$→GlcNSO$_3$-6-SO$_4$→IdoA-2-SO$_4$→GlcNSO$_3$-6-SO$_4$

Data from [32]. Substrate numbers are taken from this reference. All susceptible substrates are cleaved between the internal glucosamine and uronic acid residues.

$$\text{IdoA-2-SO}_4 \rightarrow \text{GlcNSO}_3\text{-6-SO}_4 \rightarrow \text{IdoA-2-SO}_4 \rightarrow \text{GlcNSO}_3\text{-6-SO}_4$$

sequences (substrates 4, 13, and 14). When SO$_4$ substituents in this tetrasaccharide sequence are removed from residues on either the upstream side (substrate 16) or the downstream sides (substrates 2, 15, 16, and 17) of the cleavage site, the extent of cleavage is markedly reduced.

For many years, it has been reported that lyase I could cleave heparinoids only at IdoA residues. However, heparinoids contain small amounts of GlcA-2-SO$_4$ residues (which occur in highly sulfated regions—Chapter 2), and oligosaccharides containing these residues were not available for study. Recently, however, such substrates were tested, and it was found that lyase I cleaves at GlcA-2-SO$_4$ residues, provided they are linked to a downstream GlcNSO$_3$-6-SO$_4$ residue (compound XII in Table 3.8). However, when they are linked to a downstream GlcNSO$_3$ residue (compound IX), they are not cleaved by lyase I.

To summarize, the results obtained both from examining the terminal residues of the substrates formed by lyase I cleavage and from determining the cleavage sites of purified substrates show that the enzyme prefers the most highly sulfated regions of the heparinoid chain, but can still cleave, albeit at a reduced rate, when one or more sulfates are removed from either residue at the cleavage site. However, lyase I does not distinguish between GlcA and IdoA residues at the site of cleavage.

b.   Specificity of heparin lyase II

Heparin lyase II is the least discriminating of the three lyases. As shown in Table 3.6, it cleaves all of the substrates tested and will cleave at either IdoA (substrates 2, 4, 5, and 13-17) or GlcA (substrates 3 and 18) residues with little regard to the degree of sulfation around the site of cleavage. It should be noted that, because lyase II does not distinguish between GlcA and IdoA, lyase II converts substrates 3 and 17 to the same product disaccharides. Interestingly, lyase II did not cleave at the GlcA upstream from the GlcNSO$_3$-3,6-di-SO$_4$ in the antithrombin-binding region (substrate 5 was cleaved only at the IdoA residue and yielded product 20 which was not further cleaved). However, it apparently would cleave at other positions in the antithrombin-binding region, thus destroying anticoagulant activity. Heparin lyase II would be the best enzyme to use in sequencing higher oligosaccharides, since it cleaves at most of the linkages in heparinoids, even though cleavage at some linkages may occur at somewhat reduced rates. The results in Table 3.8 are consistent with those in Table 3.6; i.e., lyase II cleaved all of the tetrasaccharides tested, including both substrates that have GlcA-2-SO$_4$ at the cleavage sites.

An interesting question remains: "what can heparin lyase II *not* cleave?" The only answer to date is the unique region from the antithrombin-binding sequence found in substrate 5 and product 20, but even this linkage may be cleaved slowly (see footnote 3, Table 3.6). Lyase I and III will not cleave at this site either. Consequently, a mixture of all three heparin lyases apparently would yield oligosaccharide 20 as a tetrasaccharide that could not be converted to disaccharides. Also, lyase II might be useful for generating GlcNSO$_3$-3,6-di-SO$_4$-containing sequences that are found outside of the antithrombin-binding sequences. As noted earlier, none of these substrates included the unsulfated blocks from heparan sulfate, so the action of lyases II and III on GlcNAc→GlcA bonds is not shown here.

c. Specificity of heparin lyase III

As shown in Table 3.7, heparin lyase III can cleave *only* at unsulfated HexA residues (substrates 3, 15, 17, and 18), especially if there are relatively few SO$_4$ groups on the adjacent residues. Cleavages occur at either GlcA (substrates 3 and 18) or IdoA (substrates 15 and 17) residues.[10] The GlcN downstream from the uronic acid at the cleavage site may be either N-acetylated (substrate 15) or N-sulfated (substrates 3, 17, and 18). Substrates 5 and 15 contain GlcNAc-6-SO$_4$→GlcA sequences, and only substrate 15 is cleaved at this site. Unfortunately, none of the other substrates contain any GlcNAc residues, even though it is clear from many other studies that GlcNAc→HexA linkages are primary cleavage sites for lyase III. Addition of a SO$_4$ to C6 of the upstream GlcN markedly reduces the rate (compare substrates 3 and 18). Like lyase II, lyase III cleaves substrates 3 and 17, converting them to the same products, so that one cannot deduce the structure of the uronic acid at the site of cleavage from the structures of the products or the specificity of the enzyme.

---

[10] The finding that lyase III attacks at IdoA is a surprising discovery that is counter to all prior dogma concerning the specificity of this enzyme.

It is especially interesting to compare substrate 15 (cleaved) with substrate 5 (not cleaved), where the only structural difference is in the extra 3-O-SO$_4$ on the reducing terminal GlcNSO$_3$-3,6-di-SO$_4$ of substrate 5. Although substrate 15 is cleaved at two sites by lyase III, the addition of the 3-O-sulfate in substrate 5 appears to prevent *both* cleavages. Since this would be the only *potential* lyase III cleavage site within the antithrombin binding sequence, these results suggest that lyase III, in contrast to lyases I and II, cannot cleave within the ATIII-binding region of heparin. *Thus, to make a low molecular weight heparin that retains maximum anticoagulant activity, lyase III may be the enzyme of choice.*

The lyase III results in Table 3.8 are consistent with those in Table 3.7; i.e., lyase III requires an unsulfated HexA at the site of cleavage, but can cleave at either GlcA or IdoA. Again, the GlcNAc→HexA linkage was not present in any of these substrates.

5.   Kinetic parameters

The need for milligram amounts of structurally defined oligosaccharide substrates has permitted only limited studies of the kinetic properties of the heparin lyases [see, for example, 283]. However, some comments on the significance of K_m and V_{max} values for heparin and heparan sulfate are in order. As we have seen, these polymers exhibit quite marked polydispersity. Thus, the molecular weight values used to calculate concentrations of these polymers are average molecular weights, and it is not clear whether chains of different lengths are cleaved at the same rates. Also, each heparinoid chain contains multiple segments that are susceptible to each enzyme, in effect increasing the "substrate concentration" over that calculated using the concentrations and the average molecular weights. Even though there may be multiple enzyme-binding sites on each heparinoid chain, steric problems may preclude enzyme binding at all potential cleavage sites. Finally, for *other* polymer-metabolizing enzymes, for which the literature contains more extensive kinetic studies, there is often a marked rise in K_m and/or fall in V_{max} as the size of the substrates become shorter [284]. All of these factors complicate the interpretation of the K_m and V_{max} values for these polymeric substrates.

6. Oligosaccharide mapping of heparin lyase digestion products

a. Introduction

One might anticipate that the information above would allow one to predict with reasonable accuracy the products that are obtained when heparinoids are treated exhaustively with each of the lyases. However, the complexities of both the heparinoid structures and the lyase specificities are such that the products that are formed when these heparinoids are converted to their limit digests with any single lyase are not obvious. This is reflected in the profiles of digestion products of these enzymes that have been used to map the structural differences among heparinoids, in much the same manner that proteins are characterized by peptide mapping. Three procedures have been used to profile oligosaccharides—high pressure liquid chromatography, polyacrylamide gel electrophoresis, and capillary zone electrophoresis [282, 285-289].

In the high pressure liquid chromatography procedure, purified lyases are used to prepare limit digests of heparin and heparan sulfate and the products are separated on strong anion exchange (SAX) columns using a NaCl gradient. In the gel electrophoresis procedure, the digestion products are separated by electrophoresis on isocratic or gradient gels. Typical high performance liquid chromatography profiles are presented in Figure 3.7. These show that the major products from the actions of each lyase on heparin or heparan sulfate are relatively few in number and are well resolved by the high performance liquid chromatography method. However, it is possible that some of these digests contain large fragments that do not emerge from the column. In fact, polyacrylamide gel electrophoresis profiles of the same digests, shown in Figure 3.8, appear to give more complex profiles, consistent with the suggestion that the HPLC profiles depict only the low molecular weight lyase fragments. Obviously, this methodology can be used to compare the action patterns of lyases from different sources against heparinoid substrates [11] [282]. The SAX high performance liquid chromatography profiles are particularly interesting. They demonstrate that the profiles are different for each enzyme and that heparin and heparan sulfate profiles for each of the enzymes are quite distinct. Many of the peaks in these profiles are not identified. Nevertheless, the HPLC profiles are quite simple, but the simplicity of the profiles may belie the fact that

---

[11] In examining the profiles from the earlier work, one must keep in mind that some of these studies may have been done with lyase preparations that contained mixtures of these enzymes. In addition to the three lyases that attack heparinoids, *Flavobacterium heparinum* also produces other lyases that attack other glycosaminoglycans that may contaminate some of the heparinoid fractions, to yield products that are not derived from heparinoids.

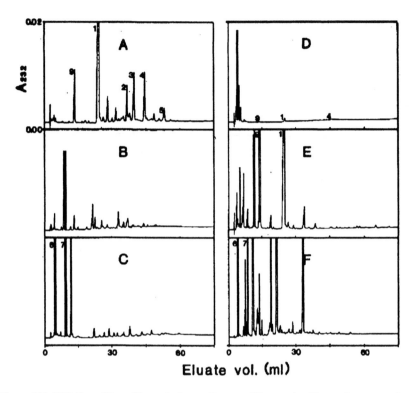

Figure 3.7. HPLC profiles of heparin lyase digests. Oligosaccharide products, numbered according to Table 3.4, are formed by (A) lyase I acting on heparin, (B) lyase I acting on heparan sulfate, (C) lyase III acting on heparan sulfate, (D) lyase III acting on heparin, (E) lyase II acting on heparin, and (F) lyase II acting on heparan sulfate. See Table 3.4 for peak identifications. Reprinted from [282] with permission, © 1990 by the American Chemical Society.

the minor, and perhaps more interesting, oligosaccharides formed by these enzymes are seen as background, or "noise," in these profiles. Amplification of the elution profiles might show a more complex pattern. This point is illustrated in the high performance liquid chromatography profiles of the nitrous acid cleavage products (Figure 3.5). In fact, a similar complexity of minor components can be seen when the lyase I high performance liquid chromatography profiles are amplified [37, 281, 286]. Furthermore, the gradient PAGE profiles in Figure 3.8 are clearly more complex than the corresponding HPLC profiles (e.g., compare the profile of the products formed by digestion of heparin with lyase I in panel A of Figure 3.7 with lane 6 of Figure 3.8, which shows the same digest). The major peaks, as well as some of the minor products, in the profiles in Figure 3.7 have been identified in several reports [278, 280, 283, 290-293]. The structures indicated by the numbered peaks in Figure 3.7 are listed in Table 3.4.

Capillary zone electrophoresis may give higher resolution of heparinoid oligosaccharides than either the high performance liquid chromatography or the PAGE methods [285-288]. To date, however, the resolution of mixtures of standard di- and oligosaccharides is considerably better than obtained with lyase digests of heparinoids.

b. Lyase I cleavage products

The primary oligosaccharides that are found in limit heparin lyase I digests of heparin in Figure 3.7 (panel A) are oligosaccharides 1-5 and 9, with the major peak being the $\Delta^{4,5}$-HexA-2-SO$_4$ $\rightarrow$GlcNSO$_3$-6-SO$_4$ disaccharide. The profile also shows several smaller peaks that are not identified. We have already noted that oligosaccharides 2 and 4 can be cleaved further by lyase I. Their accumulation is probably due to their low concentrations in these digests, which may be below the K_m values for these products. Kinetic studies of the progress of the lyase I digestion show that these five oligosaccharides all accumulate at identical rates during the incubation [283]. Linhardt *et al.* [278] have compared the amounts of the five major structures in lyase I limit digests of 20 different heparins, including antithrombin-affinity fractionated heparins. The results are compiled in Table 3.9, where values are given in percentages of the total products formed in the digests. All samples yielded all of the five major products, and, as expected, the most prominent product of lyase I digestion, in all cases, is the disaccharide

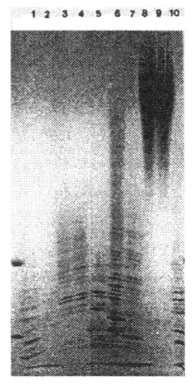

Figure 3.8. Gel electrophoresis profiles of heparin lyase digests. Lanes 1-5 show heparan sulfate products of digestion with (1) lyase I (Sigma), (2) lyase II, (3) lyase III, (4) lyase III, (5) both lyase I and lyase III. Lanes 6-10 show heparin products of digestion with (6) lyase I, (7) lyase II, (8) lyase III, (9) lyase III, (10) both lyase I and lyase III. Reprinted from [282] with permission, © 1990 by the American Chemical Society.

formed from the common IdoA-2-SO$_4$→GlcNSO$_3$-6-SO$_4$ disaccharide. However, the yields of this disaccharide, and of the other oligosaccharides, showed quite marked variations among the different heparins.

The profiles for lyase I digestion of heparan sulfate (panel B, Figure 3.7, and lane 1, Figure 3.8) are clearly different from those for heparin, and contain many unidentified peaks. This might be anticipated from the limited cleavage of heparan sulfate by heparin lyase I (Table 3.3).

c.   Lyase II and lyase III cleavage products

Figures 3.7 and 3.8 show that lyase II yields very complex profiles from both heparin and heparan sulfate. Once again, the two substrates give quite different oligosaccharide profiles. Although both heparin and heparan sulfate give multiple peaks in their profiles, only oligosaccharides 1 and 9 are identified in the heparin profile, whereas only oligosaccharides 6 and 7 are identified in the heparan sulfate profile. Lyase II yields the most complex profile. Most of the peaks have not yet been identified. Interestingly, combinations of lyases I and III do not yield the same profile as lyase II.

Lyase III yields tiny amounts of oligosaccharides 1, 4, and 9 from heparin (panel D of Figure 3.7). In fact, Figure 3.8 (lanes 8 and 9) shows that lyase III converts heparin to quite large, slow-migrating fragments. Heparan sulfate yields very small amounts of the late emerging peaks (more heavily sulfated disaccharides and tetrasaccharides); the only large peaks are disaccharides 6 and 7, and one unidentified peak in the same region (panel C).

One thing that is clear from these profiles is that they show a large number of unidentified peaks. Clearly, there is a great deal more work required to obtain a full understanding of the action of these enzymes.

7.   Heparin lyase summary

Heparin lyase I and, to a lesser extent, lyases II and III have been used extensively in the examination of the structural features of heparinoids. Their

Table 3.9. Percentages of Oligosaccharides in Heparin Lyase I Limit Digests of Different Heparins [1]

Oligosaccharide	% of Total Disaccharides
1	32 - 67
2	3 - 9
3	8 - 15
4	10 - 28
5	8 - 10

[1] Data from [278]. See Table 3.4 for oligosaccharide identification.

substrate specificities have been appreciated in general terms for many years, but their more recent testing with substrates of known structures has both refined our understanding of their specificity and confounded some of our prior conceptions. Lyase I clearly exhibits the most exacting specificity, recognizing only the most highly sulfated disaccharide units at the site of cleavage. Contrary to earlier indications, lyase I cleaves at both IdoA and GlcA residues. The percent of total uronic bonds cleaved is approximately equal to the percentages of IdoA-2-SO_4→GlcNSO$_3$-6-SO_4 disaccharides, but the yield of $\Delta^{4,5}$-HexA-2-SO_4→GlcNSO$_3$-6-SO_4 falls far short of this total. Under typical incubation conditions, some of the major tetrasaccharides that are present in "limit digests" are, in fact, substrates for the enzyme.

Lyase II has the least exacting specificity. It recognizes both GlcA and IdoA, and can cleave at disaccharides that have varying degrees of sulfation. Given this apparent lack of specificity, it remains a curious observation that, although lyase II cleaves at 85% of the uronic acid residues in heparin, it cleaves at only 39% of the uronic acid bonds in heparan sulfate. This suggests that it may not cleave within the unsulfated sequences of heparan sulfate. This could be tested using the *E. coli* K5 polysaccharide as substrate.

Lyase III cleaves at both GlcA and IdoA bonds, and favors the unsulfated disaccharides at its cleavage sites. However, it can cleave bonds in which the GlcN at the site of cleavage is both N- and O-sulfated, but it appears not to cleave bonds in which the uronic acid is O-sulfated. The fact that 94% of the uronic acid bonds are cleaved in heparan sulfate is consistent with the low content of IdoA-2-SO_4 in these polymers.

Despite considerable progress in the understanding of these enzyme activities, much uncertainty remains about the heparin lyase specificities. Nevertheless, the lyases are useful (a) for comparison of different heparinoids by the mapping procedures described, (b) for comparison of the action patterns of enzymes from different sources, (c) for the characterization of heparin and heparan sulfate structures, and (d) for preparing fragments of heparinoids that are not easily obtained by other cleavage methods.

D. Other Cleavage Methods

There are a number of emerging technologies for cleaving heparinoids that offer some potential for specific cleavages that cannot be achieved with nitrous acid or the lyases. For example, some of the true heparin hydrolases (endo-β-glucuronidases) have been purified, but their substrates specificities have received limited attention. These are the enzymes that cleave the heparinoids during their cellular metabolism (discussed in Chapter 5), and their products are apparently quite large fragments. There appear to be a variety of such enzymes that exhibit differences in specificity, but very little structural work has been carried on their specificities and their products.

There are also some further *chemical* methods for cleavage of heparinoids that have not been exploited. These approaches are discussed in Chapter 4.

E.   Analysis of Cleavage Products

1.   Quantification methods

Quantitative analysis of the individual cleavage products formed by chemical or enzyme cleavage requires a detection procedure for measuring the separated components. When lyases are used to cleave heparinoids, the lyase-derived oligosaccharides can be assayed by the 233-nm absorbency of their unsaturated uronic acid residues. Thus, passage of lyase digestion products through an HPLC column coupled to a UV detector has been used to isolate and identify separated components [278, 294]. The extinction coefficients are similar for all cleavage products, so that molar amounts of di-, tetra-, hexasaccharides, etc. can be calculated from the UV absorbency of each separated component. The nitrous acid cleavage products have no such absorbency and so must be detected by derivatization of their reducing terminal anhydromannose residues. Although it would be possible to couple various chromophores to the aldehyde group of the anhydromannose, the only approach that has been developed extensively takes advantage of the incorporation of ^3H into each product by NaB^3H_4 reduction of the anhydromannoses reducing terminals to [^3H]anhydromannitols. This reaction is stoichiometric, yielding products that all have the same molar specific radioactivity, equal to one-fourth of the molar specific activity of the NaB^3H_4 used for the reduction. Passage of NaB^3H_4 reduced mixtures through an HPLC column coupled to a radioactivity flow detector can be used to measure separated components. The number of ^3H cpm found in any separated di-, tetra-, hexasaccharide, etc. can be used to calculate the molar amount of the individual components. Note that the reducing terminal GlcN residues of the oligosaccharides formed by lyase cleavage can also be reduced with NaB^3H_4 so that radioactivity could be used for the detection of these products also. A variety of other reagents that react stoichiometrically with the reducing terminal could also be used in the detection and quantitation of these products.[12] For example, the oligosaccharides could be derivatized with

---

[12] Under alkaline conditions, 2,5-anhydromannose equilibrates with 2,5-anhydroglucose, thus converting a single compound into a mixture of two compounds. Consequently, any

reagents such as phenylhydrazine, 2-aminopyridine [295], 2-aminopyridine-biotin conjugates [296], 2-cyanoacetamide [297-299], diaminobenzoic acid [300], carboxybenzoyl-quinolinecarboxaldehyde [301], 2-aminoacridone [288], or others, all of which could be used to quantify each purified oligosaccharide. Such methods will be viable, provided these reagents yield derivatives that can be formed stoichiometrically and measured with high sensitivity. However, after the derivatization and molar response aspects of analysis are worked out for any of these chromophoric reagents, it would be necessary to develop suitable high resolution HPLC methods that would give good separation of each disaccharide. To date, only one system for separation of such derivatized oligosaccharides has been reported [288].

2. Separation methods

A number of high performance liquid chromatography procedures have been developed to separate and quantify the di- and oligosaccharides formed by deaminative cleavage of the heparinoids with nitrous acid. These include chromatography on strong anion exchange columns [28, 259], weak anion exchange columns [258], and reverse phase ion pairing columns [260]. Alternatively, some of the disaccharides can be separated by paper chromatography [259, 302, 303], paper electrophoresis [302, 303], thin layer chromatography [304], or capillary zone electrophoresis [279, 285-288]. For the lyase products, UV assays have been used for detecting oligosaccharides eluted from SAX HPLC columns with a 0-1 $\underline{M}$ linear sodium chloride gradient [278, 294] or from a sulfonated styrene-divinylbenzene copolymer column [289]. In all of these approaches, the elution positions of the di- or tetrasaccharide peaks are used to identify previously characterized oligosaccharides, and the ^{3}H cpm or total UV absorbency recovered in each peak is used to obtain ratios of disaccharides or quantitative measures of the amounts of each disaccharide present in the glycosaminoglycan. Heparinoids metabolically labeled with [^{3}H]glucosamine or ^{35}SO$_4^{2-}$ (Chapter 5) yield disaccharides with the corresponding labels, and these disaccharides from both types of cleavage can also be detected and quantified following their reduction with NaBH$_4$.

---

new method for converting oligosaccharides with reducing terminal anhydromannose residues to chromophoric derivatives should avoid alkaline conditions.

IV. DETERMINATION OF OLIGOSACCHARIDE STRUCTURES

A. Introduction

Although it is not possible to sequence the mixture of structurally varied chains typically found in heparinoids, it is relatively easy to determine the sequence of a tetra- or hexasaccharide that is obtained in a pure form. Furthermore, the general procedures useful for sequencing these smaller oligosaccharides should be applicable to higher oligosaccharides, provided such oligomers can be obtained in pure form. In fact, the isolation of such pure fragments is a bigger challenge than the sequencing! The chemical or enzymatic sequencing procedures are based on the two cleavage methods described above. It will be convenient to discuss sequencing of oligosaccharides formed by each cleavage method separately, and then to consider the problems and prospects for using these two methods together. The sequencing principles, illustrated with a few examples, can be readily extended to additional cases. Physical methods for determining sequences of heparinoid oligosaccharides are also discussed. Most notably, NMR has emerged as a prominent method for oligosaccharide sequencing.

B. Oligosaccharides Formed by Nitrous Acid Cleavage

1. Determination of reducing terminal and total disaccharide content

The principles that are described here would apply equally well to the analysis of oligosaccharides derivatized by NaB^3H_4 reduction or with one of the chromophoric reagents suggested earlier. As noted, the only systematic approach that has been developed for separation and quantification of nitrous acid-generated heparin fragments involves NaB^3H_4 reduction and analysis of the labeled products. This approach was described by Bienkowski and Conrad [28] and is illustrated by the example in Figure 3.9.

A pure NaB^3H_4-reduced oligosaccharide contains an ^3H label in its reducing terminal anhydromannitol. The labeled oligosaccharide will have a molar specific radioactivity that is one-fourth that of the NaB^3H_4 used for its reduction. When this radiolabeling approach is used in the sequencing of an oligosaccharide, it is desirable to use the same batch of NaB^3H_4 for all reductions so that all of the labeled products will have the same specific activity.

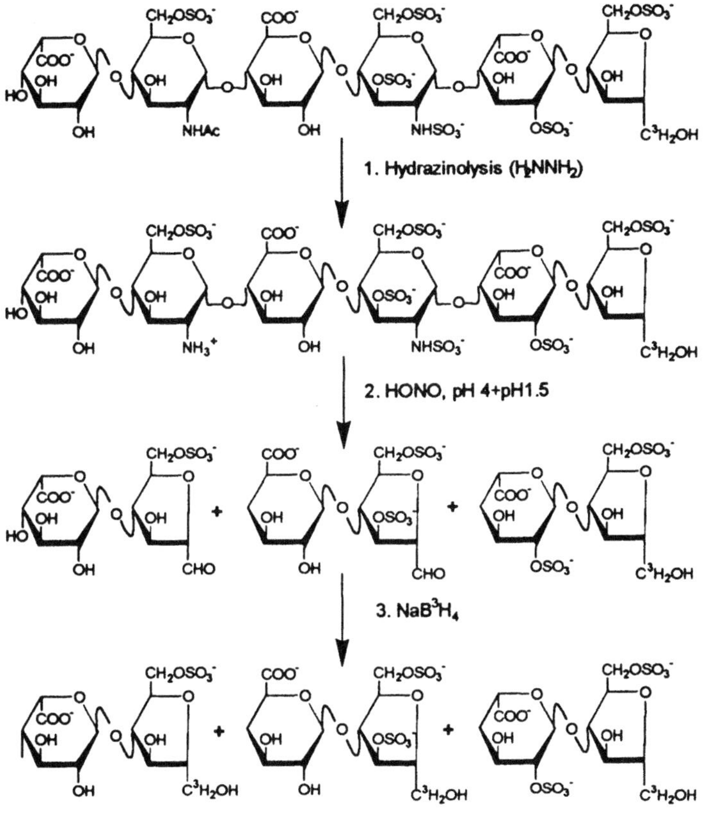

Figure 3.9. Determination of reducing terminals of oligosaccharides.

Conversion of the labeled oligosaccharide to disaccharides by nitrous acid treatment (preceded by hydrazinolysis if necessary, as in the present example) will yield a mixture in which the reducing terminal fragment is the only labeled product. By direct HPLC analysis of an aliquot of the cleavage mixture, this product can be identified. When a second aliquot of the same mixture of disaccharides is *re-reduced* with the same labeled NaB^3H_4 and analyzed, it will yield the reducing terminal fragment *plus* additional fragments in the oligosaccharide, each with a reducing terminal $[^3H]AMan_R$. Since there is a reasonable possibility that one of the disaccharides detected in the second analysis will be identical to the reducing terminal disaccharide, one may have to determine how much ^3H in addition to that from the first analysis is present in the second analysis. The ratio of total cpm in the disaccharides measured in the second analysis (i.e., including the reducing terminal disaccharide) to the cpm in the reducing terminal disaccharide alone will yield the total disaccharides per oligosaccharide.

If the oligosaccharide in question is a tetrasaccharide, then the procedure just described will yield its sequence. Also, if the oligosaccharide is a hexasaccharide or higher oligosaccharide, the sequence will be obvious if all of the disaccharides are identical. However, if the internal and nonreducing terminal disaccharides are different, additional work will be necessary to determine the order in which the latter disaccharides occur. Several approaches might be considered.

2.   Determination of the sequence of disaccharides in an oligosaccharide

a.   The case in which the oligosaccharide contains one or more N-acetylated GlcN residues

If the oligosaccharide contains GlcNAc residues, as in our example in Figure 3.9, then direct nitrous acid cleavage without prior N-deacetylation will yield a disaccharide and a tetrasaccharide. If the ^{3}H-labeled product of the nitrous acid cleavage of the ^{3}H-labeled oligosaccharide is a disaccharide, as shown for our example in Figure 3.10, then the new product that appears on re-reduction with NaB^{3}H$_4$ is a tetrasaccharide. The latter can be identified by its high performance liquid chromatography retention time.

Suppose that our example had the GlcNAc-containing disaccharide in the middle of the hexasaccharide instead of at the nonreducing terminal. In this case, the tetrasaccharide fragment formed by nitrous acid cleavage will be ^{3}H-labeled without re-reduction, giving immediate sequence information. The tetrasaccharide will, in all likelihood be among those that have already been characterized (Table 2.3), so that it can be identified by its high performance liquid chromatography retention time.

b.   The case in which the oligosaccharide contains only N-sulfated GlcN residues

If the oligosaccharide contains only N-sulfated GlcN residues, then it will be cleaved completely to disaccharides without prior N-deacetylation. If a single type of disaccharide is formed, then the sequence is obvious [see, for example, 305]. If the nonreducing terminal of an oligosaccharide formed by nitrous acid treatment is a nonsulfated uronic acid, then it could be removed by treatment with β-glucuronidase.[13] The resulting uronidase-treated oligosaccharide could

---

[13] Some commercial β-glucuronidase preparations also contain α-iduronidase. Such samples are particularly useful in identification of the nonreducing terminals, since they

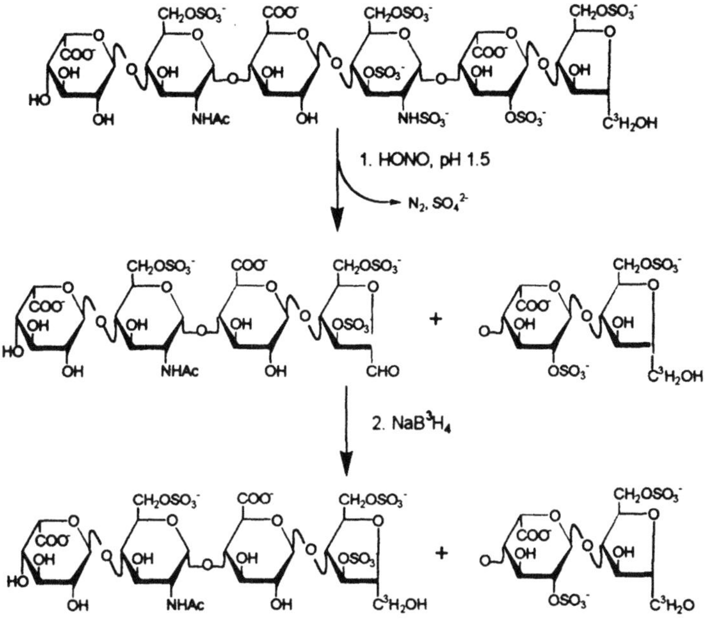

Figure 3.10. Sequencing a heparinoid oligosaccharide.

then be cleaved with nitrous acid to yield the mixture of disaccharides found in the original analysis with the nonreducing terminal GlcN (with any of the original O-sulfate groups still attached) in place of the original nonreducing terminal disaccharide. Such an analysis shows (a) that the nonreducing terminal uronic acid in the original oligosaccharide was not sulfated, (b) the identity of the nonreducing terminal disaccharide unit, and (c) for a hexasaccharide, the sequence of disaccharides. When there are two or more different internal disaccharides (octa- or higher oligosaccharides), additional work would be required, after determining the reducing terminal and nonreducing terminals as above. The more classical approach would be to obtain a partial cleavage of the labeled oligosaccharide and to isolate the reducing terminal and other oligosaccharide fragments for separate sequencing. Such fragments may be recognizable by their co-elution with previously identified oligosaccharides, or they may be isolated and sequenced individually in order to deduce the disaccharide units that are joined together in the original oligosaccharide, and the sequence in which they are joined. However, the need for these more extensive sequencing requirements has not arisen very often in the heparin field.

---

remove either GlcA or IdoA from the nonreducing terminals of oligosaccharides, provided these residues are not sulfated.

As the need arises to sequence longer oligosaccharides, the principles described here may be applied. Also, these relatively straightforward approaches could be adapted for solid phase sequencing.

C.   Oligosaccharides Formed by Lyase Cleavage

1.   Determination of reducing terminal and total disaccharide content

In principle, the same approach can be used for sequencing lyase cleavage products as described for nitrous acid cleavage products; i.e., the terminal disaccharide could be determined and the total disaccharide composition could be established after complete cleavage of the oligosaccharide to disaccharides that can be recognized by comparison of their high performance liquid chromatography retention times with those of standards. Using only the lyases in such a sequencing protocol offers some advantages over the nitrous acid approaches, but has some disadvantages as well. Among the advantages are the facts that (a) in most cases total cleavage to disaccharides can be obtained using a mixture of all three lyases, (b) selective cleavage may be obtained using individual lyases, and (c) a number of lyase products have been characterized for use as standards. A major disadvantage is that all three lyases cleave at both GlcA and IdoA residues to give nonreducing terminal $\Delta^{4,5}$-HexA's, the origin of which will be uncertain. Thus, one must obtain additional information on the uronic acid composition in the original oligosaccharide and its position in the sequence.

2.   Determination of the sequence of disaccharides in an oligosaccharide

If the oligosaccharide is generated by heparin lyase cleavage, and therefore has a nonreducing terminal $\Delta^{4,5}$-uronic acid, the $\Delta^{4,5}$-uronic acid could be removed by treatment with a glycuronidase [269] or with mercuric salts [306] to yield a product with a nonreducing terminal GlcN residue. Cleavage of this product with enzymes or nitrous acid would yield the nonreducing terminal as the *monosaccharide* in place of an equivalent amount of one of the previously detected disaccharides.

D.   Use of Both Cleavage Methods Together in Structural Analysis

In the determination of the sequence of any unknown oligosaccharide obtained from a heparinoid, it is desirable to avail oneself of all methodologies that are accessible. Thus, there may be instances in which nitrous acid cleavage offers advantages over lyase cleavages (for nitrous acid, the sites of cleavage are known, and the uronic acid residues are not altered as a result of the cleavage) and other instances in which lyase cleavage offers advantages (lyases offer some specificity in their sites of cleavage, will cleave without prior de-N-acetylation of GlcNAc residues, and will generate nonreducing terminal residues that are easy to recognize and quantify). For example, cleavage of heparinoid polymers or oligomers by nitrous acid yields a unique reducing terminal structure (AMan), whereas the cleavage with one of the lyases yields a unique nonreducing terminal structure (the $\Delta^{4,5}$-uronic acid). Thus, it seems that it should be possible to combine the two cleavage methods to obtain sequence data. Linhardt and co-workers have included a combination of protocols in an integrated approach for determining the structures of oligosaccharides formed by lyase I cleavage of heparin [280]. These combined methodologies include carbazole assays to determine the number of uronic acid equivalents, SO_4 analysis to determine the number of SO_4 groups per disaccharide, fast atom bombardment mass spectrometry for measurement of molecular weight, nitrous acid cleavage to determine the number and positions of N-sulfated GlcN residues, an indole color reaction to detect the presence of anhydromannose following nitrous acid cleavage, IO_4^- oxidation to determine whether 1,2-glycol groups (as found in unsulfated uronic acid residues) are present in the oligosaccharides, and ^{1}H-NMR and ^{13}C-NMR spectrometry to detect and quantify the amounts of specific monosaccharide residues and to determine their sequences.

One such combination of approaches that might be considered in determination of the structure of an oligosaccharide isolated from a mixture of lyase cleavage products would be to label the reducing terminal residue—a GlcN residue which may or may not be sulfated—by NaB^3H_4 reduction. This approach would result in an oligosaccharide with both nonreducing terminals and reducing terminals that are clearly identified by the $\Delta^{4,5}$-HexA and the [^{3}H]glucosaminitol residues, respectively, thus facilitating the sequencing work. Alternatively, one might choose to cleave an oligosaccharide generated by nitrous acid cleavage with one of the lyases, thus facilitating the sequencing by releasing the nonreducing terminal as the only fragment that had *no* $\Delta^{4,5}$-HexA residue. In considering such combinations of approaches, several cautions must be observed.

A first concern arises if one seeks to use lyases to cleave oligosaccharides formed by nitrous acid cleavage. It is not known which of the lyases, if any, will cleave structures that have the anhydromannose (or [^{3}H]anhydromannitol) in place of the expected GlcN residue at the reducing terminal. However, if the lyases work on these structures, they would convert internal uronic acid residues to $\Delta^{4,5}$-HexA derivatives and the nonreducing terminal disaccharide to a HexA →GlcN derivative, i.e., a product lacking the $\Delta^{4,5}$ bond. Second, $\Delta^{4,5}$-HexA's are alkali labile and may be altered during NaB^{3}H$_4$ reduction. Thus, mild conditions must be used for the reduction of $\Delta^{4,5}$-HexA-containing oligosaccharides [307]. Third, successful reduction of such oligosaccharides generates oligosaccharides that have [^{3}H]glucosaminitol residues at their reducing terminals instead of the anhydromannitol residues present after the reduction of nitrous acid-generated oligosaccharides. In contrast to anhydromannitol residues, which do not react with nitrous acid, *glucosaminitol* residues (either free or bound to an oligosaccharide at the reducing terminal) are converted by nitrous acid to mixtures of products, including glucose. Therefore, the reducing terminal structure cannot be (or, at least, has not been) identified after cleavage with nitrous acid.

   An interesting possibility that has not been used would involve direct treatment of a lyase-generated oligosaccharide with nitrous acid without prior NaB^{3}H$_4$ reduction. This would cleave the oligosaccharide and convert both reducing terminal N-sulfated GlcN residues and internal N-sulfated GlcN residues to AMan residues (the anomalous nitrous acid reaction referred to above occurs only *after* the GlcN is reduced to glucosaminitol). Nitrous acid will convert the nonreducing terminal disaccharide to a $\Delta^{4,5}$-HexA→AMan disaccharide and the internal disaccharide(s) to HexA→AMan sequences (all with the original O-sulfate groups attached). After reduction of this mixture with NaB^{3}H$_4$, all of the internal and reducing terminal disaccharides can be identified by high performance liquid chromatography as typical nitrous acid-generated disaccharides. The nonreducing terminal $\Delta^{4,5}$-HexA→AMan$_R$, with the original sulfate groups still attached, would be a new structure and would have to be characterized. However, if all such derivatives of the $\Delta^{4,5}$-HexA→AMan$_R$ were known, they could easily be identified by high performance liquid chromatography or capillary zone electrophoresis.

V. DETERMINATION OF MOLECULAR WEIGHT

   A variety of procedures have been used for determination of the molecular weights of heparin and its fragments, including end group analysis [308-311], viscosity [310], and gel permeation chromatography [312-316].

The most common procedure used currently is the size exclusion column chromatography approach. The resulting data can be used to determine the weight-, number-, and peak-average molecular weights, as well as the polydispersities of the heparinoids [313]. The elution profiles here have been monitored by the ultraviolet absorbency of the N-acetyl groups of the GlcNAc residues or by refractive index [314, 315]. Column supports used have included Sephadex 100 [317], Ultragel AcA44 [310, 318] and TSK G3000SW or TSK G2000SW [319]. In order to use the resulting column elution profiles to calculate the molecular weight of the heparinoid, it is necessary to calibrate the column using an appropriate heparin molecular weight standard whose molecular weight has been determined by an independent means. Since all means for determining the molecular weight of any heparin standard have some shortcomings, obtaining a truly valid molecular weight standard is difficult. One solution to this problem, described by Nielsen [320], is to use a heparin sample that has been partially digested with heparin lyase I as the standard for column calibration [314, 315]. Such a sample gives a series of partially resolved oligosaccharide peaks, each of which differs in size from its neighbors by one disaccharide unit (Figure 3.11). When the column effluent is monitored by a differential refractive index detector, an elution profile showing peak sizes proportional to the mass of heparinoid is obtained. However, when the effluent is monitored by an ultraviolet flow detector set at 235 nm to observe the nonreducing terminal $\Delta^{4,5}$-uronic acid residues, the elution profile shows peak sizes proportional to the number of molecules of oligosaccharide in each peak (since each lyase-generated oligosaccharide contains one nonreducing terminal $\Delta^{4,5}$-uronic acid). Thus, by chromatographing a standard di-, tetra-, or hexasaccharide on the column, one can identify the corresponding oligosaccharide size in the profile of the mixture of lyase I products and can obtain a ratio of the refractive index response to the 235-nm absorbency. This allows the identification of the size of each successive peak by counting the number of disaccharides added for each successive peak eluting before the stan- dard(s) and by calculating the ratio of the total mass in each peak to the number of reducing terminals using the refractive index and UV responses.[14] The degrees of polymerization and the elution positions can then be used to generate a standard curve (log MW vs elution volume) that can be extrapolated to cover the range of elution positions found for heparins and low molecular weight heparins. The degree of polymerization of the heparinoid in question can be multiplied by the average molecular weight of a disaccharide unit (~600) to get a reasonable approximation of the molecular weight of the heparinoid. Alternatively, if the disaccharide composition of the heparinoid is determined

---

[14] Generally, each size class elutes as a single unresolved peak containing all components of the same degree of polymerization.

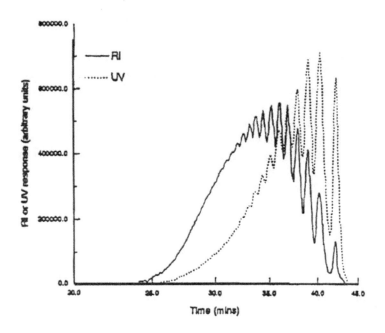

Figure 3.11. Determination of molecular weight of heparinoids. Figure from
Dr. Barbara Mulloy, with permission.

separately, the actual average molecular weight of a disaccharide can be
calculated more accurately.

The most obvious way to determine the molecular weight by end group
analysis is to measure the number of total monosaccharides per reducing
terminal residue (or nonreducing terminal residue). Reducing terminals can be
labeled by NaB^3H_4 reduction or by coupling the reducing terminal to a
chromophoric group [308, 309, 311]. The nonreducing terminal is more difficult
to derivatize, but oligosaccharides formed by heparin lyase cleavage contain a
$\Delta^{4,5}$-uronic acid residue at the nonreducing terminal that can be measured. Desai
and Linhardt [311] have described a ^{13}C NMR procedure for determining the
ratio of the number of internal monosaccharides to the number of reducing
terminal disaccharides. Here, the ratio of the total anomeric carbons to the
reducing terminal anomeric carbons gives the number of total monosaccharides
per reducing terminal monosaccharide. In this approach, the quantification of the
anomeric carbons used to calculate the degree of polymerization required time
averaging of the anomeric carbon signals over periods up to 48 hr. In addition, a
total disaccharide compositional analysis of the heparinoids was carried out in

order to obtain average disaccharide molecular weights that were necessary to convert disaccharides/reducing terminal to number average molecular weights. Although this is a unique approach to determine molecular weights by end group analysis, it has some obvious drawbacks, including the time required for each analysis. Furthermore, a number average molecular weight gives no idea of the degree of polydispersity of the heparinoid preparation. Finally, since 300,000 scans were required for time averaging for a molecular weight of 10,000, then to obtain good spectral data for a heparan sulfate with a molecular weight of 30-40,000 would require a marked extension of this time and would probably result in poor quantification of the anomeric signals.

A laser light scattering method has been applied to heparins and low molecular weight heparins [316]. This approach, referred to as high performance size exclusion chromatography-multiangle laser light scattering (HPSEC-MALLS), takes advantage of the focusing of three laser beams from different angles on the effluent of a gel filtration column. Precise molecular weights are obtained for each segment of the elution profile. This method, which has been applied to a range of different heparinoids, offers several advantages over previous methods:

- Compared to *relative* methods of molecular weight determination which rely on external standards, the HPSEC-MALLS technique is *absolute*. It derives the weight average molecular weight directly from scattered light intensity as a function of angle (as formulated by light scattering theory). No molecular weight standards are required for column standardization.

- The HPSEC-MALLS technique is insensitive to fluctuations in flow rate and the concomitant error in elution volume, an affliction of methods relying on precise column calibration.

- The presence of high molecular weight aggregates at vanishingly low concentrations in a polymer solution is readily detected by light scattering, a phenomenon that goes unnoticed by other modes of detection.

- The HPSEC-MALLS technique enables one to see how a molecule behaves in *solution*.

- Using the HPSEC-MALLS approach, one does not have to make assumptions about a molecule's conformation in solution to estimate its molecular weight and size.

VI. PHYSICAL METHODS IN HEPARINOID STRUCTURAL STUDIES

A. Introduction

In parallel with the chemical approaches described above, a number of physical methods for analysis of heparinoid structures have been developed, including ^{1}H-NMR and ^{13}C-NMR spectroscopies, mass spectrometry, and X-ray spectroscopy. For characterization of the primary structures of heparinoids and heparin fragments, the NMR and mass spectrometry methodologies have been of greatest value. NMR and X-ray spectroscopy have also yielded a great deal of information on the three-dimensional structures of these polymers. Excellent reviews of the physical methods that are used in the characterization of heparinoids are available [12, 123-126] and these approaches will be summarized only briefly here.

B. X-ray Crystallography

X-ray crystallographic studies of heparinoids have yielded basic information concerning the three-dimensional structure of heparin in the solid state. The pioneering studies of Atkins, *et al.* [128, 321, 322] on heparin films demonstrated that heparin forms crystalline regions of the most common sequences of heparin, namely, the repeating IdoA-2-SO$_4$→GlcNSO$_3$-6-SO$_4$ disaccharides. These sequences form a two-fold screw axis with four residues per turn and a disaccharide repeat period of 8.4 Å. However, the resolution obtained by these methods is not high enough for accurate deduction of ring conformations [reviewed in 124].

C. NMR

X-ray diffraction studies do not give information concerning the solution conformations of heparinoids and molecular dynamics of these polymers. Such information has been derived largely from NMR studies. The applications of these methodologies in the study of heparinoids were initiated by Perlin [323, 324] and extended by Perlin, Casu and others [11, 12, 125, 132, 138, 325-330]. The use of NMR in the characterization of heparinoids developed in step with the increasing strength of the magnets that improved the resolution of the signals [52] and with the development of two-dimensional methods [331, 332]. NMR is emerging as the method of choice for determination of heparin oligosaccharide structures.

Two nuclei, ^1H and ^{13}C, have been of value in the study of heparinoid structures. Individual ^1H and ^{13}C resonances have been identified by measuring the chemical shifts and coupling constants of model compounds—either fragments of heparin, chemically modified heparins, or synthetic structures [34, 290, 293, 330], and by two-dimensional NMR methods [331].

Figures 3.12 and 3.13 show typical ^1H-NMR [12, 32] and ^{13}C-NMR [94] spectra, respectively, of heparin with the assignments of peaks to protons or carbons in heparin. Figure 3.12 shows the proton NMR spectra of two tetrasaccharides isolated after cleaving heparin with lyase I, with the individual protons labeled. The top structure contains GlcA-2-O-SO$_4$ as the internal HexA unit, whereas the bottom structure contains IdoA-2-O-SO$_4$. These residues are readily distinguished on the basis of the chemical shifts of their anomeric protons, as well as those of their C2 protons. Also, both structures contain the nonreducing terminal $\Delta^{4,5}$-uronic acids, which are readily recognized by the downfield resonances of their C4 protons.

The ^{13}C-NMR peaks of unfractionated heparin (UH) and of antithrombin-binding (HA) and antithrombin-nonbinding (LA) fractions of the heparin, prepared by heparin affinity chromatography, are shown in Figure 3.13. Peaks are identified for the critical carbons in heparinoids [35, 333]. These include as major peaks the C2 of GlcNSO$_3$-6-SO$_4$ + GlcNSO$_3$, the C6 of GlcNSO$_3$-6-SO$_4$, and the C1 of IdoA-2-SO$_4$. Peaks are also assigned for C2 of GlcNAc + GlcNAc-6-SO$_4$, for C6 of GlcNSO$_3$, and for C1 of IdoA and GlcA. Most notably, a minor peak for C2 of GlcNSO$_3$-3-SO$_4$ + GlcNSO$_3$-3,6-di-SO$_4$, which are derived from the antithrombin-binding region, is also assigned for the unfractionated heparin, and this peak is enriched in the HA-heparin but depleted in the LA-heparin. Integration of these peaks allows quantitative analysis of the composition of the heparinoids that agrees quite well with the analyses that are obtained by high performance liquid chromatography of nitrous acid-cleaved products [94]. The chemical shifts, spin-spin coupling constants, and spin-spin splitting data from NMR spectra of heparinoids yield several types of structural information, including monosaccharide identities, monosaccharide compositions (above), monosaccharide sequences, and ring conformations. Various workers have used both ^{13}C-NMR and one- and two-dimensional ^1H-NMR for analysis of compositions, anomeric configurations, and sequences of heparin-derived oligosaccharides [34, 279-281, 291, 331, 333-337]. The two-dimensional NMR approaches allow the assignment of interacting protons on adjacent carbons in a monosaccharide or adjacent monosaccharides in an oligosaccharide, which are critical for NMR sequencing of such oligosaccharides. Evidence concerning ring

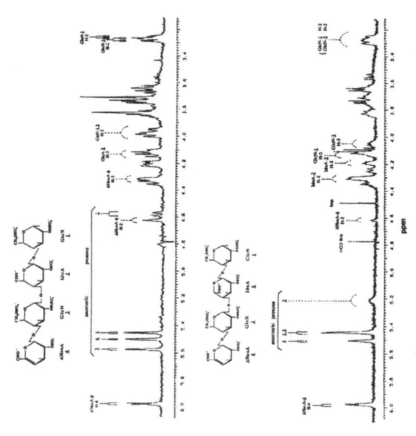

Figure 3.12. ¹H-NMR of heparinoids. Reprinted from [32], with permission.

110

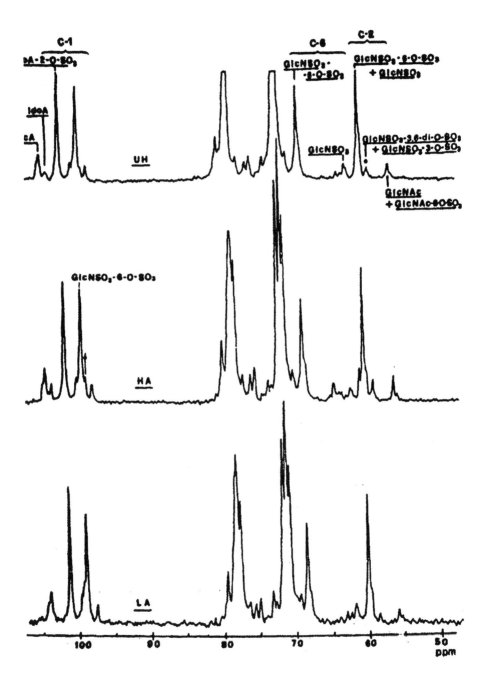

Figure 3.13. ^{13}C-NMR spectra of heparinoids. Reprinted from [94], with permission.

conformations also has been derived from ^{1}H-^{1}H coupling constants, which are related to the H-C-C-H torsional angles[11, 12, 124, 132, 138, 293, 325-328, 338]. A discussion of the conformations of heparinoids deduced from NMR studies is presented in Chapter 2.

D. Mass Spectrometry

Mass spectrometry offers a highly sensitive methodology that can be used to determine molecular weights of compounds and more detailed structural information using only microgram amounts of material. However, mass spectrometric methods that give complete structural information for heparinoids have yet to be developed. Because polymeric heparinoids, in contrast to other macromolecules, are mixtures of structurally varied, polydisperse chains, attempts to obtain structural information on polymers have not been successful [339, 340]. Even with heparin fragments, the high levels of sulfate and carboxyl groups give these molecules several negative charge sites, making observation of a singly charged molecular ion and structurally significant fragment ions difficult. However, progress has been made with pure heparinoid oligosaccharides. Early analyses of heparin oligosaccharides by ^{252}Cf plasma desorption mass spectrometry [339] using tridodecylmethylammonium chloride (TDMAC) as the surfactant gave molecular ions in which the TDMA+ ion displaced all of the alkali metal counterions of the sulfate and carboxyl groups. This blocked the negative sites by ion pairing so that an intact molecular ion could be observed. The best results were obtained for negative ion spectra in which fragmentation patterns resulted from losses of TDMA+ groups and sulfite ions from the GlcNSO3 residues, as well as losses of reducing terminal and nonreducing terminal monosaccharides.

A similar study on synthetic heparin oligosaccharides (Na$^+$ salts) was carried out by fast atom bombardment mass spectrometry (f.a.b.-m.s.) using 1-thioglycerol as the matrix [341]. The fragmentation patterns in the positive ion extraction mode showed the replacement of sodium ions by protons, and the loss of NaSO$_3$ with proton replacement, but no significant glycosidic bond fragmentation. Thus, the primary structural information obtained was the number of sulfate groups and the nature of the counterions. In the negative ion extraction mode, similar results were obtained, but the O-sulfate groups were much more stable.

A more recent study of lyase I-generated heparin oligosaccharides using negative ion f.a.b.-m.s. with triethanolamine as the matrix gave both molecular ions and more extensive fragmentations [342]. Here, loss of NaSO$_3$ was again observed in the fragment ions, but, in addition, glycosidic bond cleavage gave structurally significant fragments. In this study, the mass spectra of a disaccharide, two tetrasaccharides, and one hexasaccharide were rationalized in terms of their previously determined structures[15]. In addition, the structure of a hexasaccharide that had not been previously reported was deduced from its mass spectrum, and the deduced structure was confirmed by chemical, enzymatic, and NMR studies. This methodology was applied in subsequent work on an oligosaccharide from the antithrombin-binding sequence [281]. Similar approaches have been reported for determination of the molecular weights of size-fractionated oligosaccharides [315]. The use of mass spectrometry in structural analysis of heparin oligosaccharides is being used more frequently in conjunction with NMR and other methodologies [see, for example, 31, 343].

Dell *et al.* [344] have described a unique approach for mass spectrometry of heparinoids in which the problems associated with the sulfate groups are eliminated. Here, the heparinoid is permethylated to block all of the free hydroxyl groups; the methylated product is desulfated, exposing a new set of free hydroxyl groups where the sulfate groups had been; and these hydroxyl groups are derivatized by peracetylation. The product is then subjected to analysis by mass spectrometry.

VII. USE OF ANTIBODIES

A number of antibodies directed against heparan sulfate proteoglycan core proteins have been reported [345-347]. There have also been several reports of antibodies directed against epitopes in the polysaccharides themselves. The first monoclonal antibody against intact heparin was reported by Straus *et al.* [348]. The MAb recognized heparin but not heparan sulfate or other glycosaminoglycans. Furthermore, when the heparin was modified by de-N-sulfation, de-N, O-sulfation, or carboxyl reduction, the antibody reactivity was lost. However, although full antibody reactivity required a decasaccharide or larger oligosaccharide, the actual epitope was not further characterized.

Shibata *et al* [349, 350]. reported autoantibodies from tight skin mice and from patients with antiphospholipid antibody syndrome that were directed against heparan sulfate.

A novel monoclonal antibody that recognizes regions in heparinoids that contain N-unsubstituted GlcNs has been useful in demonstrating the presence of GlcN residues in heparan sulfate [21, 351]. Similarly, a monoclonal antibody directed against the *E. coli* K5 polysaccharide was found to recognize only N-acetylated domains of heparan sulfates [352].

---

[15] The structure of one of these tetrasaccharides was incorrectly assigned in prior work from this lab, and was not corrected until after this work was reported (see Footnote 8).

Finally, Pejler *et al.* [353] have described monoclonal antibodies that are directed against the reducing terminal sequences of fragments formed by cleavage of heparinoids with nitrous acid which require anhydromannose or its reduced form (anhydromannitol) as a part of the epitope.

ADDITIONAL REFERENCES

Malsch, R., Mrotzek, T., Huhle, G., and Harenberg, J. (1996) Purification of the monoclonal heparin antibody H-1.18, *J. Chromatog. A* 744.
Okamoto, M., Mori, S., Ishimaru, M., Tohge, H., Nakata, Y., and Endo, H. (1997) An enzyme-linked immunosorbent assay for heparan sulfate proteoglycans, *Life Sci.* 60, 1811-1819.

Chapter 4. Structural Modification of Heparinoids

I.   MODIFICATION OF POLYMERIC HEPARINOIDS

A.   Introduction

Although best known for its anticoagulant activity, heparin exhibits a number of other biological activities. Even its effects on blood coagulation are mediated or modulated through more than a dozen proteins (Chapter 8). These biological effects may be observed with isolated heparin-binding proteins, in cultured cells, or in whole animals that receive heparin or its derivatives by intravenous or subcutaneous administration. If we assume that each of these biological activities depends on the interaction of a limited oligosaccharide segment in the heparin sample (e.g., a hexasaccharide) with a heparin-binding protein, and if we recall that there are up to 5800 possible hexasaccharides in heparin, then we can begin to see the problem in identifying specific heparin sequences that mediate the activities in question. The difficulty in describing structure-activity relationships for heparin and heparin-binding proteins is compounded almost beyond comprehension when one considers the effects of heparin on a multicomponent system such as the hemostasis system, which involves many heparin-binding proteins in the blood and on the surface of the endothelial cells, each of which may recognize a different hexasaccharide sequence. Isolation of the hexasaccharide sequence responsible for the biological activity is the best way to establish the structure-activity relationships, but is difficult. An alternative approach is to prepare several structurally modified heparinoids and to determine the effects of these structural changes on the biological activity. Modification of heparinoids may result in loss of some biological activities but retention of others. Also, the testing of modified heparinoids for their binding to various heparin-binding proteins may reveal some of the structural features that are involved in the binding.

As the structural features of heparinoids have emerged, we also have come to understand the selective chemical reactivities of the functional groups in heparinoids. In addition to the sulfate groups, most of which can be removed without otherwise altering the polymer structures, the amino groups and their substituents, the carboxyl groups, and the ring structures of the monosaccharides may be altered stoichiometrically by specific chemical treatments. Both classical methodologies and methodologies that are unique to the glycosaminoglycans have been used to prepare modified heparinoids. Such procedures have been useful in obtaining heparinoids with altered biological properties.

There are additional reasons for preparing structurally modified heparinoids. One reason is that these modification procedures can be used in the further characterization of the fine structures of heparinoids. A second, more subtle reason is that some of these procedures offer new ways for selective cleavage of heparinoids. As will be discussed in Chapter 6, the isolation of heparinoid fragments that will bind specifically to a heparin-binding protein requires a method for cleaving the heparinoid polymer that *does not cleave in the middle the specific oligosaccharide segments that are required for protein binding*. Thus, the ability to cleave the polymeric heparinoid at a variety of sites is useful in identification of these specific sequences. Consequently, after outlining the structural modification procedures, we will discuss how some of these procedures may facilitate the isolation of heparinoid fragments for analysis of heparinoid fine structure and for use in studies of heparin-binding proteins.

B. Periodate Oxidation

Periodate oxidation, illustrated in Figure 4.1, is truly a classical reaction in carbohydrate chemistry. Periodate cleaves carbon-carbon bonds when each of the adjacent carbons is substituted with an hydroxyl or an amino group. Since all of the monosaccharide residues in heparinoids are linked at C4, the only candidate sites for periodate attack are the C2-C3 bonds. With only a few exceptions, all of the C2 amino groups of the GlcN residues in heparinoids are

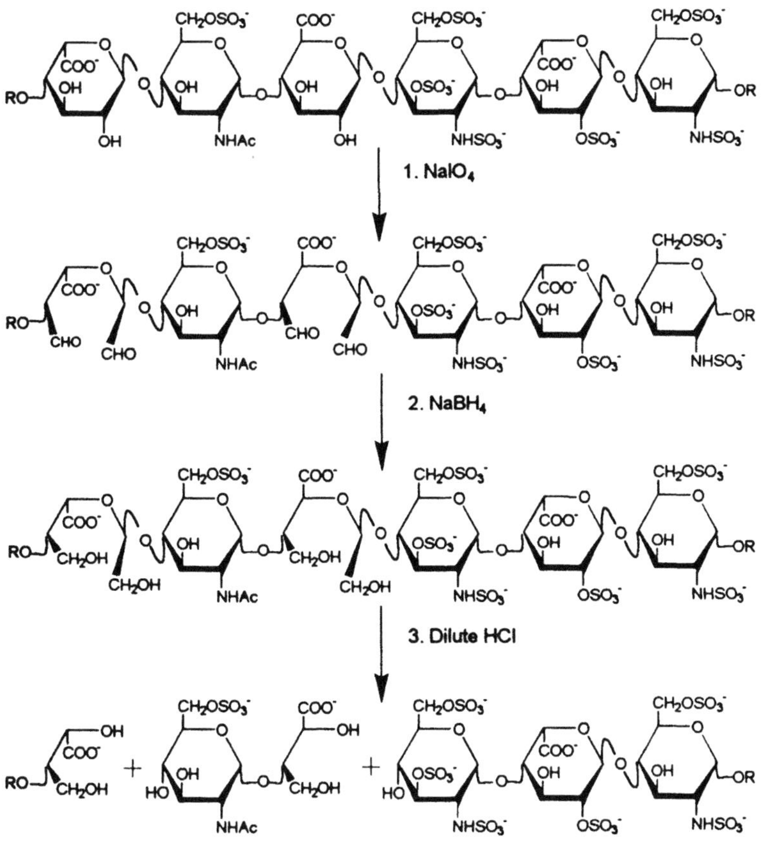

Figure 4.1. Periodate oxidation of heparin. The periodate-oxidized heparinoid is reduced with borohydride and cleaved at the oxidized residues with dilute acid.

either acetylated or sulfated (Chapter 2). Consequently, the GlcN residues are resistant to periodate. Furthermore, many of the GlcA and IdoA residues are sulfated at C2 and they, too, are resistant to periodate. Thus, the C2-C3 bonds of the *unsulfated* GlcA and IdoA residues are the only potential sites of periodate attack.

Figure 4.1 illustrates the periodate susceptibility of a segment of heparin that is involved in the specific binding of heparin to antithrombin [see 354, and references therein]. This oligosaccharide segment contains an internal GlcA that is unsulfated. Thus, periodate cleaves the C2-C3 bond of this residue and, in doing so, destroys the anticoagulant activity of heparin [355]. Of course, the periodate treatment also results in the destruction of all other unsulfated uronic acid residues in the polymer, including those in all of the GlcA→GlcNR and IdoA→GlcNR disaccharides. Interestingly, IdoA is oxidized much more rapidly than GlcA. In addition, the GlcA itself is oxidized more rapidly in the GlcA→ GlcNAc-3,6-di-SO_4 combination than in the GlcA→GlcNSO_3-3,6-di-SO_4 combination; i.e., the substitutions on the adjacent GlcN residues affect the rates of oxidation of the uronic acids [354].

Since the periodate oxidation reaction is commonly run under slightly acidic conditions (pH~4, 0-4°C), the 1,2-glycols of the unsulfated uronic acids are cleaved, but polymer remains intact, i.e., the molecular weight of periodate-oxidized heparin is essentially the same as that of the original heparin. The aldehyde groups formed by periodate oxidation have several undesirable features from the standpoint of the possible use of these products clinically, including the fact that they are quite reactive and can easily complex with other compounds, especially through Shiff base formation between the aldehydes and free amino groups of proteins and other structures. However, this reactivity can be eliminated by reducing the aldehydes to secondary alcohols with NaBH$_4$, as shown in reaction 2 of Figure 4.1. We will discuss the cleavage of the aldehyde-reduced product (reaction 3) in Section II.B., p. 129.

C. De-N-Acetylation and De-N-Sulfation

The N-acetyl groups can be removed from the GlcNAc residues in heparinoids with little, if any, further modification of the polymer, as shown in Figure 4.2. This is achieved by heating heparin with a solution of hydrazine/hydrazine sulfate at 95-100 °C for 4 to 6 hr [257, 258]. Neither the N- nor the O-sulfates are lost during this reaction.

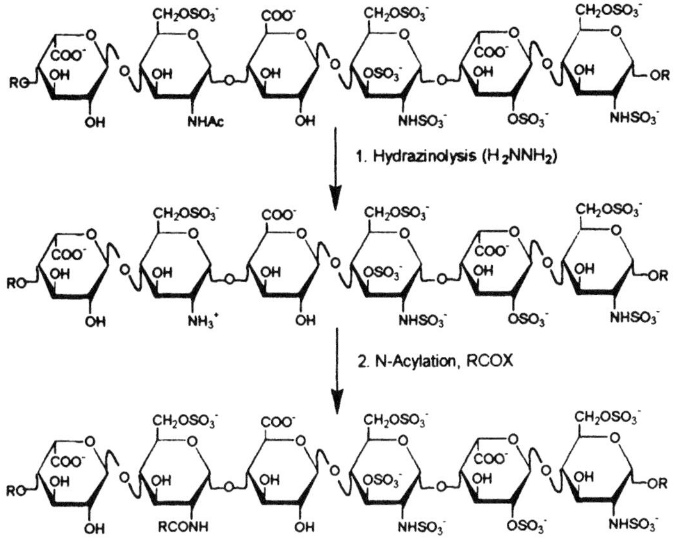

Figure 4.2. De-N-acetylation and re-N-acylation of a heparin segment.

A side reaction that occurs much more slowly than the de-N-acetylation results in the conversion of a few of the carboxyl groups of the uronic acid residues to hydrazides, as shown in Figure 4.3. Originally, the hydrazinolysis reaction was studied in anhydrous hydrazine [257], but it was found later that the rate of de-N-acetylation is increased and the rate of hydrazide formation is decreased by increasing the water content of the hydrazinolysis reaction mixture to 30% [23]. Interestingly, under both aqueous and nonaqueous reaction conditions, the rates of hydrazide formation are different for different disaccharide units in the polymer [23, 257]. When the uronic acid hydrazides are formed, they can be restored to their original uronic acid structures by treatment of the deacetylated product with I_2, HIO_3, or nitrous acid [257].[1] It may be noted that the hydrazinolysis reaction is one of the steps in total disaccharide analysis of heparinoids using nitrous acid to cleave the polymers to disaccharides (see Chapter 3).

The formation of uronic acid hydrazides, even when minimized under aqueous reaction conditions, cannot be totally eliminated. Consequently, an undesirable side reaction, namely, the cleavage of the polymer between the uronic acid hydrazide and its upstream GlcN residue, occurs to a limited extent. This is shown in reaction 2 of Figure 4.3. This is a β-elimination reaction that is

---

[1] Of course, the nitrous acid will also cleave the polymer.

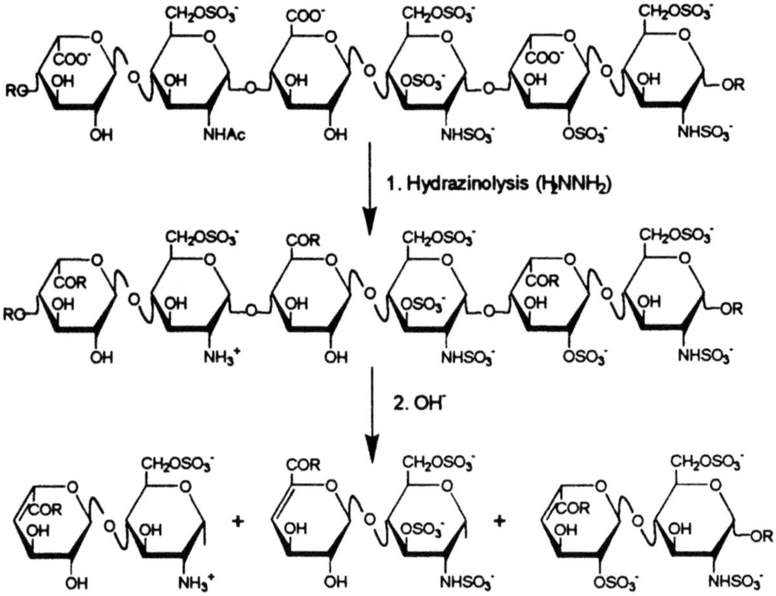

Figure 4.3. Hydrazide formation as a side reaction of the hydrazinolysis reaction. [COR = (C=O)NHNH$_3$$^+$].

initiated at C5 of the derivatized uronic acid residues. The reaction is formally identical to that catalyzed by the bacterial lyases (Chapter 3). This results in a small amount of depolymerization of the heparinoid that cannot be avoided during the de-N-acetylation.

Like the N-acetyl groups, the N-sulfate groups of heparin can be removed selectively, in this case by solvolysis. Solvolysis is usually run by heating the pyridinium salt of heparin in dimethyl sulfoxide containing a small amount of water [356]. When this reaction is run at 50°C for short times, N-sulfate groups are removed but none of the other structural features of heparinoids are modified.

Thus, both the N-acetyl groups and the N-sulfate groups of heparinoids can be removed separately and selectively.

D.   Modifications of N-Unsubstituted Heparinoids

Both the de-N-acetylated and the de-N-sulfated GlcN residues are susceptible to a variety of further conversions. Furthermore, since the N-acetyl and the N-sulfate groups can be removed separately and selectively, these further reactions can be used to modify *either* the original GlcNAc or the original GlcNSO$_3$ residues in heparinoids.

The most obvious reaction of an N-unsubstituted polymer is the re-acylation of the amino group with acetyl or other acyl groups, as shown in reaction 2, Figure 4.2. This offers the possibility of obtaining *labeled heparinoids* with little further modification of the polymer by acetylating with [^{3}H] or [^{14}C]acetic anhydride [357]. Alternatively, the amino groups can be sulfated, with or without labeled reagent, by treatment with the sulfur trioxide-triethylamine complex [358, 359]. Further conversions of the N-unsubstituted polymers are described below.

E.   2-O-Desulfation

Although N-sulfate groups are extremely labile to acid, they are resistant to alkaline conditions. The O-sulfates, however, can be removed with considerable selectivity under certain alkaline conditions. This allows alkaline de-O-sulfation under conditions that leave the N-sulfates completely intact. Two procedures have been used to remove 2-O-sulfate groups [360-364]. These are illustrated in Figure 4.4. In the first (Figure 4a), heparin is dissolved in 0.05-1.0 $\underline{M}$ NaOH (pH 12.5-12.8) and the solution is frozen and lyophilized. This treatment results in loss of 2-O-sulfate groups without any loss of 6-O-sulfates. All of the following reactions occur stoichiometrically:

IdoA-2-SO$_4$-GlcNSO$_3$-6-SO$_4$ → IdoA-GlcN-6-SO$_4$

IdoA-2-SO$_4$-GlcNSO$_3$ → IdoA-GlcNSO$_3$

GlcA-2-SO$_4$-GlcNSO$_3$-6-SO$_4$ → GlcA-GlcNSO$_3$-6-SO$_4$

GlcA-2-SO$_4$-GlcNSO$_3$ → GlcA-GlcNSO$_3$.

Under the reaction conditions, the methyl glycoside of GlcNSO$_3$-3-SO$_4$, a model compound, does *not* lose its O-sulfate group, suggesting that the important 3-O-sulfate group from the antithrombin-binding sequence of heparin remains after the lyophilization process [365]. However, it has been demonstrated more recently by disaccharide analysis that the product obtained when *polymeric heparin* is lyophilized under these alkaline conditions no longer contains GlcA →GlcNSO$_3$-3,6-di-SO$_4$ units, indicating that the 3-O-sulfate is removed in the reaction [366, 367].[2] The product of this reaction shows a markedly reduced anticoagulant activity, as expected from the loss of critical 2-O-and 3- O-sulfate substituents in the antithrombin-binding pentasaccharide (Chapter 7).

---

[2] It is clear that the 6-O-sulfates are not removed in the reaction.

A second type of alkaline desulfation reaction (Figure 4b) involves refluxing heparin in 0.1 M Na_2CO_3 for periods up to 30 hr [360-364]. This reaction also results in stoichiometric loss of 2-O-sulfate groups, but the original IdoA-2-SO_4 unit is converted to an L-galacturonic acid residue! Thus, the IdoA-2-SO_4-containing disaccharides are converted to the corresponding disaccharides

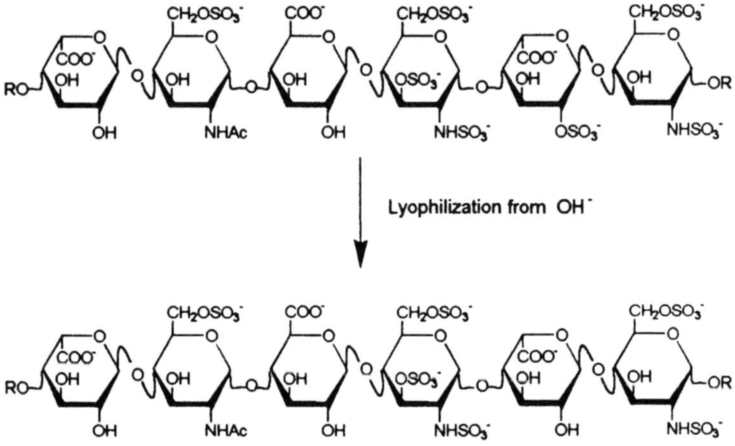

a. 2-O-desulfation without epimerization

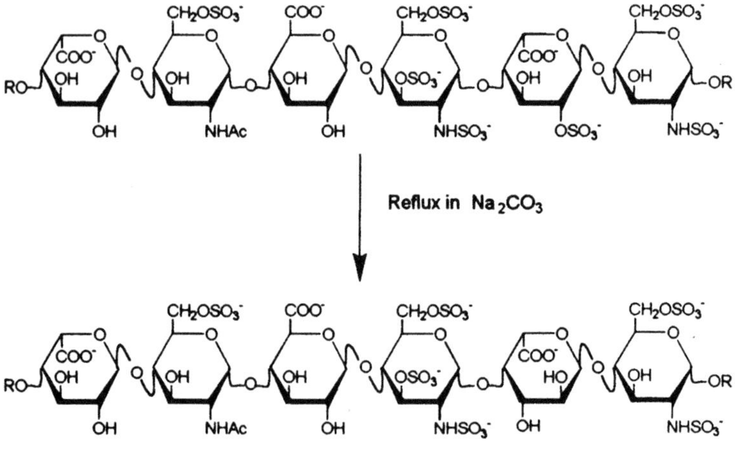

b. 2-O-desulfation with epimerization

Figure 4.4. 2-O-desulfation of heparinoids.

that contain L-galacturonic acid in place of the IdoA-2-SO$_4$. Because the GlcA-2-SO$_4$ residues are found in such minor amounts in heparinoids (Chapter 2), the effect of the reaction with Na$_2$CO$_3$ on these residues has not been demonstrated. Presumably, however, these residues would be converted to L-altruronic acid residues. The mechanism of this inversion of configuration at C2 and C3 of the IdoA has been discussed [360-364].

Obviously, these reactions result in the formation of heparinoid derivatives that may offer further possibilities for polymer modification or for testing for biological activity. The lyophilization process is particularly attractive because the product disaccharides that are formed are disaccharide units that are normally present in heparinoids. Thus, the desulfation merely alters the relative amounts of the normal disaccharides without introducing unnatural units into the heparinoid product, as occurs in the Na$_2$CO$_3$-catalyzed reaction.

F.   6-O-Desulfation

The solvolytic procedure that is used for de-N-sulfation of heparin can also be applied in 6-O-desulfation [368, 369]. When heparin is heated in dimethyl sulfoxide:water (19:1) at 90°C, the relative rates of desulfation are N-SO$_3$ >> 6-O-SO$_4$ > 2-O-SO$_4$. Thus, 6-O-desulfation occurs more rapidly than 2-O-sulfation, so that most of the 6-O-sulfates can be removed while a high percentage of the 2-O-sulfates remain. An alternate method for 6-O-desulfation has been described in which the pyridinium salt of the heparinoid is heated with N-methyl-trimethylsilyl-trifluoroacetamide in pyridine [370, 371]. Following the reaction, the GlcN residues can be re-N-sulfated (below) [358, 359].

G.   Carboxyl Reduction

The carboxyl groups of heparinoids can be reduced stoichiometrically, without depolymerization or alteration of any of the other structural features of the polymer [253, 372]. The reaction, illustrated in Figure 4.5, is carried out in aqueous media by activating the carboxyl groups of the polymer with a water soluble carbodiimide at pH ~4.75. The activated carboxyl groups can then be reduced completely by adding an excess of NaBH$_4$ (or, if a labeled product is desired, NaB^{3}H$_4$). The products of carboxyl reductions of heparinoids contain L-idose and D-glucose residues in place of the original L-iduronic acid and D-glucuronic acid residues, respectively. A complicating aspect of the reaction for the reduction of the activated carboxyls is that the carbodiimide adduct attached

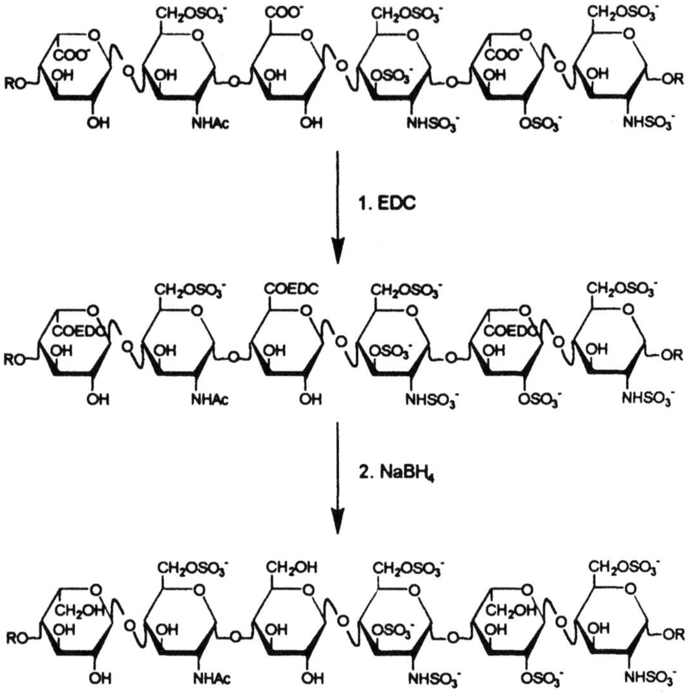

Figure 4.5. Carboxyl reduction of heparinoids. EDC = 1-ethyl-3-
(dimethylaminopropyl) carbodiimide.

to the carboxyl is labile under alkaline conditions, which must be used for the
$NaBH_4$ reduction [373], whereas the $NaBH_4$ is labile under the acidic conditions
used in the activation phase of the reaction sequence. However, the reduction
cannot be carried out with $NaCNBH_3$, which can be used under acid conditions,
since the $CNBH_3^-$ is not a strong enough nucleophile for the reduction. Thus, to
complete the reduction without loss of the activated intermediate, a large excess
of the $NaBH_4$ is added directly to the pH 4.75 activation reaction mixture. The
destruction of the BH_4^- at this acid pH results in voluminous evolution of H_2 gas,
but, as the BH_4^- is destroyed, the borate that is formed raises the pH to a level
where the $NaBH_4$ is now stabilized but still is present in such a large excess that
the reduction reaction proceeds much more rapidly than the loss of the
carbodiimide adduct. Note that use of NaB^3H_4 for the reduction would require
much too much labeled reagent if a large portion of the reagent were used

simply to raise the pH to neutrality; thus, it is preferable to raise the pH of the activation reaction mixture to neutrality before adding NaB^3H_4.[3]

Carboxyl-reduced heparinoids can be analyzed for their disaccharide content by the nitrous acid cleavage (with or without hydrazinolysis) and HPLC methods described in Chapter 3 for unmodified heparinoids. Since each disaccharide unit in these carboxyl-reduced polymers has lost one negative charge, the disaccharides emerge earlier in the elution profiles than the unmodified heparinoid disaccharides (see Figure 3.5).

Carboxyl-reduced heparinoids exhibit much greater acid lability of the glycosidic bonds of glucose and idose than the corresponding uronic acid glycosides in the starting materials (Chapter 3). Thus, the glucose residues are hydrolyzed with the ease of the glucose glycosides of amylose and glycogen. There are no analogous polymers in Nature that contain large amounts of L-idose in glycosidic linkages. However, as noted in Chapter 3, IdoA glycosidic bonds in heparin exhibit an acid lability similar to neutral hexose glycosides (e.g., D-glucose), and, in fact, limited studies have suggested that the L-idose residues in carboxyl-reduced heparins are much more acid labile than their D-glucose counterparts [374].

H.   Carboxyl Derivatization

The use of water-soluble carbodiimides to activate carboxyl groups of GlcA and IdoA residues in heparin suggests that the activation can be used for other modifications at the uronic acid carboxyl groups. After all, BH_4 is a much weaker nucleophile than other nucleophiles that might be used to react with the activated carboxyl groups. In fact, the water-soluble carbodiimide activation of carboxyl groups was first described by Hoare and Koshland [375] for derivatizing carboxylate amino acid side chains of proteins. In these studies, the activated carboxyl groups were reacted with glycine methyl ester to convert them to amides of glycine methyl ester. It has been reported that glycine methyl ester also serves as a nucleophile for converting heparin carboxyl groups to amides via the carbodiimide-activated intermediate [376]. In addition, Shaklee et al. [367, 377] have shown that aminomethylsulfonate ($^+H_3NCH_2SO_3^-$) can be used as a nucleophile to convert the carboxyl groups of heparinoids to the corresponding amidomethylsulfonated heparinoids, derivatives that contain modified uronic acid residues but retain negative charges on the sulfonate groups of the adducts. This approach offers the potential to use a broad range of

---

[3] In both cases it is necessary to carry out the reduction in a well-ventilated hood.

nucleophilic agents to convert the carbodiimide-activated carboxyl groups of heparinoids to a variety of other derivatives.

I.   Epimerization

As described earlier, when the uronic acid carboxyl groups of heparinoids are derivatized with hydrazine or carbodiimides, the C5 proton adjacent to the carboxyl derivative is labilized, and the bond between the uronic acid and its upstream GlcN residue may be cleaved under basic conditions by a β-elimination reaction (see reaction 3 in Figure 4.3). However, when a uronic acid residue at the nonreducing terminal of a heparin oligomer or polymer becomes so derivatized, there is no upstream bond to be cleaved. In this situation, the uronic acid may be epimerized at C5, thus converting an oligosaccharide containing either a GlcA or an IdoA nonreducing terminal residue to an equilibrium mixture of oligosaccharides containing both GlcA and IdoA at the nonreducing terminal. The epimerization has been particularly useful in converting disaccharides such as the relatively abundant IdoA-2-SO_4→AMan$_R$-6-SO_4 or IdoA-2-SO_4→AMan$_R$ to mixtures containing large amounts of the relatively rare GlcA-2-SO_4→AMan$_R$-6-SO_4 or GlcA-2-SO_4→AMan$_R$ [258, 378].

It has been reported that C5 epimerization of uronic acids occurs when alginates, as well as heparin and chondroitin sulfates, are treated with CO_2 under supercritical conditions [379, 380]. Some depolymerization of alginate takes place under the reaction conditions. Although no results with heparin have been reported, this type of transformation would yield polymers in which only GlcA and IdoA, the naturally occurring uronic acids, would appear in the products, although their relative proportions would be altered.

J.   Sulfate Migration

When the pyridinium salt of heparin is heated at 90° C, N-sulfate groups migrate from GlcNSO$_3$ to the C3 positions of either the original GlcNSO$_3$-6-SO_4 residue (i.e., the GlcNSO$_3$-6-SO_4 would be converted to a GlcN-3,6-diSO_4 residue) or to the upstream (presumably) IdoA-2-SO_4 (yielding IdoA-2,3-diSO_4) [381]. Thus, for example, the sequence

$$IdoA-2-SO_4→GlcNSO_3-6-SO_4→IdoA-2-SO_4$$

would be converted to a mixture of

$$IdoA-2,3-diSO_4 \rightarrow GlcN-6-SO_4 \rightarrow IdoA-2-SO_4$$

and

$$IdoA-2-SO_4 \rightarrow GlcN-3,6-diSO_4 \rightarrow IdoA-2-SO_4$$

This transformation, when followed by re-N-sulfation, results in increased anticoagulant activity. Further study of this interesting reaction has not been reported.

K.   Oversulfation

The heparinoids, as well as other polysaccharides that can be used as heparin mimics, can be oversulfated by treatment with the sulfur trioxide-trimethylamine complex at 50 °C for 24 hr [382] or by treating the tributylammonium salt of the heparinoid with anhydrous dimethylformamide and the pyridine sulfur trioxide complex [383, 384].[4] These reactions add sulfate groups randomly to free hydroxyl groups in these polysaccharides.

II.   MODIFIED HEPARINOIDS IN STRUCTURAL AND BIOLOGICAL STUDIES

A.   Introduction

In spite of our general knowledge of the structures of heparinoids, there remain many aspects of the fine structure that we do not understand. Furthermore, with the exceptions of antithrombin, the fibroblast growth factors, lipoprotein lipase, extracellular superoxide dismutase, and hepatocyte growth factor, we know almost nothing about the heparinoid structures that regulate the activities of heparin-binding proteins. The procedures outlined above, together with those described in Chapter 3, are useful in addressing these areas of incomplete understanding. However, some of these procedures have not yet been widely applied.

As has been suggested in several of the previous discussions, most of the products of the modification reactions may be tested for their structural features

---

[4] Note that heparinoids in various hydrophobic salt forms, such as the tributylamine salt, can be dissolved in nonaqueous solvents.

or their biological activities, or they may become starting materials for further modification. For example, carboxyl-reduced heparinoids can be used as starting materials for further modifications, including periodate oxidation, de-N-acetylation, desulfation, etc. In addition, heparin may be cleaved to oligosaccharides, and the heparin fragments may be used to prepare periodate-oxidized, de-N-acetylated, or desulfated products. Most of the reactions described here and in Chapter 3 give stoichiometric conversions with high yields of the desired products. Thus, a variety of well-characterized modification reactions can be used to prepare unique heparinoid structures. In addition, the degrees of conversion in these reactions may be controlled, thus permitting the preparation of a range of *partially* modified heparinoids. The extents of conversion can be controlled by limiting the reaction time, by modifying the reaction conditions, or by limiting the amounts of reactants that are consumed in the reaction. For example, nitrous acid is consumed during deaminative cleavage, and carbodiimide is consumed during the carboxyl reduction reaction; thus, partial cleavage with nitrous acid or partial carboxyl reduction can be obtained by limiting the amounts of the respective reagents added to the reaction mixtures. Furthermore, when the procedures that are used to modify glycosaminoglycans are run under conditions that limit the extent of conversion, it is usually found that the targeted groups of different monosaccharides in these polymers react at significantly different rates. Such rate differences may be inherent in the differences of the monosaccharide structures themselves. For example, D-GlcA and L-IdoA have carboxyl groups that extend from C5 of their respective pyranose rings in different orientations, and it is reasonable to expect that these two uronic acids in heparinoids would exhibit different reactivities when they are treated with the water-soluble carbodiimide. Alternatively, rate differences may be affected by differences in the substituents on the same monosaccharide ring or on adjacent residues. For example, it has been established that the de-N-acetylation of GalNAc-4-SO_4 and GalNAc-6-SO_4 in chondroitin SO_4 occurs at quite different rates [258], that uronic acid hydrazide formation rates are different for GlcA- and IdoA-containing disaccharides and their variously sulfated forms [257], that carboxyl activation with carbodiimide proceeds much more rapidly for the GlcA→GlcNSO$_3$-3-SO_4 disaccharide unit than for all of the other disaccharide units in heparin,[5] and that the rates of periodate oxidation of GlcA and IdoA residues in heparinoids depend on the structures of the surrounding residues. Thus, any of the reactions just described might be used for complete or partial modifications of heparinoids at specific types of residues. When partial modifications are carried out, one may find that these reactions exhibit some selectivity for different residues.

---

[5] Personal communication from Dr. P. N. Shaklee.

Although the creation of a modified structure that has lost one or a few of the activities of the original heparinoid may give some information about the structural features of the heparinoid that are required for the activities, more detailed information may be obtained with specific heparinoid fragments. Many of the structural modifications that have been described may be useful in the preparation of unique segments of the polymeric heparinoid. We will now turn our attention to the applications of these reactions in the preparation of unique structures and the deduction of heparinoid fine structure. We will also discuss the ways in which these reactions can be used to prepare a variety of heparinoid fragments. Despite their apparent usefulness, many of the approaches have not yet been widely exploited.

B.   Periodate-Oxidized Heparinoids

Periodate oxidation of heparinoids yields polyaldehyde structures that can be derivatized by most of the typical reactions of aldehydes, allowing the preparation of a variety of modified heparinoids. In addition, the polyaldehydes are subject to β-eliminative cleavage at room temperature *under mildly basic conditions*, as shown in Figure 4.6. When the polyaldehydes are reduced to the corresponding polyalcohols, reaction 2 in Figure 4.1, the products now become *resistant to base but quite labile to dilute acid*, again introducing the possibility of cleavage to smaller fragments. Thus, periodate oxidation followed by (a) alkali treatment, or (b) $NaBH_4$ reduction and acid treatment, yields low molecular weight heparinoid fragments. Since periodate destroys the antithrombin-binding sequence, these fragment heparinoids have very little anticoagulant activity. However, they may retain other activities.

The cleavage of periodate-oxidized heparinoids yields a mixture of fragments whose molecular weight distributions are a reflection of the distribution of unsulfated uronic acids in the heparinoid. Since the abundance of these residues in heparins is approximately one in 8-10 disaccharides (see Table 2.2), one would expect this reaction sequence to reduce the average molecular weight to one-eighth to one-tenth of the original molecular weight if the unsulfated uronic acids are evenly distributed. However, if these residues are clustered, then a higher molecular weight oligosaccharide profile would be obtained. Furthermore, the fragments that are recovered after such a treatment would be sequences of contiguous disaccharides, *all* of which contain 2-O-sulfated uronic acids.

The cleavage of heparan sulfates in this manner also gives fragments that reflect the distance between unsulfated HexAs and that contain contiguous disaccharides with 2-O-sulfated HexAs. For heparan sulfates, cleavage will

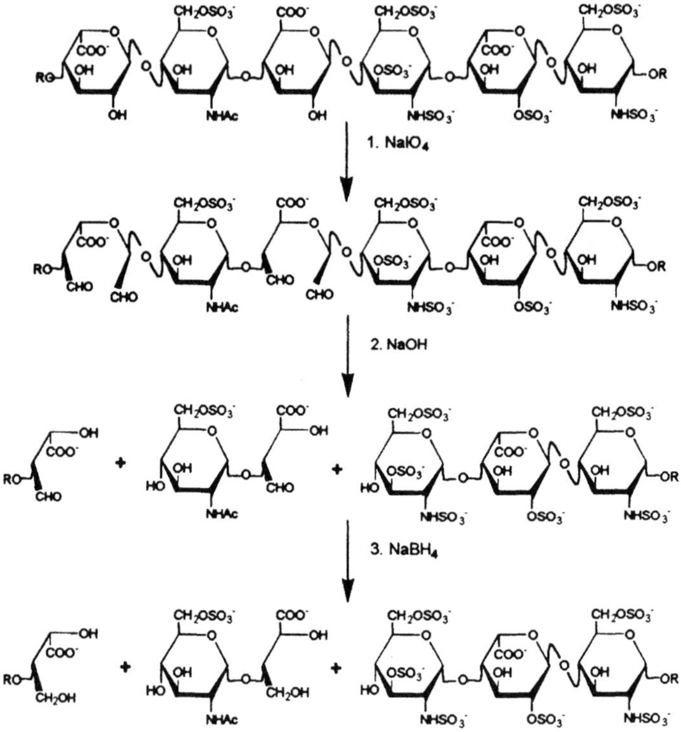

Figure 4.6. Alkaline cleavage of periodate-oxidized heparin.

occur at every disaccharide unit in the unsulfated blocks and at any unsulfated HexA that may occur within the sulfated blocks. Since the overall size of the blocks containing only GlcNSO$_3$ residues can be obtained by the de-N-acetylation/nitrous acid cleavage procedure (Section II.C., below), a comparison of the size of the overall GlcNSO$_3$-containing blocks with the size of the blocks obtained after the periodate oxidation/cleavage reaction will give an indication of the presence of unsulfated uronic acids within the sulfated blocks in the polymer. Currently, we have only limited information on the amounts of unsulfated uronic acids in the sulfated blocks of heparan sulfates from various sources.

C.   Modifications of N-Unsubstituted Heparinoids

Both the de-N-acetylated and the de-N-sulfated GlcN residues are susceptible to a variety of further conversions, including N-sulfation, N-acylation, nitrous acid cleavage, and periodate oxidation. Furthermore, since the N-acetyl and the N-sulfate groups can be removed separately and selectively, these further reactions can be used to modify *either* the original GlcNAc or the original GlcNSO$_3$ residues in heparinoids. For preparation of polymeric forms or unique fragments for testing of biological activities or fine structure, it is convenient to consider de-N-acetylated forms first and then to discuss the de-N-sulfated forms. Thus, the following modifications of de-N-acetylated heparinoids may be useful.

•   The exposed amino groups may be substituted with a variety of acyl groups, including, of course, acetate, or they may be fully sulfated.

•   After removal of the N-acetyl groups from GlcNAc residues, the C2-C3 bonds of the resulting GlcN residues can be cleaved by periodate. Of course, the unsulfated uronic acid residues are also oxidized. As described earlier, the oxidized polymers can now be cleaved at all oxidized positions by alkaline β-elimination of the polyaldehyde or by mild acid hydrolysis of the borohydride-reduced polyaldehyde.

•   The heparinoid may be cleaved with nitrous acid at pH 4 (Chapter 3).[6] The size range of the resulting fragments will reflect the sizes of the sequences between GlcNAc residues [see, for example, reference 385 for an illustration of this approach]. Furthermore, all of the GlcN residues in the fragments will be N-sulfated, allowing the isolation, characterization, and biological testing (for the first time) of the fully N-sulfated segments of these polymers. For de-N-acetylated heparin, for example, a uniform size distribution of the fragments would indicate that the GlcNAc residues are randomly distributed; thus, a polymer with 10% of its GlcN residues as GlcNAc would yield fragments with average molecular weights of one-tenth that of the starting heparinoid. For heparan sulfate, all of the GlcN residues in the unsulfated blocks would be de-N-acetylated by hydrazinolysis and would be converted to disaccharides by treatment with nitrous acid at pH 4, whereas the highly sulfated blocks would remain intact.

---

[6] If a polymer that has been de-N-acetylated by hydrazinolysis is to be cleaved with nitrous acid, it is not necessary to first destroy the uronic acid hydrazides formed in the hydrazinolysis step because the hydrazides are converted back to the original uronic acids by nitrous acid (at *either* pH 4 or pH 1.5)

Turning now to the de-N-sulfated heparinoids, the following modifications may be considered.

- The exposed amino groups may be substituted with a variety of acyl groups, as in the case of the de-N-acetylated heparinoids.

- After removal of the sulfate substituents, the C2-C3 bonds of the resulting GlcN residues may be oxidized by periodate, destroying the ring structures of the deblocked GlcN residues in addition to those of the unsulfated uronic acids. Again, the oxidized polymers can now be cleaved at all the oxidized positions by alkaline β-elimination of the polyaldehyde or mild acid hydrolysis of the borohydride-reduced polyaldehyde. This would release the isolated GlcNAc residues in heparins in mono-, di-, or trisaccharide forms, depending on whether the neighboring HexA residues were 2-O-sulfated (i.e., resistant to periodate).

- The GlcN residues may be cleaved with nitrous acid at pH 4. However, this offers no advantage for cleavage at the N-sulfated GlcN residues, since these can be cleaved with nitrous acid at pH 1.5 without prior removal of the N-sulfate groups. The size of such fragments reflects the sizes of the sequences between GlcNSO₃ residues. For heparan sulfate, such cleavages provide fragments containing only the unsulfated segments.

D.   De-O-Sulfation

Alkaline lyophilization 2-O-desulfation as well as 6-O-desulfation reactions result in the formation of heparinoid derivatives that are particularly attractive because the products contain disaccharide units that are normally present in heparinoids. Thus, desulfation merely alters the relative amounts of the normal disaccharides without introducing un-natural units into the heparinoid product. The 2-O-desulfation is selective for the 2-O-sulfates (and the 3-O-sulfates on GlcN residues), whereas the 6-O-desulfation reaction removes 2-O-sulfates at a rate that is much slower than that of the 6-O-desulfation reaction. The resulting derivatives may offer possibilities for further polymer modification, including periodate oxidation, which, after 2-O-desulfation, will now cleave at the C2-C3 bonds of *all* of the uronic acid residues in the polymer, thus offering the possibility of cleavage of the polymer completely to its GlcN monosaccharide constituents.

E. Carboxyl Group Modification

As previously noted, the glycosidic bonds of glucose and idose residues in carboxyl-reduced heparinoids are more acid labile than their corresponding uronic acids. Although it is not known whether the L-idose glycosidic bonds in carboxyl-reduced heparins are more acid labile than their D-glucose counterparts, it is clear that the presence of a sulfate at C2 of a hexose renders the glycosidic bond much more acid labile than the corresponding unsulfated hexose [386]. In fact, it has been shown that the idose-2-SO$_4$ glycosidic bond in carboxyl-reduced heparin is extremely acid labile, thus offering the possibility for selective hydrolysis of these polymers at the idose-2-SO$_4$ bonds [374].

Any modification of a uronic acid carboxyl group that renders the C5 proton adjacent to the derivatized carboxyl more acidic will offer the opportunity for a base-catalyzed elimination reaction to cleave the polymer at the glycosidic bond of the GlcN residue immediately upstream from the derivatized uronic acid residue. This base-catalyzed reaction is formally identical to the cleavages catalyzed by the bacterial heparinases (Chapter 3). We noted earlier that a side reaction in the deacetylation of GlcNAc residues by hydrazinolysis yielded uronic acid hydrazides and that heparinoids substituted in this manner were subject to a β-eliminative cleavage at these uronic acid hydrazides (Figure 4.3). Thus, heparinoids can be cleaved at these positions by hydrazinolysis; i.e., this approach would yield a low molecular weight heparin. The degree and site of cleavage would depend on the rate of hydrazide *formation*, which clearly is significantly higher under the anhydrous hydrazinolysis conditions than when water is present, and which differs for different uronic acids depending on its location in the polymer.[7] To our knowledge, this approach for cleaving heparinoids or other uronic acid-containing polysaccharides has not been exploited.

The derivatization of the carboxyl groups in heparinoids with aminomethylsulfonate has already been described (Section I.H., p. 125). These amide forms of the uronic acids markedly facilitate the β-eliminative cleavage of the glycosidic bond linked to C4 of the uronic acid.[5] Since the aminomethylsulfonate derivatization of heparinoids can be carried out stoichiometrically, it would be possible to cleave such polymers *completely* to disaccharides in alkaline solution. Also, it would be possible to control the degree of depolymerization by limiting the extent of the derivatization. These approaches have not been developed.

---

[7] For example, the $t_{1/2}$ for hydrazide formation under the anhydrous conditions is 147 min for GlcA → AMan$_R$ and 894 min for IdoA→AMan$_R$ [257].

F.   Epimerization

The potential for C5 epimerization of the uronic acid residues of heparinoids with CO_2 under supercritical conditions has not been exploited in the preparation of heparinoid derivatives. However, there are some interesting possibilities. For example, if one were able to convert the *E. coli* K5 polysaccharide into a polymer that contains similar numbers of GlcA and IdoA residues, one could obtain an interesting heparinoid analog that might be subject to further modification using the enzymes of heparin metabolism [see, for example, 384, 387].

G.   Low Molecular Weight Heparins

1.   Biological activities

The most exploited of the modified heparins are the partially depolymerized heparins. The reason for this is that high molecular weight and low molecular weight heparins exhibit differences in anticoagulant activities and biological half-lives. This will be described in greater detail in Chapter 7. For now, we may simply state that high molecular weight heparin catalyzes the antithrombin-mediated inhibition of both thrombin and Factor Xa, whereas low molecular weight heparin catalyzes the antithrombin-mediated inhibition of Factor Xa but shows a reduced activity with thrombin. The clinical advantages of low vs high MW heparins are discussed elsewhere. The now wide spread clinical availability of low molecular weight heparins prepared by a variety of means has led to the comparison of unfractionated heparins with these low molecular weight heparins for biological activities other than the anticoagulant activity.

2.   Methods of cleavage

The potential for use of low molecular weight heparins as pharmaceuticals has led to the development of unique low molecular weight heparins that are prepared in different manners so that each preparation can be protected by a unique patent. Because the biological activities of various low molecular weight heparins seem to depend more on their molecular weight ranges than on the methods used to cleave the heparin, a variety of methods have been used for their preparation. Of course, the cleavage method must not destroy the antithrombin III-binding sequence or introduce toxic side effects if the products are to be used an anticoagulants. However, for preparation of nonanticoagulant heparins, other cleavage procedures are acceptable. The most highly refined and exploited cleavage methods are partial nitrous acid cleavage and heparin lyase cleavage, but a number of other methods have been used.

One such method that has been used exclusively for the generation of low molecular weight heparin is the cleavage of heparin with hydrogen peroxide in the presence of Cu^{2+} ions [120, 388]. This procedure is not used for any other structural modification and presumably yields structurally unmodified low molecular weight heparins. However, some modifications must occur because the lowest molecular weight products of this procedure are poorly cleaved by heparin lyases [120].

The following are cleavage procedures that can be used to prepare low molecular weight heparins. Some of these would destroy anticoagulant activity.

• Partial cleavage with nitrous acid at pH 1.5.

• Partial or complete periodate oxidation, followed by direct cleavage in dilute alkali, or by mild acid (after the polyaldehyde is reduced to the polyalcohol).

• Partial or complete de-N-acetylation, followed by treatment with nitrous acid at pH 4.

• Partial de-N-sulfation, followed by treatment with nitrous acid at pH 4.

• Partial activation of carboxyl groups with (a) carbodiimides, (b) hydrazine, (c) aminomethylsulfonate, or (d) other reagents, followed by treatment with dilute alkali.

• Size fractionation of unfractionated heparin to obtain the low molecular weight range of the total material.

III. SYNTHESIS OF OLIGOSACCHARIDES

The chemical synthesis of simple disaccharides presents a number of problems. These include (a) establishing the order (from nonreducing terminal to reducing terminal) of the monosaccharide residues, (b) forming the proper anomeric configuration at C1 (α- vs β-) of each residue, and (c) forming the linkage between the anomeric carbon and the desired carbon (C4, in the case of heparinoids) on the downstream monosaccharide. These problems, which are imposing enough, are markedly compounded by the problems of obtaining a reasonable source of IdoA and getting the N- and O-sulfate groups into the appropriate positions. And, as the size of the desired product is increased, the effort is compounded further. Finally, although the products of such syntheses are hydrophilic, the reactions leading to the products are carried out in nonaqueous media. Clearly, the synthesis of a structure as basic to heparin chemistry as the antithrombin-binding pentasaccharide is a multistep process in which the yields of the final product are extremely low. Nevertheless, such work has been carried out with remarkable success in several laboratories, and the studies of the resulting products have been of enormous value in characterizing the interactions of heparin with antithrombin and other proteins.

A description of the synthetic strategies used in the preparation of heparin fragments is beyond the scope of this book. However, the approaches used and the results of these synthetic approaches have been reviewed [328, 389-392]. The 1989 review by Petitou is particularly valuable in laying out the strategies used for the synthesis of heparin fragments and the difficulties encountered. Also, some of the more recent work on the synthesis and activities of a variety of antithrombin-binding pentasaccharide structures is shown in Chapter 7.

ADDITIONAL REFERENCES

Grootenhuis, P. D. J., Westerduin, P., Petitou, M., and van Boeckel, C. A. A. (1995) Rational design of synthetic heparin analogues with tailor-made coagulation factor inhibitory activity, *Nat. Struct. Biol.* **2**, 736-739.

Liu, J., Pervin, A., Gallo, C. M., Desai, U. R., Van Gorp, C. L., and Linhardt, R. J. (1994) New approaches for the preparation of hydrophobic heparin derivatives, *J. Pharm. Sci.* **83**, 1034-1039.

Nadkarni, V. D., Pervin, A., and Linhardt, R. J. (1994) Directional immobilization of heparin onto beaded supports, *Anal. Biochem.* **222**, 59-67.

Chapter 5. The Cellular Metabolism of Heparan Sulfate Proteoglycans

---

I. OVERVIEW

As noted in Chapter 2, the basic features of the biosynthetic pathways for heparin and heparan sulfate synthesis are identical. This conclusion is based primarily on the finding that microsomal preparations from heparin-synthesizing mastocytoma cells and heparan sulfate-synthesizing hepatocytes produce similar heparin-like products under identical incubation conditions [101]. These results also indicate that the cellular controls that regulate the degree of maturation of the heparinoids in whole cells are lost in cell extracts, since the hepatocyte microsomal extracts produce a heparin-like product that is much more highly matured than the heparan sulfate produced by intact cells. The pathway of biosynthesis of heparinoids has been deduced largely through experimentation with cell extracts. This chapter integrates these cell extract results with studies that have been carried out using cultured cells and isolated cellular organelles. It details the methodologies that are most commonly used in such studies; describes the metabolic flow of the heparinoids during their synthesis, secretion, and further metabolism; and identifies areas for further studies.

Heparan sulfate proteoglycans are synthesized, secreted, endocytosed, and catabolized rapidly. Despite this dynamic processing, heparinoids are concentrated in characteristic sites in the cell. Heparan sulfate generally occurs as a proteoglycan that is concentrated on the surface of most animal cells, whereas heparin is found as free glycosaminoglycan chains in secretory granules of connective tissue mast cells and related hematopoietic cells. The cellular pathways leading to the dispositions of these heparinoids involve (a) the synthesis, in the rough endoplasmic reticulum, of the necessary core protein which serves as the heparinoid acceptor, (b) entrance of the core protein into the secretory pathway where the GlcA→Gal→Gal→Xyl linkage region is added at Ser-Gly sites on the core protein, (c) polymerization of GlcA and GlcNAc to form the initial polymeric form of the heparinoid, (d) maturation of these initial polymers by the reactions that generate the final, highly sulfated glycosaminoglycan chains (Chapter 2), and (e) the trafficking of the products to their final destinations. The trafficking steps take different routes for heparin and heparan sulfate proteoglycans. In the case of the heparin proteoglycan, the heparin chains attached to the serglycin core protein in the mast cells are cleaved by an endo-β-glucuronidase to give free heparin chains which form complexes with basic proteases and peptidases; these complexes are packaged in secretory granules. In the case of the heparan sulfate proteoglycan, the intact proteoglycan is secreted from the cell to take up residence on the cell surface. From their respective sites of concentration, both the heparan sulfate proteoglycan and the heparin chains undergo further metabolism. When mast cells degranulate, the heparin/protease complexes are released with the secretory granules and dissociate to release the enzymes. The fate of the heparin from these granules has not been described, but it is probably taken up by cells and catabolized to its monosaccharide and sulfate components that are then recycled in metabolism or excreted (Section V.A., p. 169). The cell surface heparan sulfate proteoglycan, however, is either shed [as a result of (a) proteolytic cleavage of the core protein or (b) phospholipase cleavage of the GPI anchor] and/or endocytosed and delivered either to the lysosomes for catabolism or to other intracellular targets (Section III.I., p. 160). In this chapter, we will first outline the methodology used to study proteoglycan metabolism in cell cultures and will then describe the nature of heparinoid metabolism in cells.

IL   METHODS FOR STUDYING PROTEOGLYCAN METABOLISM

A.   Metabolic Labeling

1.   Introduction

As a rule, the cellular metabolism of heparinoid proteoglycans is studied in cell cultures. Generally, radiolabeled precursors for the heparinoid or the core protein are added to the culture medium, and the appearance of the labeled product inside the cell and on the cell surface is measured. In the case of heparan sulfate, the labeled products are synthesized, secreted, and then either shed into the medium or endocytosed and catabolized. In fact, these processes result in a steady state level of heparan sulfate proteoglycan on the cell surface. In order to account for all of the product that is being processed by the cells, it is necessary to measure the proteoglycans and the free heparan sulfate chains in (a) the culture medium pool, (b) the cell surface pool, and (c) the intracellular pool. For cultures of attached cells, the culture medium pool is readily obtained, since the medium can be removed while the cells are still attached to the culture dish. The cell surface pool is released from washed cells by treatment with low levels of isotonic trypsin. In some cases, this can be done without releasing the cells from the dish, but usually a suspension of cells is obtained which yields the cell surface pool in the supernatant after the cells are pelleted by centrifugation. Finally, the material that can be extracted from the pelleted cells is the intracellular pool. It may be noted that any proteoglycan form of the heparan sulfate that is present on the surface of the cells is converted to free glycosaminoglycan chains (with short peptides attached) by the trypsin treatment, thus yielding a product that does not reflect the true nature of the proteoglycan on the cell surface. However, the intracellular pool can be protected from the trypsin by washing the trypsin away and by adding trypsin inhibitors before breaking the cells. The extraction of heparan sulfate proteoglycan from the lysed cells *without* trypsin treatment yields the *total* cell surface plus intracellular pool (the "cell-associated pool"). After characterization of the intracellular pool from the trypsin-treated cells and the total cell-associated pool, one can subtract the amount of material in the intracellular pool from the total cell-associated pool to obtain a reasonable characterization of the actual nature of the cell surface fraction (proteoglycan, free heparan sulfate chains, etc.). Further approaches for characterization of these fractions will be described below.

Depending on the domains of the proteoglycan that one wishes to study, one may use labeled precursors for the core protein (e.g., cysteine, methionine, or leucine), for the glycosaminoglycan chains (glucosamine, sulfate, galactose, xylose, acetate), or for the N- or O-linked oligosaccharides or GPI anchor that may be attached to the core protein (glucosamine, mannose). Since the labeling of *proteins* in such studies is such a common procedure, this will not be discussed further here.

2.   Labeled glucosamine

D-Glucose normally is a precursor of the amino sugars that are incorporated into glycosaminoglycans. As shown in Figure 5.1, the branch point in glycolysis leading to the amino sugars occurs at the level of D-fructose-6-PO_4, which is aminated by glutamine and acetylated by acetyl CoA to yield N-acetyl-D-glucosamine-6-PO_4. The latter is ultimately converted to UDPGlcNAc, which equilibrates with UDPGalNAc. When exogenous D-glucosamine is supplied to cultured cells, it enters the cells and is phosphorylated to join the endogenous pathway at GlcN-6-PO_4. In a pathway not shown in Figure 5.1, exogenous glucosamine also is incorporated into CMP-sialic acid, which is the donor for the sialic acid that appears in N- and O-linked oligosaccharides and in glycolipids. Thus, when [^{3}H]glucosamine is supplied to cells (one may also use [^{14}C] glucosamine, but it is more expensive), the label ends up in all products that contain GlcNAc, GalNAc, sialic acid, or their derivatives, but in no other products. The labeled products include the heparinoids, the chondroitin sulfate/dermatan sulfate group of glycosaminoglycans, hyaluronic acid, keratan sulfate, many glycolipids, GPI anchors, and the N- and O-linked oligosaccharides that are attached to the core proteins and other glycoproteins.

For cultured cells, the specific radioactivities of the UDPHexNAc pools are controlled by the proportions of the glucose and the [^{3}H]glucosamine that enter the UDPHexNAc pools. These proportions depend, in turn, on the *relative concentrations* of the glucose and GlcN in the growth medium, which are constantly changing during a labeling experiment. This is illustrated in Figure 5.2 from a study of the incorporation of labeled glucosamine into chondroitin

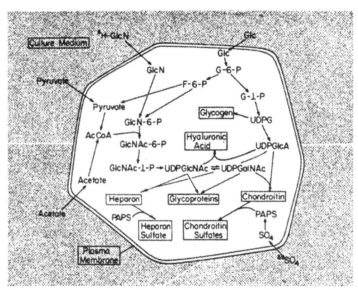

Figure 5.1. Uptake and metabolism of glucosamine and sulfate by cells.
Reprinted from [393], with permission.

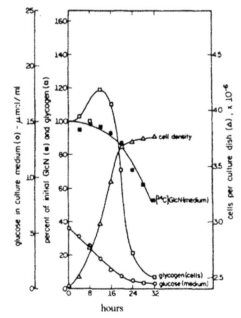

Figure 5.2. Glycogen and glucosamine metabolism in
cultured chondrocytes. Reprinted from [393], with
permission.

sulfate by chick embryo chondrocytes [393]. When fresh culture medium is added to a culture of cells (time 0), the glucose concentration of the medium is high, and cells incorporate some of this glucose into glycogen and metabolize the rest to support cell growth. As the glucose concentration drops, the cells begin to draw on their glycogen stores and, depending on the cell density, may reduce both the glucose in the culture and the glycogen to quite low levels before the medium is changed again. While the glucose and glycogen levels are relatively high, labeled GlcN disappears from the culture quite slowly, but as these levels drop, the [³H]GlcN is consumed more rapidly. The specific radioactivity of the UDPHexNAc pool is relatively low while glucose and glycogen are high, but increases continuously as these levels drop, approaching that of the [³H]GlcN that is supplied in the medium when glucose and glucosamine are fully depleted. Thus, one cannot obtain a linear rate of incorporation of *label* into glycosaminoglycans if the labeling period is extended over more than a few hr.

Usually the concentration of the labeled GlcN supplied in a labeling experiment is very low and is chosen on the basis of the number of μC_i of high specific activity [³H]GlcN that are required for the degree of labeling desired (frequently related to the *cost*/μC_i). Thus, for optimum incorporation of label into glycosaminoglycans, the glucose concentration in the culture medium at the beginning of a labeling experiment should be no greater than 1 g/liter and the labeling period should be limited to a few hr. It is important to remember that the glucose will be consumed at a rate that depends on the number of cells in the culture dish. Consequently, the *rate of increase of the specific activity* of the UDPHexNAc pool will be higher in high density cultures than in low density cultures. Shortly after the glucose disappears, the labeled GlcN also is totally consumed, and the incorporation of label ceases. Proteoglycan synthesis also ceases, since the glucose is the ultimate source of carbon and energy for proteoglycan biosynthesis. From this description, it can be appreciated that it is important to choose the labeling times and the culture medium composition to maximize the amount of label incorporated, and to maintain (as nearly as possible) a linear rate of incorporation of label (which occurs while the glucose concentration in the medium is still relatively high (e.g., 0.5 to 1 g/liter). When [³H]glucosamine is used as a labeled precursor for heparinoids, *every* disaccharide unit in the main polymer chain will become labeled.

3.   [³⁵S]Sulfate

Use of $^{35}SO_4^{2-}$ as a metabolic precursor for heparinoid synthesis in cultured cells offers several advantages over labeled GlcN, but also has some disadvantages. If the cell type that is being studied does *not* express the metabolic pathways leading from sulfur-containing amino acids to inorganic sulfate (e.g., chondrocytes, [394]), the specific activity (in cpm/μmole of SO_4) of the sulfate substituents in the glycosaminoglycans synthesized by the cells will be identical to that of the $^{35}SO_4^{2-}$ supplied in the culture medium. Thus, if one isolates a labeled heparinoid from such cells and converts it by treatment with heparin lyase(s) or nitrous acid to fragments that can be separated and identified, one can use the number of cpm in the fragment to calculate the actual number of moles of the fragment. Since both the unlabeled SO_4^{2-} in the culture medium and the $^{35}SO_4^{2-}$ enter the cell in constant proportion on the sulfate transporter and are converted directly to 3'-phosphoadenosine-5'-phosphosulfate, the donor of the $^{35}SO_4$ for the heparinoid chains, the specific radioactivity of the incorporated $^{35}SO_4$ does not change during the labeling period. Note, however, that when the glucose in the culture is exhausted (again, at a rate that depends on the number of cells in the culture), $^{35}SO_4^{2-}$ incorporation ceases, not because the $^{35}SO_4^{2-}$ in the culture medium is used up, but because the carbon source required to synthesize the heparinoid chains is no longer available [393]. As discussed below, when $^{35}SO_4^{2-}$ incorporation ceases, the $^{35}SO_4$-labeled heparinoids rapidly disappear as they are endocytosed and catabolized. One disavantage of using $^{35}SO_4^{2-}$ as a labeling agent is that it only labels sulfated disaccharides; thus, for heparan sulfate, the large blocks of unsulfated disaccharides do not become labeled, and any change in the relative proportions of sulfated and unsulfated blocks might be (incorrectly) interpreted as changes in total heparan sulfate synthesized.

Many animal cell types are able to convert the sulfur atoms of cysteine and methionine into inorganic sulfate, which can be incorporated into heparinoids. As illustrated in Figure 5.3, there are several routes by which cysteine is catabolized, but only the route leading to β-sulfinylpyruvate appears to lead to the generation of sulfite [395]. A number of cell types contain sulfite oxidase. Thus, administration of sulfite to cells leads to its conversion to sulfate, which is incorporated into glycosaminoglycans [396].

The ability of animal cells to utilize cysteine and methionine as a source of sulfate for glycosaminoglycan synthesis was not generally recognized in the proteoglycan field until the report by Esko *et al.* [397] that certain CHO cell mutants that had lost the capacity to transport SO_4^{2-} into the cells were still able to synthesize sulfated glycosaminoglycans. These mutants incorporate very little $^{35}SO_4^{2-}$ from the culture medium into glycosaminoglycans, but the heparinoids produced by the mutant when [^{3}H]GlcN is used as the precursor show the same degree of sulfation that is observed in wild type cells. Also, addition of [^{35}S]cysteine to the culture medium results in the formation of [$^{35}SO_4$]heparan sulfate. The sulfur of methionine is also converted to SO_4^{2-} but not as efficiently. Further examination of a variety of cell types has shown that the capacity of

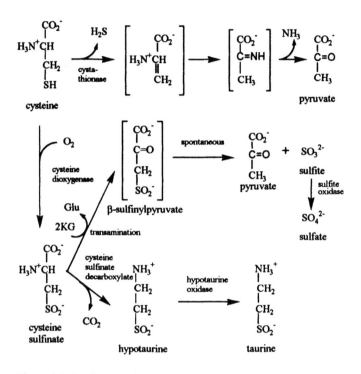

Figure 5.3. Catabolism of cysteine.

cells to convert cysteine to SO_4^{2-} varies considerably [394, 396, 398-402]. In CHO cells, increasing the level of cysteine in the culture medium reduces the amount of $^{35}SO_4^{2-}$ incorporated into glycosaminoglycans. Thus, in these cells the two sources of $^{35}SO_4$ seem to generate a single pool of 3'-phosphoadenosine-5'-phosphosulfate. However, in hepatoma cells, an increase in the concentration of unlabeled cysteine in the culture medium has no effect on the amount of label from $^{35}SO_4^{2-}$ incorporated into heparinoids; likewise, an increase in the amount of unlabeled SO_4^{2-} in the culture medium has no effect on the amount of label from [^{35}S]cysteine incorporated into heparinoids.[1] If both labeled precursors are supplied at the same time, the amount of $^{35}SO_4$ in the heparinoid products is additive. Apparently the cysteine metabolism in hepatoma cells is more complex than in CHO cells. In fact, multiple pathways for cysteine catabolism have been reported in hepatocytes [395]. It should be obvious, by analogy with the discussion of glucose and glucosamine incorporation into heparinoids, that in cells that can convert cysteine into SO_4^{2-}, the specific radioactivity of $^{35}SO_4$ in

---

[1] Unpublished work of Y.-C. Guo and H. E. Conrad

heparinoids formed from $^{35}SO_4^{2-}$ as the lone labeled precursor will not be equal to that of the SO_4^{2-} supplied in the culture medium (and *may* increase as cysteine is consumed). A study of mutants in the $Cys \rightarrow SO_4^{2-}$ pathway when the cells are incubated with $^{35}SO_4^{2-}$ might be informative, since these would yield $^{35}SO_4$ substituents in the heparinoids with a specific activity equal to that of the $^{35}SO_4^{2-}$ supplied in the medium.

In addition to using mutants that cannot transport SO_4^{2-} into the cells, tests of the contribution of exogenous SO_4^{2-} can be studied in wild-type cells using various stilbene derivatives that are inhibitors of sulfate transport [see, for example, 403].

4.   Other precursors (galactose, xylose, acetate, mannose)

Although radiolabeled galactose, xylose, mannose, or acetate precursors of glycosaminoglycans can be used for special labeling purposes, they are of much less value that the precursors described above. Xylose and galactose are incorporated into the linkage region. Labeled galactose also is incorporated into the uronic acid units of heparinoids, although not very efficiently. [2-^{3}H]Mannose may be used to label the mannose and fucose residues that appear in N- and O-linked oligosaccharides or in GPI anchors. Mannose is phosphorylated, and mannose-6-PO_4 is converted to GDP-mannose and GDP-fucose to label these residues, but it is also converted to fructose-6-PO_4, and can thus enter the glycolytic pathway and potentially label a great variety of metabolites. Therefore, it is customary to use [^{3}H]mannose labeled at C2 as the precursor of complex carbohydrates, because any mannose-6-PO_4 that flows into glycolysis will lose the ^{3}H as it is converted to fructose-6-PO_4. Although it will become incorporated into the N-acetyl groups of GlcNAc residues, acetate is such a non-specific precursor that it is of little value in the study of glycosaminoglycan metabolism.

B.   Extraction of Proteoglycans

As a rule, the conditions requried for solubilization of proteoglycans and free heparan sulfate chains depend on the type of experiment being performed. Generally work on the metabolism of heparan sulfate proteoglycan is carried out using cultured cells. Such cells usually contain the desired products in several pools. For convenience, one can distinguish experimentally the pool of proteoglycans in the culture medium (which clearly are already solubilized), the pool of proteoglycan associated with the surface of washed cells, and the heparan sulfate present inside the cells. The culture medium pool and the intracellular pool are readily soluble in water and can be treated as any other solubilized cell products. The cell surface pool, however, is usually anchored (a) by a core protein that spans the plasma membrane or (b) by a GPI anchor that is intercalated into the outer leaflet of the membrane. The latter pools are extracted from cells with buffers containing dilute detergents [404, 405]. The transmembrane syndecan proteoglycans are all quantitatively extracted in cold neutral buffers that contain Triton X-100 or other detergents and can easily be separated from other proteins and from nonhydrophobic proteoglycans by ion exchange chromatography and intercalation in liposomes. It is more difficult to separate them from each other as they have similar overall size, charge densities, and affinities for a variety of ligands [146, 406].

Epithelial GPI anchored proteins are associated with complexes that are insoluble in cold solutions of nonionic detergents such as Triton X-100, but are extracted in these buffers at 37 °C. Efficient extractions of glypican may also be obtained in cold solutions of octyl glucoside, a nonionic detergent with a structure that resembles that of a glycolipid and that therefore may solubilize the lipid patches more efficiently than other detergents [146]. Another method for extraction of GPI anchored proteins involves the use of 4'-amino-benzamido-taurocholic acid for selective extraction [407].

C.   Separation and Analysis of Proteoglycans

In addition to the proteoglycans, all glycoproteins become labeled when [^{3}H]glucosamine is supplied to cells. $^{35}SO_4^{2-}$ labels a more restricted group of polymers, but can supply $^{35}SO_4$ to tyrosine residues in some proteins, to sulfatides, or to N-linked oligosaccharides of certain glycoprotein hormones [408]. Consequently, to distinguish the labeled heparinoids from other labeled polymeric products, one must either isolate the heparan sulfate proteoglycans or apply unique heparinoid chemistry (nitrous acid) or enzymology (heparin lyases) to characterize the heparinoid component in the mixture of labeled products (Chapter 3). Because of the high negative charge associated with the glycosaminoglycan chains, these proteoglycans are easily separated from most other labeled products by anion exchange chromatography [see, for example, 225, 404]. Glycoproteins, which elute from DEAE-cellulose columns in low salt, are completely separated from the proteoglycans, which elute with high salt. However, the different types of proteoglycans may be difficult to separate by anion exchange chromatography, even with quite shallow gradients. Thus, additional approaches may be required. For example, the chondroitin sulfate/dermatan sulfate group of glycosaminoglycans may be removed by digestion with chondroitinases. Alternatively, specific proteoglycans can be selectively precipitated with antibodies directed against their core proteins (Chapter 3).

To determine whether an isolated heparinoid is in a proteoglycan form or a free heparan sulfate chain, advantage may be taken of the fact that serine-linked glycosaminoglycans are readily released from their core proteins by an alkali-catalyzed β-elimination reaction (below). N-linked oligosaccharides remain attached to the core protein when the reaction is run at room temperature, but any O-linked oligosaccharides that are present on the core protein will be released along with the glycosaminoglycan chains. This reaction will result in a marked reduction in the molecular weight of a proteoglycan, but will have no effect on the molecular weight of a free glycosaminoglycan chain. The procedure also can be used for the isolation of the free glycosaminoglycan chains, the O-linked oligosaccharides, and the core protein (still substituted with any N-linked oligosaccharides that were present originally).

Complete release of the O-linked glycosaminoglycans, as well as any O-linked oligosaccharides, can be obtained in 24 hr at room temperature without the removal of N-linked oligosaccharides [409, 410]. At higher temperatures (35-50°C) the rate of the reaction is increased, but there is loss of some sulfate residues, some cleavage of N-linked oligosaccharides [411], and some cleavage of the polypeptide chain. Under milder conditions, the serine residues to which the O-linked carbohydrates are linked are converted to dehydroalanine residues which may be reduced to alanines with $PdCl_2$ [410], but the core protein is not cleaved. Thus, the molecular weight of both the core protein and the released glycosaminoglycan chains can be estimated by gel electrophoresis or gel filtration [404, 412]. $NaBH_4$ is included in the reaction mixture to convert the newly formed xylosyl reducing terminals to xylitols, thus preventing alkaline peeling from the reducing terminal of the chain. If desired, NaB^3H_4 can be used in the reaction to label the reducing terminal xylose of the glycosaminoglycan chains (as well as the reducing terminal residues of the O-linked oligosaccharides). It is advisable to free the proteoglycan from oxyanions such as phosphate prior to reaction, since these will catalytically destroy borohydride [413]. If NaB^3H_4 is used in the reaction to label the reducing terminal xylitol, the ratio of total disaccharides in the glycosaminoglycan chains to [^{3}H]xylitol can be used to estimate the average size of the glycosaminoglycan chains. In addition, the *linkage region* can be recovered from an ^{3}H-labeled glycosaminoglycan chain by trimming away the main chain with lyases or nitrous acid (Chapter 3), leaving the labeled linkage region plus any uncleaved disaccharides.

III. METABOLISM OF HEPARAN SULFATE PROTEOGLYCANS

A. Core Protein Synthesis

As described in Chapter 2, a number of proteins have been identified as specific core proteins, or carriers, of heparin or heparan sulfate chains. These core proteins must be synthesized by the cell in order for heparinoid synthesis to proceed; i.e., the core proteins serve as primers for heparinoid synthesis, and, in their absence, the heparinoid chains cannot be formed. Thus, the regulation of the expression of genes for these core proteins represents a first stage in controlling the *amount* of heparinoids synthesized by cells (Section IV.B., p. 162). One might expect that the structures of the heparinoid chains added to different core proteins would vary. However, the limited information that is available indicates that the structures of the heparinoid chains are not influenced by the core protein(s). Instead, these structures apparently are controlled by the unique cellular regulation of the synthesis of the heparinoid chains in each cell type. For example, Shworek *et al.* [155] have shown that all three glycosaminoglycan attachment sites on ryudocan (syndecan-4) receive chondroitin sulfate and heparan sulfate in equal proportion and that the structures of the glycosaminoglycan chains at each site are essentially identical. A number of other reports also indicate that the heparan sulfate structures are cell type-specific and not core protein-specific [154, 156, 157, 206, 414, 415]. Thus, the observation that syndecan 3 from neonatal rat brain carries unique heparan sulfate chains that bind basic fibroblast growth factor, but not other heparin-binding proteins [416], may reflect the fact that these cells synthesize one major core protein and that the control mechanisms in these cells for heparan sulfate chain synthesis yield a product that has selective bFGF binding capacity; i.e., there is probably no selectivity of *syndecan* proteoglycans for basic fibroblast growth factor. All indications show that if the same cells produced several types of heparinoid core proteins, all would carry heparan sulfate structures with identical selectivities for bFGF.

Core protein synthesis is cell type-specific and is developmentally regulated [144, 158-167]. Since the amount of core protein that is synthesized limits the amount of heparan sulfate that can be formed, it is clear that the steady state levels of heparan sulfate proteoglycans that appear on the surface of different cell types may vary significantly. However, as documented in Section VI.B., p. 173, many cell types express several core proteins, so that the quantification of the amount of heparan sulfate on any single core protein that a cell produces may not represent the total heparan sulfate produced by the cell.

Although it is clear that the expression of the several heparan sulfate core proteins is highly regulated in development, a clear picture of how gene expression of the core proteins is controlled is still emerging. For example, cDNAs for serglycin have been obtained from rat [179, 417], mouse [179, 182, 185], and human [180, 183, 418-421], but the entire gene structure has been reported only for human serglycin. Also, the entire genes have been described for mouse syndecan [198, 199], and human perlecan [232]. The serglycin gene [186] has a 5'-flanking region that contains cis-acting elements as well as elements that interact with trans-acting factors [185]. It has also been shown that several multipotent hematopoietic cell precursors can be induced to differentiate and modulate the expression of serglycin by treatment with 12-O-tetradecanoylphorbol-13-acetate [422]. However, the mechanisms that control the normal physiological regulation of serglycin expression have not yet been established.

The mouse syndecan gene [198, 199] has five exons that code for the signal sequence, the N-terminal glycosaminoglycan attachment region, the main part of the extracellular domain, the protease-susceptible site, and the transmembrane and cytoplosmic domains, respectively. The 5'-flanking region contains TATA and CAAT boxes and other potential binding sites for Sp1 (GC box), NF-kB, MyoD (E box), and Antennapedia. The structure of the promoter region suggests that the control of syndecan-1 expression is both constitutive and developmentally regulated.

The gene for human perlecan [232] is quite complex, with 94 exons and a 5'-flanking region that lacks TATA or CAAT boxes but contains several GC boxes with binding sites for transcription factors SP1 and ETF. The gene contains multiple transcription initiation sites distributed over 80 base pairs of genomic DNA.

B. Early Glycosylation

Although different signal sequences may play specific roles in the targeting of secreted proteins [423], all core proteins have signal sequences that target these proteins to the lumen of the rough endoplasmic reticulum where they are further modified. The core proteins lose their signal sequences and become N-glycosylated at their glycosylation recognition peptide sequences (Asn- X-Ser or Asn-X-Thr). For the glypicans, the initial transcripts contain two signal peptides: the one at the N-terminal that directs the polypeptide into the lumen of the rough endoplasmic reticulum and a C-terminal peptide that directs the addition of a preformed GPI anchor (see Chapter 2) to the C-terminal amino acid of the matured core protein.

Pathways for the synthesis and addition of the GPI anchors, shown in Figure 5.4, have been reviewed [424-430]. The anchor is synthesized by the stepwise addition of monosaccharides to phosphatidylinositol. The initial step, the addition of GlcNAc from UDPGlcNAc, is followed by de-N-acetylation of the GlcNAc residue to yield a unique N-unsubstituted GlcN that is retained in the final structure. The inositol is fatty acylated and the mannose residues are added from dolichol-P-mannose, a step that presumably occurs in the lumen of the rough endoplasmic reticulum [430]. The final step in the synthesis of the preformed anchor is the transfer of ethanolamine phosphate from phosphatidylethanolamine to the nonreducing terminal mannose, forming a phosphodiester linkage. Further modifications of this common structure occur in some anchors and may include the addition of side chain monosaccharides or ethanolamine phosphates, fatty acylation of inositol ring hydroxyl groups, and transacylation of the glycerol moiety of the phosphatidyl inositol unit. Within minutes of completion of the synthesis of the core protein, the

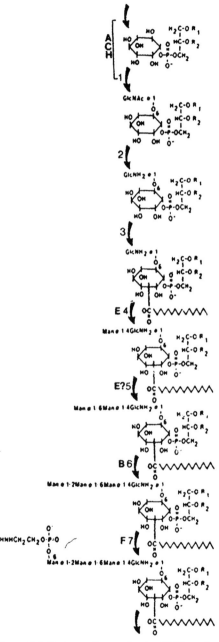

Figure 5.4. Pathway for synthesis of GPI anchors. Reprinted from [426], with permission.

preformed anchor is attached to the protein by a putative transamidation reaction in which a specific C-terminal peptide sequence is displaced from the protein as the ethanolamine phosphate moiety of the anchor takes its place. The peptide signal for GPI anchor attachment has been described [431-435]. It is made up of a hydrophobic amino acid sequence at the C-terminal end of the protein and a processing site containing two to three small amino acids (Ser, Ala, Gly, Asp, Asn, or Cys) positioned 10-12 residues upstream from the hydrophobic domain. The C-terminal hydrophobic domain temporarily anchors the completed protein to the membrane of the endoplasmic reticulum, and the transamidase cleaves at the small amino acid sequence and attaches the GPI anchor on the luminal side of the endoplasmic reticulum.

C.   Addition of the Heparinoid Chains

1.   Site of synthesis

The pathway for heparinoid chain biosynthesis has been described in Chapter 2. Briefly, the initial stage of biosynthesis of both the heparinoids and the chondroitin sulfate/dermatan sulfate group is the addition of the linkage region to the Ser→Gly sequences on the acceptor core proteins. This is followed by the addition of the repeating disaccharide GlcA→GlcNAc to form the [GlcA →GlcNAc]$_n$ chain. Finally, maturation reactions take place. These include de-N-acetylation/N-sulfation of some of the GlcNAc residues, epimerization of GlcA to IdoA, 2-O-sulfation of both GlcA and IdoA residues, 6-O-sulfation of the GlcNAc or GlcNSO$_3$ residues, and 3-O-sulfation of some of the GlcNSO$_3$ residues. All of these reactions occur in the endoplasmic reticulum or the Golgi prior to secretion from the cell. Also, during this process, both N- and O-linked oligosaccharides may be added to the core proteins, provided the appropriate peptide acceptor sequences are present. The structures of the N- and O-linked oligosaccharides attached to the core proteins have received little attention, but their biosynthesis presumably follows the well-established pathways. It should be noted, however, that virtually all of the radiolabeled precursors that are used to study heparinoid proteoglycan metabolism may be incorporated into these non-glycosaminoglycan constituents of these proteoglycans.

2. Heparinoid chain synthesis

The initial step in addition of the heparinoid chains is the formation of the linkage region. To attach the GlcA→Gal→Gal→Xyl linkage region, separate enzymes are required for each of these monosaccharides. Esko *et al.* [436] have shown that CHO cells mutants lacking the xylosyl transferase and the galactosyl transferase I do not synthesize either heparan sulfate or chondroitin sulfate. This indicates that the same enzymes are used for the synthesis of the linkage regions of both heparan sulfate and chondroitin sulfate. The galactosyl transferase II and the GlcA transferase I, an enzyme that is distinct from the one that participates in the synthesis of the main [GlcA→GlcNAc]$_n$ chain of the heparinoids, are also used for both heparan sulfate and chondroitin sulfate/dermatan sulfate synthesis. Thus, as described in Chapter 2, the pathways of heparan sulfate and chondroitin sulfate synthesis are identical up to the branch point that occurs at the addition of the first HexNAc residue. The mechanisms that control the reactions at this branch point are not well understood, but it has been shown that, in heparan sulfate synthesis, the GlcNAc transferase that adds the first GlcNAc to the linkage region is a different enzyme from the GlcNAc transferase that participates in the main chain elongation [73, 74]. Also, it appears that the Ser-Gly acceptor sequences for heparan sulfate chain addition (a) occur in clusters and (b) are flanked by acidic amino acid residues and an adjacent tryptophan [150, 151]. The xylosylation reaction has been localized to the late endoplasmic reticulum and the *cis*-Golgi [437, 438], whereas the addition of the rest of the linkage region occurs in the Golgi [437-439].

The synthesis of the repeating disaccharide chains of the heparinoids takes place in the Golgi. At least two of the enzymes in heparinoid biosynthesis are single proteins that catalyze concerted reactions. The enzyme that synthesizes the [GlcA→GlcNAc]$_n$ polymer is apparently a single protein that has both glucuronosyl transferase and N-acetylglucosaminyl transferase activities [81, 440]. The second protein that catalyzes two reactions is the enzyme that carries out the de-N-acetylation/N-sulfation reactions [104, 105, 441]. This enzyme has been cloned from two sources and, for both proteins, the two activities can be demonstrated in a single protein [107-109, 442]. However, whereas the apparent catalytic domains of the two enzymes show high sequence similarity, the N-terminal domains are different. It has been suggested that there may be multiple forms of this enzyme with different activities [20, 117]. A third case in which a single protein may catalyze multiple activities has been suggested for the 2-O-sulfotransferase and the 6-O-sulfotransferase from mastocytoma cells [112]. However, these activities are clearly found in separate enzymes in CHO cells [113, 115]. Finally, it has been reported that a CHO cell mutant that lacks 2-O-sulfotransferase produces heparan sulfate chains lacking both IdoA-2-SO$_4$ and

GlcA-2-SO$_4$ residues. This suggests that a single enzyme may catalyze sulfation of both uronic acid residues [114].

A further indication of the localization of the various maturation reactions along the passageway through the secretory apparatus is the time required for labeled core protein and the ^{35}SO$_4$-labeled products to reach the cell surface following the addition of the respective labeled precursors. The amount of heparan sulfate proteoglycan on the cell surface can be measured by the release of labeled heparan sulfate from intact cells with isotonic trypsin. The transit time for labeled core protein is 30-40 min after a labeled amino acid precursor is supplied, indicating the total processing time for proteoglycan synthesis. In contrast, ^{35}SO$_4$-labeled proteoglycan appears on the cell surface almost immediately after the addition of ^{35}SO$_4^{2-}$ to the culture medium [404, 443, 444], indicating that the sulfation reactions occur in the *trans*-Golgi, just prior to secretion of the completed proteoglycan to the cell surface.

The donors of the monosaccharides (UDPGlcA and UDPGlcNAc) and the sulfate (3'-phosphoadenosine-5'-phosphosulfate) for heparinoid synthesis are all synthesized in the cytoplasm of the cell and must be transported into the lumen of the endoplasmic reticulum or the Golgi. Membrane transport systems have been described for 3'-phosphoadenosine-5'-phosphosulfate [445-447] and the UDP sugars [445, 448]. This work has been reviewed [449].

3.   Regulation of heparinoid structure

Both the structural features and the amounts of heparinoids produced by different cell types can vary considerably. It has also been shown that, in a given cell type, the structures of heparan sulfate chains synthesized by cultured cells are different in different metabolic pools [404, 450, 451] and may change when growing cells reach confluence [412]. Although the mechanisms that control these changes are not understood, several controls on heparan sulfate structure can be demonstrated in whole cells. As originally shown in studies with cell-free systems, the degree of maturation of the [GlcA→GlcNAc]$_n$ chains depends on the extent of the initial de-N-acetylation/N-sulfation reactions, which might be controlled by SO$_4^{2-}$ availability (Chapter 2). Mutations in the N-sulfotransferase in CHO cells result not only in the reduction of the percentages of the glucosamine residues that are N-sulfated, but also in reduced levels of IdoA and O-sulfate groups in the final products [452, 453]. It has been shown that the transport of UDP-galactose into the Golgi exerts a limitation in the rate at which Gal becomes available in the Golgi for glycosaminoglycan linkage region and main chain (keratan sulfate) biosynthesis [448].

It has also been demonstrated in cell extracts that the chain length of the heparinoid chains can be regulated. Lidholt *et al.* [454, 455] have found that the heparinoid chains formed when 3'-phosphoadenosine-5'-phosphosulfate is present at the beginning of the reaction are longer than those formed when PAPS is omitted. Thus, the chain length of the polymer may also be controlled by early (N-sulfation) reactions. This also raises the possibility that the rate of transport of PAPS into the Golgi might play a role in the control of heparan sulfate structure.

Finally, apparent mechanisms for control of the structures of the heparinoids have been deduced from studies of biosynthetic enzyme specificities. Regulation may result from the fact that certain sequences in the maturing heparinoid chains are substrates for more than one of the maturation enzymes, such that the ultimate structures of these sequences may depend on the effectiveness of the two enzymes in their competition for a given substrate sequence. Such competitions have been described in Chapter 2. The examples cited are (a) the GlcA 2-O-sulfotransferase and the GlcA epimerase, both of which can act on the polymer chain immediately after the de-N-acetylation/N-sulfation reaction, and (b) the IdoA 2-O-sulfotransferase and the GlcNAc (GlcNSO$_3$) 6-O-sulfotransferase that can act on the product that is present immediately after the GlcA residues are epimerized to IdoA residues. In effect, the conversion by one of these pairs of competing enzymes determines the ultimate structure of the affected disaccharide units. Thus, the structure of the final heparinoid product may depend on the amounts, the kinetic properties, and/or the regulation of each competing enzyme.

D.   Trafficking to the Cell Surface

After proteoglycan synthesis has been completed in the Golgi apparatus, the products migrate to the cellular sites where they are normally concentrated. Thus, heparan sulfate proteoglycans are secreted to the cell surface where they can interact with a variety of heparinoid-binding proteins, whereas heparin chains are cleaved from their proteoglycans in mast cells and packaged in secretory granules in complexes with basic proteases and peptidases. Heparan sulfate proteoglycans with transmembrane core proteins can interact both with extracellular heparin-binding proteins and with intracellular components that bind to the cytoplasmic domain of the core protein. Proteoglycans with transmembrane core proteins may be spread over the total cell surface in stratified epithelia and in isolated epithelial and fibroblastic cells, but are found only on the basolateral surfaces of polarized simple cuboidal epithelia [144].

GPI-anchored heparan sulfate proteoglycans are anchored in the outer leaflet of the cell membrane bilayer and can interact only with extracellular proteins. GPI anchored proteins are targeted to the apical surface of a variety of epithelia, perhaps because they are sorted together with glycosphingolipids, which are also delivered to the apical surface [456-458]. However, although apical sorting of GPI anchored proteins seems to be the general rule, the GPI anchored proteins in a rat thyroid cell line are sorted to the basolateral surface [459]. Furthermore, one GPI anchored protein is symmetrically distributed in these cells. Thus, generalization may be premature. In fact, it has been shown that the apical sorting of glypican is lost as the degree of substitution of the core protein with heparan sulfate chains is increased [460]. Prevention of chain sulfation in glypican that is fully substituted with heparan sulfate chains does not restore apical sorting.

In addition to the membrane-*anchored* proteoglycans, there are membrane-*associated* proteoglycans or free heparan sulfate chains. Most often these have been recognized as proteoglycans that are displaced from the cell surface by heparin, suggesting that such structures are attached to cell surface "heparin-binding proteins" and that exogenous heparin competes with the heparan sulfate chains of the displaceable proteoglycans for the heparin-binding proteins [404, 461]. However, other agents have also been found to displace cell surface heparan sulfate proteoglycan, the most notable being a variety of inositol phosphate derivatives [222, 462]. In some cases, heparin and phytic acid (inositol hexaphosphate) have been found equally effective in replacing cell surface heparan sulfate proteoglycan, but in others the phytic acid and other inositol phosphate derivatives have been found to be more effective [222].

E.   Metabolic Turnover: Shedding and/or Endocytosis

Perhaps the most unique aspect of heparan sulfate proteoglycan metabolism is the fate of these proteoglycans after they reach the cell surface. Kraemer was the first to study heparan sulfate metabolism in cultured cells [463-468]. This work demonstrated (a) that cultured cells produce heparan sulfate; (b) that the heparan sulfate is found in the culture medium, on the cell surface, and in the intracellular pools; and (c) that a portion of the cell surface pool can be displaced from the cells by heparin. It also demonstrated the block structure of the heparan sulfate chains (Chapter 2) and suggested that all animal cells produce heparan sulfate proteoglycans. Finally, these studies showed that the cell surface heparan sulfate proteoglycan participates in a dynamic metabolism that is cell cycle dependent.[2] Kraemer's work set the stage for more recent studies that have confirmed that although heparan sulfate proteoglycan is *concentrated* on the cell surface, this cell surface material represents a transient and dynamic pool of proteoglycan.

When cells are labeled continuously over a 15- to 20-hr period, the amount of cell surface heparan sulfate proteoglycan reaches a steady state [404, 450, 469]. When the labeling precursor is removed from the culture medium of these cells, the cell surface pool disappears with a half-life of 3-8 hr [216, 404, 405, 444, 461, 469-475]. This is a remarkable rate of turnover when one considers that the doubling time for most cells in culture is at least 20-24 hr; thus, the turnover of cell surface heparan sulfate proteoglycan proceeds through three to five half-lives during a single cell cycle! The fate of the proteoglycan varies with the cell type and the type of core protein. Available data show that the metabolic flow of cell surface heparan sulfate proteoglycan takes different routes in different cells. In all cases, some of the material is shed (desquamated![2]) into the culture medium, but a significant fraction is internalized by the cells and transported to the lysosomes, where it is catabolized, or to other cellular compartments.

The trafficking of the heparan sulfate proteoglycans may depend on the type of core protein in these structures, although proteoglycans with the same type of core protein may be processed differently in different cell types. These differences have been reviewed [146, 476]. The proteoglycans may be shed from the cell surface, perhaps by proteolytic cleavage of the core protein, by phospholipase cleavage of the GPI anchors, by endogenous heparanase cleavage [477], or by the loss of proteoglycans from intracellular pools that occurs at mitosis (Section III.H., p. 160). Alternatively, they may be internalized, either via coated pits or by nonclathrin-coated membrane invaginations. Interestingly, full accounting for the total cell surface heparan sulfate proteoglycan often shows that the amount of material that is shed into the culture medium may be much less than the total amount of proteoglycan present on the surface of the cells at the beginning of a chase period, suggesting that the pericellular matrix pool of proteoglycan is, in large part, internalized. The possibility remains that some of the material that is internalized by the cells is partially degraded inside the cell and is then released into the culture medium, albeit by unknown mechanisms [474, 475, 478].

---

[2] The title of Kraemer's 1972 paper, "Cell-cycle dependent desquamation of heparan sulfate from the cell surface" is one of the classic titles in heparinoid literature, and one that sent many workers in the field scrambling for their dictionaries!

F.   Pre-lysosomal Processing of Internalized Proteoglycan

The portion of the heparan sulfate proteoglycan that is internalized is partially processed before it reaches the lysosomes. For example, whereas the cell surface heparan sulfate is present as a proteoglycan, the internalized material is so rapidly processed to free heparan sulfate chains that very little of the proteoglycan form can be detected in the intracellular pool [470, 479]. The extra-lysosomal conversion of the internalized proteoglycan may result from protease cleavage of the core protein and/or endo-β-glucuronidase cleavage of the heparan sulfate chain [404, 480]. Endo-β-glucuronidases have been characterized in several laboratories [64, 97, 98, 471, 478, 481-491], but the cellular localization of these enzymes has not been firmly established. However, there is clear evidence that these are not lysosomal enzymes [469]. Bame [492] has presented evidence that there are multiple species of heparinases in CHO cells.

The processing of internalized proteoglycan is a multistep pathway [reviewed in 476] and proceeds by different routes for GPI anchored proteoglycans and transmembrane proteoglycans in rat ovarian granulosa cells [223]. In the latter cells, all of the GPI anchored proteoglycan is internalized by the cells, whereas, in human lung fibroblasts, most of the GPI anchored proteoglycan is shed into the culture medium [215]. For the internalized heparan sulfate proteoglycan, the pre-lysosomal metabolism, in which the heparan sulfate chains are progressively converted to smaller oligosaccharides, may occur over periods ranging from a few minutes up to 3 hr. [484, 488]. Ultimately, most of the internalized heparan sulfate chains enter the lysosomes, where they are converted to monosaccharides and free sulfate (below).

G.   Lysosomal Catabolism

Fragments formed by the prelysosomal endogluronidase(s) enter the lysosomes where they are converted to GlcNAc, GlcA, IdoA, and free SO_4^{2-} by the concerted action of nine lysosomal exo-enzymes. The pathway for catabolism of heparinoids has been established largely by the work of Hopwood and colleagues who have presented comprehensive reviews on the subject [493, 494]. The pathway is shown in Figure 5.5.

The heparinoid chains that result from the cleavage by endo-β-glucuronidase(s) (step 1 in Figure 5.5) have reducing terminal GlcA residues and nonreducing terminal GlcN residues. The latter may be N-acetylated, N-sulfated, and/or O-sulfated. These oligosaccharides are degraded by the stepwise action of a series of exo-enzymes acting from the nonreducing end of the oligosaccharide. A general feature of this sequence is that, as each nonreducing terminal residue is exposed, there is an initial removal of any O- and/or N-sulfate groups to produce nonsulfated nonreducing terminal monosaccharides that are then removed by glycosidases. The lysosomal enzymes that are involved in the catabolism of heparinoids include those that hydrolyze the $N\text{-}SO_3$ that is

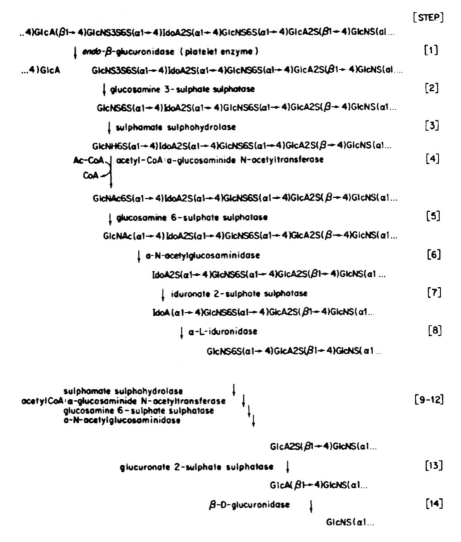

Figure 5.5. Catabolism of heparinoids. Reprinted from Nature [reference 493], with permission. © 1989, Macmillan Magizines Ltd.

present in $GlcNSO_3$, the five different types of *ester* SO_4, ($GlcNAc$-6-SO_4 , $GlcNSO_3$-6-SO_4, $GlcNX$-3-SO_4, $IdoA$-2-SO_4, and $GlcA$-2-SO_4), and the three types of glycosidic bonds (β-$GlcA$, α-$IdoA$, and α-$GlcNAc$). In intact cells, lysosomal catabolism and, to some extent, endosomal catabolism, are blocked by NH_4Cl or chloroquine (Section IV.F., p. 166).

In addition to the sulfatases and glycosidases required to degrade these oligosaccharides, there is a particularly interesting acetyl-CoA:α-glucosamine-N-acetyltransferase that plays a role in the conversion of N-sulfated glucosamine residues to hexosaminidase-susceptible substrates. As shown in step 3 of Figure 5.5, the N-sulfate is first removed by a sulfamidase, leaving an N-unsubstituted GlcN residue. Such a residue is not a substrate for the hexosaminidase, which recognizes only GlcNAc (or GalNAc) residues. To deal with this problem, there is a membrane-bound N-acetyltransferase that catalyzes the transfer of acetyl groups from acetyl-CoA to the free amino group, thus generating a viable substrate for the hexosaminidase [494, 495]. This represents a rare example in which acetyl-CoA is a critical participant in a catabolic pathway.

The enzymes in this pathway have been purified. Many of them have been cloned, and their substrate specificities and kinetic properties have been reported [summarized in 493, 494]. A particularly interesting property of these catabolic enzymes is that the substrates for each of the enzymes appear to be inhibitors for the other enzymes in the pathway. Perhaps this results from the fact that there are only subtle structural differences between successive intermediates in the pathway so that the substrate binding sites of the catabolic enzymes show limited discrimination between substrates and products. However, it is difficult to understand how these inhibitory actions may be of physiological significance. Hopwood has proposed a model for heparan sulfate catabolism in which the entire catabolic process is represented as a highly organized system of membrane-bound enzymes, a suggestion based on the fact that these enzymes tend to copurify. One might speculate that such a complex could limit the accessibility of enzymes in the pathway to the heparinoid fragments that may inhibit their activities.

Except for GlcN 3-O-sulfate sulfatase and the GlcA 2-O-sulfate sulfatase, a genetic deficiency in humans of any one of these enzyme activities interrupts the progression of steps in Figure 5.5 and leads to the accumulation of heparan sulfate fragments, often with dire physical and mental consequences. These genetic defects are referred to collectively as *mucopolysaccharidoses*. Interestingly, several of the heparinoid-degrading enzymes act on other glycosaminoglycan substrates as well. For example, iduronate 2-O-sulfate sulfatase acts on both heparin and dermatan sulfate, and its deficiency (Hunter's syndrome) results in the accumulation of oligosaccharides from both glycosaminoglycans. Similarly, GlcN 6-O-sulfate sulfatase cleaves sulfates from α-linked GlcN-6-SO$_4$ residues on heparan sulfate and β-linked GlcNAc-6-SO$_4$ residues on keratan sulfate.

The monomeric products of the catabolic pathway are transported out of the lysosomes by specific transport systems for the IdoA, GlcA, GlcNAc, and SO$_4^{2-}$ [496-501].

H.   Cell Cycle Aspects of Heparan Sulfate Metabolism

Certain aspects of heparan sulfate proteoglycan metabolism are cell cycle dependent. Kraemer and Tobey [466] first showed that there is a pre-mitotic loss of heparan sulfate proteoglycan from cells. This observation has been repeated and extended by the demonstration that, just prior to mitosis, there is an abrupt loss of heparan sulfate and heparan sulfate proteoglycan from all cell surface and intracellular pools [502, 503]. These results may be related to those of Preston *et al.* [504] who found a four-fold lower rate of glycosaminoglycan synthesis at mitosis than during interphase and suggested that the mitotic depression af glycosaminoglycan synthesis resulted from the generalized cessation of membrane vesicle transport during mitosis [505-507]. It has also been shown that smooth muscle cells synthesize and secrete sulfated proteoglycans in the G1 and G2 phases of the cell cycle but that the incorporation of ^{35}SO$_4^{2-}$ stops at the beginning of the S phase and during mitosis [508]. Finally, it has been demonstrated that exogenous heparan sulfate proteoglycan is taken up by a hepatoma cell line and blocks cell division late in the G1 phase of the cell cycle [502, 503]. Exogenous heparin also blocks the cell cycle of smooth muscle cells in late G1 [509] (Section V.A., p. 169).

I.   Nuclear Heparan Sulfate

A particularly interesting example of intracellular processing and trafficking is the flow of a portion of the pericellular matrix heparan sulfate proteoglycan into and back out of the nuclei of hepatoma cells [222, 412, 451, 469, 503]. In these cells heparan sulfate proteoglycan is secreted into the pericellular matrix and is then endocytosed and cleaved to free chains, a portion of which are transported to the nuclei via a path that does not pass through the lysosomes. In the continuous presence of ^{35}SO$_4^{2-}$, the nuclear pool reaches a steady state level, but labeled heparan sulfate can be chased out of the nuclei by incubating the cells in media containing unlabeled SO$_4^{2-}$. The amount of heparan sulfate associated with the nuclei increases three-fold as growing cells become confluent, and the level of nuclear heparan sulfate shows an inverse correlation with the rate of cell division. Furthermore, when heparan sulfate proteoglycan from confluent cells is added exogenously to growing cells, cell division is inhibited as the heparan sulfate is taken up and becomes associated with the nucleus. The heparan sulfate proteoglycan from log-phase cells is taken up less effectively and has no effect on cell division. Again, the rate of cell division decreases with increasing levels of nuclear heparan sulfate.

There is a significant precedence for the appearance of glycosaminoglycans in the nuclei of mammalian cells [510-516]. Furthermore, there are many suggestions in the literature that heparan sulfate plays a role in the regulation of cell growth [492, 517-541]. However, fluorescein-labeled heparin is taken up by rat vascular smooth muscle cells and appears around—but not in—the nucleus [542].

IV.   PERTURBATION OF HEPARINOID METABOLISM

A.   Introduction

There are a number of inhibitors of proteoglycan metabolism, some which are relatively specific for glycosaminoglycans and others which may alter both proteoglycan metabolism and other cellular pathways. For example, inhibitors of protein synthesis prevent all protein synthesis, including that of the core proteins. When core protein synthesis is blocked, the acceptors for heparinoid chain synthesis are not available and heparinoid synthesis is also prevented. Somewhat more specific inhibitors are those that prevent GPI anchor synthesis. Such inhibitors specifically block the synthesis of *all* GPI anchored proteins, including the glypicans. Interference with specific reactions, such as the sulfation reactions, is quite specific for all of the sulfated glycosaminoglycans (heparan sulfate, chondroitin sulfate, dermatan sulfate, keratan sulfate), but will also prevent sulfation of tyrosines in proteins [543] or certain N-linked oligosaccharides [408]. Finally, one might inhibit the catabolism of heparan sulfate as well as other catabolites that enter the lysosomes using lysosomotrophic agents, or inhibitors of specific catabolic enzymes, provided such inhibitors can be delivered to the lysosomes in whole cells. The inhibitors that are useful for altering heparan sulfate metabolism in whole cells are described here.

B.   Beta-Xylosides

p-Nitrophenyl-β-xyloside and 4-methyl-umbelliferyl-β-xyloside have been used for a number of years as artificial acceptors for the synthesis of glycosaminoglycans that are formed on a typical GlcA→Gal→Gal→Xyl→Ser linkage region. As has been noted, these glycosaminoglycans include heparin and heparan sulfate, as well as chondroitin sulfate and dermatan sulfate. Addition of one of these artificial acceptors to the culture medium at concentrations up to 1 mM results in the artificial "priming" of the biosynthesis of the linkage region and the subsequent addition of the glycosaminoglycan chains that are attached to the linkage region. The resulting products are high molecular weight chains that have the original β-xylosides at their reducing terminals. For efficient usage of the β-xylosides in intact cells, the aglycon must be sufficiently hydrophobic to allow the β-xyloside to diffuse across the plasma membrane and the Golgi membranes to get to the site of action of the enzymes that carry out the synthesis of the linkage region and the glycosaminoglycan chains. These artificial acceptors will compete with the xylosyl→core protein for the enzymes that add the monosaccharides of the linkage region and the glycosaminoglycan chain. The extent of diversion of the glycosaminoglycan chains from the core proteins depends on the concentration of the xyloside added to the culture medium. The xylosides will also prime glycosaminoglycan synthesis when protein synthesis is inhibited or in mutants that lack the xylosyl transferase. Even when no inhibitor of protein synthesis is present, high concentrations of p-nitrophenyl-β-D-xyloside may intercept the entire glycosaminoglycan chain biosynthetic machinery. This has been observed in chondrocytes, which produce primarily chondroitin sulfate proteoglycan. The xylosyl-core protein is still produced in the presence of the β-xyloside and is secreted to the pericellular matrix without attached chondroitin sulfate chains [544].

It is clear that the addition of one or two of the linkage region galactose residues to the β-xyloside will render these β-xylosides membrane-impermeable. Nevertheless, the products built on β-xylosides (which themselves are membrane impermeable) progress, by unknown mechanisms, through the Golgi stack with full access to all of the biosynthetic enzymes. Furthermore, the glycosaminoglycan chains are not retained in the plasma membrane when they reach the cell surface. They are secreted exclusively into the culture medium. Another interesting observation is that the biosynthetic maturation of the glycosaminoglycan chains is significantly altered in the presence of β-xylosides, yielding glycosaminoglycan structures having demonstrable differences from the structures produced in the absence of the β-xyloside [545, 546].

The two β-xylosides referred to earlier are the ones that have been used routinely in most glycosaminoglycan work. For reasons that are not entirely understood, these xylosides prime the synthesis of chondroitin sulfate much more efficiently than they prime the synthesis of heparan sulfate, even in cells that normally produce much more heparan sulfate than chondroitin sulfate [see, for example, 451]. Furthermore, a unique β-xyloside, naroparcil, specifically primes only dermatan sulfate synthesis *in vivo* [547]. Esko and collaborators have developed a series of β-xylosides with different aglycons that are much more effective in the priming of heparan sulfate synthesis, although they also prime the synthesis of chondroitin sulfate chains [546, 548]. In these studies, a variety of aglycones (5,6,7,8-tetrahydro-2-naphthol-, *cis/trans*-decahydro-2-napthol-, 2-naphthol-, 3-

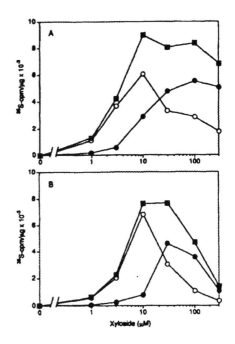

Figure 5.6. Effect of β-xyloside concentration on glycosaminoglycan synthesis. 2-Naphthol-β-D-xyloside (A) or 3-estradiol-β-D-xyloside (B) was administered to CHO cells deficient in xylosyl transferase at the concentrations indicated, and the amounts of chondroitin sulfate (open circles), heparan sulfate (closed circles) and total glycosaminoglycan (closed squares) were measured after a 4- to 5-hr incubation. See text for further explanation. Reprinted from [546] with permission.

estradiol-, 6-ethoxy-2-naphthol-2-(2-naphthoxy)-1-ethyl-, 6-butoxy-2-naphthol-, 4-(2-naphthoxyl)-1-butyl-, 6-hydroxyquinoline-, 5-hydroxyindole-,4-n-butylphenyl-, 9-phenanthrol-, 6-hydroxyindole-, 4-phenylphenol-) have been coupled in β-linkage to xylose and each has been tested for its ability to prime the synthesis of both heparan sulfate and chondroitin sulfate in CHO cells. As shown in Figure 5.6, increasing the concentrations of xylosides gives marked increases in the synthesis of total glycosaminoglycans (closed squares), but the relative proportions of chondroitin sulfate (open circles) and heparan sulfate (closed circles) change as the concentration of the β-xyloside is increased, with chondroitin sulfate production peaking at relatively low β-xyloside concentrations and heparan sulfate synthesis peaking at higher concentrations. It is important to note that xylosides are consumed during glycosaminoglycan synthesis in cultured cells. Therefore, after the initial β-xyloside concentration is optimized for heparan sulfate priming, the β-xyloside concentration will drop throughout the incubation period, and the level of the β-xyloside in the medium may shift from one that is optimum for heparan sulfate synthesis toward one that is optimum for chondroitin sulfate synthesis. Consequently, for maximal priming of heparan sulfate synthesis, timing is very important.

Similar priming profiles are observed for the other β-xylosides, but the concentrations of β-xylosides that give peak productions of chondroitin sulfate and heparan sulfate vary with the different β-xylosides. Also, for a given β-xyloside, the optimum priming concentrations vary with the cell type. As a general rule, the maximum rates of synthesis of glycosaminoglycan chains in the presence of β-xylosides are far greater than the rates of synthesis in the absence of the β-xyloside, indicating that *the rate of core protein synthesis is normally the rate-limiting step in proteoglycan synthesis*; i.e., the capacity of cells to synthesize glycosaminoglycan chains is much greater than that required to fully glycosylate the core proteins that are normally produced by the cells. Thus, the mechanisms that control the synthesis of core proteins regulate the amount of heparan sulfate produced by the cells.

Priming of glycosaminoglycan chain synthesis is quite specific for β-xylosides. β-Glycosides of other pentoses do not prime. Furthermore, β-xylosides with various hydroxyl group substitutions on the xylose residue do not prime well [549]. Attempts to intercept heparinoid chain synthesis by artificial acceptors containing monosaccharides other than xylose have not been particularly successful, although some priming has been obtained with β-galactosides [480]. Analogous approaches have been taken in the synthesis of O-linked oligosaccharides, which can be primed with N-acetyl-α-D-galactosaminides [550], and in the synthesis of N-linked oligosaccharides, which can be primed with hydrophobic peptides containing the Asn-X-Ser/Thr recognition sequence [551].

C.   Chlorate and Selenate

The interactions between heparinoids and heparin-binding proteins are generally dependent on the distribution of the sulfate substituents along the heparinoid chains. Thus, if cells are prevented from producing a fully sulfated heparan sulfate, heparin-binding proteins cannot bind to their surfaces. Chlorate and selenate are agents that inhibit sulfation but do not seem to have other effects on the synthesis of glycosaminoglycans or other sulfated polymers [552-

557]. Concentrations of chlorate as high as 30 mM have been used with no apparent harm to the cells. Chlorate behaves as a sulfate analog that replaces sulfate in the synthesis of 3'-phosphoadenosine-5'-phosphosulfate to yield 3'-phosphoadenosine-5'-phosphochlorate, which either is itself unstable or yields heparinoids with unstable chlorate groups that are spontaneously hydrolyzed [553].[3] Heparinoid chains are still synthesized in the presence of these specific sulfation inhibitors, but they are markedly reduced in sulfate substituents.[4] Consequently, these inhibitors are excellent agents for determining whether cell surface heparan sulfate proteoglycans participate in the responses of cells to growth factors and other agents on cells. For example, it is known that cell surface heparan sulfate proteoglycan must interact with basic fibroblast growth factor in order for the growth factor to bind to its receptor [559-561]. In the presence of chlorate, unsulfated heparan proteoglycan is found on cell surfaces, and basic fibroblast growth factor cannot bind or exert its effects on cells (Chapter 9). Some caveats with respect to the use of chlorate are noted in Chapters 9 and 11.

D.   Inhibitors of GPI Anchor Synthesis

Failure to attach GPI anchors to proteins that are normally processed with these C-terminal structures results in an alteration in their metabolism, due in some cases to the retention of the proteins within the endoplasmic reticulum and/or Golgi apparatus, and eventual catabolism. Such proteins do not appear on the cell surface [reviewed in 539]. Failure to attach the GPI anchor occurs naturally in paroxysmal nocturnal hemoglobinuria patients [562, and references therein]. A number of mutants in the biosynthesis of GPI anchors are known [426, 563, 564], but the heparan sulfate metabolism of such mutants has not been studied. There are also inhibitors that can be added to cells to block GPI anchor synthesis and thus prevent cell surface expression of such proteins, including glypicans. These include mannosamine [214, 565-567], 2-fluoro-2-deoxy-D-glucose [568, 569], and phenylmethylsulfonyl fluoride [570, 571]. Such inhibitors lack specificity, as they interfere with synthesis of O- and N-linked oligosaccharides as well, and they block the normal metabolism of *all* GPI anchored proteins. However, if the toxic effects of these inhibitors are not severe in short-term experiments, such inhibitors would allow the selective inhibition of glypican migration to the surface of cells that produce both

---

[3] This would be analogous to the arsenate/phosphate competition in the synthesis of ATP.

[4] Interestingly, the heparan sulfate chains of heparan sulfate proteoglycan have been shown to increase in length when sulfation is inhibited by chlorate [558].

glypican and proteoglycans with transmembrane core proteins. Such experiments have not been reported. It has been shown that 2-fluoro-2-deoxy-D-glucose is incorporated by chondrocytes into keratan sulfate in place of D-galactose and into chondroitin sulfate. In chondroitin sulfate synthesis, the 2-fluoro analog of D-glucuronic acid is found as the UDP-activated derivative and is incorporated into chondroitin sulfate in place of GlcA [572]. If this occurs in other cell types or for other glycosaminoglycans, then one might find that cells will synthesize a modified heparan sulfate containing fluorodeoxy-D-glucuronic acid in place of GlcA. Such studies have not been reported.

E.   Transition State Analogs

Glucuronic acid is utilized in metabolism as a building block for chondroitin sulfate/dermatan sulfate and the heparinoids, and also in detoxification reactions in which aromatic substrates are "glucuronosylated" and excreted. In the catabolic hydrolysis of GlcA-containing metabolites, a glucuronopyranose residue passes through a half-chair intermediate during glycoside bond hydrolysis. A classical inhibitor of these cleavage reactions is saccharolactone, a transition state analog for the half-chair form of GlcA, which binds tightly to the glucuronidase and inhibits the reaction. In biosynthetic reactions, glucuronic acid is transferred from UDPGlcA to the appropriate acceptors. Transition state analogs for UDPGlcA have been obtained [573]. These have been found to work quite well as inhibitors of the glucuronosyl transferases involved in detoxification, but have not been studied as inhibitors of glycosaminoglycan biosynthesis.

F.   Inhibition of Catabolism

Endosomes and lysosomes are acidified, membrane-enclosed compartments in cells that contain enzymes that have acidic pH optima. Thus, it has been possible to prevent, or slow, the action of these enzymes by adding agents, such as NH_4Cl, chloroquine, or tilorone which enter these organelles and raise the pH above the optima for most of the enzymes [404, 469, 470, 574]. Also, lysosomal catabolism can be specifically prevented by lowering cell incubation temperatures. It is well established that endosomes cannot fuse with lysosomes at temperatures below 20 °C [575]. This blocks heparan sulfate catabolism [469]. One additional agent that inhibits heparan sulfate proteoglycan catabolism is leupeptin, which blocks core protein cleavage immediately following endocytosis [576].

G. Mutants in Proteoglycan Synthesis

As in other fields of biology, the use of mutants to study heparan sulfate proteoglycan metabolism and biological activities has been extremely useful [70, 577]. This approach has been explored by Esko and co-workers who mutagenized CHO cells and selected, by a replicate plating technique, mutants showing a diminished ability to produce $^{35}SO_4$-labeled colonies [578, 579]. Wild type CHO cells produce both chondroitin sulfate and heparan sulfate proteoglycans. Table 5.1 lists a variety of mutants in proteoglycan metabolism that have been characterized. These include several mutants that lack the glycosyl transferases required to synthesize the linkage region [436, 580], and a mutant that is unable to transport SO_4^{2-} across the plasma membrane [397], as well as mutants in the heparinoid chain polymerization reactions [440], in the N-sulfotransferase [79, 452, 453], and in 2-O-sulfotransferase [114]. Mutants lacking the xylosyl and galactosyl transferases required for linkage region synthesis produce neither chondroitin sulfate nor heparan sulfate, demonstrating that the same enzymes are utilized to synthesize the linkage regions of both glycosaminoglycans. In the presence of β-xylosides the xylosyl transferase mutant can synthesize both chondroitin sulfate and heparan sulfate. The mutant that lacks the SO_4^{2-} transport system [581] has been discussed in Section II.A.3., p. 142. This mutant cannot utilize SO_4^{2-} from the culture medium for synthesis of heparan sulfate proteoglycan but can convert cysteine to SO_4^{2-} for use in the sulfation reactions. The N-sulfotransferase mutants in both CHO cells [from references 70, 79, 452, 453] and COS cells [582] produce heparan sulfate proteoglycans with reduced sulfation. Interestingly, the mutant that lacks N-acetylglucosaminyl transferase also lacks the glucuronosyl transferase. Since the mutation is a point mutation, it appears that the polymerization of GlcA and GlcNAc to form the initial $[GlcA{\rightarrow}GlcNAc]_n$ chain is catalyzed by a single enzyme. Also, these mutants continue to produce chondroitin sulfate, indicating that a separate set of activities are required for chondroitin sulfate polymerization. In a similar manner, it has been found that the enzyme that catalyzes the N-sulfation of GlcN residues also carries the de-N-acetylase activity. The mutant that lacks the 2-O-sulfotransferase produces a heparan sulfate that lacks both IdoA-2-SO_4 and GlcA-2-SO_4, suggesting that both uronic acid residues are 2-O-sulfated by the same enzyme [114]. Rosenberg and co-workers have described a particularly interesting group of mouse LTA cell mutants that produce normal levels of an altered heparan sulfate with markedly lowered ability to bind to antithrombin III [583]. Finally, a now-classic series of Thy-1 mutants include those with defects in GPI anchor synthesis [reviewed in 426], but these have not been examined for their effects on proteoglycan metabolism.

Table 5.1. Mutants in Glycosaminoglycan Biosynthesis

Cell Type	Mutant	Defective Activity	Phenotype	References
CHO	pgsA	xylosyl transferase	proteoglycan-deficient	[436, 580]
CHO	pgsB	galactosyl transferase	proteoglycan-deficient	[436, 580]
CHO	pgsC	sulfate transport	normal proteoglycans	[from reference 70, 397, 581]
CHO	pgsD	glucuronosyl/N-acetyl-glucosaminyl transferases	heparan sulfate-deficient	[from reference 70, 440]
CHO	pgsE-206	N-sulfotransferase	undersulfated heparan sulfate	[from reference 70, 79, 452, 453]
COS		N-sulfotransferase	undersulfated heparan sulfate	[582]
CHO	ldlD	UDPGlcNAc 4-epimerase	chondroitin sulfate-deficient	[from reference 70]
CHO	pgsF--17	2-O-sulfotransferase	undersulfated heparan sulfate	[114]
LTA	VI-7	Unknown regulatory protein	reduced antithrombin-binding	[385]

Such mutants have been useful for confirming the pathways of heparinoid biosynthesis, for establishing the points of branching in chondroitin sulfate and heparinoid biosynthesis, for detecting regulatory genes that determine the structures of the glycosaminoglycan chains, for isolating genes corresponding to the mutations, and for deducing the physiological roles of the heparinoids. Some of these uses of mutants are discussed in more detail elsewhere, but several examples of the value of such mutants are cited here.

Mutants that cannot produce heparan sulfate proteoglycan fail to bind to cell adhesion molecules such as type V collagen [584] and fibronectin [585]. They also fail to bind ligands such as basic fibroblast growth factor [114, 194, 559, 586] or thrombospondin [538]. They do not bind certain viruses, such as Herpes simplex virus or HIV [587-589]. Perhaps the most interesting result is that CHO cells require heparan sulfate proteoglycan to grow as tumors in athymic mice; those mutants that cannot produce heparan sulfate proteoglycan cannot produce tumors, a result not found in mutants lacking chondroitin sulfate proteoglycan alone [590].

V.   METABOLISM OF EXOGENOUS HEPARINOIDS

A.   Metabolism in Cell Culture

Although exogenous heparin has a broad range of effects on a variety of cell types, the metabolism of exogenous heparin by different cells has received limited attention. Vascular smooth muscle cells from the rat aorta [542, 591] bind [^{3}H]heparin at 4 °C in a saturable manner with a K_d of 10^{-9} $\underline{M}$. These cells have approximately 100,000 binding sites per cell. When heparin-saturated cells are warmed to 37°C, 80% of the bound heparin is shed into the culture medium and the remainder is internalized by the cells in a biphasic manner. Fifty % of the internalized heparin is taken up with a $t_{1/2}$ of 15-20 min, whereas the remainder is internalized with a $t_{1/2}$ of 1-2 days. Fluorescein-labeled heparin is also internalized by these cells and concentrated in the perinuclear region. The internalized heparin is catabolized, but details of the site, kinetics, or extent of catabolism have not been described. In our discussion of heparin metabolism in mast cells, it was noted that heparin may be released from heparin/protease complexes after degranulation of the cells. This released heparin would be expected to follow a similar endocytosis/catabolism route (by some as yet unidentified cells in the mast cell environment). Thus, metabolism of endogenous heparin is probably a true physiological event, at least for some cell types.

Heparin inhibits the growth of vascular smooth muscle cells. A comparison of heparin-sensitive and heparin-resistant cells has shown that both cell types bind and internalize heparin and degrade it in lysosomes [592, 593]. However, during the incubation with heparin the sensitive cells show an increased binding and internalization of heparin, whereas the resistant cells exhibit reduced binding.

B. Metabolism *in Vivo*

1. Introduction

Study of the *in vivo* metabolism of heparinoids is more difficult than the study of metabolism in cell cultures for several reasons. First, one may use isotopically labeled heparinoids in some animals but not in humans and, even when such experiments are done in animals, the amount of labeled material required is quite large. Second, without labeled heparinoids, it is difficult to measure actual amounts and structures of the metabolized heparinoids. Consequently, most studies have been limited to the examination of the anticoagulant activity in the circulation following the administration of exogenous heparins or low molecular weight heparins to animals or humans. Such studies are driven by the fact that these heparinoids are used clinically as drugs which, in appropriate doses, are effective anticoagulants. However, when heparins are given in excess, serious bleeding occurs. Thus, a primary emphasis in studies of the levels of heparin and its metabolism *in vivo* has been on the changes in anticoagulant activity at intervals following heparin administration. Reviews of this literature, which is quite extensive, may be consulted for further reference to the primary literature [594-597]. As pointed out in these reviews, observing changes in the anticoagulant activity in the circulation following administration of exogenous heparin is not the same as following the actual metabolic changes in amounts and structures of heparins as catabolism proceeds. Chapters 7 and 8 show that the anticoagulant activity depends on a variety of factors, including the molecular weight of the heparin and the effects of heparin on antithrombin and other heparin-binding proteins in the vasculature.

2. Pharmacokinetics

The anticoagulant activity of heparin in the circulation has been studied following intravenous, subcutaneous, intramuscular, intraperitoneal, and oral administration. In clinical practice, heparins are given either intravenously or subcutaneously. These drugs can be given as a bolus or by continous infusion. As might be anticipated, uptake, drug effect, and metabolism are very dependent on the mode of delivery.

Intravenous heparin becomes immediately available. When given as a bolus, its anticoagulant activity appears rapidly and then is lost rapidly by two metabolic routes. First, it is taken up by the endothelium and catabolized [598]. Second, it is excreted by the kidneys [599, 600]. The endothelial uptake of heparin occurs most rapidly, but is saturable. Thus, low doses of heparin saturate the endothelium whereas higher doses first saturate the endothelium and then must be catabolized and excreted via the slower process that takes place in the kidneys. As a consequence, the half-life of heparin in the circulation depends on the dose of heparin administered and ranges from 30 min at a dose of 25 units/kg to 150 min with a dose of 400 units/kg. Furthermore, the higher molecular weight portions of the heparin are cleared most rapidly by the endothelium. In any event, the disappearance of a bolus dose of heparin is rapid. Consequently, when immediate and continuous anticoagulant effects are required, intravenous heparin is given by continuous infusion. Low molecular weight heparin may also be used. One of the main advantages of low molecular weight heparin over unfractionated heparin is that it does not bind to the endothelial cells or to some of the other heparin-binding proteins in the circulation. Consequently, it exhibits a half-life of approximately two- to four-fold longer than heparin [for example, see 601].

When heparins are given subcutaneously, the anticoagulant activity in the blood rises much more slowly, appearing only after 1-2 hr, then peaks and falls. Because the bioavailability is reduced, higher bolus doses can be given intermittantly, and these have an extended anticoagulant effect.

3.   Oral administration of heparinoids

Although there have been no heparin products that have been developed for oral administration, there have been a number of literature reports of oral uptake of heparin [602-609]. In fact, it has been reported that orally administered heparin exhibits a dose-dependent antithrombotic activity [609]. In a particularly interesting report, Larsen et al. [610] showed that when a single dose of [$^{35}SO_4$]heparin or [^3H]heparin is administered orally to rats, a series of labeled heparin oligosaccharides are observed in the blood for at least 12 hr. Oligosaccharides with anti-Xa activity (Chapter 8) are observed early and are degraded to smaller fractions, including disaccharides, which are excreted in the urine. Thus, heparin is progressively broken down to smaller and smaller fragments before excretion. The catabolism of internalized heparin takes place in the liver and kidney, and in the vascular endothelium [611]. Interestingly, when heparin fragments formed by bacterial heparin lyase treatment (Chapter 3) are injected intravenously, the fragments are excreted unchanged; i.e., they are resistant to the catabolic effects that are observed for the *hydrolytic* fragments that are formed from oral heparin [612]. This is consistent with the known pathways for catabolism of heparin oligosaccharides in which the fragments are degraded sequentially from the nonreducing terminal (Section II.G., p. 157). The heparin lyase products have a nonreducing terminal $\Delta^{4,5}$-uronic acid that is not recognized by enzymes in animal tissues.

VI. DYNAMIC ASPECTS OF HSPG METABOLISM—A KEY TO ITS FUNCTION?

A. Introduction

For many years, scientists interested in the biology of carbohydrates in general, and glycosaminoglycans in particular, have wrestled with the question, "What is the physiological role of these structures?" The answers have continued to be elusive, but the speculations have improved. We do not have to answer these questions for the entire carbohydrate community here, but we should try to deal with this question as it pertains to the heparinoids. Our knowledge of the structure and metabolism of these polymers, as incomplete as it is, offers both suggestions for their physiological role and experimental approaches for testing these hypotheses. For example, the ubiquity of the heparinoids, especially the heparan sulfate proteoglycans, in the animal kingdom compel us to believe that these structures are critical for the well-being of the animal organism. We are struck by the enormous cellular machinery devoted to the biosynthesis of the heparinoids, the complex trafficking of these structures to the cell surface, and the fact that they have no sooner taken their place of the cell surface than they are internalized and catabolized. Surely, the dedication of such extensive cellular machinery to heparan sulfate proteoglycan metabolism must indicate the essentiality of the heparinoids. However, we are restrained by results that have shown that mutations in the enzymes of heparan sulfate proteoglycan metabolism, or specific perturbations in heparan sulfate metabolism, result in little change in the well-being of cells in culture (see Section IV.G., p. 167). Perhaps we should be considering tissues and whole organisms in order to get a true picture of the role of heparan sulfate proteoglycan. Indeed, Murch et al. [8] have shown that genetic deficiency in heparan sulfate synthesis can be lethal. In any event, we must collect our hypotheses concerning the role of the heparinoids so that we can test them. As detailed earlier, the most unusual aspects of the heparan sulfate proteoglycans that may give some clue to their role are (a) the multiplicity of core proteins and (b) their dynamic metabolism. These features of heparan sulfate proteoglycans are a starting point for the formation of our hypotheses.

B. Variation in Core Protein Expression

A complicating factor in describing heparan sulfate proteoglycan metabolism is the observation that a given cell line may produce several different core proteins. Although our present knowledge is limited, production of multiple species of heparan sulfate proteoglycans by cells appears to be quite common. Certainly this possibility must be asssessed in the examination of any cell line that has not been previously described. A few of the best studied examples will suffice to illustrate this point.

- Human fibroblasts express four different heparan sulfate proteoglycans (in addition to two different chondroitin sulfate proteoglycans) [200, 406].

- A mouse mammary epithelial cell line produces two core proteins: syndecan and a 38-kDa core protein that has not yet been characterized [191].

- A rat osteosarcoma cell line produces two heparan sulfate proteoglycans, one that has a GPI anchor and another that has a transmembrane core protein [225].

- Similar results have been found in ovarian granulosa cells [613] and vascular endothelial cells [193].

- Rat aortic vascular smooth cells synthesize both syndecan-1 (syndecan) and syndecan-2 (fibroglycan), the relative proportions of which are altered by various growth factors [614].

In addition to the expression of multiple species of heparan sulfate proteoglycans, it is clear that proteoglycan expression is developmentally regulated [144, 158-167].

What little we know at present suggests that the heparan sulfate glycosaminoglycan structure is not controlled by the core protein [157, 206]; i.e., presumably a given cell type will attach the same mixture of heparan sulfate chains to one core protein that it will attach to all other acceptor core proteins that the cell produces. This needs to be examined more carefully to be sure that the conclusion holds for a variety of cells. However, once the proteoglycans reach the cell surface, there may be some influence of the core proteins on the further metabolism of the heparan sulfate proteoglycan.

C.   Dynamics of the Cell Surface Pool of HSPG

The cell surface heparan sulfate proteoglycan is a dynamic pool that turns over constantly, with half-lives in various cell lines that range from one-third to one-eighth of the time required for the cell cycle. The structures of the heparan sulfate chains of these proteoglycans are readily altered in response to a wide range of agents added to cultured cells (Section VI.D., below). Thus, it has become apparent that the structures of the heparan sulfate chains presented on the surfaces of cells can change extremely rapidly. In those cells whose metabolism has been examined in detail, it appears that there is a steady state level of heparan sulfate proteoglycan in the pericellular matrix that can be observed when cells are incubated continuously with labeled precursors. This steady state occurs when the rate of heparan sulfate proteoglycan loss from the cell surface by endocytosis and/or shedding is equal to its rate of replacement. The mechanisms that control the rate of synthesis and secretion of the heparan sulfate proteoglycan are better understood than those that control its disappearance. Determining the fate of the extracellular proteoglycan is complicated by the fact that there are multiple core proteins in different cells and that these core proteins appear to exert some control over the trafficking of the extracellular heparan sulfate proteoglycan. A proteoglycan arriving at the cell surface may undergo further metabolism before it is released from the surface and/or internalized by the cell. Such metabolism might include the proteolytic cleavage of the core protein or the cleavage of the GPI anchor by phospholipases C or D or the internalization of transmembrane core proteins in response to a signal from the cytoplasmic domains of the core protein. Furthermore, it appears that the metabolic flow of cell surface heparan sulfate proteoglycan shows significant differences from one cell type to another. These are complex issues that will require more study.

D.   Dynamic Changes in HSPG Structures

The fact that there is a steady state level of heparan sulfate proteoglycan on the cell surface does not mean that there is a constant range of structures of the heparan sulfate chains in this pool. Changes in the structures of the heparan sulfate chains in response to changes in the physiological environment of the cells may be a critical aspect of the behavior of heparan sulfate proteoglycan in the fulfillment of its normal physiological role. Because of the rapid turnover of cell surface proteoglycan, a rapid change in the structures of the heparan sulfate chains could take place in response to various extracellular signals that are transmitted to the sites of heparan sulfate proteoglycan synthesis. Furthermore, the protective effects of heparan sulfate-bound proteins might play some role in control of the structures of heparan sulfate oligosaccharides after they are internalized (discussed below). Not much attention has been devoted to the structural characterization of the heparan sulfate chains under diverse culture conditions, but there are reports showing that these structures undergo changes under a variety of conditions. These changes are observed as differences in the oligosaccharide composition of the heparan sulfates. At the outset, it may be noted that the compositions of heparan sulfate from the culture medium, the cell surface, the intracellular, and the nuclear pools in a rat hepatoma cell line are clearly different [404, 412]. Furthermore, the *amount* of heparan sulfate as well as its composition in these pools changes when growing cells reach confluence [412]. Similar results have been reported for human neuroblastoma cells [615]. In fact, the suggestion from these results is that as the relative amounts of heparan sulfate in the various cellular pools are altered, the structures of the heparan sulfate chains are also altered, but this suggestion will require further study. In any event, it is clear that the amounts of heparan sulfate in different cellular pools can be perturbed by a variety of agents, including β-xylosides, glucose levels, catechin, TGF-β, inositol, 12-O-tetradecanoylphorbol-13-acetate, sodium butyrate, dibutyryl cAMP, insulin, retinoic acid, estradiol, basic FGF, and hydrocortisone [451, 545, 546, 616-619].

How are these structures controlled? During their biosynthesis, a mixture of heparan sulfate chains is added to core proteins in the secretory organelles. The structural features of the heparan sulfate chains that are necessary for binding to proteins are controlled by a group of maturation enzymes in the Golgi apparatus. The simplest and most obvious kinds of controls that may be exerted during the biosynthesis of heparan sulfate are those resulting from varying the relative levels of the maturation enzymes and the metabolic precursors. We have noted in Chapter 2 that there are branches in the biosynthetic pathway in which pairs of maturation enzymes may compete for the same heparinoid sequences, and that the final structures that are produced may depend on the relative proportions of such sequences that take one branch of the pathway or the other. In a similar manner, the *regulation* of enzyme activity can play a role in directing the synthesis down one branch of a pathway rather than another. We do not know anything about the mechanisms that control these various reactions *in vivo*. Aspects as simple as the availability of 3'-phosphoadenosine-5'-phosphosulfate or the UDP sugars, the amounts of metal ions, or the pH of the Golgi lumen may play some role. There may also be effects of starvation on heparan sulfate structure like those that have been reported for N-linked oligosaccharides [620, 621]. The concentration of glucose in the culture medium has, in fact, been found to alter heparan sulfate proteoglycan metabolism in a hepatoma cell line [451]. Also, the maturing heparinoid chains may exert some feedback regulation of enzyme activity, but if this occurs, the experimental dissection of such mechanisms is almost too complex to contemplate!

It has been suggested that the control of heparan sulfate structures does not end with its biosynthesis. These derive from the finding that different structures are found in the culture medium, cell surface, and intracellular pools. The heparan sulfate in all of these pools is derived from the cell surface pool of heparan sulfate proteoglycan.[5] Thus, one might conclude that the metabolism of the heparan sulfate proteoglycan *after it is secreted from the cells* results in fractionation of the heparan sulfate chains! One way in which these structures might be controlled after the secretion of the heparan sulfate proteoglycan is suggested in Section VI.E., below.

E.   Intracellular Trafficking Dynamics

1.   Introduction

The number of examples in which the intracellular trafficking of endocytosed heparan sulfate chains has been characterized is too small to allow us to generalize, but this aspect of metabolism may turn out to be the most variable, yet the most important, feature of heparinoid metabolism. The results that are available suggest a scenario in which one or more heparin-binding proteins are bound to specific sequences in the heparan sulfate chains on the cell surface and internalized in these complexes with the heparan sulfate chains. The complexes are attacked by proteolytic enzymes and endo-β-glucuronidases that remove the core protein, leaving enzyme-resistant complexes composed of the heparan sulfate fragments and heparin-binding proteins (or protein fragments). These complexes are targeted to various cellular destinations. They may be transported to the lysosomes for catabolism or directly recycled to the plasma membrane. They may be targeted to the Golgi and reprocessed (which may

---

[5] Note should be taken of the fact that the intracellular pool size of the newly synthesized heparan sulfate chains is quite small compared to that of the chains that have re-entered the cells by endocytosis. Also, the transit time of newly synthesized heparan sulfate proteoglycan in the secretory pathway is short compared to the residence time of endocytosed heparan sulfate. Therefore, the labeled heparan sulfate in the intracellular pool is composed primarily of heparan sulfate chains that have been endocytosed by the cells, especially when its presence is measured by $^{35}SO_4$-incorporation, which occurs just prior to heparan sulfate proteoglycan secretion.

involve some alteration of the heparan sulfate chain structure) and re-secreted.[6] They may be transported to the nucleus (e.g., in bFGF/heparan sulfate complexes, see Chapter 9) where the complex dissociates and one or both members of such complexes exert regulatory controls. The actual targeting of the cell surface proteoglycan may be controlled initially by a specific interaction between the core protein and a receptor (which, for the transmembrane core proteins, could be on the cytoplasmic side of the plasma membrane). However, the core protein is rapidly removed from these chains immediately after internalization. Therefore, neither the core protein nor its receptor would be expected to have any further influence on the routing of the heparan sulfate chains. Consequently, further routing would have to depend on the structural features of each particular heparin-binding protein (which might carry a targeting sequence or present a heparan sulfate-induced conformation that is necessary for the targeting) or the ligand receptor. This process is complex enough even if a single heparin-binding protein is internalized, but the competition of multiple heparin-binding proteins for the available cell surface proteoglycans and the simultaneous internalization of several such complexes, could increase the complexity considerably.

Although there is a great deal of speculation in these few sentences, there is also a significant body of evidence to support these concepts, some of which are reiterated below and expanded in other chapters.

2. Protection of protein-heparin complexes from enzymes

In heparinoid/protein complexes, the protein is protected from proteases and the heparinoid is protected from heparin-cleaving enzymes. For example, there have been reports showing that basic fibroblast growth factor, when complexed with heparin, is protected from the action of trypsin and other proteases [622]. Similarly, when heparin is complexed with antithrombin, the segment of heparin that contains the specific antithrombin-binding sequence is specifically protected from heparin lyase I [623]. In fact, this was one of the methods that was used to isolate the antithrombin-binding heparin oligosaccharide. Thus, when one considers the large number of heparin-binding proteins that are found in the extracellular environment of cells, the opportunities for the formation of complexes between cell surface heparan sulfate and proteins, each of which may recognize a different sequence in the

---

[6] Although the appearance of free heparan sulfate chains in the culture medium pool of cultured cells is usually attributed to direct shedding from the cell surface, re-secretion could be an alternative mechanism for the release of these free chains from the cell.

heparan sulfate, are many. Consequently, there is the possibility that each of these proteins protects a specific heparinoid sequence during the pre-lysosomal metabolism of the internalized heparan sulfate. Alternatively, some heparin-binding proteins may recognize similar heparinoid sequences, and these proteins may compete for binding to these sequences.

3.   Heparan sulfate can also bind to heparin-binding proteins

A number of examples show that many of the heparin-binding proteins can recognize sequences in both heparin and heparan sulfate. In each case, the heparin-binding ability of these proteins was recognized first, but heparan sulfate also was found to contain the necessary sequences for binding to the same proteins. Examples include basic fibroblast growth factor [624-628], antithrombin [629-632], thrombospondin [633], and lipoprotein lipase [634, 635]. There are other examples, and it may be anticipated that more will appear when such comparisons are made.

4.   The degree of interaction depends on amount of heparan sulfate on the cell surface and on the amount of heparinoid-binding protein in surroundings

Variables in the binding of various proteins in the extracellular millieu to cell surface heparan sulfate sequences are the amount and structure of heparan sulfate on the surface of the cell and the number and amounts of heparin-binding proteins that can come into contact with the cell. Since cells usually have a greater capacity to synthesize heparan sulfate chains than is required to substitute the core protein(s) that the cell produces (Section IV.B., p. 162), the *amount* of heparan sulfate on the cell surface is controlled by the expression of the core proteins. In addition, in cell culture, the total amount of the heparan sulfate proteoglycan exposed to the culture medium depends on the number of cells in the culture, a number that is increases during the time course of many experiments. Of course, we know almost nothing about how the structures of the heparan sulfate chains are controlled. In cell culture, where serum is almost always a component of the culture medium, there is limited control over the number and amounts of heparin-binding proteins surrounding the cells, and we do not know what heparan sulfate structures are required for binding to most of the heparin-binding proteins. However, it is likely that there are usually sufficient amounts of heparin-binding proteins in the serum to occupy a significant proportion of the cell surface heparan sulfate proteoglycan.

5.   Are heparan sulfate/protein complexes internalized as such?

There is a body of literature that indicates that heparin-binding proteins are internalized as heparan sulfate/protein complexes. A number of examples are cited elsewhere, but a short list includes thrombospondin [633], basic fibroblast growth factor [636], and lipoprotein lipase [634, 635].

6.   Heparan sulfate oligosaccharides and heparin-binding proteins persist in the intracellular pool

The discussion of processing of internalized heparan sulfate proteoglycan in Section III.E., p. 155, describes the findings that the heparan sulfate chains spend up to 3 hr in pre-lysosomal compartments. Furthermore, it has been shown that internalized basic fibroblast growth factor is not degraded until approximately 2 hr after internalization by endothelial cells, apparently because it is complexed with heparan sulfate [637]. The compartments in which the heparan sulfate and heparin-binding proteins reside prior to their delivery to the lysosomes have not been well characterized. However, it is easy to see how protection of specific segments of the heparan sulfate chains from enzymatic breakdown by such complexations might result in the appearance of a subset of heparan sulfate chains in the intracellular pool that has a composition that is different from that of the heparan sulfate chains on the cell surface.

7.   The role of core protein

Limited data available suggest that, steric constraints aside, the interactions of heparin-binding protein with heparinoids on the surface of *any given cell type* would be the same regardless of the core protein that carries the chain. Thus, for heparin-binding proteins that normally interact with mast cell heparin or with cell surface heparan sulfate, the *amounts* but not the *types* of core proteins that are synthesized by a given cell type control the extent of interaction between the heparin-binding protein and the cell.[7] A corollary to this statement is that cells may control their interactions with heparin-binding proteins by the types of heparinoid chain structures that they synthesize, no matter which of the core proteins are present to prime heparinoid synthesis.

---

[7] This has not been tested directly, but the present information suggests that this is true.

The internalized heparan sulfate proteoglycan appears within minutes as free heparan sulfate chains, recognized as such by the fact that the molecular weights of $^{35}SO_4$-labeled heparan sulfate chains are not reduced when treated with alkaline borohydride, a reaction that releases glycosaminoglycan chains from their core proteins (Section II.C., p. 146). It follows then that the role of the core protein in the processing of heparan sulfate chains must be complete immediately after the endocytosis event.[8] Thus, the metabolic role of the core protein may be to recognize receptors for endocytosis or to serve as a site of cleavage by cell surface proteases or phospholipases (for the glypicans). Consequently, the core protein's main role during the endocytosis might to direct the heparan sulfate chains into unique sets of endosomes. Also, a role for the core proteins in proteoglycan trafficking through the secretory pathway cannot be ruled out.

8.   Basic fibroblast growth factor and heparan sulfate are found in the nucleus

Basic fibroblast growth factor [638-640], fibroblast growth factor receptor [641, 642], and heparan sulfate [469] have been reported in the nuclei of cells. Furthermore, it is clear that heparan sulfate and FGF enter the cell together with an FGF receptor in a ternary (or quaternary) complex (Chapter 9). However, the trafficking of the heparan sulfate and basic fibroblast growth factor together, with or without the FGF receptor, has not been examined. Similarly, the trafficking of other heparin-binding proteins with heparan sulfate has received little attention.

9.   Summary

We have just enough knowledge to appreciate that a number of facts support each of the proposed events in heparan sulfate trafficking following its endocytosis, but not enough knowledge to be sure that these facts fit together in the manner suggested. One reason for the gaps in our knowledge is that those who study proteoglycan metabolism usually do not study heparin-binding proteins, and those who study heparin-binding proteins do not study the metabolism of proteoglycans. Furthermore, only a handful of scientists in either

---

[8] It is possible that the internalized heparan sulfate and cleaved core protein remain in a vesicle which also contains the receptor itself. Such a vesicle may retain the targeting sequences of the receptor or the core protein.

group (or in any group, for that matter!) study heparan sulfate structures. Nonetheless, heparan sulfate proteoglycans are uniquely poised on the cell surface to interact with heparin-binding proteins, and everything we know about the metabolic flow of endocytosed proteoglycans is in accord with the suggestion that both the protein and the heparan sulfate in protein/heparan sulfate complexes are trafficked together. A corollary to this suggestion is that both heparan sulfate chains and heparin-binding proteins are present in the same pre-lysosomal compartments, and an attempt to understand the metabolism of either the heparan sulfate chains or the heparin-binding proteins must take into account both components of such complexes. The problem is further complicated by the possibility that several heparin-binding proteins may be internalized at the same time, perhaps on different segments of the same heparan sulfate chain (or even competing for the same segments). With these complexities in mind, one can see that any attempt to understand the control of the metabolic flow of the heparan sulfate chains would be greatly facilitated (a) if cells that produce a single core protein were chosen, (b) if the number of heparin-binding proteins in each study were limited to one or two, and (c) if the trafficking and catabolism of both heparan sulfate and a single heparin-binding protein were followed together.

ADDITIONAL REFERENCES

Bame, K. J., and Robson, K. (1997) Heparanases produce distinct populations of heparan sulfate glycosaminoglycans in Chinese hamster ovary cells, *J. Biol. Chem.* **272**, 2245-2251.

Bossuyt, X., and Blanckaert, N. (1997) Carrier-mediated transport of uridine diphosphoglucuronic acid across the endoplasmic reticulum membrane is a prerequisite for UDP-glucuronosyltransferase activity in rat liver, *Biochem. J.* **323**, 645-648.

Brinkmann, T., Weilke, C., and Kleesiek, K. (1997) Recognition of acceptor proteins by UDP-D-xylose proteoglycan core protein β-D-xylosyltransferase, *J. Biol. Chem.* **272**, 11171-11175.

Carey, D. J., Bendt, K. M., and Stahl, R. C. (1996) The cytoplasmic domain of syndecan-1 is required for cytoskeleton association but not detergent insolubility, *J. Biol. Chem.* **271**, 15253-15260.

Dolan, M., Horchar, T., Rigatti, B., and Hassell, J. R. (1997) Identification of sites in domain I of perlecan that regulate heparan sulfate synthesis, *J. Biol. Chem.* **272**, 4316-4322.

Klaassen, C. D., and Boles, J. W. (1997) The importance of 3'-phosphoadenosine 5'-phosphosulfate (PAPS) in the regulation of sulfation, *FASEB J.* **11**, 404-418.

Leppä, S., Vleminckx, K., Van Roy, F., and Jalkanen, M. (1996) Syndecan-1 expression in mammary epithelial tumor cells is E-cadherin-dependent, *J. Cell Sci.* **109**, 1393-1403.

Pollard, V. W., Michael, W. M., Nakielny, S., Siomi, M. C., Wang, F., and Dreyfuss, G. (1996) A novel receptor-mediated nuclear protein import pathway, *Cell* **86**, 985-994.

Romarís, M., Bassols, A., and David, G. (1995) Effect of transforming growth factor-β1 and basic fibroblast growth factor on the expression of cell surface proteoglycans in human lung fibroblasts, *Biochem. J.* **310**, 73-81.

Uhlin-Hansen, L., Kusche-Gullberg, M., Berg, E., Eriksson, I., and Kjellén, L. (1997) Mouse mastocytoma cells synthesize undersulfated heparin and chondroitin sulfate in the presence of brefeldin A, *J. Biol. Chem.* **272**, 3200-3206.

Chapter 6. Heparinoid/Protein Interactions

I. ASPECTS OF HEPARINOID/PROTEIN INTERACTIONS

A. Introduction

We have discussed the structures of heparin and heparan sulfate, and have alluded to their complexes with heparin-binding proteins. We turn now to a discussion of the interactions of these structures with the heparin-binding proteins. Both heparins and proteins are large molecules. The interactions between a protein and heparin or heparan sulfate take place over rather restricted domains on the surfaces of the two interacting polymers. In the discourse below, the nature of these interactions and the experimental approaches that are used to address them are described.

It is easy to observe that a protein will bind to a heparinoid. It is much more difficult to characterize the molecular contacts at the binding domains of the protein and the heparinoid. In fact, the molecular interfaces between heparinoids and their binding proteins have been described for only a few such interactions [15, 643]. We will discuss these cases in more detail in later chapters, but it may be useful here to summarize the results for several of the most-studied cases so that the nature of the binding between protein and heparinoid may be appreciated in the later discussion.

B. Antithrombin

Antithrombin is a serine protease inhibitor that inactivates several members of the blood coagulation cascade, the most important of which are thrombin and factor Xa. Antithrombin is not an effective inhibitor of these proteases unless it first binds to a heparinoid; i.e., the heparinoid "activates" antithrombin. The smallest heparin-derived oligosaccharide that will activate antithrombin is a pentasaccharide with a unique structure (Chapter 7). This pentasaccharide apparently binds, largely via electrostatic bonding, between two α-helices, both of which are enriched in arginine and lysine residues. Some of the anionic groups on the pentasaccharide bind to one of the two α-helices, while other anionic groups bind to the second. As a result of the binding, antithrombin undergoes a change in conformation that converts it to a suicide substrate for thrombin or factor Xa.

The pentasaccharide/antithrombin complex is very reactive with its factor Xa target, but is not reactive with its other target, thrombin. In order for antithrombin to inactivate thrombin, a longer segment of heparin is required. This segment must contain the same pentasaccharide sequence, but must extend beyond the surface of the antithrombin so that it can bind to thrombin as well. Thus, the heparin serves two roles in the antithrombin inactivation of thrombin: (a) it activates the antithrombin by binding a pentasaccharide portion of its sequence to the heparin-binding site on antithrombin, and (b) it presents a surface that can bind both the antithrombin and the thrombin side by side. As implied, thrombin is also a heparin-binding protein.

There are a number of heparin-binding proteins whose activities are activated by heparin binding. However, as for the thrombin/antithrombin reaction, the activity of many of these requires a heparinoid sequence long enough to bind both the heparin-activated protein and a second protein.

C. Fibroblast Growth Factors

The fibroblast growth factors (FGFs) represent additional examples in which heparinoids play a template role in the activation of heparin-binding proteins. The FGFs are a family of heparin-binding growth factors that modulate cellular activities on binding to cell surfaces. Although several different cellular responses to FGFs may occur, the most prominent is mitogenesis. For stimulation of mitogenesis, FGFs must bind simultaneously to heparan sulfate proteoglycan and a fibroblast growth factor receptor on the surface of the cell, or to two molecules of FGF. Thus, as in the antithrombin case, the heparinoid serves as a template that brings two heparin-binding proteins together on the cell surface.

Both the FGFs and the receptor proteins have clusters of basic amino acid side chains on the protein surfaces that participate in the heparinoid binding. In contrast to the pentasaccharide binding site on antithrombin, which is made of two α-helices in apposition to each other, the heparin-binding regions of FGF appear to be made up of β-strands and β-bends. Thus, heparin-binding domains are not restricted to specific types of secondary structures.

The heparin sequences that bind to FGFs have been partially characterized. Hexasaccharide and octasaccharides bind well to FGFs alone, but decasaccharides or dodecasaccharides are required for binding to both FGFs and their receptors and for eliciting the mitogenic responses. The FGF-binding heparinoids need not be so highly sulfated as the antithrombin-binding pentasaccharide. One thing that is clear from the studies of the FGFs and a number of other heparin-binding proteins is that the antithrombin-binding pentasaccharide shows no specific binding to the FGFs or to other proteins. It is also clear that not all members of the mixture of heparan sulfate chains from any given tissue bind to all FGFs. Thus, the idea that specific, although not necessarily unique, heparinoid sequences are required to mediate the activities of individual heparin-binding proteins is emerging.

II.   HEPARINOIDS AND PROTEINS: PARTNERS IN BINDING

A.   Introduction

Heparinoids are ubiquitous in multicellular organisms [3]. Heparin and heparan sulfate are unique among the glycosaminoglycans in their ability to bind to a large number of different (primarily extracellular) proteins. Although the current list of heparin-binding proteins numbers more than 100 (Appendix), the list is growing rapidly. Of course, 100 proteins represent a small fraction of the total proteins in eukaryotic cells, and it is possible that the ability to bind heparin is simply a gratuitous property of these proteins. However, the binding of heparinoids by many of these proteins results in a modification of the protein's activities or metabolism, suggesting that heparinoids have evolved and persisted through all levels of animal phylogeny for some common purpose. In one sense, the heparin-binding proteins represent a "family" of proteins that bind to heparin and heparan sulfate. The members of this family are not related by the sequence similarities that are found in a typical family of proteins; they do not even seem to show a very high degree of similarity in the domains that bind to the heparinoids, although this conclusion may be premature in this young field.

What distinguishes a protein that binds to heparinoids from one that does not? The structural features of the heparinoids suggest that the peptide sequences in proteins that bind to heparinoids must be rich in basic amino acids, and indeed this generally turns out to be the case. As will be discussed in greater detail below, these proteins have surface domains enriched in clusters of basic amino acids, especially arginines and lysines, that bind to heparinoids in large part through multiple contacts with the negatively charged sulfate and carboxyl groups (which are fully ionized under physiological conditions [124, 644]) of the heparinoids. The fact that heparin-binding proteins have domains rich in basic amino acids is not necessarily reflected in the isoelectric point of these proteins. For example, acidic fibroblast growth factor (Chapter 9) and extracellular superoxide dismutase (Chapter 10) have pI's in the range of 4.5 to 6, but both bind tightly to heparin.

Heparinoids are generally restricted to animal cells. Some bacteria produce polysaccharides that have the same, or similar, monosaccharide sequences, but lack the sulfate substituents (Chapter 2). Most of the heparin-binding proteins in our list also are derived from animal cells, although it is clear that some viruses and bacteria produce heparin-binding proteins and use them to gain entry into animal cells, taking advantage of the fact that these host cells are coated with heparan sulfate proteoglycans. Heparinoids have not been identified in plants. In addition to the heparinoids, most animal cells produce other sulfated polysaccharides—most notably chondroitin sulfate, dermatan sulfate, and keratan sulfate, all of which occur in proteoglycan forms. However, these polymers show relatively little tendency to bind to proteins and influence their activities.

We will now examine the structural features, first of the heparinoids, and then of the proteins, that influence the binding affinities of these two partners.

B.   The Heparinoid Partner in Binding

1.   The structural features of heparinoids that result in protein binding

As described in Chapter 2, the heparinoids exhibit quite linear, helical tertiary structures—not the type of structure that one might anticipate would show remarkable ability to bind to proteins. Linear three-dimensional structures are also observed for the other glycosaminoglycans found in animal tissues [123, 645], and these fulfill our expectations that they bind poorly to proteins. Why then do heparinoids bind to proteins while other glycosaminoglycans do not?

The most obvious explanation is that the biosynthesis of IdoA-containing polymers, in contrast to the other glycosaminoglycans, leads to structures that are highly sulfated. Thus these polymers can present multiple, closely arrayed, charged sulfate and carboxyl groups for polyvalent association with positively charged domains on the surface of proteins. But superimposed on this basic structural feature is the ability of the IdoA rings to undergo rapid conformation changes that allow the heparinoids to present a variety of *arrangements* of these charged groups, adding further versatility for the binding of these polymers to proteins. This is illustrated in Figure 6.1, which shows two solution conformations of a heparin dodecasaccharide. In the left structure all of the IdoA residues are in the $^1C_4$ conformation; in the right structure all of the IdoA residues are in the $^2S_0$ conformation. Both sequences run from the nonreducing terminal at the bottom to the reducing terminal at the top. The sulfates on the C2 of the GlcN, the C2 of the IdoA, and the C6 of GlcN are labeled A2, I2, and A6, respectively, and the carboxyl groups are denoted by asterisks. It can be seen (a) that the N-sulfate, the 2-O sulfate, the 6-O-sulfate, and the carboxyl substituents on each successive $GlcNSO_3$-6-SO_4→IdoA-2-SO_4→$GlcNSO_3$-6-SO_4→IdoA-2-SO_4 sequence form clusters, first on one side of the helix and then on the other; and (b) that the shift from the $^1C_4$ conformer to the $^2S_0$ conformer of IdoA results in quite significant changes in the relative positions of the sulfates and carboxyls in such clusters. Thus, when heparinoids interact with proteins that have surface domains that are rich in arginine and lysine, the protein may induce changes in the ring conformations of the IdoA residues that allow the array of anionic groups on the heparinoid to become juxtaposed to the array of cationic charges on protein. In such protein/heparinoid complexes, some of the IdoA residues may take the $^1C_4$ conformation whereas other IdoA residues may be in the $^2S_0$ conformation; i.e., the distribution of arginine and lysine residues on the surface of the protein may control the ring conformations. Furthermore, it seems quite possible that, under the influence of a protein, some IdoA residues may take a $^4C_1$ conformation, even though this equi-energetic ring form does not normally occur in internal positions of heparin in free solution (see Chapter 2).

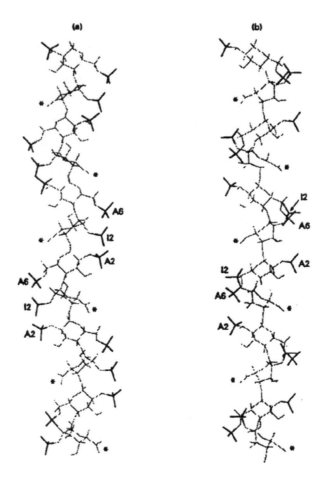

Figure 6.1. Solution conformations of heparin. Reprinted from
[132], with permission.

Superimposed on the flexibility that results from changes in the IdoA ring
conformation are the rotations about the glycosidic linkages, even though the
extent of these rotations is rather limited (Chapter 2). Together, the multiplicity
of negatively charged sulfate and carboxyl substituents, the facile changes in
ring conformation that allow re-positioning of these anionic groups, and the
rotations at the glycosidic linkages seem to account for the remarkable
ability of heparinoids to fit themselves to proteins. According to this concept,
GlcA-containing glycosaminoglycans (and indeed GlcA-containing segments of

heparinoids), despite glycosidic bond flexibilities that are comparable to that of
IdoA-containing glycosaminoglycans, do not bind to proteins because (a) their
biosynthetic pathways do not lead to such highly sulfated sequences, and (b) the
GlcA residues are locked in the $^4C_1$ conformation. However, GlcA residues may
make unique contributions to protein binding, both by supplying some rigidity to
parts of the protein-binding heparinoid sequences and by offering an orientation
of their carboxyl groups that is different from that of the IdoA carboxyl group.

Several aspects of the three-dimensional structures of heparinoids as they
interact with proteins must be considered. First, heparin itself has long stretches
of highly sulfated IdoA-containing disaccharides, whereas heparan sulfate has a
block structure in which highly sulfated blocks contain variable (but generally
small) numbers of contiguous IdoA-containing disaccharides. Thus, it would
appear that, over its entire length, heparin would exhibit the maximal ability to
undergo an induced conformational adaptation as this polymer interacts with
proteins and that heparan sulfate would have the same capacity, but over the
shorter IdoA-containing segments of the polymer [132, 329]. Interestingly, when
heparan sulfate serves as a template for binding two proteins together, as in both
the thrombin/antithrombin and the FGF/FGF receptor cases, it is possible that
there may be intervening unsulfated oligosaccharide blocks between the sulfated
blocks that bind to each of the proteins.

Further experimental data are necessary to address several questions: (a)
what are the three-dimensional structures of the heparinoid chains when they are
bound to proteins? and (b) what effect does the covalent attachment of one, or
several, heparinoid chains to their core proteins have on the interactions of
heparanoid chains with proteins? These two circumstances are difficult to study
and have received very little attention.

2. Other aspects of heparinoids that must be considered in
heparinoid/protein interactions.

Because of its common use as an anticoagulant, heparin is readily
available and inexpensive. Some structurally modified heparins can also be
obtained readily (Chapter 4). Consequently, most of the studies of heparin-
binding proteins have utilized heparin and its derivatives. However, the
occurrence of activating oligosaccharide sequences in heparinoid polymers that
modulate the activities of heparin-binding proteins must be examined in light of
several points that we have addressed earlier: (a) it is the heparan sulfate chains
of the heparan sulfate proteoglycans that mediate most of the biological
activities of the heparinoids, (b) heparan sulfate structures produced by different
tissues are distinct mixtures of structures (thus, one cannot extrapolate results

obtained with heparan sulfate from one cell type to all other cell types), and (c) heparin and heparan sulfate chains may contain similar disaccharide sequences for short stretches (e.g., tetra- or hexasaccharides), but the arrays of unique structures in each of these polymers diverge as the sequences become longer. It is important to understand that many of the disaccharide sequences found in heparan sulfate are not found in heparin; similarly, many of the disaccharides sequences found in heparin are not found in heparan sulfate. For example, a common sequence in *heparin* may be represented by the following:

$$[IdoA-2-SO4 \rightarrow GlcNSO3-6-SO4]_n,$$

where n may be as high as 4-6 (perhaps higher). Such a highly sulfated sequence presumably would become rare in most heparan sulfates as $\underline{n}$ is increased above 2 or 3. Thus, if this type of sequence were required for the biological activity of a particular protein, we might get a biological response with heparin that would be achieved rarely with heparan sulfates. In addition, the response of cells to this particular protein would be greatest in a tissue that produced a more highly sulfated heparan sulfate than found in most others.

Conversely, a sequence such as

$$[GlcA \rightarrow GlcNAc]_2 \rightarrow GlcA \rightarrow GlcNAc-6-SO_4 \rightarrow IdoA-2-SO_4 \rightarrow GlcNSO_3-6-SO_4$$

might be quite common in a *heparan sulfate* but would be rare in heparin. Thus, if such a sequence were required for biological activity, it is possible that heparin would not elicit an activity that would be readily observed with most heparan sulfates. Since virtually all studies of heparin-binding proteins have used heparins, it is even possible that some heparan sulfate-binding proteins have not yet been discovered. Unfortunately, structurally modified heparins, such as those described in Chapter 4, are not true mimics of heparan sulfate structures.

There is a related point. As stated earlier, heparinoids are conformationally adaptable in regions that are enriched in IdoA residues, but are much less adaptable in regions enriched in GlcA residues. Consequently, the limited length of IdoA-containing sequences in heparan sulfates reduces the ability of these polymers to bind to proteins; i.e., in contrast to heparins that are conformationally adaptable throughout their full length, the heparan sulfates exhibit the capacity to become fitted to a protein in the IdoA-rich regions, but are less adaptable in the GlcA-rich regions. Thus, the induced conformational fit that can occur in the IdoA-rich regions depends on just how long these regions are.

For some proteins, a range of heparin oligosaccharides with significantly different structural motifs may bind with similar affinities. In such cases, it is possible to identify a structurally modified heparin oligosaccharide that contains the minimum structural features required for binding (requirement of GlcA,

IdoA, GlcA, GlcNSO$_3$, or GlcNAc in certain positions; required positions of sulfates; etc.). However, beyond the requirements in the minimum structure, it may be possible to interchange GlcA and IdoA in some positions or to add further sulfates without altering the binding significantly. Thus, the heparinoid sequences that bind to proteins may be described in terms of either their *minimal* or their *acceptable* structural features required for binding to a protein. One reason that heparin binds to so many proteins may be that it contains sequences that represent acceptable heparinoid structures that will bind to these proteins; i.e., the extra sulfates or IdoA residues, which may contribute little to binding, do not prevent binding. Of course, the most important question here is whether the *acceptable* structural features are present in a heparan sulfate produced by a tissue that may respond to the heparin-binding protein.

In contrast to the proteins, in which the heparin-binding domains may be made up of groups of amino acid residues that are not contiguous in the primary sequence of the polypeptide, the sequences in heparinoids that bind to proteins seem to be single oligosaccharide segments within the heparinoid chain. This is undoubtedly a result of the fact that heparinoids do not fold back on themselves as readily as polypeptides. However, it is clear that a single heparinoid chain may contain multiple binding sites that can accommodate the heparin-binding protein. Thus, a single heparinoid chain may bind several molecules of the same protein, provided these heparinoid binding sequences are located at a sufficient distance from each other so that steric interference does not block protein binding. Such interactions result in protein oligomerization induced by the polysaccharide. Similarly, one heparinoid chain may bind two different proteins side by side, as we have described earlier for the thrombin/antithrombin and the fibroblast growth factor/fibroblast growth factor receptor cases. Furthermore, several such pairs of proteins may bind to a single heparinoid chain, provided the chain is long enough. Again, this binding utilizes linear, contiguous oligosaccharide sequences for the two proteins.

As a general rule, the minimal size of a heparinoid sequence that is required for binding to a monomeric protein falls in the range of hexa- to octasaccharides. However, when the biological activity that is mediated by the heparinoid requires that the heparinoid bind to *two* proteins, the minimal length of heparinoid required for such a template role is longer. It may be as small as a decasaccharide or a dodecasaccharide, although somewhat longer segments may be required for maximal activity. An 18-mer is required for the antithrombin-mediated inactivation of thrombin. Although the number of proteins for which the structures of these oligosaccharides have been studied is small, it is generally found that both IdoA residues and sulfate substituents are required for binding and activity.

It now clear that there are no consensus sequences in the heparinoid chains that are utilized for binding to *all* heparin-binding proteins. One manifestation of this fact is that when heparin is fractionated by affinity chromatography on several different immobilized proteins, different fractions of the total heparin are obtained. For example, affinity fractionation of heparin on immobilized antithrombin gives one fraction of heparin that binds to antithrombin and one that does not. The former has anticoagulant activity; the latter does not. However, affinity chromatography of heparin on every other heparin-binding protein that has been tested to date gives bound and unbound fractions, both of which have very similar anticoagulant activities.

Finally, it must be emphasized that binding of heparinoids to proteins *in vivo* occurs in the presence of physiological levels of salts and metal ions. Consequently, these conditions must be mimicked as closely as possible in *in vitro* studies in order to establish the potential physiological significance of the binding. For example, tissue plasminogen activator binds heparin *in vitro* at low salt concentrations, but when the salt is raised to physiological levels, the binding is lost. Thus, early proposals that attributed physiological significance to the binding of heparinoids to tissue plasminogen activator had to be revised (Chapter 8).

C.   The Protein Partner in Binding

1.   Consensus primary amino acid sequences for heparin binding

At the outset, the discussion of the peptide sequences that bind to heparinoids will consider only the primary structures of these peptides, a level of understanding that is, at the same time, our most comfortable view, and yet only a partial view. When limited information was available on the sequences in proteins that bound to heparin, it seemed that the sites of heparinoid binding could be reasonably described by primary sequences of appropriately spaced amino acids. Unfortunately, this concept is an oversimplification.

In a paper that has been widely cited, Cardin and Weintraub [646] examined peptide stretches that are enriched in basic amino acids in a large number of heparin-binding proteins and suggested that the following represent two general types of "consensus sequences" for heparin binding:

$$X \underline{B} \underline{B} X \underline{B} X$$

$$X \underline{B} \underline{B} \underline{B} X X \underline{B} X$$

In these sequences, B is a basic amino acid and X is any hydropathic amino acid. These sequences represented the most common positions of the basic amino acids in peptides that were found in proteins that bind to heparin. The identification of these sequences was based on the experimental demonstration that heparin-binding peptides conforming to these structures had been reported in vitronectin, platelet factor 4, apolipoprotein B (apoB), and apolipoprotein E (apoE). Although heparin-binding peptides in other heparin-binding proteins were not available at the time, a search for similar sequences in a broad range of proteins that had been reported to bind heparin led to the conclusion that these proteins all contained sequences that would fit one or both of the consensus sequences.

A comparison of the consensus sequences in the variety of heparin-binding proteins that were examined led to several generalizations:

- Since these are *consensus* sequences, not every residue that occurs in a "B" position is necessarily a basic amino acid and not every residue that occurs in an "X" position is necessarily a non-basic amino acid.

- Arginine and lysine are the most abundant basic amino acids in these sequences; histidine occurs infrequently. Acidic amino acids are not present.

- Helical projections of these sequences place the basic amino acid residues in clusters on one side of the helix.

- Proteins that show high degrees of *overall* homology with these heparin-binding proteins, but that do not bind to heparin, lack basic amino acids in some or all of the "B" positions in the corresponding sequences.

Peptides that contain these consensus sequences, on binding to heparin, may undergo changes in secondary structure. Some increase in helical structure, some increase in β-structure, and some exhibit very little change in conformation. Some heparin-binding proteins contain more than one consensus sequence. Two or more of these may participate in heparin binding. These sequences are not necessarily contiguous in the primary structure since they may be brought into proximity by the protein folding. Also, such pairs of sequences that may be involved in cooperative binding to heparinoids may not have the same types of secondary structure.

Although the Cardin and Weintraub treatment has been extremely useful in identifying heparin-binding peptides in proteins, several limitations should be noted.

- Although a large number of heparin-binding proteins were described in the Cardin and Weintraub paper, heparin binding peptides had been identified for only four of these. Thus, the actual database used for the generalization was small.

- There were no considerations of the structures of the heparinoids, especially with regard to the distances between sulfate and carboxyl substituents that would be expected to interact with the arginine and lysine side chains. It takes two partners for the binding, and only one partner was addressed in this treatment. One should note, however, that (a) the amount of information that was available before 1989 was quite small compared with the amount available today, and (b) the analysis would have been much more complex if the heparinoid partner had been considered as well.

- The peptides that were chosen on the basis of these consensus sequences as the most likely heparin-binding domains for some of the proteins (thrombospondin, antithrombin, heparin cofactor II, protein C inhibitor) have turned out to be the actual heparin-binding domains, but for other proteins (the FGFs) the chosen peptides are now known not to participate in heparin binding; i.e., the generalizations led to some incorrect peptides assignments.

- Despite the known high sequence similarity of FGF-1 and FGF-2, both in primary and tertiary structure, the analysis led to the selection of *different* heparin-binding peptides from nonhomologous regions of the two proteins, neither of which is now recognized as a prominent part of the heparin-binding domain of these proteins.

- It is difficult to match some of the heparin-binding peptides from proteins characterized since the 1989 analysis to either of the two consensus sequences

Although Cardin and Weintraub describe the heparin-binding consensus peptides with the usual N- to C-terminal directionality, there is no discussion of the directionality of the heparin segment that binds (nonreducing terminal to reducing terminal, or vice versa). It would be particularly important to determine whether distances between positive charges in the peptides can be fit to the distances between negatively charged sulfates and carboxyl groups in the heparinoid. In fact, some workers have concluded that the role of clusters of basic amino acids in heparin binding is more subtle than this treatment suggests [647].

2. Secondary and tertiary structures for heparin binding

Cardin and Weintraub proposed that the consensus heparin-binding sequences might occur in either helices or β-strands. In fact, this prediction appears to be true. The heparin-binding peptides in antithrombin are α-helices, whereas the heparin-binding peptides in the FGFs are β-strands. It has also been demonstrated that when heparin-binding peptides bind to heparin, they may undergo a change in their secondary structures [648, 649].

Cardin and Weintraub also suggested that two (or more) regions in a protein may be brought together by the folding to form a cleft for heparinoid binding. This has been amply demonstrated in several cases. Despite this, it remains an interesting fact that when a cleft containing two heparin binding elements is formed by protein folding, one of the elements seems to contain most of the basic amino acid residues involved in heparin binding. It is unclear whether there is any *a priori* reason why this should be so. Perhaps linear amino acid sequences are best adapted to binding to linear heparinoid sequences.

A second approach for deduction of the protein structures that bind to heparinoids has been reported by Margalit *et al.* [650] who took advantage of the fact that 18 heparin-binding sequences had been identified 4 years after the Cardin and Weintraub report. These workers modeled all of these sequences according to their known or proposed secondary structures—in some cases α-helices and in others β-strands—and noted that all of these peptides contained two basic amino acids that were about 24 Å apart. Furthermore, any other basic amino acids in these sequences were generally on one side of the secondary structure, opposite the side that is enriched in hydrophobic amino acids. They proposed that such structures may represent a common motif for heparin-binding peptides. This motif is simpler than the Cardin/Weintraub consensus sequence approach since there are only two basic amino acids required, and the only provision is that they be ~24 Å apart. There are, however, some problems with this conclusion.

The number of amino acids in these peptides that are terminated by the two basic amino acids at the ends of the motif depends on the type of secondary structure: in the β-strands the peptide sequences are 8-mers; in the α-helices, they are 14-mers. In the analysis, one of the heparin-binding peptides from FGF-2 fits the motif, but the other, which is the most critical one for binding (Chapter 9), does not. Furthermore, four of the other peptides listed by Margolit *et al.* as heparin-binding peptides do not fit the motif no matter which secondary structure they take.

The primary sequence in antithrombin that contains most of the heparin contact amino acid residues shows a great deal of β-structure in the isolated peptide and is induced to form a higher percentage of β-structure when it binds to heparin [648, 649]. However, this same peptide appears to be an α-helical structure in the intact protein (Chapter 7).

Once again, very little direct attention has been given to the distribution of charges in the heparinoid partner that would bind to this common structural motif.

3.   Conclusions

In the identification of the heparin-binding regions of a protein whose sequence is known, it is useful, as a first step, to examine the sequence for the presence of the consensus sequences described by Cardin and Weintraub. However, when such a sequence is present, it is not necessarily implicated as a portion of the heparin-binding domain of the protein, nor does the absence of such a sequence ensure that the protein will not bind to heparin; further experimentation is necessary. One problem with consensus sequences in proteins is that they imply some uniformity of structure in the heparinoids that bind to these proteins. Present data suggest that this is not the case. It is interesting to compare the attempts to find consensus sequences for heparinoid binding with the much more extensive attempts to find consensus peptides for nucleic acid binding. Nucleic acids are more ordered structures than heparinoids, and, while there are several motifs for protein sequences that bind to nucleic acids, the location of these motifs within a polypeptide sequence does not assure that such sequences will bind to nucleic acids [651, 652]. If there are no consensus peptide sequences for nucleic acid binding, perhaps we should not place too much emphasis on the proposed consensus sequences for heparinoid binding.

To summarize, it is premature to try to predict which peptide sequences of a heparin binding protein participate in heparin-binding based on either primary or secondary structural considerations. It is much easier to do this after the crystal structure is known. Unfortunately, in this young field there are only a few heparin-binding proteins for which crystal structures are available.

III.   THE BINDING OF METAL IONS BY HEPARINOIDS

Heparin and heparan sulfate bind strongly to divalent metal ions. Thus, the question arises, "do physiological concentrations of ions such as Ca^{2+}, Zn^{2+}, Mg^{2+}, Mn^{2+} or other metal ions affect protein-heparinoid interactions? One question is whether metal ion binding alters the ring conformations or overall conformations of these polymers and whether such alterations have some effect on binding to proteins. However, most of the studies of the binding of heparin, heparan sulfate, or heparinoid fragments to proteins and heparin-binding peptides have not addressed this question.

The binding of metal ions to heparinoids has been studied by a number of workers [reviewed in 124]. These studies have included a variety of mono- and divalent ions, as well as a number of cationic dyes and synthetic polycationic peptides and protamine. Metal binding has been measured by infrared spectroscopy [653], NMR [654, 655], potentiometric titration [656, 657], circular dichroism [654], and polarography [658]. However, all of these studies have been confined to heparin and most have used metal ion concentrations that are not at physiological levels. From the standpoint of our main considerations here, it would be rather easy to determine the effects of increasing concentrations of pertinent metal ions (within the physiological ranges) on heparinoid/protein interactions, but such studies have not been done [see, however, reference 659].

IV.   METHODOLOGY

A.   Introduction

The chapters that describe the individual heparin-binding proteins cite many examples of the methodology that is unique to this area of study. However, it may be useful here simply to outline some of the more salient aspects of the measurement of binding and the identification of the domains of proteins and heparinoids that are directly involved in protein/heparinoid interactions.

B.   Measurement of Binding and Biological Activities

The strength of binding of a heparinoid to a protein is expressed in terms of the dissociation constant, K_d, a ratio of the product of the *molar* concentrations of the two partners in binding to the *molar* concentration of the associated complex at equilibrium. For heparin-binding proteins, the K_d's are always dependent on the concentration of NaCl, with higher salt concentrations raising the K_d's. *Under physiological salt conditions*, a K_d value of 10^{-7}-10^{-9} $\underline{M}$ represents strong binding, whereas a K_d value of 10^{-4}-10^{-5} $\underline{M}$ represents weak binding. In heparinoid studies, the K_d must be interpreted carefully. When K_d measurements are made in the laboratory, the protein is usually pure but the heparinoid is not. The impurity of the heparinoid may be reflected in its range of molecular weights, in its range of structures, or, more commonly, both. In the case of molecular weight polydispersity, it is possible that lower molecular weight species may bind with affinities that are different from those of the higher molecular weight species. In the case of structural variability, one can anticipate that only a fraction of the components of mixtures of heparinoids—even those of a uniform molecular weight class—will bind to a protein. Furthermore, several molecules of some proteins will bind to a single heparinoid chain. These aspects of heparinoid structure complicate the calculation of the molecular weight of the heparinoid that actually binds to the protein. Usually, an *average* molecular weight is chosen to obtain the molar concentration for use in the K_d calculation, in which case the results may not be very meaningful.

The binding of pure heparin pentasaccharide to antithrombin (Chapter 7) represents a case in which both pure protein and pure heparinoid have been used [391]. In this case a valid K_d value can be obtained. However, when peptide fragments of antithrombin, or any other heparin-binding protein, are used to measure K_d values, these values must be interpreted carefully. In the antithrombin case, two problems appear. First, the antithrombin-binding site for the pentasaccharide is formed by two helices which form a cleft into which the pentasaccharide fits. One of these segments contains most of the basic amino acids that participate in the pentasaccharide binding. Synthetic peptides, or cleaved fragments of the protein that contain this domain, also bind the pentasaccharide. However, the full binding affinity of the protein cannot be represented by only one of the two peptides that are involved in pentasaccharide binding to antithrombin. Furthermore, in the antithrombin case, the synthetic peptide takes a β-strand conformation rather than the helical form found in the protein.

Binding affinities can be measured in several ways. Perhaps the best way is by isothermal titration calorimetry [660-663]. Alternatively, *relative* affinities of different heparinoids for a given protein are easily measured by observing the NaCl concentration required to displace the heparinoid from a protein affinity matrix (or to displace the protein from a heparin affinity matrix). It has been demonstrated that the NaCl concentrations are directly proportional to the K_d values [662]. When NaCl concentrations above 1 $\underline{M}$ are required for elution, strong binding is indicated, whereas elution at concentrations below 0.3-0.4 $\underline{M}$ reflects relatively weak binding.

In another approach, one can measure the concentration of heparinoid required to displace either the protein or a standard heparinoid (either of which may be labeled) from a protein/heparinoid complex [627].

Finally, for some proteins, the binding of heparin causes a change in conformation that can be measured as a change in UV absorbency or in the fluorescence or CD spectrum. Thus, observing such changes as a function of heparinoid concentration can yield K_d values. Biological assays can also be used. For example, the concentration of the heparinoid necessary to obtain a change in binding of the protein to a cell surface can be measured.

An alternative approach for measuring the selectivity and affinity of heparinoid binding to proteins has been described by Lee and Lander [664]. In this procedure, referred to as affinity co-electrophoresis, mixtures of agarose containing decreasing concentrations of the binding protein are prepared and allowed to gel in side-by-side lanes in the electrophoresis apparatus. Then a fixed concentration of an ^{125}I-labeled heparin is electrophoresed through each lane. Following electrophoresis, the position of the heparin is detected by autoradiography. Since the rate of electrophoretic migration of proteins under the conditions used is much lower than the rate of heparin migration, the migration of the heparin through the lanes is retarded by the binding protein in the lane, with the degree of retardation being dependent on the concentration of the protein in the gel and the affinity of the protein for the heparin. At the highest protein concentrations, heparin migration is completely retarded; at the lowest concentrations, heparin migration is unretarded; and at intermediate concentrations of the protein, intermediate rates are observed. Those components of the heparin mixture that do not bind to the protein are not retarded at any protein concentration. This method, which has been applied to antithrombin, basic FGF, acidic FGF, fibronectin, nerve growth factor, and serum albumin, permits the calculation of the K_d for heparin binding as well as the separation of small amounts of those species of the heparin that bind to the protein from those that do not bind.

Several practices found in the literature on heparin-binding proteins are not advisable. *Expression of heparinoid concentrations in terms of international units of heparin is not useful.* With the exceptions of a few of the proteins of the coagulation cascade (Chapter 8), the anticoagulant activity of heparin is not related to binding or biological activities of heparin-binding proteins. Furthermore, anticoagulant activities vary for different preparations of heparin, so that the *weight* of heparin or the number of moles of heparin used cannot be calculated from the number of units. A second problem is that the *displacement of a protein from a protein/heparinoid complex by the addition of basic proteins or peptides does not necessarily imply that there is specific binding of the displacing peptide to the heparinoid.* For example, protamine is used clinically to neutralize the anticoagulant activity of heparin, but it does not bind specifically to the antithrombin-binding pentasaccharide.

C.   The Identification of Heparinoids that Bind to Proteins

1.   Cleavage of heparinoids

The structural variability of heparinoids makes it impossible to identify the protein-binding domains of the heparinoid without converting the polymeric heparinoid to oligosaccharides. For example, when it became clear that heparin could be separated into antithrombin-binding and antithrombin-nonbinding fractions by affinity chromatography on immobilized antithrombin, it appeared that it would be easy to identify the antithrombin-binding heparin sequence by measuring the disaccharide compositions of the two heparin fractions. However, the differences were so obscured by the "sameness" of heparin structural elements that had nothing to do with antithrombin binding that it was impossible to get much new information from analyses of the polymeric forms. Even if significant structural differences between the binding and nonbinding fractions were obvious, it would not have been possible to determine the disaccharide sequences of the two polymeric fractions (see Chapter 3). When heparin was first fragmented and then subjected to affinity chromatography on antithrombin columns, the hexa- and octasaccharide fractions that bound to the column contained the unique sequences, which could then be isolated and characterized.

The identification of oligosaccharides that bind to proteins requires that fragments as small as hexa- or octasaccharides be obtained. The limited experience to date suggests that the fragmentation of heparinoids yields such a complex mixture of oligosaccharides that the desired fragment will be only a small proportion of the total mixture of fragments. Therefore, it is desirable to maximize the yield of the binding fragments. Heparinoids can be cleaved by nitrous acid, heparin lyases, or other methods described in Chapters 3 and 4. Since different cleavage methods cleave at different sites, *some methods may cleave the heparinoid at sites in the middle of the desired sequence*. Thus, when the nature of the binding sequence is not known, it is desirable to do an initial experiment to determine which cleavage method gives the maximum yield of the desired fragments.

2.   Fractionation of heparinoid oligosaccharides

All of the cleavage methods for fragmenting heparinoids yield mixtures containing thousands of oligosaccharide species that vary in both size and structure (Chapter 2). It might seem that it would be easy to isolate the desired heparinoid fragments by affinity chromatography on the immobilized protein.

However, affinity matrices usually have quite small capacities, and there are often multiple structural variants of oligosaccharides in this mixture that have different degrees of affinity for the protein. Therefore, initial fractionation procedures that can accommodate relatively large amounts of material and that can eliminate most of the background oligosaccharides are appropriate. The most common approaches include fractionation according to size (gel filtration) and charge (chromatography on DEAE-cellulose or other ion exchangers).

Procedures that have relatively low sample capacity but high resolution include high pressure liquid chromatography on anion exchange or ion pairing columns, as well as zone electrophoresis (Chapter 3). These low capacity methods can be used in conjunction with affinity chromatography to yield highly purified oligosaccharides. We have pointed out elsewhere that the complexities of the mixtures of oligosaccharides increase rapidly with increasing sizes of the oligosaccharide mixtures.

D.   The Identification of Heparin-Binding Domains of Proteins

The methodology for preparing peptide fragments of proteins is much more generally appreciated than the heparinoid methodology outlined above. The proteins themselves can be highly purified by affinity chromatography on heparin affinity columns [665], and their fragments or desired synthetic peptides can be purified in the same manner.

As a first step in the identification of the peptide sequence that is responsible for heparin binding, the protein can be cleaved and the resulting peptide mixture can be fractionated by chromatography on heparin affinity matrices. *Obviously enzymes that cleave at basic amino acid residues should be avoided so that cleavages do not occur in the middle of potential heparin-binding domains.* Interestingly, fragmentation of a pure heparin-binding protein with proteolytic enzymes or agents such as CNBr yields a mixture of peptides that is much less complex than the oligosaccharide mixtures obtained when heparinoids are cleaved.

Tight-binding peptides can be eluted from heparin affinity matrices with salt and sequenced in an attempt to identify amino acids that participate in the binding. If these peptides are involved in heparin binding, they should block the formation of protein/heparin complexes, as should antibodies directed against them. However, as noted earlier, the fact that a peptide blocks formation of the protein/heparin complex formation does not necessarily imply that the peptide is specific for the binding. Thus, the final implication of such peptides in heparinoid binding requires a demonstration that the peptides give *specific* inhibition of binding.

There are other approaches that have been used with varying degrees of success. For example, in heparin/protein complexes, derivatization of amino acid side chains in the heparin-binding domain often effectively blocks heparin binding. The best way to assess the importance of individual amino acids in heparin binding is by *single* replacement of amino acids in the potential heparin-binding domains by site-directed mutagenesis. Many examples of this approach are cited in later chapters.

ADDITIONAL REFERENCES

Mikhailov, D., Mayo, K. H., Pervin, A., and Linhardt, R. J. (1996) ^{13}C-NMR relaxation study of heparin-disaccharide interactions with tripeptides GRG and GFG, *Biochem. J.* **315**, 447-454.

Stearns, N. A., Prigent-Richard, S., Letourneur, D., and Castellot Jr., J. J. (1997) Synthesis and characterization of highly sensitive heparin probes for detection of heparin-binding proteins, *Analyt. Biochem.* **247**, 348-356.

Chapter 7. Antithrombin, the Prototypic Heparin-Binding Protein

---

I. THE HEPARIN/ANTITHROMBIN INTERACTION

A. The Serpin Family

Antithrombin is a member of the serpin superfamily of proteins that includes more than 40 proteins [666-670]. With only a few exceptions, the members of the serpin superfamily are <u>serine</u> protease <u>inhibitors</u>.[1] Each serine protease inhibitor plays a role in inhibiting excessive action of its specific target protease(s). The serpins combine with proteases to form inactive complexes, which are then cleared from the circulation. The serpins are, in fact, *suicide substrates* for their target proteases; they interact with the active sites on the proteases and are cleaved. The bonds that are cleaved are referred to as the *active centers*, or *reactive centers (or sites)*, of the serpins, and these active centers, plus the surrounding amino acid sequences, determine the specificities

---

[1] The proteins in this family that have no protease inhibitory function include thyroxine-binding globulin, cortisol-binding globulin, ovalbumin, angiotensinogen, and maspin [671]. There are, in addition, a group of uterine proteins that are members of the serpin family that inhibit aspartate proteases rather than serine proteases [672].

of the serpin-protease interactions.[1] The amino acid pair that forms the peptide bond at the site of cleavage is designated P1-P1', with the surrounding region designated ...P3-P2-P1-P1'-P2'-P3'..., etc. [673]. On cleavage by the protease target, P1 becomes a new C-terminal amino acid for the sequence ...P3-P2-P1, and P1' becomes a new N-terminal amino acid for the sequence P1'-P2'-P3'... . The cleaved serpins remain in inactive complexes with their proteases. In some cases, these complexes break down slowly but, in others, they are not readily dissociated, even in detergent, urea, or guanidine. These complexes are apparently stable acyl enzyme intermediates [668, 669]. Although most serpins will combine rapidly and directly with their target proteases, several of them, including antithrombin, heparin cofactor II, protease nexin I, and protein C inhibitor, are poor inhibitors, but become quite excellent inhibitors when they are complexed with heparin. The heparin causes a conformation change in the serpin that facilitates the protease/serpin binding and/or exposes the active center of the serpin to the protease. The present discussion will deal primarily with antithrombin, for which the target proteases are thrombin and factor Xa, and perhaps factors XIa, IXa, and XIIa [674-678].

Since the role of a serpin is to prevent the harmful action of its target protease, it must inactivate the protease rapidly enough under physiological conditions to fulfill this function. In the laboratory, individual serpins may inhibit several different serine proteases, some of which are not *physiological* targets for the serpin. This is because the rates of reaction with susceptible proteases may not be fast enough with some proteases to be of physiological significance. For example, in the laboratory, antithrombin is an inhibitor of trypsin and plasmin [679], in addition to the proteases cited above, but the reaction of trypsin and plasmin with antithrombin is too slow *in vivo* to be of physiological importance. In fact, although antithrombin inhibits trypsin, this inhibition is not stimulated by heparin. Travis and Salvesen [680] have offered criteria for distinguishing the physiologically important targets of protease inhibitors from those that are merely incidental laboratory serpin targets. The primary criterion is the rate at which the inactivation occurs, expressed as the half-time of association of proteases with their protease inhibitors under physiological conditions (i.e., at the physiological concentrations of the reactants). According to these criteria, a $t_{1/2}$ must be less than 100 msec for effective control of the protease activity. The validity of this criterion is confirmed for most of the serpins by the findings that genetic defects in the serpins reflect the failure to regulate the target protease(s) identified by this standard. Interestingly, when this criterion is applied, none of the coagulation factors that are inactivated by antithrombin would be physiological targets for antithrombin in the absence of heparin. When heparin is present, the rates of inactivation are increased 1000-fold or more, thus raising the rates to ones that meet the criterion. However, Travis and Salvesen suggest that even in the presence of heparin, the only protease that reacts rapidly enough to qualify as a target protease for antithrombin is thrombin. The other antithrombin-susceptible serine proteases in the coagulation cascade, including factor Xa, still do not qualify! Nevertheless, *in vivo* results show clearly that factor Xa *is* a target for antithrombin.

B. Antithrombin Inhibits Blood Coagulation

As detailed by Rosenberg [681] and by Björk and Lindahl [682], it was recognized even before the discovery of heparin that thrombin slowly loses its activity when added to plasma or serum, and it was thought that serum contained a specific "anti-thrombin" that inactivated thrombin. After the discovery of heparin by McLean [45] and the demonstration of its potent anticoagulant activity, it was found that heparin anticoagulant activity required an accessory protein in the blood which was called "heparin cofactor" [683]. Ultimately it was found that the anti-thrombin of serum and the heparin cofactor were the same protein, which has come to be known as "antithrombin III," or simply "antithrombin."[2] Most workers consider that the target proteases for antithrombin are the serine proteases of the coagulation cascade, namely thrombin and factors IXa, Xa, XIa, and XIIa. Thus, antithrombin is an inhibitor of blood coagulation.[3] Its essential role in the control of hemostasis is most clearly established by the propensity for thrombosis in patients with genetic defects that reduce antithrombin activity [684-686]. Since its early isolation and characterization [675, 687], routine methods have been developed for the isolation of antithrombin [688]; the pure protein has been characterized in detail [675, 689-691]; and its human gene has been cloned [692, 693].

C. Reactions of Heparin with Antithrombin and its Target Proteases

1. The reaction sequence

The reaction of heparin with antithrombin and proteases proceeds through several stages, which are illustrated in Figure 7.1. Initially, heparin binds to antithrombin (reaction 1), causing a conformation change. This change allows further interactions between antithrombin and the bound heparin which result in

---

[2] In the literature both "anthrombin" and "antithrombin III" are used to refer to the same protein. Björk and Lindahl [682] suggest that only the former name be used.

[3] For a description of the blood coagulation system, see Chapter 8.

$$H + AT \rightarrow H/AT \rightarrow H/AT^* \overset{T}{\searrow} H/AT^*/T \rightarrow H/AT^{**}\text{-}T \overset{H}{\swarrow} AT^{**}\text{-}T \rightarrow AT^{**} + T$$

Reaction   1   2   3   4   5   6

Figure 7.1. The reaction steps in the heparin-induced inactivation of thrombin by antithrombin. H = heparin; AT = antithrombin; T = thrombin; AT* = antithrombin that has undergone a conformation change as a result of heparin binding; AT** = antithrombin that has been cleaved at its active center, with its two fragments still joined by a disulfide bond. Noncovalent interactions are indicated by a slash (/), whereas covalently linked species are indicated by a hyphen (-).

stronger binding (reaction 2). The heparin/antithrombin complex then reacts with a target protease, in this case thrombin, forming a ternary complex (reaction 3). In this complex the protease cleaves the antithrombin at its active center near the C-terminal (reaction 4), yielding a new complex that contains the cleaved, and conformationally altered, antithrombin *covalently* bound to the active site of the protease (an acyl enzyme intermediate).[4] The C-terminal peptide that is cleaved from the antithrombin remains covalently bound to the larger fragment of the antithrombin via a disulfide bond. As a result of the antithrombin cleavage, the antithrombin affinity for the bound heparin is markedly diminished [694, 695] and so the heparin dissociates from the complex *in an unaltered form* (reaction 5), free to catalyze further reactions between antithrombin and its target proteases; i.e., heparin is a catalyst in these reactions. The remaining acyl enzyme complex dissociates very slowly (reaction 6) to yield the unaltered protease and the cleaved antithrombin, which no longer has either serpin activity or heparin-binding ability [696, 697]. In the laboratory, the covalently linked complex between the cleaved antithrombin and the thrombin is stable in 1% sodium dodecyl sulfate or guanidinium chloride at pH 7.5 and 37 °C [675, 697-701], but undergoes a slow dissociation with a $t_{1/2}$ of 5.7 days (k = 1.4 x 10^{-6} sec^{-1}) [702, 703], a rate that increases markedly at pH's above 8.5 or in the presence of nucleophiles [698, 700, 703-705], i.e., under nonphysiological conditions. The individual steps in this reaction sequence are now described in greater detail.

2.   The reaction of heparin with antithrombin

---

[4] For the indication of various complexes between heparin and proteins, or between two proteins, the system that is adopted throughout this treatise is that *non-covalent complexes* will be designated by a slash between the two components of a complex (e.g., heparin/antithrombin), whereas *covalent complexes* will be designated by a dash between the two components of the complex (e.g., thrombin-antithrombin).

The reaction of heparin with antithrombin in the absence of target proteases (reactions 1 and 2 in Figure 7.1) represents the first stage of the overall process of the heparin-mediated serpin reaction. This bimolecular reaction has been studied using pure antithrombin and unfractionated heparin or heparin fractions that have been obtained by affinity chromatography of heparin on immobilized antithrombin [706-708]. Unfractionated heparin has an anticoagulant activity of 150-180 units/mg, whereas the heparin that binds to antithrombin with high affinity (HA-heparin) has elevated anticoagulant activity (3-400 units/mg) and the unbound heparin (LA-heparin) has a very low anticoagulant activity (20 units/mg). Both the binding constants for association of heparin and antithrombin and the rates of complex formation have been measured. These studies have taken advantage of changes in the spectral properties of antithrombin that occur when it binds heparin. The initial observation of the spectral change by Villanueva and Danishefsky [709] demonstrated that the binding of heparin causes an altered UV absorbency of antithrombin, signaling a conformation change induced by the heparin [709]. This change is also observed as an increase in fluorescence emission and a change in the near-UV circular dichroism [703, 709-713]. These spectral changes are elicited by both HA-heparin and LA-heparin, but only the changes induced by HA-heparin, which are much larger, correlate with the critical conformation change that accompanies the appearance of the serpin activity of antithrombin [714]. Heparin and the antithrombin-binding pentasaccharide prepared from heparin (Section I.G., p. 221) give similar changes [715]. Thus, the heparin-induced change in conformation of antithrombin appears to be the primary effect of heparin in the activation of the serpin activity [716-718].

Table 7.1 shows a comparison of the dissociation constants for antithrombin with HA-heparin and LA-heparin. These results show that the K_d for binding of HA-heparin to antithrombin is two to three orders of magnitude lower than that for LA-heparin, and that low molecular weight fractions show

Table 7.1. Binding of Heparins to Antithrombin

Antithrombin plus...	Molecular weight	K_d ($\underline{M}$)	Reference
HA-heparin	11,000	0.5×10^{-6}	[719]
HA-heparin	15,000	1.25×10^{-6}	[703, 714, 720]
HA-heparin	6,000	1×10^{-7}	[721]
HA-heparin		2×10^{-8}	[720-722]
Unfractionated heparin		5.74×10^{-8}	[723]
LA-heparin	15,000	0.2×10^{-4}	[714, 719]
LA-heparin	6,000	1×10^{-4}	[721]

somewhat diminished affinity for antithrombin. The dissociation constant depends on the ionic strength (μ) and the pH [724, 725]. The K_d is 100-fold higher at $\mu = 0.5$ than at physiological μ, and increases 25-fold as the pH is increased from 5.5 to 8.5. These observations suggest that the binding of heparin to antithrombin is due primarily to electrostatic interactions, a conclusion that is confirmed by other studies.

The spectral changes that occur when heparin binds to antithrombin are attributed largely to an alteration in the environment of a buried tryptophan, [711, 726]. These have been used to measure the kinetics of heparin binding. [711, 718, 722, 726-728]. The results demonstrate that the reaction of antithrombin with heparin and pentasaccharide occurs in two stages, one in which the heparin is bound relatively weakly ($K_d = 4.3 \times 10^{-5}$ $\underline{M}$), causing the conformation change, and a second rapid stage ($k = 10.4$ sec^{-1}) in which the heparin binding becomes much stronger ($K_d = 6 \times 10^{-8}$ $\underline{M}$) [729]. The conformation change has been calculated to increase the affinity for heparin by >300-fold.

3.   The reaction of the heparin/antithrombin complex with target proteases

When heparins or heparin fragments that have been fractionated according to molecular size are tested for their ability to induce the inhibition of factor Xa by antithrombin, neutralization of the protease activity is achieved with heparin segments containing as few as five monosaccharide units, i.e., with the pentasaccharide fragment of heparin that is described in detail in Section I.G.3., p. 222, Thus, antithrombin is "activated" by the pentasaccharide. When the same tests are carried out with *thrombin*, it is found that heparin segments having less that 18 monosaccharide units fail to give effective thrombin neutralization [317, 715, 730-734]. The explanation for this difference is that although the activation of antithrombin by a short heparin segment (pentasaccharide) is sufficient to render the pentasaccharide/antithrombin complex reactive with its target proteases, the neutralization of thrombin requires that the heparin segment be of sufficient length to bind to *both* antithrombin and thrombin, i.e., they must be brought together side by side on the heparin template in order to react, an effect referred to as "approximation." Thrombin alone reacts poorly with the activated antithrombin unless it can sit down beside antithrombin on an extended heparin template. factor Xa, however, exhibits no such requirement for this approximation. The approximation effect is also observed for factors IXa and XIa [317, 681, 682, 735], even though the latter are probably not significant *in vivo* targets for antithrombin. The approximation effects have been discussed in the literature [714, 736, 737].

4.   The interaction of heparin with thrombin

In the approximation reaction, heparin must bind directly to thrombin. Studies of this binding have indicated that there is little specificity in the heparin sequence required for thrombin recognition [738]. All heparin fractions bind to thrombin [736, 739-744]. For example, HA-heparin and LA-heparin bind to thrombin with the same affinity. Furthermore, chromatography of heparin on a thrombin affinity matrix gives some fractionation [738], but does not give fractions with significant differences in binding to thrombin. Thrombin binds to heparin primarily through nonspecific electrostatic association with any three contiguous disaccharides of the heparin chain [722, 745]. The heparin binds to anion exosite II on thrombin (Section I.F., p. 219) that is distant from its active site [746]. On binding, there is no spectral change in the thrombin [747]. The K_d for this interaction has been reported as 1.7×10^{-9} $\underline{M}$ [748] and 1×10^{-9} $\underline{M}$ [749].

At optimum concentrations of HA-heparin, the rate of the reaction of thrombin with antithrombin is increased 2000-fold by the heparin [721]. However, at higher heparin concentrations, different heparin molecules can bind separately to antithrombin and thrombin, thus preventing the binding of both thrombin and antithrombin to the same heparin chain. Thus, heparin concentrations greater than needed to maximize the reaction rate result in a decrease in the rate of inactivation of thrombin by antithrombin [40, 711, 714, 721, 724, 726, 750-754].

D.   The Structure of Antithrombin

1.   Primary structure and cleavage site of antithrombin

Antithrombin is present in human serum at a concentration of approximately 2.5 $\mu\underline{M}$ (125-200 μg/ml) [691, 755, 756]. It is a single-chain protein with a molecular weight of 54,000-68,000, depending on the degree of carbohydrate substitution [675, 682, 689]. Human antithrombin has been sequenced and cloned [692, 757, 758]. After removal of a 32 amino acid signal sequence, the secreted protein contains 432 amino acids and has disulfide bonds at Cys 8-Cys 128, Cys 21-Cys 95, and Cys 247-Cys 430 [704]. Antithrombin contains ~10% carbohydrate, with N-linked oligosaccharide attachment sites at Asn 96, Asn 135, Asn 155, and Asn 192. The degree of N-glycosylation at these positions varies; for instance, there are two human isoforms that differ in glycosylation at Asn 135 [759, 760]. The glycosylation patterns of antithrombin have significant effects on the affinity of the protein for heparin [759-763]. When antithrombin is bound to heparin, it is cleaved specifically at Arg 393-Ser 394 (P1-P1') by thrombin and by the other serine proteases of the coagulation cascade [682, 764, 765], forming stable covalent acyl enzyme complexes in which the cleaved 39 amino acid C-terminal peptide remains covalently attached to the complex by the Cys 247-Cys 430 disulfide bond [682, 766, 767]. Variants that have amino acid substitutions at Arg 393 do not neutralize the target proteases, but do retain their heparin-binding properties [768].

2.   The three-dimensional structure of antithrombin

Although the primary sequence of human antithrombin has been known for some time, the crystal structure has only recently become available [767, 769-771]. In the early considerations of the three-dimensional structure of antithrombin, the high degree of homology among members of the serpin family [690, 716, 772] led to strong arguments that the crystal structure of α_1-protease inhibitor, which serves as an archetype of all serpins,[5] is an appropriate mimic of the three-dimensional structures of other serpins, including antithrombin [773-775]. Because of the strong sequence similarities among members of the serpin family, the projection of the amino acid sequence of antithrombin onto the crystal structure of α_1-protease inhibitor yields a three-dimensional structure of antithrombin that is consistent with an extensive body of literature (Figure 7.2). The same type of projection has been made for the other heparin-activated serpins, including heparin cofactor II, protease nexin I, and protein C inhibitor [776-778] (see Chapter 8). Furthermore, as X-ray crystallographic structures of other members of the serpin family, including α_1-antichymotrypsin [779], leukocyte elastase inhibitor [780], the latent form of plasminogen activator inhibitor [781], and the cleaved [782] and uncleaved forms of ovalbumin [783], became available, a picture consistent with that of the three-dimensional structure of α_1-protease inhibitor has emerged for all serpins. As discussed below, the actual crystal structure of antithrombin is completely consistent with these earlier deductions.

α_1-Protease inhibitor that has been cleaved at its active center crystallizes in three different forms, the three-dimensional structures of which are very similar. These crystal structures have been solved and used for the modeling of other serpin structures [774, 778, 781, 784, 785]. Eighty percent of the amino acids in α_1-protease inhibitor are in eight well-defined helices (A-H) and three large β-sheets (A-C). Figure 7.2 shows the corresponding structural features of antithrombin. When the primary sequences of antithrombin and other serpins are

---

[5] Although the α_1-protease inhibitor is an an inhibitor of trypsin and is often referred to as α_1-antitrypsin, its physiological target is leukocyte elastase, which, in contrast to trypsin, meets the reaction rate criteria of Travis and Salvesen for a physiologically significant target protease. α_1-Protease inhibitor is *not* one of the heparin-activated serpins.

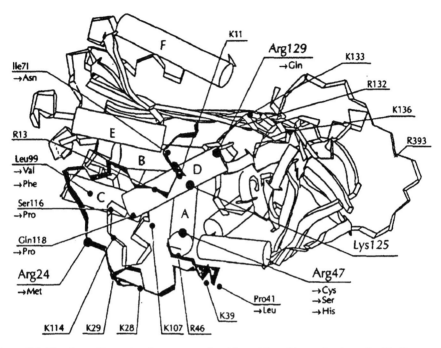

Figure 7.2. The three-dimensional structure of antithrombin, with emphasis on the binding sites for the heparin pentasaccharide. Reprinted from [767], with permission.

aligned with that of α_1-protease inhibitor, 28% of the residues are identical, and a much higher percentage of the residues are similar. The most highly conserved amino acid residues are found in the regions of secondary structure. Insertions of 1, 2, or 3 amino acid residues occur in the random coil regions of the chains that connect the secondary structures. Thus, such insertions do not affect the tertiary structure. Considerable variations are observed in the lengths of the N- and C-terminals of members of the serpin family. Since both the N- and C- terminals project into the solution, such extensions do not perturb the tertiary structures. Among the serpins, the shortest analogs begin with residue 23 of the α_1-protease inhibitor and end with residue 391, which are the first and last buried residues in the globular structure. Antithrombin has a relatively unstructured N-terminal extension of 45 residues that is linked to the C and D helices by disulfide bonds. Although α_1-protease inhibitor contains no disulfide bonds, antithrombin has Cys 247→Cys 430 in a conserved area, and, if α_1-protease inhibitor contained Cys residues in these positions, this bond could be formed without significant perturbation of the tertiary structure. The glycosylation sites are on the surface [786]. Consequently, the N- linked oligosaccharides project into solution.

3.   The active center of antithrombin

A large body of evidence describes the structural features of the active centers of serpins and the changes in serpin structure that take place when the P1-P1' bonds are cleaved to form the stable serpin-protease complexes [666, 668-670, 787, 788]. The active center of α_1-protease inhibitor is found at Met 358-Ser 359. In antithrombin, the homologously aligned amino acids are Arg 393-Ser 394, which is the active center of antithrombin. In a pathological variant of α_1-protease inhibitor in a child with a bleeding disorder, Met 358 is replaced by Arg, a change that converts α_1-protease inhibitor into an inhibitor of thrombin [774]. This observation, which demonstrates the apparent positional identity of the active centers of α_1-protease inhibitor and antithrombin, is one of many indications of the homologies among members of the serpin family.

Until recently, the only crystal structures that were available were those of the cleaved α_1-protease inhibitor and uncleaved ovalbumin (a member of the serpin family that does serve as a protease inhibitor). These structures plus other data led to some general conclusions about the active centers of serpins. These centers are found in a loop that extends from the main body of the serpin and that connects strands of β-sheets A and C, shown at the tops of the structures in Figure 7.3. Thus, the active centers of serpins extend from the main structures of these proteins and have mobilities that allow them to become seated in the active sites of their target proteases. The helical portion of the loop at the top of the putative active center loop of ovalbumin restricts the mobility of the loop— probably preventing ovalbumin from acting as a serpin [6]. Interestingly, when the peptides bound at the active centers of α_1-protease inhibitor and other serpins are cleaved by their target proteases, the amino acid residues that had formed the P1-P1' bond move to opposite ends of the cleaved structure, approximately 70 Å apart. This is illustrated for α_1-protease inhibitor in Figure 7.3. The cleaved α_1-protease inhibitor structure shows the newly formed N-terminal amino acid (Ser 359, P1') at the top of the structure and the newly formed C-terminal (Met 358, P1) at the bottom. In addition to the large distance between the residues of the cleaved bond, the reactive site loop that originally appeared to protrude from the main body of the serpin has now become a new strand of β-sheet A (the strand terminating in P1). Thus, the cleavage is accompanied by a significant change in conformation that converts the serpins from "stressed" to "relaxed" forms, consistent with the observation that they become much less temperature sensitive on cleavage. It is proposed that the insertion of the cleaved loop into the A sheet results in the stabilization of the serpin-protease complex [668, 788-

---

[6] Although ovalbumin is a member of the serpin superfamily, it is not cleaved by any target protease. Thus, its crystal structure yields an approximation of the structure of uncleaved serpins.

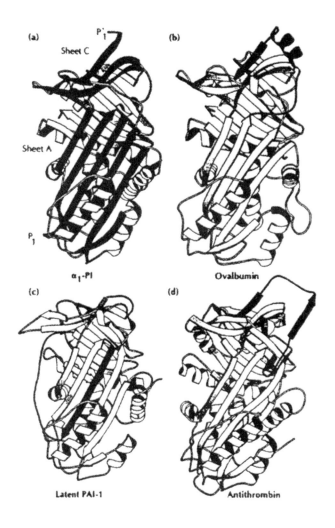

Figure 7.3. Structures of cleaved α_1 protease inhibitor (α_1-PI) and uncleaved ovalbumin, latent plasminogen activator inhibitor-1 (PAI-1), and antithrombin. Reprinted from [666], with permission.

792] [but see 793]. In the orientation shown in Figure 7.3, the protease cleavage in the reactive center loop occurs at the top of the serpin molecule and the active site serine of the protease remains covalently linked to the new C-terminal, which ends up at the bottom of the serpin. Thus, as the cleaved loop slides into the A sheet, the two covalently linked proteins must undergo a marked change in orientation with respect to each other. In the case of antithrombin, cleavage at the active center results in loss of the capacity to bind heparin, presumably due to this conformation change.

Strong support for this model in the antithrombin case comes from the recent description of the structure of crystals of uncleaved, cleaved, and latent forms of antithrombin [767, 769-771]. The actual crystal structure of the uncleaved antithrombin depicted in Figure 7.3 generally conforms to that deduced from the projection of the antithrombin sequence onto α_1-protease inhibitor. In the cleaved antithrombin the new C-terminal peptide becomes fully inserted into β-sheet A. This gives the same separation of P1 and P1' as observed for α_1-protease inhibitor. The inserted β-strand also results in the loss of the functional heparin-binding site.

Interestingly, the active center loop of the *uncleaved* antithrombin, which lacks the short helical structure seen in the ovalbumin loop, still protrudes from the sheet structure of the antithrombin in the manner suggested by the ovalbumin structure, but a small segment of the loop is inserted into the β-sheet A (Figure 7.3). The loop is quite flexible and can be displaced in several ways. For example, the binding of heparin apparently expels this loop to the render it susceptible to thrombin and factor Xa [see 794 and references therein]. In addition, under mild denaturing conditions, uncleaved antithrombin is converted into a latent form in which the loop is fully inserted into the sheet, yielding an inactive, or latent, form of antithrombin [795]. Such a latent form of a serpin, in which the active center loop is fully inserted into the β-sheet, was originally described for plasminogen activator inhibitor-1 (the dark β-strand in latent PAI-1 in Figure 7.3.) [781] and, more recently, for α_1-protease inhibitor [796]. We will discuss these serpins in greater detail in Chapter 8.

E.   Sequences in Antithrombin that Bind to Heparin

1.   Introduction

The deduction of the structural features involved in heparin/protein interactions requires structural characterization of both the heparin and the protein. Furthermore, since the interactions between two polymers take place over limited, and relatively small, regions of the total structures, it is necessary to determine which regions of each polymer are involved in the interaction. The binding of heparin to antithrombin is the most extensively studied of all the heparin/protein interactions and represents the prototype for other heparin/protein complexes. Because the experiments that have led to an understanding of the heparin/antithrombin interaction serve as a model for similar studies with other heparin-binding proteins, they are discussed in some detail here.

The heparin segment that binds to antithrombin was characterized relatively early, and only more recently has there been some agreement in the literature on which peptide regions of antithrombin recognize heparin. The available information supports the view of the binding of heparin to antithrombin that is illustrated in the two panels of Figure 7.4 [from 774, 797]. Here, both the specific heparin pentasaccharide sequence and the antithrombin polypeptide sequences that participate in the interaction are illustrated, and the docking of the pentasaccharide between two peptide segments of the antithrombin is shown. The interactions between the heparin pentasaccharide and the protein are electrostatic and involve (a) the negatively charged sulfate and carboxyl groups of the pentasaccharide and (b) specific positively charged Lys and Arg side chains on the protein. Notably, *two* peptide segments that are brought together by the folding of the antithrombin chain appear to interact with the charged groups on opposing faces of the pentasaccharide. The experimental results leading to these conclusions are outlined below.

2.   Two primary peptide sequences are involved in binding of heparin by antithrombin

a.   Introduction

The identification of the amino acid sequences of antithrombin that are involved in the specific binding of heparin has evolved slowly. Present data now seem to focus on two peptide sequences that are well removed from each other in the primary sequence: one within the A124-R145 sequence, and the other within the P41-W49 sequence [57, 713, 718, 798-811]. These two segments are held in proximity for heparin binding by the Cys 8-Cys 128 disulfide bond, selective reduction of which results in loss of heparin-binding and anti-factor Xa activity [812]. Other residues in antithrombin also seem to participate in heparin binding (below).

b.   The A124-R145 sequence

The A124-R145 sequence, found in the D helix of antithrombin, is rich in basic amino acids, including lysines 125, 133, 136, and 139 and arginines 129, 132, and 145, as shown:

A K L N C R L Y R K A N K S S K L V S A N R
 125   130   135   140   145

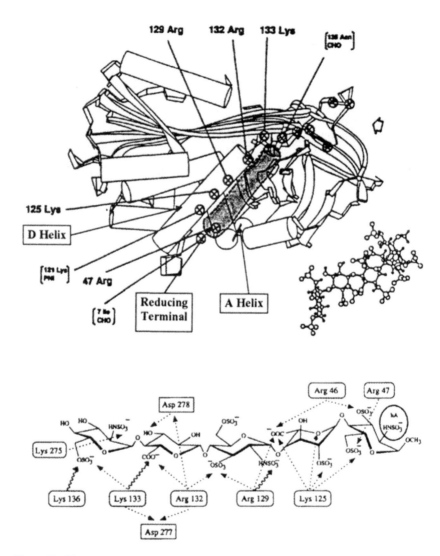

Figure 7.4.The binding of the heparin pentasaccharide to antithrombin. In the top structure, the stick figure structure and the shaded cylinder represent the heparin pentasaccharide described below. Reproduced from references [774] (top) and [797] (bottom).with permission. © 1989 and 1991, respt., American Chemical Society.

In fact, this sequence contains the highest positive charge density in antithrombin. This peptide occurs in an α-helical structure [696, 800, 813]. The implication of this sequence in heparin binding has been based largely on the characterization of hereditary defects in antithrombin that result in loss of heparin binding and on indirect probes of the structure [see 643]. Chemical modifications have implicated Lys 125 [801, 814, 815], as well as Arg 129 and Arg 145 [804] in the binding of heparin. Furthermore, Chang [801] has shown that heparin specifically protects Lys 125 and Lys 136, as well as Lys 107, from labeling with 4-N,N-dimethylazobenzene-4'-isothiocyano-2'-sulfonic acid, while facilitating the derivatization of Lys 236, apparently a manifestation of the conformation change. This sequence has also been implicated by the isolation of heparin-binding antithrombin fragments containing these amino acids. For example, cleavage of antithrombin with cyanogen bromide [802] or *Staphylococcus aureus* V8 protease [816] yields heparin-binding peptides 104-251 and 114-156, respectively. Furthermore, replacements of single basic amino acids in this sequence with Ala result in markedly reduced heparin binding [663], and antibodies directed against peptide 124-145 block heparin binding to antithrombin [803]. Finally, hereditary mutations in this region also affect heparin binding. As one example, an abnormal antithrombin with a mutation of Arg 129 to Gln fails to bind heparin [804].

Both the crystal structure studies on antithrombin and the projection of the antithrombin sequence on the α_1-protease inhibitor crystal structure show that the 124-145 peptide forms an α-helix. However, two groups have now reported on the structural features and heparin-binding properties of synthetic peptides containing this sequence and have demonstrated that the isolated peptide, in solution, forms primarily a β-strand rather than an α-helix [648, 649]. These synthetic peptides bind heparin and compete with antithrombin for heparin binding, but it is not yet clear whether the peptides recognize the antithrombin-binding pentasaccharide *specifically*, although they do block heparin binding to antithrombin (but see Section 4.D., Chapter 6). These results raise the possibility that the binding of heparin may induce a conformation change in the D helix. However, this suggestion is not compatible with studies on the docking of the heparin pentasaccharide to antithrombin, which implicate the α-helical structure in heparin binding (see Section I. H., p. 230). Thus, it appears that the 124-145 peptide does not occur as a β-strand in intact antithrombin and that the α-helical structure is stabilized, perhaps via hydrophobic interactions [817]. This problem highlights a point made in Chapter 6, namely that the information gained from the study of heparin-binding peptides may be very useful, but must be examined together with the three-dimensional structure of the same sequence in the intact protein.

c.   The P41-W49 sequence

The P41-W49 sequence, part of which is found in the A helix of antithrombin, appears to be essential for heparin binding [806, 811]. It contains Arg 46 and Arg 47 as the only basic residues.

P E A T N R R V W

41   45   49

Pro 41 [805], Arg 47 [806], and Trp 49 [57, 713, 807, 808] have been identified as critical residues in the binding of heparin to antithrombin. Hereditary mutations in this region that affect heparin binding include Pro 41→Leu, Arg 47→Cys, Arg 47→His, and Arg 47→Ser [805, 806, 809-811]. It appears that Arg 46 and Arg 47 play direct roles in heparin binding, while the nearby residues play critical roles in the positioning of the arginines.

d.   Other sequences

The apparent involvement of amino acid sequences 41-49 and 124-145 in heparin binding is supported by many reports, but other sequences in antithrombin have also been implicated. For example, it has been shown that hereditary mutations such as Ile 7→Asn, Met 20→Thr, and Arg 24→Cys, or substitutions in C-terminal residues 402-407 decrease heparin binding [682]. Also, lysines 107 and 114, which lie close to the 124-145 sequence, and Leu 99 have been implicated in heparin binding [801, 814]. Indeed, mutations of Arg 393→His or →Pro at the active center of antithrombin have been shown to affect heparin binding [818, 819]. Other peptide regions have also been suggested for heparin binding, e.g., 286-297 [820], lysines 26, 29, 62, 234, and 241 [681], and 290-300,[7] NMR studies confirm that Lys 125 is important for heparin binding, but give no evidence for a role of Lys 290, Lys 294, or Lys 297 [821].

Finally, the glycosylation patterns of antithrombin have significant effects on the affinity of the protein for heparin [759-763]. Changes in glycosylation of antithrombin may have either positive or negative effects on heparin binding. For example, an antithrombin variant in which there is no glycosylation at Asn 135 has an enhanced affinity for heparin [760]. Conversely, a variant that contains an additional carbohydrate site because of conversion of Ile 7 to an Asn has diminished affinity for heparin [786].

e.   Location of the heparin-binding sites in antithrombin

At this point, it is reasonable to ask where the critical heparin-binding 124-145 and 41-49 amino acid sequences are found in the three-dimensional structure of antithrombin? Gratifyingly, these residues form a band of positively charged arginines and lysines from the base of the A helix across the underside of the adjacent D helix (Figure 7.4). This band of basic amino acids

---

[7] J. D. Esko, personal communication.

accommodates the critical heparin pentasaccharide sequence that is involved in the binding (Section I.H., p. 230). The critical role of Arg 24, implied by the studies outlined above, suggests that the N-terminal extension of antithrombin, depicted by the dark strand in Figure 7.2, is folded to accommodate the end of the pentasaccharide as it binds to the A and D helices. Before discussing the docking of heparin to antithrombin, we must describe in more detail both the heparin binding site on the surface of thrombin and the heparin sequence that binds to antithrombin.

F.   Sequences in Thrombin that Bind to Heparin

As noted in Section I.C.3., p. 208, the heparin-mediated neutralization of thrombin by antithrombin requires that heparin bind to both antithrombin and thrombin. Thus, thrombin must have a heparin-binding site.

The structure and activities of thrombin have been described in several reviews [681, 812, 822-824]. Thrombin is synthesized in the liver as a zymogen, prothrombin, which is glycosylated at three sites and γ-carboxylated at a number of Glu sites in its N-terminal region. Human prothrombin has a 36 amino acid signal propeptide. It is secreted as a 72 kDa protein with 579 amino acids, having the active serine protease domain in the C-terminal portion of the chain [see 825 and references therein]. The conversion of prothrombin to active thrombin by factor Xa involves cleavage at Arg 271-Thr 272 and Arg 320-Ile 321 to yield α-thrombin containing two chains that are joined by a disulfide bond. The cDNA sequences for α-thrombins from a number of species have been characterized [826, 827]. Human thrombin originally has 49 amino acids in the A chain and 259 amino acids in the B chain [828]. X-ray crystallographic structures have been reported for a number of derivatives of thrombin, including the human PPACK-thrombin (D-Phe-Pro-Arg chloromethylketone-thrombin), an active site inhibited form [829]. Also, many surface residues have been mapped by site-directed mutagenesis [830, 831]. Several important surface features of the ellipsoid-shaped protein (approximately 45 x 45 x 50 Å) are illustrated in Figure 7.5, which shows the typical serine protease active site catalytic triad as a closed triangle in the center (equivalent to His 57, Asp 102, and Ser 195 in chymotrypsin).[8] The directionalities in this illustration are designated "north,"

---

[8] The residues in Figure 7.5 are numbered according to the equivalent numbers in chymotrypsinogen; i.e., the active site His, Asp, and Ser residues in prothrombin and chymotrypsinogen are aligned so that the corresponding residues in chymotrypsinogen (His 57, Asp 102, and Ser 195) and prothrombin (His 363, Asp 419, and Ser 525) are matched. Then the amino acids in the prothrombin sequence are numbered with the chymotrypsin numbers, with appropriate allowances for inserted sequences in the prothrombin [see 823].

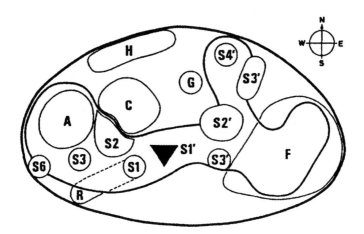

Figure 7.5. The surface structure of thrombin. The regions of the protein are designated as follows: ▼ , catalytic triad; A, aryl binding site; F, fibninogen recognition exosite (exosite I); G, glycosylation site; S1-S6, P1-P6 binding sites; S1'-S4', P1'-P4' binding sites; H, heparin binding site (exosite II); C, chemotactic site, R, RGD sequence. Reprinted from Thrombsis Research, reference [823], © 1993, with kind permission from Elsevier Science, Ltd, The Boulevard, Langford Lane, Kidlington OX5 1GB, UK.

"south," "east," and "west" as an aid in the description of the different binding regions of the enzyme [823]. The substrate binding site is formed by a crevice that runs from "west" to "east" all the way across the surface of the protein, with the active site Asp, His, and Ser residues in the middle of the crevice. The crevice is indicated by substrate recognition sites, designated by S1, S2, etc. (binding sites for P1, P2, etc.) and S1', S2', etc. (binding sites for P1', P2', etc.). It is in this crevice that the active center loop of antithrombin must bind [832, 833].

Two regions on the surface of thrombin constitute patches of positively charged amino acids; these are designated "anion exosite I" and "anion exosite II." Anion exosite I, indicated by F on the "east" side of the surface, has been identified as the recognition site for fibrinogen, thrombomodulin, and heparin cofactor II, whereas anion exosite II, H, on the "northwest" surface, has been identified as the heparin-binding site [see 823], based on chemical modification and mutation data [746, 834-836]. Anion exosite II is made up of a number of basic amino acids, all in the B chain, including arginines 93, 97, 101, 126, 165, 173, 175, and 233, and lysines 169, 235, 236, and 240. As described below, all of these appear to be positioned so that they can bind to heparin. Many of these residues are found in segments Arg 165-Arg 175 and the helical sequence, His 230-Ile 242.

$$\underline{R} \ P \ V \ C \ \underline{K} \ D \ S \ T \ \underline{R} \ I \ \underline{R}$$

165   170   175

$$H \ V \ F \ \underline{R} \ L \ \underline{K} \ \underline{K} \ W \ I \ Q \ \underline{K} \ V \ I$$

230   235   240

These peptides (a) appear on opposite sides of a groove that is 10-12 Å wide and 20-25 Å long, and (b) could accommodate a highly-charged heparin hexasaccharide as required for the heparin/antithrombin-mediated neutralization of thrombin [824]. This heparin binding site is blocked in prothrombin. Therefore, prothrombin cannot bind heparin [834, 837] and does not respond to heparin-activated antithrombin.

G.   The Sequence in Heparin that Binds to Antithrombin

1.   Introduction

The availability of pure antithrombin in the mid-70's permitted its use in the affinity fractionation of commercial heparin [706-708]. This early work demonstrated that only one-third of commercial hog mucosa heparin displays high affinity for antithrombin. Ultimately it was found that the antithrombin-binding heparin contains a well-defined pentasaccharide that binds specifically to antithrombin and induces its serpin activity. However, the initial work was complicated by the fact that in a polymer containing an average of 20-30 disaccharide units, the 2½ disaccharide sequence (i.e., the pentasaccharide) that bound to antithrombin was difficult to recognize by comparing the overall compositions of the antithrombin-binding (HA-heparin) and nonbinding (LA-heparin) heparin fractions. Consequently, it was necessary to isolate much shorter fragments of the antithrombin-binding heparin in order to deduce the structure.

2.   Assays

The isolation of any structure with a unique biological activity requires one or more bioassays that reflect the desired activity. Although the induction of the antithrombin activity requires only a short (pentasaccharide) segment of heparin, the pentasaccharide-activated antithrombin inactivates factor Xa but not thrombin (see Section I.C.3., p. 208). This was recognized (but not well understood) early in the work on the identification of the active segment of heparin. Thus, the heparin-mediated inactivation of factor Xa was used as an assay for the identification of the shortest specific segment of heparin that is required to activate antithrombin. A second assay for antithrombin activation was based on the spectral changes in antithrombin that result when heparin, or an active heparin fragment, binds to the protein (Section I.C.2., p. 206). The spectroscopic assay measures only the heparin-antithrombin interaction; it may not necessarily reflect the *specific* binding that is required for the biological activity or the formation of a heparinoid/antithrombin complex capable of neutralization of the protease. Thus, the spectroscopic assay represents a simple measure of a two-component reaction rather than the three-component reaction measured in the factor Xa neutralization assay, but the latter reflects the true biological activity assay. Both assays were used to identify the specific heparin sequence that forms the active heparin/antithrombin complex.

3.   The minimal heparinoid sequence for antithrombin activation

Identification of the minimal sequence in heparin required for binding to antithrombin was carried out concurrently in several laboratories [90, 317, 326, 623, 674, 675, 677, 838-847]. This work involved the isolation of short fragments formed by the partial cleavage of heparin with nitrous acid or heparin lyase I, as well as chemical synthesis of pentasaccharide analogs [35, 325, 848-852]. Because this work has been summarized elsewhere [391, 681, 682, 684, 797, 853], the details will not be repeated here. These studies relied on affinity chromatography of heparin fragments on immobilized antithrombin and assay of the separated oligosaccharides by the procedures referred to above. The results have shown that the smallest oligosaccharide that binds with high affinity to antithrombin is the pentasaccharide illustrated in Figure 7.6. Following the designations used by Grootenhuis and van Boeckel [797], removal of each group marked with a double exclamation mark (!!) leads to a loss of more than 95% of the anti-Xa activity, whereas removal of each group marked with a single exclamation mark (!) leads to a loss of about 75% of the activity [846, 847, 854]. Thus, these groups are critical for antithrombin binding, and their role in binding must be reconciled with the positions of the antithrombin basic amino acid side chains that form the counterions for the electrostatic-binding interactions.

Several features of the pentasaccharide structure deserve comment:

- The numbering of the monosaccharide units in the pentasaccharide is a relic of earlier work in which a *hexasaccharide* was believed to be the smallest oligosaccharide that bound to antithrombin. The residues in the

UNITS   **2**   **3**   **4**   **5**   **6**

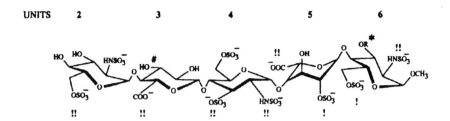

Figure 7.6. Structure-activity relationships of heparin pentasaccharide analogs. Reprinted with permission from [797]. Copyright 1991, American Chemical Society.

hexasaccharide were designated unit 1 (nonreducing terminal) to unit 6 (reducing terminal). With the demonstration that the non-reducing terminal residue (unit 1) could be removed from the hexasaccharide without significant reduction in binding to antithrombin, the pentasaccharide structure became the smallest antithrombin-binding oligosaccharide, but the numbering of the monosaccharide units that was used for the hexasaccharide was retained.

- The pentasaccharide contains a unique 3-O-sulfated $GlcNSO_3$ at unit 4 that is a minor monosaccharide component of heparin (see Tables 2.2 and 2.3).[9] The discovery of the 3-O-sulfate group was a result of serendipity. In research directed toward developing a simple assay for the enzyme that removes the N-sulfate groups from heparin, Leder discovered that human urine contained a sulfatase that would remove the sulfate group from $GlcNSO_3$-3-SO_4 [855]. At that time, no naturally occurring 3-O-sulfated groups were known in heparin or in any other natural product. However, when this enzyme was tested on heparin fragments, it was discovered that it would release free sulfate from these fragments, provided the GlcN residue containing the 3-O-sulfate was N-sulfated [855], suggesting for the first time that heparin contained 3-O-sulfate groups. This was confirmed by chemical [623] and NMR studies [840, 843, 844].

- The pentasaccharide contains one rigid GlcA residue at unit 3 [138, 325, 326]. Apparently this rigid ring structure between the upstream GlcN residue that carries a critical 6-O-sulfate (unit 2) and the downstream $GlcNSO_3$ that carries the critical 3-O-sulfate group (unit 4) is required for

---

[9] Although it was originally believed that the 3-O-sulfate group was found *only* in the antithrombin- binding pentasaccharide, it is now cear that it also occurs in other parts of heparin chains (48, 62, 115, 119, 185).

224 Heparin-Binding Proteins

antithrombin binding [327, 856]. Also, the specific orientation of its carboxyl group may be important.

- The pentasaccharide with an N-acetylated GlcN in unit 2 is the most commonly isolated form. However, the N-acetyl group can be removed and replaced with an N-sulfate group without any marked alteration in the binding and activity of the oligosaccharide [735]. The frequent occurrence of GlcNAc at unit 2 may be related to the necessity for the GlcA residue in position 3, since, when GlcNAc lies just upstream from a GlcA residue, the epimerase that interconverts GlcA and IdoA cannot act on the GlcA (see discussion of the biosynthesis of heparin in Chapter 2). Thus, a GlcNAc residue in the antithrombin-binding pentasaccharide may simply optimize the biosynthetic chances of retaining the critical GlcA residue at unit 3. This argument has been advanced by Lindahl *et al.* [90], who have also taken note that the procedure for obtaining antithrombin-binding heparin fragments (cleavage of heparin with nitrous acid) can cleave at many of the $GlcNSO_3$ residues that may have been present in unit 2 of the antithrombin-binding pentasaccharides in the heparin. In fact, Lindahl and co-workers have demonstrated that there are N-sulfated GlcN residues present in the antithrombin-binding oligosaccharides that are obtained by affinity chromatography of heparin fragments [90].

- Although the pentasaccharide sequence occurs in only one-third of the chains in hog mucosa heparin, it is now recognized that the amounts of this sequence in various heparins and heparan sulfates are quite variable [26, 57, 206, 632, 753, 857-860].

4.   Effects of structural modifications of the pentasaccharide on it activity

The identification of the natural pentasaccharide as a sequence that is part of some heparin chains and the development of the chemical synthetic procedures for synthesis of such structures [733, 797, 861] set the scene for further studies of the structure-activity relationships of various derivatives of the pentasaccharide. Figure 7.7 shows the activities of a variety of chemically synthesized variants of the pentasaccharide [391]. The K_d values for pentasaccharide binding to antithrombin and the activity in the factor Xa assays are shown. Several points are of note.

- All of these structures contain the same monosaccharide units in the same sequence that is found in the natural pentasaccharide. With the exception of pentasaccharide 1, all are methyl glycosides. The activities of pentasaccharides 1 and 2, which differ only in the α-methyl glycoside

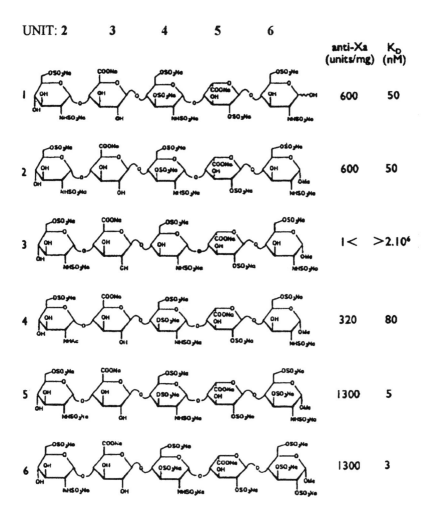

Figure 7.7. Biological properties of synthetic pentasaccharides. Activities shown are the anti-Xa activity of the pentasaccharide and the K_d for the dissociation of the pentasaccharides from antithrombin. Reprinted from [391], with permission.

substitution at the reducing terminal of pentasaccharide 2, are identical, indicating that the methyl glycoside has no effect on the activity.

- Pentasaccharide 4 is the most commonly occurring sequence found in antithrombin-binding heparin fragments, whereas pentasaccharide 2 is the same structure in which the N-acetyl group on unit 2 is replaced by an N-sulfate group. Both of these structures are quite active in both assays, with pentasaccharide 2 being somewhat more active than pentasaccharide 4.

- Pentasaccharides 2 and 3 are identical except that unit 4 in pentasaccharide 3 lacks the 3-O-sulfate. The removal of the 3-O-sulfate gives total loss of activity.

- Pentasaccharides 5 and 6 represent an improvement on Nature. In both of these structures, an additional 3-O-sulfate group is added to the reducing terminal residue (unit 6). Comparison of structure 5 with the natural structure (pentasaccharide 2) shows a doubling of anti-Xa activity and a 10-fold lowering of the K_d. The increased activity is due to the interaction of the extra 3-O-sulfate with Arg 46 on the "northern side" of the pentasaccharide (below).

- Pentasaccharide 6 is identical to pentasaccharide 5, but with a replacement of the N-sulfate group on the reducing terminal residue with an O-sulfate group. The activities of 5 and 6 are essentially identical.

Not shown in Figure 7.7 is another interesting result of the synthetic work in which an additional 3-O-sulfate group is added to unit 2. This reduces the activity by 60%. This substitution appears to cause steric hindrance of the binding, ascribed to the electrostatic repulsion of this 3 O-sulfate by the Asp 278 [797].

5.   The structure surrounding the antithrombin-binding pentasaccharide

Although it was originally believed that the 3-O-sulfate group was found only in the antithrombin-binding pentasaccharide, it is now clear that it also occurs in other parts of heparin chains [94, 117, 354, 862, 863]. As described in Chapter 2, a number of heparin fragments that contain 3-O-sulfated GlcN residues have been characterized. These are abstracted from Table 2.3 and presented in Table 7.2 so that we can see the variations in the monosaccharide sequences that occur around the 3-O-sulfated GlcN residues and try to identify sequences that contain 3-O-sulfated GlcN residues but show no antithrombin-binding capacity. The oligosaccharides in Table 7.2 are fragments formed by cleaving heparin with nitrous acid or heparin lyase I, and a few are purified directly from the lower molecular weight side fractions of heparin. As isolated, the nitrous acid products have reducing terminal AMan residues derived from N-sulfated GlcN residues in the polymer, and these are depicted as the original GlcN residues in bold at the reducing terminals. The heparin lyase products have $\Delta^{4,5}$-uronic acids at their nonreducing terminals, and these are depicted as the original IdoA residues in bold (it is assumed that these were all derived from IdoA rather than GlcA residues). The oligosaccharides for which neither the reducing terminal nor the nonreducing terminal is shown in bold are those that were isolated directly from the lower molecular weight fraction of heparin. When examined in this fashion, there are several cases in which nitrous acid cleavage and lyase cleavage yield the same fraction. For example, oligosaccharides 15 and 22 are the same, as are oligosaccharides 49 and 51 and oligosaccharides 12 and 18.

The monosaccharide units are numbered as in Figure 7.6, and the critical structural features for antithrombin-binding from Figure 7.6 are indicated in the top row of the table. As shown, some of the GlcN residues are both 3-O-sulfated and 6-O-sulfated, whereas others are only 3-O-sulfated. The latter have been grouped at the bottom of the table. The activities of these structures have been measured in anti-Xa assays, or simply by their tight binding to antithrombin. A number of the oligosaccharides have not been tested for either activity. A compilation similar to Table 7.2 has been presented by Toida et al. [863].

Units 2, 3, 4, 5, and 6 represent the antithrombin-binding pentasaccharide. Several points are of particular interest:

- Unit 2 always contains the critical 6-O-sulfate, but the GlcN residues may be either N-acetylated or N-sulfated.

- Unit 3 is invariably a GlcA residue. Presumably a GlcA residue in this position is essential for 3-O-sulfation of the downstream GlcN.

- Unit 4 always has both of the critical substituents for antithrombin binding, namely the N-sulfate and the 3-O-sulfate. In most cases this GlcN residue is also 6-O-sulfated, but there are several examples in which the 6-O-sulfate is missing.

- Unit 5 is always an IdoA residue and, in only one case (oligosaccharide 39) does the IdoA lack the essential 2-O-sulfate. As expected, the latter oligosaccharide is almost inactive.

- In unit 6 the GlcN residue is always N-sulfated, and in only one case (oligosaccharide 50) is the critical 6-O-sulfate missing.

Interestingly, the two oligosaccharides, 39 and 50, that lack structural features that are necessary for productive antithrombin binding are both of sufficient length to contain all five of the residues of the antithrombin-binding pentasaccharide. Oligosaccharide 39 shows only minimal activity, and, although no activity measurements were made on oligosaccharide 50, it also is presumably inactive. Thus, these two oligosaccharides represent sequences that contain 3-O-sulfated GlcN residues but lie outside of the antithrombin-binding regions. Three of the four oligosaccharides that lack the nonessential 6-O-sulfate on unit 4 do not extend to units 5 and 6. It is interesting to consider whether the

Table 7.2. Sequence of Oligosaccharides Surrounding the

No.[1]	Ref.	Activity [2]	-2	-1	1	2	3
			CRITICAL SUBSTITUENTS [3]			**6-O-SO₄**	**COO⁻**
15	[28]	NA			IdoA	GlcNAc-6-SO₄	GlcA
22	[30]	NA			**IdoA**	GlcNAc-6-SO₄	GlcA
27	[31]	NA			**IdoA**	GlcNSO₃-6-SO₄	GlcA
35	[34]	<1 u			**IdoA -2-SO₄**	GlcNSO₃-6-SO₄	GlcA
36	[35]	B			IdoA	GlcNSO₃-6-SO₄	GlcA
41	[30, 36]	68 u	**IdoA -2-SO₄**	**GlcNSO₃-6-SO₄**	IdoA	GlcNAc-6-SO₄	GlcA
39	[37, 863]	8 u			**IdoA -2-SO₄**	GlcNAc-6-SO₄	GlcA
47	[38]	B	IdoA-2-SO₄	GlcNSO₃-6-SO₄	IdoA	GlcNAc-6-SO₄	GlcA
49	[38, 39]	B			IdoA	GlcNAc-6-SO₄	GlcA
51	[41]	240 u, B			**IdoA**	GlcNAc-6-SO₄	GlcA
48	[35]	B			IdoA	GlcNSO₃-6-SO₄	GlcA
52	[37, 863]	NA	**IdoA -2-SO₄**	**GlcNSO₃-6-SO₄**	IdoA	GlcNAc-6-SO₄	GlcA
54	[37, 863]	180 u	**IdoA -2-SO₄**	**GlcNSO₃-6-SO₄**	IdoA	GlcNAc-6-SO₄	GlcA
55	[35]	B	IdoA-2-SO₄	GlcNSO₃-6-SO₄	IdoA	GlcNSO₃-6-SO₄	GlcA
12	[28]	NA			IdoA	GlcNAc-6-SO₄	GlcA
18	[31]	NA			**IdoA**	GlcNAc-6-SO₄	GlcA
38	[36]	NA	**IdoA -2-SO₄**	**GlcNSO₃-6-SO₄**	IdoA	GlcNAc-6-SO₄	GlcA
50	[40]	NA	**IdoA -2-SO₄**	**GlcNSO₃**	IdoA	GlcNAc-6-SO₄	GlcA

[1] The oligosaccharides are given the same numbers that are found in Table 2.3. These oligosaccharides were obtained as heparin fragments formed by cleavage of heparin with either nitrous acid or heparin lyase I. Consequently, the nitrous acid products had anhydromannose residues on the reducing terminal, whereas the lyase I products had a Δ4,5 uronic acid residues on the nonreducing terminals. In order to observe the original sequences, the anhydromannose residues

3-O-Sulfated Glucosamine Residues in Heparin

Unit				
4	**5**	**6**	**7**	**8**
NSO₃, 3-O-SO₄	**COO⁻** **2-O-SO₄**	**NSO₃** **6-O-SO₄**		
GlcNSO₃-3,6-di-SO₄				
GlcNSO₃-3,6-di-SO₄				
GlcNSO₃-3,6-diSO₄				
GlcNSO₃-3,6-di-SO₄				
GlcNSO₃-3,6-diSO₄	IdoA-2-SO₄	GlcNSO₃-6-SO₄		
GlcNSO₃-3,6-di-SO₄				
GlcNSO₃-3,6-diSO₄	IdoA	GlcNSO₃-6-SO₄		
GlcNSO₃-3,6-diSO₄	IdoA-2-SO₄	**GlcNSO₃-6-SO₄**		
GlcNSO₃-3,6-diSO₄	IdoA-2-SO₄	GlcNSO₃-6-SO₄	IdoA-2-SO₄	**GlcNSO₃-6-SO₄**
GlcNSO₃-3,6-diSO₄	IdoA-2-SO₄	GlcNSO₃-6-SO₄	IdoA-2-SO₄	GlcNSO₃-6-SO₄
GlcNSO₃-3,6-diSO₄	IdoA-2-SO₄	GlcNSO₃-6-SO₄	IdoA-2-SO₄	**GlcNSO₃**
GlcNSO₃-3,6-diSO₄	IdoA-2-SO₄	GlcNSO₃-6-SO₄		
GlcNSO₃-3,6-diSO₄	IdoA-2-SO₄	GlcNSO₃-6-SO₄	IdoA-2-SO₄	GlcNSO₃-6-SO₄
GlcNSO₃-3,6-diSO₄	IdoA-2-SO₄	GlcNSO₃-6-SO₄	IdoA-2-SO₄	GlcNSO₃-6-SO₄
GlcNSO₃-3-SO₄				
GlcNSO₃-3-SO₄				
GlcNSO₃-3-SO₄				
GlcNSO₃-3-SO₄	IdoA-2-SO₄	GlcNSO₃		

are converted to GlcNSO3 residues, and the $\Delta^{4,5}$ uronic acid residues are converted to iduronic acid residues. Where these substitutions have been made, the terminals are indicated in bold type.

[2] Activity was assayed by several procedures by the different authors. Anti-xray assays are indicated by the number of units(u). Binding to antithrombin is indicated by B. NA = not assayed..

lack of 6-O-sufates on units 4 and 6 might be biosynthetically coupled. If so, oligosaccharides 12, 18, and 38 may also be derived from non-antithrombin-binding regions of heparin.

H. The Antithrombin/Heparin Interaction

Grootenhuis and van Boeckel [797] have described the docking of the heparin pentasaccharide to antithrombin. For this study, the antithrombin was modeled on the structure of α_1-protease inhibitor as described in Section I.D.2., p. 210, to obtain a three-dimensional structure of antithrombin in which the two heparin-binding peptide sequences form a pocket with the Pro 41-Trp 49 sequence of helix A opposite the Ala 125-Arg 145 sequence on helix D (Figure 7.4). The pentasaccharide used for the docking study is the most active and tight-binding species described to date, namely structure 5 in Figure 7.7, in which an additional 3-O-sulfate is added to unit 5 of the naturally occurring pentasaccharide. Based on conformation studies [138, 325, 326], the pentasaccharide is projected as a three-dimensional structure with six of the critical negatively charged groups on one side (referred to as the "south side") of the molecule and three of the critical negatively charged groups on the other (the "north side"). The location of the essential groups of the pentasaccharide on opposite sides of the structure ("south" and "north") is completely consistent with both the positions of the charged groups and the directionality of the heparin chain in the three-dimensional solution conformation of heparin proposed by Mulloy et al. [132] (Figure 6.1). Furthermore, this depiction of the structure suggests that it would interact through two different faces of antithrombin. Thus, the two negatively charged faces of the pentasaccharide can be juxtaposed with the two positively charged amino acid sequences of the antithrombin to maximize the complementarity between the lysines and arginines on the protein and the carboxyl and sulfate groups on the pentasaccharide. This is accomplished by pairing the six negative charges of the "south side" of the pentasaccharide with Lys 125, Arg 129, Arg 132, Lys 133, and Lys 136 of the D helix, and the three negative charges of the "north side" of the pentasaccharide with Arg 46 and Arg 47 of the A helix (Figure 7.4). The authors suggest that the initial step in the binding is the interaction between the "south side" and the D helix, which is then followed by the conformation change that permits the interaction between the "north side" and the A helix,[10]

---

[10] Carlson et al. [864] proposed earlier that heparin binds initially to antithrombin at a position near the Trp 49, giving the conformation change, and that further binding occurs in the Lys- rich peptide from residues 124 to 145.

Several additional features of the docking results are noted.

- The presence of two separate binding sites is in agreement with kinetic studies revealing that binding of heparin to antithrombin is a two-step process [726]. The exact location of the aminosulfate residue at unit 6 in the field of the dipole of helix A not only enhances the interaction of the pentasaccharide with antithrombin [854, 865], but also suggests the participation of the A helix in the transmission of the conformational change in the heparin-binding site toward the active center of antithrombin.

- The enhancement of the pentasaccharide activity by the addition of the extra 3-O-sulfate at C3 of residue 6 [327, 328] is due to the formation of an additional salt bridge between this 3-O-sulfate and Arg 47.

- Lys 275, which is initially distant from the pentasaccharide docking site, moves toward the 6-O-sulfate group of unit 2 so that this interaction can contribute to the binding. In fact, it has been shown that when the 6-O-sulfate is replaced with a hydroxyl group, about 30% of the activity of the pentasaccharide is lost [797].

- Lys 136, known to be involved in binding [801], moves to the essential 6-O-sulfate of unit 2.

- Carrell et al. [767] suggest further that the N-terminal extension of antithrombin, particularly Arg 24, plays a role in the pentasaccharide binding (see Figure 7.2).

I.   Structure of the Heparin/Antithrombin/Thrombin Complex

1.   The antithrombin/thrombin interaction

Since the crystal structures of antithrombin and thrombin are now established and the critical interacting sites of antithrombin (the reactive center) and thrombin (the active site) are known, it is possible to dock these two proteins to obtain an approximation of their alignment as they react. Schreuder et al. [769] have developed a model for the docking of these two proteins. In this model, in the presence of heparin the active center loop of the antithrombin projects from the A and C β-sheets and binds to the active site of the thrombin. In this alignment, the heparin-binding regions of antithrombin and thrombin are juxtaposed so that the heparin sequences containing at least 18 monosaccharides can bind to the two proteins, serving as a template to facilitate their interaction. The only structure that remains to be approximated is that of the antithrombin-thrombin complex that is formed *after* thrombin cleaves the active center of the antithrombin. This structure presumably involves the antithrombin portion of the complex with the active center loop inserted into β-sheet A and with the carboxyl group of the newly formed C-terminal arginine carboxyl group in ester linkage to the thrombin active site serine.

2.   Orientation of heparin in the heparin/antithrombin complex

On the surface of the antithrombin molecule shown in Figure 7.4, the bound heparin pentasaccharide extends from the bottom of the A helix to the top of the D helix. The active center of the antithrombin, where cleavage by thrombin or factor Xa will occur, is shown by the arrow at the top of the figure. In the docking studies, the orientation of the pentasaccharide that maximizes the complementarity of the contact sites on the protein and the pentasaccharide results in the placement of the *downstream* (reducing terminal) end of the pentasaccharide opposite the A helix while the *upstream* (nonreducing terminal) end binds to the D helix and projects to the left of the active center loop of the antithrombin.[11] With longer heparin chains, in which the pentasaccharide sequence is extended on both ends, the heparin sequences downstream from the pentasaccharide project into the solution off the A helix, whereas the sequences upstream from the pentasaccharide lie across the rest of the surface of the antithrombin. These upstream disaccharide units may interact with some of the other antithrombin amino acids that have been implicated in heparin binding. If they are extended further, they also bind to thrombin anion exosite II.

Octadecasaccharides (18-mers) that have been tested for their thrombin-neutralizing capacity are mixtures of heparin fragments in which the specific pentasaccharide occurs at different positions in each chain, but the *active* components of these mixtures presumably have the pentasaccharide at one end, with an extension of 13 additional upstream monosaccharides that includes, and terminates with, the hexasaccharide sequence that binds to the thrombin. Thus, when thrombin interacts with antithrombin in the presence of a mixture of 18-mer heparin fragments, those heparin fragments that bind these two proteins together must have the antithrombin-binding pentasaccharide on the downstream end and the thrombin-binding hexasaccharide on the upstream end, with 7 additional monosaccharides in between (giving the necessary 18 monosaccharides required to activate the thrombin-antithrombin reaction).

---

[11] As described in Chapter 2, it is customary to draw oligosaccaride structures with the nonreducing terminal (upstream) end on the left and the reducing terminal (downstream) end on the right. In Figure 7.2, the docked pentasaccharide sequence is shown in the opposite orientation.

II.   THE ROLE OF ANTITHROMBIN *IN VIVO*

A.   Introduction

This elegant picture of the mechanism of the heparin-mediated antithrombin neutralization of thrombin and factor Xa is derived from the work of many researchers over a period extending back to the 70's and earlier. We must now ask whether this mechanism is a true representation of the events that actually occur *in vivo*. A close examination of the *in vivo* control of blood coagulation raises many questions. The initial problem is that *heparin* is not found in circulation [684] [12] and therefore does not participate in hemostasis. Consequently, the heparin-mediated mechanisms that we have described may be operative only when heparin is administered clinically. However, it is clear that antithrombin does play a role in hemostasis, even when heparin is not administered, as genetic defects in antithrombin predispose patients to thrombosis [reviewed in 684, 685, 686]. Thus, there are two distinct circumstances in which the role of antithrombin may be considered: (a) that in which *exogenous* heparin is used clinically as an anticoagulant/antithrombotic, and (b) that in which antithrombin participates in *endogenous* mechanisms that regulate hemostasis. We will address these aspects below, but first we must resolve the dilemma of how antithrombin can exert its activity *in vivo* in the absence of heparin.

B.   The Activation of Antithrombin by Heparan Sulfate Proteoglycan

To this point, we have shown that antithrombin, in the absence of heparin, reacts with thrombin at a rate that is insufficient to block clotting *in vivo* but that this rate does become physiologically relevant when heparin is present. However, heparin occurs only in mast cells [867, 868], which are found beneath the endothelium, and there is some question as to whether heparin from mast cells ever gets into the blood [12] [684]. How can antithrombin exert its biological activity in the absence of heparin? This question has been addressed by Rosenberg and Marcum and their collaborators in a series of reports that have shown that the endothelial lining of blood vessels contains a cell surface heparan sulfate proteoglycan that has sufficient antithrombin-activating activity to accommodate the requirement for thrombin and Xa neutralizing activity in the vascular system [629, 630, 859, 869-871]. These studies, which have been reviewed in detail [684], have shown that mast cell-free endothelial cell

---

[12] However, see reference [866].

preparations, as well as cloned endothelial cells from a variety of species [206], produce heparan sulfate proteoglycans that have the capacity to bind to antithrombin affinity matrices and to inactivate thrombin and factor Xa in the presence of antithrombin. Of the total heparan sulfate proteoglycan derived from these various sources, only 0.3 to 10% adheres to antithrombin affinity matrices; i.e., the antithrombin-binding heparan sulfate proteoglycan represents only a small fraction of the total endothelial cell heparan sulfate. The antithrombin-binding portion contains all of the 3-O-sulfated glucosamine units of the heparan sulfate and has antithrombin activities ranging from 11 to 37 USP units/mg. Furthermore, perfusion of rat and mouse hind limbs with thrombin plus antithrombin gives 10- to 16-fold stimulation of the amount of the antithrombin-thrombin complex formed compared to nonperfused controls. This activity is completely abolished by prior perfusion of the hind limbs with purified heparin lyase. Perfusion with buffer alone fails to wash any heparinoids into the effluent, indicating that the heparan sulfate proteoglycan is tightly bound to the endothelium. The same levels of activity are found in normal and mast cell-deficient mice, indicating that mast cell heparin is not involved in these responses. These results lead to the following conclusions with respect to the *in vivo* activity of antithrombin.

- It is the endothelial cell heparan sulfate proteoglycan that activates antithrombin *in vivo*, not heparin. Sufficient activity is present in the vasculature to accommodate the total need for activated antithrombin in hemostasis.

- Antithrombin in the circulation becomes bound to the active species of heparan sulfate proteoglycan on the endothelial cell surface and is thus continuously available *in an activated state* to inhibit circulating thrombin or factor Xa.

- Only a subset of the endothelial cell heparan sulfate proteoglycan, generally representing less than 10% of the total, contains all of the antithrombin-activating activity. The large reservoir of the endothelial cell surface heparan sulfate proteoglycan that cannot bind antithrombin remains available for the binding of other heparin-binding proteins in the vasculature; i.e., antithrombin binding leaves most of the endothelial cell surface heparan sulfate proteoglycan exposed for other possible interactions (see Chapters 8-11).

- Although the endothelium can recruit only a small fraction of the circulating antithrombin (<1 %), this amount is sufficient to control thrombosis. Thus, there remains a large reservoir of antithrombin in the circulation to replace the antithrombin that is consumed in the neutralization of thrombin and factor Xa.

- Since the heparinoids that activate antithrombin serve as catalysts in the neutralization of thrombin and factor Xa, the endothelial cell heparan sulfate proteoglycan can act repeatedly in the interactions of antithrombin with its target proteases. Furthermore, as noted in Chapter 5, the cell surface heparan sulfate proteoglycan pool is in a constant state of metabolic renewal. Thus, any endothelial cell surface heparan sulfate proteoglycan that might be lost from the surfaces of these cells would be rapidly replaced by newly secreted heparinoid. It is also possible that the endothelial cells may respond to certain regulatory mechanisms to produce different levels of anticoagulantly active heparan sulfate proteoglycan.

To summarize the events that take place on the surface of endothelial cells *in vivo*, we may note that a small fraction of the antithrombin in the vasculature is bound to endothelial cell surface heparan sulfate proteoglycan in an activated state. Circulating thrombin and factor Xa bind to the activated antithrombin and cleave the antithrombin at its active center, forming a covalent protease-antithrombin complex. On cleavage, the antithrombin loses its affinity for the heparan sulfate proteoglycan and so the protease-antithrombin complex dissociates from the endothelial cell heparan sulfate and is catabolized (Chapter 8). The endothelial cell heparan sulfate proteoglycan may then bind to and activate more of the antithrombin from the circulating reservoir for further protease neutralization reactions.

As has been discussed, factor Xa neutralization by antithrombin requires only the specific antithrombin-binding pentasaccharide. Such sequences are clearly present in endothelial cell heparan sulfate. However, thrombin neutralization by antithrombin requires that the heparan sulfate contains the specific antithrombin-binding pentasaccharide and at least 13 additional monosaccharide residues placed upstream from the pentasaccharide and terminated at the upstream end (the position of thrombin binding) by a sulfated hexasaccharide sequence. Such structural features are common in highly sulfated heparins, but the endothelial cell heparan sulfate, the true inducer of antithrombin activity *in vivo*, may contain very limited amounts of these longer oligosaccharide sequences. In heparan sulfate the necessary antithrombin-binding pentasaccharide units clearly are present, and highly sulfated thrombin-binding hexasaccharides may also be present. However, our knowledge of the structures of heparan sulfate (Chapter 2) would suggest that these two structural features would occur to a very limited extent in a single block of sulfated disaccharides; i.e., the blocks of sulfated disaccharides in heparan sulfates are usually not as long as the required 18-mer. The possibility exists that heparan sulfate may have the pentasaccharide segment in one sulfated block and the thrombin-binding hexasaccharide segment in another, and that these two segments would be separated by a block of unsulfated disaccharides. Such structures have not yet been demonstrated in endothelial cell heparan sulfate, and one might expect that such structures would be much less common than the pentasaccharide alone. Such logic would suggest that the endothelial cell heparan sulfate would be more effective in the neutralization of factor Xa than in the neutralization of thrombin. In fact, heparan sulfate isolated from human aorta has been shown to be four-fold more effective in the antithrombin-mediated inhibition of factor Xa than in the inhibition of thrombin [872]. This is discussed in greater detail in Chapter 8.

III. ANTITHROMBIN AND THROMBIN AS PROTOTYPES FOR HEPARIN-BINDING PROTEINS

Although antithrombin is clearly the prototype by which other heparin-binding proteins may be judged, we have discussed *two* heparin-binding proteins in this chapter—antithrombin and thrombin. It is of some value to compare what we have learned about the similarities and differences in the interactions of these two proteins with heparin and heparan sulfate. The sequence that binds to antithrombin is clearly a very explicit structure that can tolerate only minor structural alterations without loss of its activity and that occurs rarely in most heparins. We know less about the hexasaccharide sequence that binds to thrombin, but it is a highly sulfated sequence and is a much more common structure in heparins. Clearly, the specific anticoagulant action of heparin is expressed in large part through the heparin/antithrombin complexation which leads to the inactivation of both thrombin and factor Xa. The *direct* binding of heparin to thrombin may also inhibit the activity of thrombin, albeit without permanent inactivation of the protease. The *in vivo* inactivation of thrombin and factor Xa clearly is mediated by the specific antithrombin-binding pentasaccharide in endothelial cell heparan sulfate proteoglycan and not by direct and reversible inhibition of thrombin by heparan sulfate. As we examine other heparin-binding proteins, we may consider antithrombin as the prototype for *specific* heparin- or heparan sulfate-mediated effects. The thrombin model of *nonspecific* heparin- and heparan sulfate-mediated modulation of protein activity may also be found and may be just as common, or perhaps an even more common, mode of heparan sulfate modulation of protein activities *in vivo*.

Studies on the heparin interactions with antithrombin and thrombin are instructive in another way. They demonstrate that a wide range of methodologies have been required to characterize the antithrombin/heparin/thrombin interactions. Physical techniques have included (a) NMR characterization of protein and oligosaccharide structures, (b) X-ray crystallography for determining the three-dimensional structure of the heparin-binding proteins, (c) calorimetry and circular dichroism for observing the interactions between proteins or peptides and heparin oligosaccharides, and (d) molecular modeling to deduce the three-dimensional structure of antithrombin and the docking of heparin pentasaccharide to antithrombin. Chemical approaches have been extremely useful in the synthesis of both peptides and oligosaccharides. A variety of biochemical approaches have contributed to our understanding of the mechanism of the heparin anticoagulant activity, including (a) cloning of the cDNAs for the proteins so that amino acid sequences could be deduced and site-directed mutagenesis could be used to identify critical amino acids for heparin binding, (b) biological assays to measure heparin and heparin oligosaccharide activities *in vitro*, (c) affinity chromatography for the isolation of specific antithrombin-binding heparin oligosaccharides and specific heparin-binding peptides, and (d) measurement of binding and kinetic constants to observe the interactions between heparin and antithrombin or peptides. Finally, whole animal experiments have been of value in determining the biological activity of endothelial cell heparan sulfate *in vivo* and the effects of genetic defects in antithrombin on normal hemostatic mechanisms. The time period over which the understanding of the heparin/antithrombin interaction has grown to its present stage has been more than 35 years, and during this time important new technologies have emerged. Contributions have come from many labs. All of the approaches that have been useful in the heparin/antithrombin studies point the way for studies of other heparin-binding proteins.

We will return to the question of the activity of antithrombin *in vivo*. However, a further description of the role of antithrombin in hemostatic metabolism requires that we first consider, in Chapter 8, the reactions that take place in coagulation and fibrinolysis. There is an additional motive for embarking on a discussion of the hemostatic pathways, namely, that these pathways involve a large number of heparin-binding proteins.

ADDITIONAL REFERENCES

Cadène, M., Duranton, J., North, A., and Si-Tahar, M. (1997) Inhibition of neutrophil serine proteinases by suramin, *J. Biol. Chem.* **272**, 9950-9955.

Chang, W.-S. W., and Harper, P. L. (1997) Commercial antithrombin concentrate contains inactive L-forms of antithrombin, *Thromb. Haemostas.* **77**, 323-328.

Fan, B., Turko, I. V., and Gettins, P. G. W. (1994) Antithrombin histidine variants. ^{1}H NMR resonance assignments and functional properties, *FEBS Lett.* **354**, 84-88.

Hopkins, P. C. R., Chang, W.-S. W., Wardell, M. R., and Stone, S. R. (1997) Inhibitory mechanism of serpins. Mobility of the C-terminal region of the reactive-site loop, *J. Biol. Chem.* **272**, 3905-3909.

Kridel, S. J., Chan, W. W., and Knauer, D. J. (1996) Requirement of lysine residues outside of the proposed pentasaccharide binding region for high affinity heparin binding and activation of human antithrombin III, *J. Biol. Chem.* **271**, 20935-20941.

Kridel, S. J., and Knauer, D. J. (1997) Lysine residue 114 in human antithrombin III is required for heparin pentasaccharide-mediated activation, *J. Biol. Chem.* **272**, 7656-7660.

Meagher, J. L., Huntington, J. A., Fan, B., and gettins, P. G. W. (1996) Role of arginine 132 and lysine 133 in heparin binding to and activation of antithrombin, *J. Biol. Chem.* **271**, 29353-29358.

Olson, S. T., Frances-Chmura, A. M., Swanson, R., Björk, I., and Zettlmeissel, G. (1997) Effect of individual carbohydrate chains of recombinant antithrombin on heparin affinity and on the generation of glycoforms differing in heparin affinity, *Arch. Biochem. Biophys.* **341**, 212-221.

O'Malley, K. M., Nair, S. A., Rubin, H., and Cooperman, B. S. (1997) The kinetic mechanism of serpin-proteinase complex formation. An intermediate between the Michaelis complex and the inhibited complex, *J. Biol. Chem.* **272**, 5354-5359.

Petitou, M., Barzu, T., Herault, J.-P., and Herbert, J.-M. (1997) A unique trisaccharide sequence in heparin mediates the early step of antithrombin III activation, *Glycobiology* **7**, 323-327.

Skinner, R., Abrahams, J.-P., Wisstock, J. C., Lesk, A. M., Carrell, R. W., and Wardell, M. R. (1997) The 2.5 A structure of the antithrombin indicates a conformational change at the heparin binding site, *J. Mol. Biol.* **266**, 601-609.

Stone, S. R., and Le Bonniec, B. F. (1997) Inhibitory mechanism of serpins. Identification of steps involving the active-site serine of the protease, *J. Mol. Biol.* **265**, 344-362.

Tsiang, M., Jain, A. J., and Gibbs, C. S. (1997) Functional requirements for inhibition of thrombin by antithrombin III in the presence and absence of heparin, *J. Biol. Chem.* **272**, 12024-12029.

Chapter 8. Heparin-Binding Proteins in Hemostasis

I. HEMOSTASIS

A. Introduction

The hemostatic mechanisms that regulate blood clotting result in a delicate balance between clot formation (coagulation) and clot dissolution (fibrinolysis or thrombolysis). Clots (thrombi) are formed rapidly in the initial stage of repair of injuries to the vasculature. The extent of thrombus formation and growth is under exacting control, and, as secondary tissue repair mechanisms come into play, clots are dissolved. The balance between coagulation and fibrinolysis is critical; if the clotting stage is not controlled, then excessive clot formation will yield lesions that restrict blood flow (thrombosis). However, if coagulation is inhibited, as when exogenous heparin is administered, excessive bleeding occurs. Furthermore, there are physiological conditions which may predispose to excessive clotting or to excessive bleeding, and clinical intervention is often necessary to "assist" the hemostatic mechanisms in achieving the appropriate balance between coagulation and fibrinolysis. Coagulation and fibrinolysis are complex processes that have been discussed in detail in excellent monographs by Halkier [873] and Sherry [874]. The Halkier book is a source of information on the cloning, structures, and biochemistry of the components of the coagulation and fibrinolysis systems, as well as the details of their interactions and the regulation of these interactions. Similar information is presented in the Sherry monograph, but the emphasis here is more heavily weighted toward the biological and clinical aspects of hemostasis and the conditions under which the balances between coagulation and fibrinolysis are maintained or perturbed. In addition, there are excellent reviews that the reader may consult [681, 718, 875-879]. As the description of hemostasis unfolds below, a surprising number of heparin-binding proteins emerge as critical players in hemostasis. Our immediate purpose here is restricted to an examination of the effects of heparinoids on the activities of antithrombin and the other heparin-binding proteins that play a part in hemostasis. Consequently, only a brief review of these hemostatic mechanisms is presented here, so that the roles of antithrombin and the other heparin-binding proteins in these processes can be placed in perspective. The review literature cited above may be consulted for details and for references to the primary literature. Following the description of the hemostatic mechanisms, each of the heparin-binding proteins will be described in greater detail.

B.   Coagulation

The events that lead from a triggering signal for clotting to the final formation of a highly cross-linked fibrin clot are shown in Figure 8.1. The initiating event in blood coagulation leads to a succession of reactions in which a series of proenzymes (factors IX, VII, X, XI, XII, XIII, prekallikrein, and prothrombin) are converted into active serine proteases. The activated forms of the enzymes and cofactors are designated by the letter "a" following the Roman numeral. Each activated protease, in turn, converts its successor proenzyme into an active protease. This results in a cascade of conversions, with each activated enzyme amplifying the signal from the prior protease in the cascade. The cascade leads eventually to the conversion of prothrombin (factor II) to

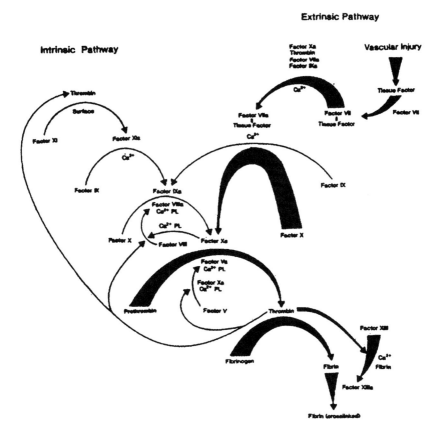

Figure 8.1. Coagulation pathways. Reprinted from [880] with permission, © 1991, American Chemical Society.

thrombin [1] (factor IIa), which initiates clot formation by converting fibrinogen to fibrin. The fibrin is then cross-linked by factor XIIIa, a transglutaminase, to give a very strong fibrin clot.

Classically, two coagulation pathways have been distinguished. The *extrinsic* pathway is initiated by tissue factor (TF, also called tissue

---

[1] Thrombin, heparin cofactor II (HCII), von Willebrand factor (vWF), plasminogen activator inhibitor-1 (PAI-1), vitronectin, activated protein C inhibitor, tissue factor pathway inhibitor (TFPI), platelet factor 4, histidine-rich glycoprotein, fibrin, and thrombospondin are heparin-binding proteins. Tissue plasminogen activator (t-PA) and urokinase (u-PA) are also heparin-binding proteins, but only under nonphysiological conditions.

thromboplastin), a membrane-anchored glycoprotein that is located in *extra*vascular tissues, but not on the endothelium or in the circulation. Tissue factor comes into contact with blood only when vascular injury occurs. Tissue factor forms a complex with factor VIIa in which the capacity of factor VIIa to convert factor X to factor Xa is stimulated ~1000-fold. Factor Xa, in turn, converts prothrombin to thrombin. In contrast to the extrinsic pathway, which is initiated by contact of blood with extravascular tissue factor, the *intrinsic* pathway represents a more extensive cascade which involves only proteins circulating in the plasma, and, in fact, utilizes several of the coagulation proteases that are part of the extrinsic pathway. The two pathways become one at the point where factor X is converted to factor Xa. The intrinsic pathway sustains the coagulation system *in vivo*. The current concept is that the extrinsic pathway is the initiator of clot formation, which is "turned on" by tissue factor and "turned off" by tissue factor pathway inhibitor (TFPI - see Section II.K., p. 281); i.e., the triggering event is a transient one and its effect is sustained by the subsequent events in the coagulation pathway. Following the triggering event and the initial formation of thrombin, the intrinsic pathway amplifies the signal from the extrinsic pathway and controls the growth and maintenance of thrombi, whether formed as a result of vascular injury or the abuse of the endothelium that occurs during the aging process [880, 881].

Two of the factors in the coagulation pathways, factor VIIIa and factor Va, are nonenzymatic proteins which serve as cofactors for factor IXa (which catalyzes X→Xa) and factor Xa (which catalyzes prothrombin→thrombin), respectively. Although factors VIIIa and Va are not proteases, they are required for the activities of their respective proteases. Both exist in pro-forms and must be activated by proteolysis for maximal activity, as shown in Figure 8.1. Interestingly, factors VIII and V, the pro forms, only attain maximal cofactor activities after they are activated by thrombin [see, for example, 882]. Consequently, small amounts of thrombin formed in the early phases of the coagulation cascades can give remarkable feedback activation of thrombin production as a result of thrombin's role in the production of factors VIIIa and Va. Thus, it is clear that the original concept of separate linear cascades of coagulation factors must now accommodate the fact that the active enzyme end-product of the pathways, thrombin, in fact can stimulate its own production by catalyzing the activation of some of the proenzymes and cofactors that lead to its own production [880]. In this manner, the extrinsic and intrinsic pathways act cooperatively. As we shall see, thrombin can also turn *off* its own production.

C.   Fibrinolysis (Thrombolysis)

Fibrinolysis, as the name implies, is the dissolution of clots that results from extensive enzymatic cleavage of the cross-linked fibrin in clots. The reactions in fibrinolysis are shown in Figure 8.2. The enzyme that is directly involved in proteolysis of fibrin is plasmin, which circulates in the blood as a zymogen, plasminogen. Plasmin is a very potent and nonspecific enzyme that is capable of cleaving a variety of proteins in the circulation. However, plasminogen has a strong affinity for both fibrinogen and fibrin, and so becomes tightly bound to fibrinogen and remains bound to the fibrin that is formed by the action of thrombin on fibrinogen. This complexation to fibrinogen, and then fibrin, focuses the activity of plasmin at the site of the fibrin clot. However, plasminogen is inactive in its proenzyme form. It is converted to plasmin by two plasminogen activators circulating in the plasma. One of these, tissue plasminogen activator (t-PA),[1] like plasminogen, has a strong affinity for fibrinogen and fibrin and so binds to fibrinogen and remains bound to the fibrin that is formed. The activity of t-PA on plasminogen is markedly stimulated

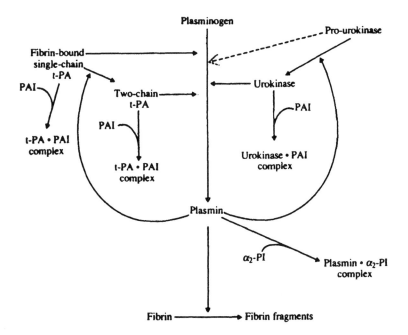

Figure 8.2. Fibrinolysis pathways. Reprinted from [873], with permission.

when fibrinogen is converted to fibrin. Thus, on the surface of the fibrin, t-PA converts the plasminogen to plasmin, thus initiating clot dissolution. In fact, the t-PA activation of plasminogen is *dependent* on the binding of t-PA to fibrin; apparently the fibrin provides an activating template on which t-PA and plasminogen can come into contact. Interestingly, once some of the plasminogen is activated by t-PA, the resulting plasmin can convert the single chain form of t-PA to a two-chain form, which has increased activity for the plasminogen → plasmin conversion. This appears to be a feedback amplification for the clot dissolution similar in kind to the feedback activation of the coagulation pathway by thrombin.

The second plasminogen activator, urokinase (u-PA),[1] circulates in the blood as a proenzyme, pro-urokinase, which is often referred to as "scu-PA" (single chain urokinase-type plasminogen activator). Pro-urokinase has little, if any, activity, but the single chain protein is converted to its active two-chain form (tcu-PA) by plasmin. However, neither pro-urokinase nor urokinase has any special affinity for fibrin; thus, urokinase becomes associated with plasmin on the fibrin clot only if it interacts with fibrin-bound plasmin as the pro-urokinase is activated. Although u-PA can be utilized clinically in the dissolution of clots, this may not be its primary role *in vivo* (see Section II.G., p. 267).

A potential problem in fibrinolysis results from the activity of lipoprotein (a) [Lp(a)], which circulates in the plasma at a concentration that ranges from levels below detection limits up to ~3 mg/ml. The effects of Lp(a) on fibrinolysis have been reviewed by Edelberg and Pizzo [883, 884]. Lp(a) is a low density lipoprotein that contains a lipid core, apoprotein B, and apoprotein(a). Apoprotein(a) is found in six isoforms, all of which exhibit strong homology with plasminogen. These homologies in apoprotein (a) are expressed in the ability of Lp(a) to bind to fibrinogen and fibrin *at the same site where plasminogen binds*. In this manner, Lp(a) competes with plasminogen for binding sites on the fibrinogen and fibrin. Since Lp(a) has no protease activity, it cannot convert plasminogen to plasmin. However, Lp(a), when it displaces plasminogen from fibrinogen, prevents plasminogen activation and thereby inhibits fibrinolysis. Individuals with elevated levels of Lp(a) (>3 mg/ml) exhibit strong tendencies to develop atherosclerotic lesions, apparently as a result of the inhibition of fibrin clot dissolution.

D.   Regulatory Mechanisms in Hemostasis

1.   Localization of hemostatic mechanisms to the sites of vascular injury

The most notable mechanisms that minimize the spread of the thrombus from the site of vascular injury to the rest of the endothelium are the mechanisms that limit coagulation and fibrinolysis to specific cell surfaces. Although almost all of the coagulation proteins circulate in the plasma, they are recruited to negatively charged surfaces of cells where the coagulation reactions take place. Coagulation can occur on the surfaces of platelets, endothelial cells, or peripheral white cells. Restriction of clotting to injured vascular sites is due to a combination of the pro-coagulation properties of blood platelets and the anticoagulant properties of other cell surfaces. An injury to the vasculature results in exposure of the subendothelial layer of smooth muscle cells which triggers the binding of platelets to the injury site and the initiation of the clotting response. This is illustrated in Figure 8.3. Under the rapid blood flow conditions found in the arterial circulation, a circulating protein, von Willebrand's factor (vWF),[1] assists in platelet adhesion by binding to the subendothelial layer of the vessels. On binding, vWF gains the capacity to bind tightly to blood platelets via the platelet surface glycoprotein Ib (Figure 8.3, injured vessel). Platelets thus become tethered by vWF to sites of arterial injury to initiate the formation of a loose "platelet plug" (1° hemostatic plug). The binding of platelets to vWF results in platelet activation, one result of which is the exposure of the platelet cell surface integrin receptor complex, GP-IIb/IIIa, which promptly binds fibrinogen, as well as fibronectin,[1] thrombospondin (TSP-1),[1] and/or vWF. These proteins cross-link the agglutinated platelets and recruit additional platelets from the circulation, resulting in the formation of a platelet aggregate and yielding a much tighter platelet plug (2° hemostatic plug), the formation of which represents the initial step in the repair of the injury to the vascular wall. Under conditions of low blood flow in the venous circulation, platelet adhesion is mediated by receptors of the integrin family and extracellular matrix proteins, including fibronectin, collagen, and laminin [885]. As the platelet aggregate is formed, the platelets are activated both by vWF and by thrombin that is formed early in the cascade, and platelet receptors for factors V, VIII, IX, and X are exposed. Some of these coagulation factors are released from the activated platelets themselves, and others are recruited from the circulation. They bind to the surface of the platelets and are thus positioned for their role in solidification of the platelet plug by the events leading to the formation of the highly cross-linked fibrin. These events focus the coagulation reactions at the sites of vascular injury. A further mechanism for stabilization of the platelet plug involves TSP-1,[1] which is released from activated platelets. Thrombospondin forms complexes with fibrin, fibrinogen, plasminogen, histidine-rich glycoprotein,[1] and heparan sulfate proteoglycan, all of which may be associated with the platelets. Thus, thrombospondin may play a role in stabilization of the platelet aggregate by forming additional cross-links between the platelets. Finally, it should be noted that factor XIIIa, a transamidase that catalyzes the cross-linking of the fibrin formed by thrombin, also has a high affinity for fibrinogen and so circulates as a fibrinogen/factor XIIIa complex. Thus, the binding of fibrinogen to the cell surfaces where clotting occurs brings along the factor XIIIa so that the latter is positioned for its role in the fibrin cross-linking reaction. As fibrinogen

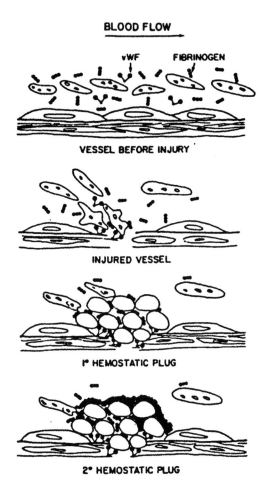

Figure 8.3. Action of vWF in platelet plug formation. Reprinted from [876], with permission.

is converted to fibrin, the fibrin adsorbs thrombin. In this bound form, thrombin (a) retains its full activity on fibrinogen, and (b) is protected from inactivation by heparin/antithrombin. Thus, the procoagulant activity is localized to the clot and thrombin may act as a regulator of clot growth [reviewed in 886].

Fibrinolysis, as already noted, is an event that is focused at the site of clot formation as a result of the tight binding of plasminogen and t-PA to fibrinogen and fibrin. Because fibrinogen represents a part of the platelet plug, fibrin is formed on the surface of platelets and other cells by coagulation reactions.

Consequently, fibrinolysis, like coagulation, takes place on these cell surfaces. Other mechanisms operate to control the *interval* between clot formation and clot dissolution.

The platelet plug, which contains all of the components of both the coagulation and the fibrinolysis systems, is rather impervious to other components of the plasma. Thus, circulating heparin and antithrombin cannot enter the plug [887, 888], nor can α_2-plasmin inhibitor (Section I.D.3., p. 248). t-PA, which is used clinically to hasten the dissolution of the plug following heart attacks and strokes, must activate plasminogen on the surface of the plug before it can enter more deeply as the surface fibrin is broken down and the plug is loosened. It is clear that the anticoagulant activity of heparin cannot be expressed at the site of vascular injury *after* the clot has been formed. However, when t-PA is used, heparin is administered as well in order to prevent *re-formation* of the thrombus at the original occlusion site or at other sites in the vasculature. Also, in surgical procedures, where vessel injury can cause serious thrombus formation, heparin is administered prior to (and during) the surgery to prevent the formation of thrombi.

2. Hemostatic mechanisms on the surfaces of endothelial cells

Although clots, or thrombi, are most commonly formed at sites of vessel injury, normal endothelium also presents a surface on which both coagulation and fibrinolysis can occur. Endothelial cells have receptors that bind the clotting factors necessary to initiate the intrinsic coagulation pathway. Also, endothelial cells bind plasminogen via receptors on their surfaces [see, for example, 889, 890], and they produce t-PA and u-PA, which can bind to the endothelial cell surface. Thus, the endothelial surface is coated with the components required for both coagulation and fibrinolysis. As described in Chapter 7, antithrombin binds to the endothelial cell heparan sulfate proteoglycan so that it is activated to turn off the coagulation pathways by inhibiting factor Xa and thrombin activities. Thus, endothelial cell surfaces have their own hemostatic mechanisms. Under normal circumstances, coagulation and fibrinolysis are balanced on the endothelial cell surface and the endothelium generally presents mechanisms that prevent formation and spreading of thrombi (see below).

3.   **Prevention of the spread of thrombosis by serpins: the role and metabolic fate of antithrombin and other serpins**

The localization of thrombus formation and dissolution to cell surfaces at the sites of vascular injury gives initial assurance that a more generalized thrombosis does not occur. However, the production of highly active, broad spectrum proteases in the platelet plug can result in potentially harmful effects if some of these enzymes escape from the plug or are generated in other regions of the vasculature. Thus, additional mechanisms operate to prevent these damaging effects of tissue degradation and thrombosis at non-injured sites in the vascular wall. This brings us finally to a statement concerning the primary role of antithrombin in hemostasis, namely the neutralization of thrombin and factor Xa in order to prevent the formation or extension of a clot to uninjured sites in the vasculature. Antithrombin is synthesized by endothelial cells and by the liver [630, 631, 697] and is secreted into the blood. A small fraction of the circulating antithrombin becomes bound to those endothelial cell surface heparan sulfate proteoglycan species that contain the antithrombin-binding pentasaccharide. The bound antithrombin is thus activated and primed for rapid reaction with any thrombin or factor Xa that appears in the general circulation, effectively neutralizing these proteases and preventing their initiation of unneeded and harmful fibrin clot formation. The inactivation of thrombin prevents (a) the conversion of fibrinogen to fibrin, and (b) the effects of thrombin in amplifying its own production by activating cofactors V and VIII. The latter result of antithrombin-mediated inactivation of thrombin appears to play a major role in the prevention of thrombin-induced coagulation [718, 891, 892]. Thus, antithrombin functions in the prevention of the spreading of the thrombus.

Interestingly, although endothelial cell heparan sulfate-activated antithrombin might be expected to control the levels of both thrombin and factor Xa (such that a genetic defect in antithrombin would result in an increased level of *both* enzymes), congenital deficiencies in antithrombin seem to result in elevated levels of factor Xa, but *not* elevated levels of thrombin [684]. This surprising result may be due to the fact that there are several other antithrombin-independent mechanisms that control thrombin levels, but that have no effect on factor Xa levels [for example, the protein C system (thrombomodulin) and the protein C inhibitor, below], i.e., there are a number of other ways to control the levels of thrombin, but factor Xa levels are controlled primarily by the antithrombin mechanism. Therefore, under conditions in which thrombin levels *tend* to become elevated (as when the levels of functional antithrombin are reduced), these alternate mechanisms lower the levels of thrombin but not those of factor Xa. Another explanation for the preferential elevation of factor Xa when antithrombin is defective is that the inactivation of thrombin by antithrombin occurs at an 8- to 10-fold faster rate than the inactivation of factor Xa. Therefore, when antithrombin levels are reduced (but not to zero), the levels of thrombin will be preferentially reduced.

There is one final mechanism by which endogenous heparan sulfate-activated antithrombin might be more critically involved in controlling factor Xa levels than thrombin levels. As described in Chapter 7, the inactivation of factor Xa by antithrombin requires heparin sequences no longer than a pentasaccharide, whereas inactivation of thrombin requires at least an 18-mer heparin segment. Furthermore, the 18-mer must have the antithrombin-binding pentasaccharide on one end of its sequence and a highly sulfated thrombin-binding hexasaccharide on the other (Chapter 7). Because of the block structure of heparan sulfate (Chapter 2), the necessary positioning of the pentasaccharide and a highly charged hexasaccharide may occur rather infrequently in heparan sulfate. Therefore, the endothelial cell heparan sulfate may be much more effective in factor Xa neutralization than in thrombin neutralization, and genetic deficiencies that result in loss of heparin-binding capacity of antithrombin may elevate factor Xa levels much more than thrombin levels. In other words, it may be that evolution has provided a variety of ways for controlling thrombin levels and that antithrombin is primarily a mechanism for controlling factor Xa levels.

In addition to antithrombin, other serpins control the activities of the proteases of the hemostatic pathways. The actions of some of these serpins are shown in Figure 8.2. Plasmin is rapidly inactivated by α_2-plasmin inhibitor (α_2-PI), and t-PA and u-PA are inactivated by plasminogen activator inhibitors (PAIs). There are at least four PAIs, but only PAI-1 [1] seems to be involved in the inactivation of both t-PA and u-PA. PAI-1 is not very stable and is stabilized in the plasma by forming a complex with vitronectin.[1] Other serpins include activated protein C inhibitor [1] (also called PAI-3) and heparin cofactor II (HCII),[1] a serpin that inactivates thrombin but not factor Xa. With the exception of the HCII, activated protein C inhibitor, and plasminogen activator inhibitor, all of these serpins are active in the absence of heparin and their activities are not stimulated by heparin. All of these serpin reactions generate relatively stable covalent complexes between the target proteases and the cleaved forms of the respective serpins. These complexes are thus waste products of the hemostatic metabolism and are cleared from the circulation by receptor-mediated endocytosis and lysosomal degradation, primarily in hepatocytes [893-897]. For antithrombin and other serpins there is an additional mechanism for clearance that is mediated through vitronectin (see Section II.I.2., p. 277).

Finally, α_2-macroglobulin, a proteinase inhibitor that acts through the unique mechanism of entrapping proteinases for endocytosis and catabolism (see Chapter 11), plays a major role in the clearance of thrombin and other proteases from the circulation [898, 899]. α_2-Macroglobulin represents 8-10% of the total protein in human serum.

4.   Prevention of the spread of thrombosis by proteolytic degradation of factors Va and VIIIa, the protein C system

We have noted (Section I.B., p. 240) that thrombin can stimulate its own production by converting factors V and VIII to their active cofactor forms, Va and VIIIa. When factors Va and VIIIa are absent, thrombin cannot be formed and coagulation cannot occur. This leads us to an additional mechanism which prevents coagulation from taking place on the surface of normal endothelial cells. A primary system exists for the proteolytic degradation of factors Va and VIIIa, namely the activated protein C system, which degrades and thus inactivates these cofactors. Protein C is a zymogen that must be activated by proteolytic cleavage to yield activated protein C (designated APC, or Ca), a reaction that is catalyzed by thrombin. However, thrombin (which will catalyze the *formation* of Va and VIIIa from factors V and VIII, respectively; Section I.B., p. 240) cannot normally activate protein C. In order to gain the capacity for the protein C activation, thrombin must bind to thrombomodulin,[1] a thrombin receptor on the surface of endothelial cells. In the thrombomodulin/thrombin complex the specificity of thrombin is markedly altered so that it *loses* the capacity to activate factors V and VIII and *gains* the capacity to cleave protein C to APC. APC itself is rather inactive in the cleavage of factors Va and VIIIa, but its activity is markedly stimulated by another protein in the circulation, protein S [see, for example, 900]. It is rather paradoxical that when thrombin is present in the platelet plug, it can stimulate its own synthesis by *activating* factors V and VIII, whereas when thrombin escapes from the plug and binds to thrombomodulin on the endothelial cell surface, its activity is changed and it can totally *inhibit* its own synthesis by activating protein C, which then catalyzes the proteolytic *inactivation* of factors Va and VIIIa. Several further points concerning the activated protein C activity are of note:

- APC in the presence of protein S turns off the coagulation pathway by destroying factors Va and VIIIa. Thus, if protein S was present in excess, exogenous APC would be expected to act as an anticoagulant.

- Activated protein C activity is neutralized by a specific serpin, protein C inhibitor[1] (actually "activated protein C inhibitor"). The activity of this serpin is heparin-dependent [901, 902], as discussed is Section II.J., p. 279.

- The thrombin/thrombomodulin complex on endothelial cell surfaces represents another way in which the levels of thrombin may be controlled. Thrombin released from the platelet plug may be immediately recruited by the endothelial cell surface thrombomodulin so that the pro-coagulant activity of thrombin would be lost. Thus, endothelial cell surface thrombomodulin represents a latent anticoagulant that diverts thrombin from its coagulant role. Consequently, thrombomodulin represents an additional means by which the endothelial cell surface is rendered anticoagulant.

• In the platelet plug, factors Va and Xa form a complex on the platelet surface. Factor Va can be inactivated by thrombin only if the Va is not complexed to factor Xa. Consequently, the inactivation of Va cannot occur in the plug unless this complex becomes dissociated. Thus, the relative rates of coagulation and fibrinolysis may depend on the rate of formation/dissociation of the factor Xa/factor Va complex.

5.   Prevention of the spread of thrombosis by neutralization of tissue factor

One final regulatory mechanism is able to turn off the extrinsic coagulation system at the initiation step by the neutralization of tissue factor (see Figure 8.1). This system utilizes a new protein, variously called TFPI (tissue factor pathway inhibitor) [1], LACI (lipoprotein-associated coagulation inhibitor), or EPI (extrinsic pathway inhibitor). TFPI combines with factor Xa to give a TFPI/factor Xa complex. This complex then combines with the tissue factor/factor VIIa complex to yield a quaternary complex in which the tissue factor/factor VIIa cannot activate factor X or factor XI, preventing further initiation of the cascade (this mechanism is detailed in Section II.K.2., p. 283). Thus, as soon as coagulation is initiated by tissue factor, the tissue factor activity is turned off by TFPI, limiting the triggering event. The small amount of thrombin resulting from this triggering event then plays a major role in sustaining the coagulation process. It activates factor Xa and also activates platelets to expose receptors for the coagulation factors. However, TFPI prevents further tissue factor-initiated coagulation and turns over the further enhancement of the coagulation systems to the thrombin-activated part of the coagulation pathway. Control of the *extent* of coagulation now defers to antithrombin and the protein C system [903].

II.   INTERACTIONS OF HEPARINOIDS WITH THE HEPARIN-BINDING PROTEINS IN HEMOSTASIS

A.   Introduction

It is clear that the discussion above is much too brief to give the reader an in-depth understanding of hemostatic mechanisms. The subject is much too complex! However, in this short survey, a number of new proteins that are critical for hemostasis have been introduced, and note has been taken that many of these are heparin-binding proteins.[1] For some of these e.g. fibrin,[1] the binding to heparin is quite weak [904, 905], and for those that bind more tightly, it is not clear whether there are critical interactions between the protein and cell surface heparan sulfate proteoglycan in their normal roles in hemostasis. However, when *exogenous* heparin is administered to patients to target the antithrombin-mediated anticoagulant activity, heparin also alters activities of these other heparin-binding proteins. Before discussing each of these proteins and their responses to exogenous heparin, some general remarks concerning the effects of exogenous heparin are in order.

First of all, it must be noted that all of the proteins that will be discussed bind to exogenous heparin. Thus, although the primary goal of heparin administration is to prevent coagulation by activating antithrombin for the neutralization of thrombin and factor Xa, all of these other heparin-binding proteins bind their share of the circulating heparin. The actual amount of heparin bound by each protein depends on the amount of the protein available, the binding constant of the protein for the heparin, and any selectivities that these proteins may have for specific sequences in heparin [discussed in 906, 907, 908]. To further complicate matters, there are, in addition to the heparin-binding proteins that play a *direct* role in hemostasis, many other heparin-binding proteins in the circulation that also bind to exogenous heparins, and perhaps to endogenous heparan sulfates (e.g., lipoprotein lipase, superoxide dismutase, apolipoproteins B and E, and various growth factors). Exogenous heparin triggers responses from many of the proteins that are involved in hemostasis. Some of these effects are procoagulant and some are anticoagulant. Normally one can monitor only one, or a few, of these responses in any treatment or clinical trial, but the responses that are *not* measured are still taking place. A number of general factors affect the responses of these different heparin-binding proteins. For example:

- Some of the activities mediated through heparin-binding proteins respond only to relatively high molecular weight heparin (the thrombin/antithrombin reaction is one example), whereas others respond to both high and low molecular weight heparin (the factor Xa/antithrombin reaction is one example).

- It is generally observed that antithrombin-binding (HA-heparin) and antithrombin-non-binding (LA-heparin) heparin fractions are effective in binding to and activating all other heparin-binding proteins, i.e., the heparin sequences that bind to *antithrombin* are not the same sequences that are recognized by other heparin-binding proteins.

- While heparin remains in the circulation, it is progressively catabolized to smaller and smaller oligosaccharides before it is excreted via the kidneys with a $t_{1/2}$ of approximately 30 min (for a bolus IV dosing, see Chapter 5). Also, as catabolism and excretion proceed, the concentration of total heparinoids falls. Therefore, when heparin is first administered, it has a relatively high molecular weight and high concentration, and this exogenous heparin may alter the activities of *all* of the heparin-binding proteins. However, as the heparin is broken down to smaller fragments, those proteins whose heparin responses require high molecular weight heparin cease to respond to the heparin, whereas those that can respond to both high and low molecular weight forms of heparin continue to respond to the heparin fragments, with the response of all proteins diminishing as the total heparinoid concentration falls.

- Low doses of exogenously supplied heparin are taken up by the endothelium, thus disappearing from the circulation. However, when higher doses of heparin are administered, the endothelium becomes saturated and excess heparin remains in the circulation and is excreted via the kidneys (Chapter 5). Therefore, the amount of heparin available to the circulating heparin-binding proteins may be quite variable depending on the dose of heparin given.

- Since the amounts of these proteins found in the blood may vary from one individual to another, the dosing of heparin required to optimize the degree of antithrombin-mediated inhibition of coagulation must be monitored during treatment to achieve the desired anticoagulant effect and limit hemorrhage.

- Interactions between proteins and heparin may result in a change in the amounts of heparin-binding proteins. As one example, a dose of heparin may cause a transient drop in the concentrations of antithrombin, thrombin, and factor Xa, as the antithrombin reacts with these coagulation proteins. A more interesting example is the case in which the levels of plasminogen activator inhibitor I (PAI-1) may be depressed after heparin administration. Since PAI-1 controls the levels of t-PA and u-PA, a fall in PAI-1 may result in a rise in t-PA and u-PA (see Section II.H.2., p. 271) and an increase in the rate of fibrinolysis.

The end result of all of these effects is that the responses to exogenous heparin that are measured immediately after heparin is administered will be different from the effects that are observed after (a) the heparin is taken up by the endothelium, (b) the heparin is partially catabolized to smaller fragments, and (c) changes occur in the plasma concentrations of heparin and the various heparin-binding proteins. Stated another way, the overall effects of heparin on hemostasis are in a constant state of change following either bolus or continuous doses of exogenous heparin. Furthermore, the dynamics of these processes when heparin is administered intravenously will be different from those observed when heparin is administered subcutaneously, for several reasons: (a) the intravenous and subcutaneous heparins encounter different cell types, and the different cell types will have surface heparan sulfate proteoglycans which bind different heparin-binding proteins; (b) when heparin is administered subcutaneously, the lower molecular weight portions of this polydisperse drug will be taken up more rapidly than the mid- to high-molecular weight portion; and (c) the subcutaneous heparin will appear in and disappear from the blood with a kinetic pattern that is quite distinct from that of the intravenous heparin.

We will turn now to a discussion of the actions of each of these heparin-binding proteins in hemostasis. For each protein there is an extensive literature. Many of the proteins that are described here interact with other proteins and with the surfaces of platelets or endothelial cells. Although our emphasis will be primarily on the interaction of heparin and heparan sulfate with these proteins, one must be aware that there are also interactions between integrins and proteins that are both qualitatively and quantitatively similar to the heparan sulfate interactions with proteins [see for example 909]. A few of the proteins that bind to heparinoids also interact with integrins, and these integrin-mediated interactions will be described here as well. However, the discussion below will focus primarily on the normal heparinoid interactions of each protein in the hemostasis pathways and the alterations in their activities that occur in response to exogenous heparin.

B.   Antithrombin

Antithrombin is present in the blood at a level of ~ 2.5 $\mu\underline{M}$. Only about 1% of this is bound to the endothelium via the cell surface heparan sulfate proteoglycan. Thus, in normal hemostasis, only this tiny fraction of the total available antithrombin is activated. Added heparin binds to the circulating antithrombin and almost certainly displaces the small amount of surface-bound antithrombin from the endothelium. Thus, the amount of the available antithrombin that is activated is increased 100-fold by exogenous heparin and the activated antithrombin remains high as long as the heparin concentration is maintained. However, the heparin/antithrombin complex cannot penetrate the platelet plug after a clot has formed. Therefore, exogenous heparin activates antithrombin for the prevention of the formation of *new* thrombi that may tend to occur after heparin is administered. We discussed in Section I.D.3., p. 248., the possibility that the endothelial cell surface heparan sulfate may be much more effective in factor Xa neutralization than in thrombin neutralization.

However, exogenous heparin will be more effective against thrombin because thrombin reacts 8-10 times faster with the heparin/antithrombin complex than does factor Xa. Immediately following the heparin dosing, before significant heparin catabolism has taken place, the heparin-activated antithrombin will neutralize both thrombin and factor Xa. After the catabolism of heparin reduces the size of the heparin fragments below the 18-mer size that is required for thrombin neutralization, then the activity of antithrombin will neutralize only factor Xa (which should prevent the formation of more thrombin—see Figure 8.1).

C. Thrombin

1. Structure and biological activity

Thrombin, because of its central role in hemostasis, interacts with many proteins in the circulation and on cell surfaces. For example, thrombin must have specific interaction sites on its surface for binding to fibrinogen, antithrombin, factors V, VIII, XI, and XIII, thrombomodulin, protein C, and other proteins in the circulation, as well as receptors on the surface of platelets and endothelial cells. It also binds exogenous heparin. The structure of thrombin and the various regions for binding of substrates and heparin were described in Chapter 7. Because there are multiple mechanisms for rapid clearance of thrombin from the vasculature, the levels of circulating thrombin are very low. Prothrombin, however, is secreted by the liver and circulates at a plasma concentration of 100 µg/ml. The complex series of events in the conversion of prothrombin to thrombin [823] takes place on cell surfaces in the presence of Ca^{2+} and factors Va and Xa (the complex of Ca^{2+}, the cell surface phospholipids, and factors Va and Xa is called prothrombinase). The many biological roles of thrombin that result from its interactions with fibrinogen, thrombomodulin, protein C, antithrombin, α_2-macroglobulin , HCII, platelets, endothelial cells, and factors V, VIII, XI, and XIII have been described above.

2. Effect of exogenous heparin

As discussed in Chapter 7, heparin binds strongly to thrombin, and at concentrations of exogenous heparin that result in the binding of separate chains of heparin to thrombin and antithrombin, the approximation effect of heparin in the heparin-mediated antithrombin inhibition of thrombin is lost. Even heparin fractions that are depleted of the specific antithrombin-binding heparin pentasaccharide sequence may block the neutralization of thrombin by antithrombin. The concentrations of heparin that are required for interference of the antithrombin-thrombin reaction are on the high end of the usual therapeutic levels (10^{-5} $\underline{M}$) [see 828, 910]. It would appear likely that smaller fragments of heparin would be as effective as larger heparin chains in the direct inhibition of thrombin (but the effect of oligosaccharides on thrombin activity has not been reported).

D.   Heparin Cofactor II

1.   Structure and heparin binding

Heparin cofactor II (HCII) has been described by Tollefsen [875, 911]. The human full-length cDNA codes for a protein with 499 amino acids, 19 of which form a signal sequence that is lost prior to secretion [912]. HCII is synthesized by hepatocytes, but not by endothelial cells, and is secreted as a glycoprotein with about 10% carbohydrate by weight (three potential sites for N-linked oligosaccharides). It is a member of the serpin superfamily and, as such, has high homology with antithrombin and the other serpins (25-30% identity). In contrast to antithrombin, it specifically neutralizes thrombin and chymotrypsin. It does not inactivate other proteases of the coagulation pathway. Both thrombin and chymotrypsin cleave HCII at Leu 444-Ser 445, near the C-terminal [913, 914], forming covalent complexes that are not disrupted by heat and sodium dodecyl sulfate. The HCII activity against thrombin is stimulated 1000-fold by either heparin or dermatan sulfate, but its activity against chymotrypsin is not affected by either glycosaminoglycan. HCII is the only serpin known that can be activated by dermatan sulfate. The activation of the HCII by either heparin or dermatan sulfate is molecular weight dependent, requiring heparinoid sequences with 20 or more monosaccharides [915]. Similarly, minimal sequences of 12-14 residues are required for activation by dermatan sulfate. Evidence suggests that both glycosaminoglycans bind both HCII and thrombin. The HCII binds to thrombin anion exosite I while the glycosaminoglycan binds to thrombin anion exosite II [823], in a manner similar to the approximation effects described for the reaction of thrombin with antithrombin (Chapter 7). Consistent with this conclusion is the finding that at concentrations of heparin that are high enough so that separate heparin chains can bind to, and block, the heparin-binding sites on HCII and thrombin, heparin inhibits the thrombin-HCII reaction.

Like antithrombin, the amino acid sequence of HCII can be projected onto the α_1-protease inhibitor crystal structure to obtain an approximation of its three-dimensional structure, which is remarkably similar to the antithrombin structure [774, 778], as shown in Figure 8.4. Comparison of the heparin-binding domains of antithrombin with the corresponding sequences of HCII [778, 916] show the involvement of Arg 103 of HCII, which is equivalent to Arg 47 of antithrombin, and a 21 amino acid sequence beginning at Lys 173, equivalent to Lys 114 of antithrombin, in which 9 residues are identical to the antithrombin sequences and 5 of the 6

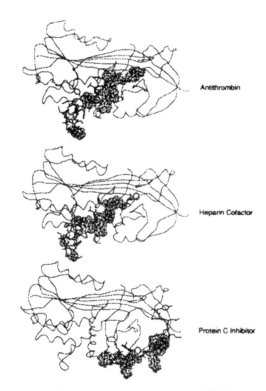

Figure 8.4. Three-dimensional structure of HCII; comparison with antithrombin and protein C inhibitor. Reprinted from [778], with permission.

basic residues are identical. Also, the reactive centers of the two serpins are in similar positions.

```
       165        170      175      180      185      190          195
HCII: K D F V N A S S K Y E I T T S H N L F R K L T H R L F R R - N F
AT:   K - F D T I S E K T S D Q - I H F F F A K L N C R L Y R K A N K
```

2. Biological activity

HCII circulates in the blood at a concentration of ~1.2 µM, approximately half that of antithrombin. In contrast to antithrombin, HCII does not bind to endothelial cell surface heparan sulfate proteoglycan and is not activated by the endothelium [917]. This is a particularly interesting observation because it shows, in a convincing fashion, that the endothelial cell heparan sulfate proteoglycan exhibits selectivity among the various heparin-binding proteins in the circulation. Since antithrombin can bind to a limited number of heparan sulfate proteoglycan sites and since HCII has a three-dimensional structure that closely mimics that of antithrombin, it is likely that this selectivity is controlled by structures of the heparan sulfate chains in the endothelial cell heparan sulfate proteoglycan. Thus, HCII does not appear to play a role in normal hemostasis. However, HCII is activated by surface structures on fibroblasts and smooth muscle cells [918]. For these cells, it is the cell surface *dermatan sulfate* that stimulates the activity of the HCII. Whinna *et al.* [919] have shown that the dermatan sulfate attached to decorin and biglycan core proteins can activate HCII, both when free in solution and when bound to collagen. This has led to the suggestion that the normal role of HCII is to prevent excessive coagulation in extravascular tissues when vascular injury occurs [912, 913, 920-923]; i.e., HCII is activated only when the blood comes in contact with the dermatan sulfate found in extravascular tissues.

3.   Effect of exogenous heparin

Although HCII is not activated by heparan sulfate proteoglycan in the vasculature, exogenous heparin activates both antithrombin and HCII, which form distinctive complexes with thrombin. The relative proportions of the thrombin-HCII complex and the thrombin-antithrombin complex that are formed depend on the concentration of heparin in the circulation. The thrombin-antithrombin complex predominates at heparin levels of 0.07-33 µg/ml (0.01-5 units/ml), whereas the thrombin-HCII complex predominates at concentrations of 33-670 µg/ml (5-100 units/ml). Dermatan sulfate alone at a concentration of 67 µg/ml can block coagulation. Thus, although HCII does not appear to be involved in normal hemostasis within the vascular system, it becomes a significant contributor to the inhibition of coagulation when exogenous heparin or dermatan sulfate is administered. Since the effect of heparin on the reaction of HCII with thrombin depends on the molecular weight of the heparin, the activity of the HCII-mediated reaction will fall as heparin is catabolized. Interestingly, histidine-rich glycoprotein (Section II.M., p. 287), present in blood at ~100 µg/ml, binds to heparin and decreases the amount of heparin available to activate HCII, but has no effect on the dermatan sulfate stimulation. However, platelet factor 4 (Section II.L., p. 285), which is released from activated platelets, binds to both heparin and dermatan sulfate and decreases the amount of both that is available to activate HCII [924].

E. von Willebrand Factor

1. Structure and heparin binding

The structure and biology of von Willebrand factor (vWF) have been described in several reviews [925-931]. Human vWF is formed by multimerization of a single basic subunit composed of 2050 amino acids. The subunit is rich in cysteine (8.3% of the total amino acids) and is heavily glycosylated (10 to 19% of the weight), with both N-linked and O-linked oligosaccharides. It is synthesized with a 22 amino acid signal sequence and a 741 amino acid propeptide (von Willebrand antigen II), both of which are lost by processing. Multimers are formed by dimerization of the basic subunit by disulfide cross-linking, and then further disulfide cross-linking that leads to multimers of the dimers. These vary in molecular mass from 500 to 10,000 kDa. vWF circulates as a complex with factor VIII. The propeptide and the mature subunit are made up almost entirely of four types of domains, designated A to D, with each type repeated several times in the polypeptide chain, as illustrated in Figure 8.5. The following structural features are of particular interest from the standpoint of the present discussion: (a) the factor VIII-binding domains, located in the N-terminal region (below); (b) the heparin- and GPIb-binding domain, located in the A1 sequence; and (c) the GPIIb/IIIa-binding domain, which contains an RGD integrin ligand in the C-terminal portion of the molecule. All of these domains are of importance in the biological function of vWF, but only the heparin-binding domain is discussed further in this section.

The binding of heparin to vWF is dependent on the molecular mass of the heparin, with the K_d increasing two orders of magnitude as the heparin molecular mass drops from 12 kDa to 4 kDa [932]. Heparin that binds to vWF has the same antithrombin-binding activity as heparin that does not [932]. vWF can be purified by affinity chromatography on heparin-Sepharose [933-935]. Two heparin-binding regions have been identified in vWF, a Ca^{2+}-dependent binding domain in the N-terminal region (amino acids 1-298) and a tighter-binding region in the A1 domain, which forms a loop that is closed by a disulfide bridge between Cys 509 and Cys 695 (Figure 8.5). Using peptide fragments of vWF, the heparin-binding sequence in the A1 domain has been traced to residues 449-728 and 480-718 [935-937]. The heparin-binding region was narrowed down to amino acids 512-673 [938], and ultimately to residues 569-580, in the N-terminal half of the A1 loop [660, 939]:

$$\underset{570}{\text{K}}\ \text{D}\ \underset{}{\text{R}}\ \underset{}{\text{K}}\ \underset{575}{\text{R}}\ \text{P}\ \text{S}\ \text{E}\ \text{L}\ \underset{580}{\text{R}}\ \text{R}\ \text{I}$$

The binding of heparin to this peptide has been studied in detail [660, 939]. The peptide binds to heparin with an affinity ($K_d = 3.5 \times 10^{-7}$ $\underline{M}$) comparable to that of vWF ($K_d = 4.5 \times 10^{-7}$ $\underline{M}$) and competes with vWF for heparin binding. Chromatography of heparin on a peptide-Sepharose affinity column yields a heparin fraction that shows a seven-fold increase in capacity to inhibit vWF-mediated platelet agglutination. Isothermal ligand titration calorimetry shows that the binding between the peptide and heparin is enthalpy driven, with the formation of 8 to 15 ionic bonds on binding. Circular dichroism spectroscopy shows that the peptide possesses 60% β-strand and only 10% α-helix, and that the binding of heparin causes a shift in the CD spectrum without much change in the content of β-strand and helix.

Although these data suggest that residues 569-580 play a prominent role in heparin binding to vWF, site-directed mutagenesis in which residues 569-573 of vWF are all converted to Ala residues yields a modified vWF in which heparin binding is not diminished [940]. Furthermore, conversion of the sequence K^{642} K K K^{645} to all Ala results in a 70% reduction of heparin binding. However, the heparin-binding measurements in these studies were carried out at sub-

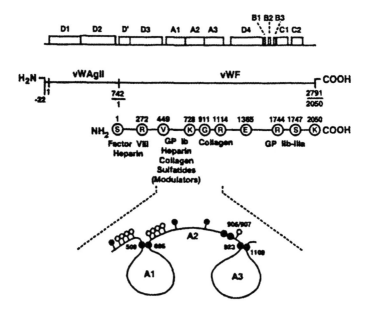

Figure 8.5. Domain structure of vWF. Reprinted from [931], with permission.

physiological ionic strengths.

2. Biological activity

vWF circulates at a plasma concentration of about 10 μg/ml (~40 n$\underline{M}$, for the monomer). The gene expression of vWF, which is synthesized only by endothelial cells and megakaryotes, has been studied [see 941, and references therein]. Endothelial cells secrete vWF constituitively to their subendothelial matrix or store it in Weibel-Palade bodies for later, regulated secretion in response to various agonists. Platelets (formed from megakaryotes) store vWF in α-granules and release it on platelet activation. vWF plays two roles in hemostasis. First, it circulates in the plasma as a complex with factor VIII, thus stabilizing factor VIII and protecting it from inactivation by activated protein C (Sections I.D.4., p. 250, and II.J.2., p. 280). When factor VIII is not in such complexes, it has a t$_{1/2}$ *in vivo* of only minutes. Second, under high sheer blood flow, vWF plays a primary role in platelet agglutination and aggregation at sites of vessel injury (Figure 8.3), an effect that may depend on the perturbation of its conformation that occurs under conditions of high shear [942]. Here, either circulating vWF or endothelial cell surface vWF adheres to the subendothelium that is exposed on tissue injury, binding to collagen or to other matrix components. The bound vWF then recruits and binds circulating platelets. This binding activates the platelets and concentrates them at the injury site, thus initiating platelet plug formation. Finally, vWF released from the activated platelets binds to the integrin receptor, GPIIb/IIIa (integrin $\alpha_{IIb}\beta_3$), to cross-link the platelets and stabilize the aggregate.

The correlations between the vWF sequences involved in each of these binding events and the heparin-binding sequence are of particular interest:

- The initial binding of vWF to subendothelial collagen appears to involve the amino acid sequences 542-622 and 948-998 within the A1 and A3 loops, respectively (Figure 8.5). The sequence in the A1 loop overlaps the putative heparin-binding sequence. However, the A1 loop can be deleted from vWF without interfering with subendothelial adhesion, and there remains some uncertainty as to whether collagen is the functional ligand for the initial vWF binding.

- The initial binding between vWF and the platelet GPIb, a receptor that is exposed on the surface of platelets before activation, involves a sequence within the A1 loop that has been traced to proteolytic fragments 440-728, 273-511, and 674-728. It is the current belief that the major vWF peptide fragment that binds to GPIb is 524-542 [929], which lies close to the heparin-binding sequence (569-583).

- Finally, the platelet cross-linking sequences appear to be the RGD sequences in the C-terminal region of the monomeric vWF. These bind to GPIIb/IIIa which becomes accessible after platelet activation. The RGD sequence is not near the heparin-binding sequence.

The binding of the vWF to the subendothelium and to the platelet GPIb both may involve sequences close to the heparin-binding region in the A1 loop. The interaction that is perhaps most important from the standpoint of the vWF-heparin interactions is the binding of vWF by GPIb. Platelets do not normally interact with vWF in the circulation. The binding of vWF to the subendothelium "activates" the vWF, making it reactive toward the platelet GPIb receptor. The activation of vWF, presumably a conformation change, can be induced *in vitro* by ristocetin, allowing the examination of the vWF/platelet binding in the absence of the subendothelial inducers. When platelets are mixed with vWF in the *absence* of ristocetin, no reaction takes place. However, when ristocetin is added, platelet aggregation can be observed in the test tube, a result of a sequence of reactions in which (a) platelets bind to the ristocetin-activated vWF, (b) the platelets are activated to expose GPIIb/IIIa, and (c) the activated platelets are then cross-linked by vWF (or other integrin-binding proteins in the plasma) through GPIIb/IIIa. When heparin is added, it binds to vWF and blocks the vWF/platelet interaction [932]. This effect apparently results from the binding of heparin to its binding site in the A1 loop, thus sterically preventing the interaction between the vWF and platelets. The heparin inhibition of platelet aggregation at sites of vascular injury can also be demonstrated *in vivo* [932]. It is not known whether there is a specific sequence in heparin that is involved in the vWF binding, but it is clear that heparin fractions that have been depleted of antithrombin-binding sequences by affinity)) chromatography on antithrombin-Sepharose are just as effective in blocking vWF-mediated platelet aggregation as those fractions that bind tightly to antithrombin. Consequently, one effect of exogenous heparin is to block platelet aggregation, thus inhibiting coagulation. This anticoagulant effect of heparin is quite independent of the antithrombin-mediated anticoagulant activity of heparin. In fact, periodate-oxidized heparin, which no longer binds to antithrombin, will also block platelet aggregation [943].

A role for the heparin-binding capacity of vWF in normal hemostasis has not been reported, nor have there been studies to determine whether heparan sulfate (especially the heparan sulfate found on the surfaces of endothelial cells) will bind to vWF and prevent vWF-platelet interactions *in vivo*. If this does in fact take place, then this would represent a means for preventing normal platelet adhesion to the endothelium, yet another mechanism for preventing thrombus formation on normal endothelia.

3.   Effect of exogenous heparin

It is clear from the above discussion that exogenous heparin may play a role as an anticoagulant by preventing the very critical step in vWF binding to platelets and platelet plug formation [930]. In fact, it has been demonstrated that intravenous administration of heparin inhibits the platelet agglutination activity of vWF [932].

Heparin also causes an additional anticoagulant effect that is only indirectly related to vWF. As described earlier (Section I.B., p. 240), the full expression of the coagulation pathway depends on the conversion of factor VIII to factor VIIIa, a cofactor for the activation of factor X by factor IX (Figure 8.1). The conversion of factor VIII to factor VIIIa is catalyzed by thrombin, and, at physiological concentrations of thrombin and factor VIII, the VIII→VIIIa reaction is inhibited by therapeutic levels of heparin [944], a direct effect of the binding of heparin to thrombin (Section I.C., p. 243). As noted earlier, factor VIII circulates as a complex with vWF, which prolongs the factor VIII half-life. The inhibition of the thrombin conversion of VIII to VIIIa occurs both in the presence and in the absence of vWF.

F.   Tissue Plasminogen Activator

1.   Structure and heparin-binding

Tissue plasminogen activator (t-PA) has been described as a heparin-binding protein, but, as we shall see, its heparin-binding capacities are limited and may be of no physiological significance. The structure of t-PA has been described in detail [945, 946]. Human tissue plasminogen activator, with a molecular mass of 70 kDa, is a single-chain glycoprotein composed of 527 amino acids that circulates in the blood at a concentration of 70 pM. The domain structure of the protein, shown in Figure 8.6 [947, 948], consists of an N-terminal finger region (homologous to the fibronectin finger region structures), followed by an EGF-like domain, two kringle domains,[2] and finally

---

[2] Single or multiple kringle domains occur in a number of other proteins, including urokinase, coagulation factor XII, human plasminogen, hepatocyte growth factor, prothrombin, and apolipoprotein (a), where they play a role in protein-protein interactions. These domain structures contain approximately 80 amino acids, including cysteine residues that form three disulfide bonds that stabilize their tertiary structures.

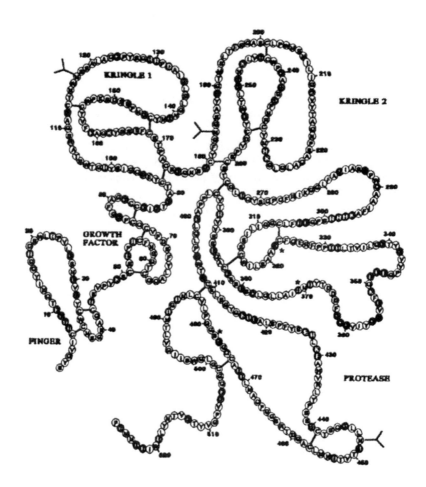

Figure 8.6. Domain structure of t-PA. Reprinted from [952], with permission.

the serine protease domain that lies in the C-terminal half of the sequence.

Although t-PA is active in its single-chain form, it is converted by plasmin cleavage at Arg 275-Ile 276 to a more active two-chain form in which the two chains are held together by a disulfide bond; i.e., in contrast to most serine proteases, t-PA is active in its "pro-enzyme," or zymogen, form [949]. Heparin and heparin oligosaccharides bind to t-PA and stimulate its activity. Heparin can be fractionated by affinity chromatography on t-PA-Sepharose columns [950, 951] to obtain a portion of the original heparin that exhibits a four-fold greater activity than the starting heparin. As is commonly found, the fractionation of the heparin into t-PA-binding and t-PA-nonbinding fractions does not fractionate heparin according to its antithrombin-binding activity; i.e., the t-PA-binding and t-PA-nonbinding fractions have similar antithrombin-mediated thrombin-neutralizing activities.

Despite extensive work on the structural details of t-PA and on the effects of replacement of one or more amino acids in various domains on its biological activities [see, for example, 952, 953-964], the polypeptide region of t-PA that binds to heparin has been a focus of only a few studies. These have suggested that the heparin binding domain lies primarily in the finger region with some contribution from the kringle-2 domain [965, 966]. However, a search for an amino acid sequence rich in arginine and lysine in these domains, or at any other regions of the primary sequence, reveals sequences with only a limited number of basic residues. The sequences that are the richest in lysine and arginine residues are the following:

Finger domain:

$$\text{W L }\underline{\text{R}}\text{ P V L }\underline{\text{R}}\text{ S N }\underline{\text{R}}\text{ V E Y C W C N S G }\underline{\text{R}}$$
$$21 30 40$$

Kringle-2 domain:

$$\text{L G}\underline{\text{K}}\text{ H N Y C }\underline{\text{R}}\text{ N P D G D A }\underline{\text{K}}\text{ P W C H L }\underline{\text{K}}\text{ N }\underline{\text{R}}\text{ }\underline{\text{R}}\text{ L}$$
$$230 240 250$$

In these sequences the lysine and arginine residues are widely spaced. Thus, it is only by default that these sequences can be implicated as heparin-binding sequences—there are no other candidates! This dilemma is aggravated by the finding that heparin oligosaccharides as short as disaccharides apparently bind to t-PA and stimulate its activity (below). The lack of a typical heparin-binding sequence raises the possibility that the heparin-binding domain may be composed of amino acid residues that are brought together by the tertiary folding of the molecule. If this is the case, it would seem that modest perturbation of the tertiary structure might reduce its affinity for heparin. However, such experiments have not been described. t-PA is the first example that we have encountered of a heparin-binding protein that seems to lack the usual clustering of basic amino acids that is generally involved in heparin binding. Consistent with this finding, it is questionable whether there is any significant interaction between t-PA and heparinoids *in vivo* (below).

2. Biological activity

In normal hemostasis, both t-PA and plasminogen bind with high affinities, through their kringle domains, to fibrinogen and fibrin, and the activation of plasminogen by t-PA takes place right on the fibrin clot; i.e., the fibrin serves as a template for plasminogen activation by t-PA and markedly stimulates the reaction. Similarly, if fibrin appears on the endothelium, t-PA and plasminogen would be expected to bind to it as well and to digest the fibrin. Thus, t-PA may play a role in maintaining the anticoagulant surface of the endothelium.

In the absence of fibrin(ogen), heparin can also stimulate plasminogen activation by binding to and activating t-PA [950, 965, 966], but heparan sulfate is not very effective [966, 967]. It has been reported that heparin oligosaccharides as small as disaccharides can stimulate t-PA activity [968].[3] In fact, di- and tetrasaccharides prepared from heparin increased the k_{cat} and lowered the K_m for the t-PA-mediated plasminogen activation, whereas higher oligosaccharides only increased the k_{cat}. In striking contrast, it has been reported that low molecular weight heparin *cannot* stimulate the activity of t-PA [969]. This apparent contradiction seems to be due to the difference in the assay procedures that have been used to study the heparinoid effects on t-PA activity, and particularly to the differences in the ionic strength of the assay media. After the earlier reports on heparin activation of t-PA, it was found that the ability of heparin to activate plasminogen by t-PA is a complex function of the ionic strength and the presence of divalent cations and chloride ions [967, 970]. The most significant conclusion from these demonstrations of the salt effects is that heparin has little effect on the t-PA-mediated activation of plasminogen *at a physiological ionic strength*, but does stimulate the reaction at lower salt concentrations. Because the earlier studies of the effects of heparin, low molecular weight heparin, and heparin oligosaccharides were carried out in the presence of different, but sub-physiological salt concentrations, different degrees of stimulation of t-PA activity were observed.

3. Effect of exogenous heparin

When t-PA is administered clinically to induce thrombolysis, heparin is also given to patients to prevent re-thrombosis at the thrombogenic site where clot lysis will occur and to prevent further thrombosis. A number of authors

---

[3] This may be similar to the effects of heparin disaccharides on FGF dimerization (Chapter 9).

have addressed the question of whether heparin has some direct effect on the efficacy of t-PA *in vivo* [see 971, 972, 973, 974, and references therein]. It seems clear that heparin enhances the re-canalization events during t-PA treatment; what is not clear is whether the heparin effects are exerted directly on the activities of t-PA or whether they are mediated through one or more of the other heparin-binding proteins in the vasculature. The latter possibility seems most likely. For example, it has been shown that heparin and low molecular weight heparin have no direct effect on clot lysis [971] but that heparin is clearly beneficial in thrombolysis when administered with t-PA [972-974]. However, the *in vitro* demonstrations that physiological salt concentrations eliminate the heparin effect on t-PA activity suggest that heparin could have no direct effect on t-PA activity *in vivo*. In the absence of t-PA, exogenous heparin cannot penetrate the clot where the plasminogen activation normally takes place, and therefore cannot directly influence or modulate the effects of endogenous t-PA. It is possible that as t-PA induces fibrin breakdown, the thrombi would become more porous and the heparin might be able to enter the platelet plug and exert an antithrombotic effect. In fact, heparin acts on many hemostatic proteins, and the effect of heparin in t-PA therapy can result from its effects at many levels [975]. It is interesting to speculate that if heparin *could* bind to t-PA in the blood, it would stimulate the conversion of circulating plasminogen to plasmin. The plasmin thus formed would be neutralized almost immediately by α_2-antiplasmin. If this were to happen, the effect of the heparin would be to catalyze a futile reaction sequence in which plasminogen is converted to inactive plasmin/α_2-antiplasmin complexes. The end result would be the heparin-catalyzed depletion of plasminogen and α_2-antiplasmin from the plasma— potentially disarming of the fibrinolytic pathway—just the opposite of the effect that is desired in the clinical administration of exogenous heparin. The fact that such a sequence of events has not been observed is consistent with the conclusion that heparin has no direct effect on t-PA activity *in vivo*.

G. Urokinase

1. Structure and heparin binding

The structure and activities of urokinase-type plasminogen activator (urokinase, u-PA) have been reviewed [946], and the three-dimensional structure of u-PA has been modeled on the chymotrypsin crystal structure [976]. u-PA appears in the circulation as a single chain glycoprotein (scu-PA) that contains 411 amino acids and has a molecular mass of 54 kDa. The single-chained protein (Figure 8.7) has an N-terminal EGF-like domain, but it has no finger domain and only one kringle, which apparently plays no protein-binding role in u-PA activities. The C-terminal region of the molecule contains the protease domain [977]. Scu-PA has very little activity, but acquires the capacity to activate plasminogen when it is cleaved by plasmin at Lys 158-Ile 159 to generate a two-chain u-PA (tcu-PA) in which the N-terminal portion of the protein containing the kringle domain remains linked to the C-terminal protease domain by a single disulfide bond. A further cleavage of tcu-PA yields a low molecular mass (32 kDa) form of u-PA. Heparin binds to u-PA and stimulates its activity [950, 978-980]. The heparin-binding domain has been localized to the kringle region of the protein [980], where there is found the following sequence within the kringle inner loop that is somewhat rich in arginine and lysine:

$$N\ Y\ C\ \underline{R}\ N\ P\ D\ N\ \underline{R}\ \underline{R}\ \underline{R}\ P\ W\ C\ Y\ V\ Q\ V\ G\ L\ \underline{K}$$
$$\quad\ \ 100\qquad\quad 105\qquad\quad 110\qquad\quad 115\qquad\quad 120$$

2. Biological activity

u-PA is synthesized and secreted as the single chain form by endothelial cells and a number of other cell types. Endothelial cells also secrete t-PA, PAI-1, and receptors for u-PA, plasminogen, and t-PA. u-PA binds through its EGF-like domain to its specific cell surface receptor [981-986], which is a GPI-anchored protein in the endothelial cell membrane. The receptor is a heavily glycosylated glycoprotein, but is *not* a proteoglycan. Thus, heparan sulfate proteoglycan is not involved in the binding of u-PA to cell surfaces. It has been demonstrated that suramin (Figure 8.8), which has been considered a low molecular weight heparin analog [987, 988], inhibits the binding of u-PA to the cell surface receptor. However, neither heparin nor heparan sulfate exhibit the same activity [989]. Thus, the suramin effect does not seem to be heparan sulfate-related.

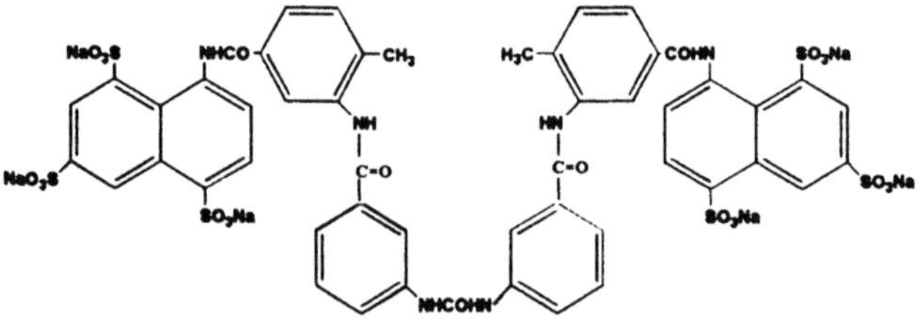

Figure 8.8. Structure of suramin, a heparin mimic.

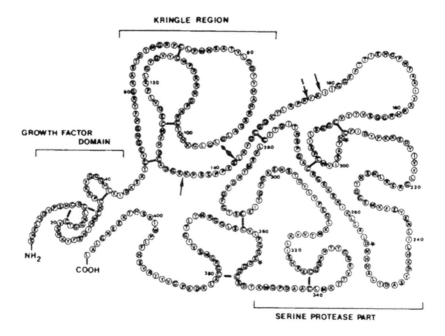

Figure 8.7. Domain structure of scu-PA. Reprinted from [946], with permission.

The secretion of u-PA by endothelial cells and its receptor-mediated binding to these cells localizes its effect to the cell surface, where the scu-PA is converted to tcu-PA [990]. The tcu-PA converts surface-bound plasminogen to plasmin, which may prevent thrombus formation on the endothelium and modulate the capacity of the cells to migrate. In fact, the ability of u-PA to stimulate cell migration suggests that it may play a significant role in tissue remodeling and metastasis [991-993]. The activity of u-PA appears to be regulated by its own rate of synthesis and turnover, and by the rate of synthesis of its receptor [992, 994-997].

3.   Effect of exogenous heparin

The effect of heparin on two-chain u-PA is very similar to its effect on t-PA; i.e., the heparin stimulates the conversion of plasminogen to plasmin by the tcu-PA. The plasmin, in turn, converts more of the inactive single-chain u-PA to the active two-chain form. A study of the binding of heparin to the isolated kringle domain of u-PA has shown that the heparin binding to the kringle is salt sensitive, but demonstrates that, at physiological ionic strengths, the K_d is ~ 8 x 10^{-5} M, suggesting that there may be modest binding of exogenous heparin to u-PA *in vivo* [980]. However, studies with whole u-PA and heparin have shown that there is an insignificant effect of heparin on u-PA activity at physiological ionic strength [967, 970]. As in the t-PA case, the demonstrations that heparin stimulates u-PA activity were carried out at sub-physiological ionic strengths.

H.   Plasminogen Activator Inhibitor-1

1.   Structure and heparin binding

The activities of t-PA and u-PA are neutralized by plasminogen activator inhibitor-1 (PAI-1), a member of the serpin family [reviewed in 906]. The full-length cDNA for PAI-1 codes for a pre-PAI-1 containing 402 amino acids and having extensive homology with other members of the serpin family. The secreted protein is a 50-kDa glycoprotein with 379 amino acids and about 13% carbohydrate [998, 999]. The active center, Arg 346-Met 347, is cleaved extremely rapidly by t-PA and u-PA to give the corresponding inactive covalent complexes of these proteases with PAI-1 [781, 998-1000]. The crystal structure of PAI-1 has been determined [781] and has been compared with that of α_1-protease inhibitor and ovalbumin, as shown in Figure 8.9 [781, 784]. PAI-1 can exist in three interchangeable conformations—an active form with an exposed active center loop, a latent form in which the active center loop is inserted into β-sheet A (as in Figure 8.9), and a substrate-like form that can be cleaved and released by its target proteases [see 1001, and references therein]. The P1 and P1' residues in the cleaved PAI-1 are found ~70 Å apart [1001], as for other cleaved serpins (see discussion of serpin reactive center loops in Chapter 6).

A comparison of the heparin-binding A and D helices of antithrombin with the corresponding structures of PAI-1 shows that PAI-1 has two basic amino acids, His 10 and Lys 28, in its A helix and six basic amino acids in its D helix. Measurements of heparin binding to PAI-1 and to mutant forms in which each of these basic amino acids is replaced singly with Ala [1003] have shown that the A helix of PAI-1 does not play a role in heparin binding but that D helix residues Lys 65, Lys 69, Arg 76, Lys 80, and Lys 88 play major roles in heparin binding, with Arg 76, Lys 80, and Lys 88 being of critical importance. Thus, the heparin-binding region of PAI-1 has been localized to the sequence of basic amino acids near the C-terminal [719, 784, 1003]:

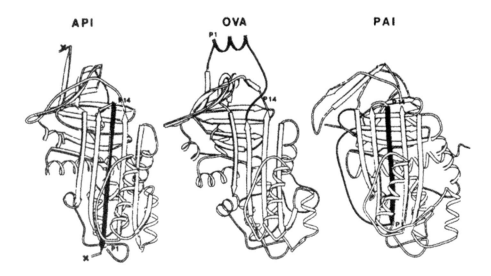

Figure 8.9. Comparison of 3D structures of α_1-protease inhibitor, ovalbumin and PAI-1. Reprinted from [1002], with permission.

K I D D K G M A P A L R H L Y K E L M G P W N K
65   70   75   80   85

Interestingly, this sequence does not conform very well to the consensus heparin-binding sequences described by Cardin and Weintraub (see Chapter 6).

2.   Biological activity

PAI-1 is synthesized by endothelial cells, smooth muscle cells, hepatocytes, various malignant cells [reviewed in 1004], and granulosa cells [784]. The rates of synthesis of PAI-1 by the various cell types are regulated by a number of growth factors, including acidic fibroblast growth factor, transforming growth factor-β, glucocorticoids, tumor necrosis factor-α, platelet-derived growth factor, tumor necrosis factor, and vascular endothelial growth factor, and by other agents including lipopolysaccharide, phorbol 12-myristic-13-acetate, endotoxins, and factors that are secreted by smooth muscle cells [999, 1005-1013]. In cell culture, PAI-1 is secreted into the medium in an active form that is rapidly converted to an inactive or "latent" form that can be reconverted to an active form by negatively charged phospholipids or denaturants. The structural basis for this latency has been established by

Mottonen *et al.* [781], who have demonstrated that the reactive center loop of PAI-1 is partially incorporated into the β-sheet A in the latent protein. PAI-1 is found in the blood at a concentration of 25-30 μg/ml and in the extracellular matrix of the endothelial cells. Regardless of its location, PAI-1 is always found in a noncovalent complex with *vitronectin* (Section II.I., p. 275), and its biological activity, which is primarily directed toward the control of plasminogen activators, is also intimately coupled to that of vitronectin. Ninety percent of the PAI-1/vitronectin complex in blood is found in the α-granules of the platelets [1012]. Neither PAI-1 nor vitronectin appear to be synthesized by platelets, and the pathways leading to their accumulation in platelets are unknown [see 1014].

In vivo, newly secreted PAI-1 immediately forms a complex with vitronectin, which stabilizes the PAI-1 by increasing its *in vitro* half-life at 37 °C° two to three-fold [998, 999, 1004, 1015, 1016]. The interaction between PAI-1 and vitronectin involves the N-terminal region of the PAI-1, a site totally separated from its heparin-binding domain. The PAI-1 binding site on the vitronectin apparently overlaps the vitronectin heparin-binding sequence [784, 1016], although some evidence has suggested that PAI-1 interacts with the N-terminal (heparin-binding) domain of vitronectin [1015, 1017]. Interestingly, heparin seems to compete with PAI-1 for binding to vitronectin when the vitronectin is layered onto plastic dishes, but not when the vitronectin is in solution [1015, 1016, 1018]. This raises the possibility that heparin *in vivo* can release free PAI-1 from its PAI-1/vitronectin complex, causing labilization and loss of activity of the PAI-1.

PAI-1 is the major serpin that neutralizes the activities of t-PA and u-PA *in vivo*. Although four distinct plasminogen activator inhibitors have been identified, PAI-1 is the only cleaved serpin that is found in plasma in an inactive (covalent) complex with the t-PA [999]. Thus, PAI-1 plays a major role in the control of fibrinolysis. High levels of PAI-1 are correlated with myocardial infarction, deep vein thrombosis, re-occlusion after angioplasty, and coronary artery disease [cited in 1007, 1012]; i.e., when PAI-1 levels are excessive, the activities of t-PA and u-PA are neutralized and fibrinolysis cannot proceed at the rate that is necessary to maintain hemostasis. PAI-1 reacts extremely rapidly, and in a *heparin-independent* manner, with t-PA and u-PA, with a second order rate constant of 10^7-10^8 M^{-1} sec^{-1} and a K_d of less than 10^{-12} M for the t-PA/PAI-1 complex [998]. Curiously, the heparin-binding capacity of PAI-1 does not seem to play a role in its direct endogenous activities, but see Section II.I., p. 275.

Many of the activities of PAI-1 have been studied using the pure, active form of PAI-1. However, the form of PAI-1 that is found in plasma, in platelets, and in the endothelial extracellular matrix is the PAI-1/vitronectin complex. The complexation with vitronectin has no effect on the reactivity of PAI-1 with t-PA and u-PA. However, vitronectin does give a 1000-fold stimulation of the rate of PAI-1-mediated neutralization of *thrombin* [998, 1000, 1019, 1020]. Thus, the PAI-1/vitronectin complex is effective in the neutralization of t-PA and u-PA and controlling thrombolysis, but it is also effective in the neutralization of thrombin, thus blocking thrombus formation. These complementary activities would seem to prevent both fibrinolysis and coagulation at the same time!

The regulation of the activities of t-PA and u-PA in the general circulation appears to be controlled by the PAI-1/vitronectin complex in the endothelial cell extracellular matrix. As noted earlier that the PAI-1/vitronectin interaction blocks the heparin-binding domain of the vitronectin, but leaves the heparin-binding domain of the PAI-1 exposed. This raises the possibility that the cell surface heparan sulfate proteoglycan may bind the complex to the endothelial cells. However, the PAI-1/vitronectin complex cannot be dissociated from the extracellular matrix of the endothelial cells with high salt or with heparin [cited in 998], observations that appear to rule out the involvement of the heparan sulfate. Perhaps vitronectin RGD (integrin-binding) sequences are involved in the binding. In any event, the PAI-1/vitronectin complex on the surface of the endothelial cells reacts with circulating t-PA or u-PA, neutralizing their activities. Once the inactive plasminogen activator-PAI-1 complexes have been formed, they lose their affinity for vitronectin and are released from the cell surface [998, 1004, 1018]. The released t-PA-PAI-1 complex is rapidly cleared from the plasma by receptor-mediated endocytosis and catabolism, again with no apparent involvement of cell surface heparan sulfate proteoglycan [see 1021, 1022, 1023]. As discussed in Chapter 11, the receptors for clearance of the complexes appear to be members of teh LDL receptor family.[4]

The PAI-1/vitronectin complex is involved in several activities in cells [1017, 1024], but its activity in the platelets is certainly one of its more important roles. As noted earlier, 90% of the PAI-1 in the blood is found in the platelet α-granules. When the platelets are activated by thrombin in the initial stages of coagulation, PAI-1/vitronectin and growth factors are released and the releasates block the initiation of clot lysis by t-PA in two ways: (a) the PAI-1/vitronectin complex released from the platelets inhibits the activation of plasminogen by t-PA so that the clot can form [1025], and (b) the transforming growth factor-β and epidermal growth factor-like activities released from the platelets can stimulate PAI-1 synthesis by endothelial cells and hepatocytes [1026] so that there is a delayed PAI-1 response (approximately 2 hr). This suggests that PAI-1/vitronectin (and perhaps just vitronectin alone—see Section II.I., p. 275) may prevent plasminogen activation until after the coagulation

---

[4] t-PA is also cleared from the circulation via PAI-1-independent mechanisms (see Section II.F.).

cascade can generate an effective procoagulant response and may also retard or prevent fibrinolysis at sites that are removed from the site of vascular injury.

3.   Effect of exogenous heparin

There is little direct evidence in the literature concerning the effect of heparin on the activity of PAI-1. PAI-1 is clearly a heparin-binding protein, and in the PAI-1/vitronectin complex its heparin-binding region is apparently not blocked. However, an *in vivo* role for this heparin-binding property has not been established. It has been shown that exogenous heparin stimulates thrombin inhibition when PAI-1 is in excess but that it enhances the rate of PAI-1 inactivation when thrombin levels are increased [1027]. Heparin binds to PAI-1 and may displace it from the PAI-1/vitronectin complex. Thus, in the presence of exogenous heparin, the PAI-1/vitronectin complex may be replaced by a PAI-1/heparin complex. Heparin has little effect on the reactivity of PAI-1 with t-PA and u-PA, but stimulates the PAI-1-mediated neutralization of thrombin 100-fold [784, 1015, 1016, 1028], although the stimulated rate is considerably lower than the reaction rates for antithrombin or HCII with thrombin. If complexation with vitronectin stimulates the thrombin-neutralizing activity of PAI-1 1000-fold, while its complexation with heparin gives only a 100-fold stimulation of the same activity, then, if heparin displaces PAI-1 from the PAI-1/vitronectin complex, it seems likely that heparin will *lower* the thrombin-neutralizing activity of PAI-1. This heparin effect on thrombin neutralization is molecular weight dependent, suggesting that heparin serves as a template for binding the thrombin and PAI-1 (approximation). The involvement of endothelial cell heparan sulfate proteoglycan in the activation of PAI-1 toward thrombin, if any, has not been described. A further point relates to the well-recognized stabilization of PAI-1 by vitronectin. It is not known whether PAI-1 is also stabilized in a PAI-1/heparin complex. If it is not, it is possible that heparin displacement of PAI-1 from vitronectin may result in a destabilization of PAI-1.

To summarize, PAI-1 circulates in the plasma in a complex with vitronectin, and it is also found in this complex in the α-granules of the platelets and in the subendothelial extracellular matrix. Vitronectin stabilizes PAI-1 in the circulation without altering its plasminogen activator neutralizing capacity and it converts PAI-1 into a serpin that will also neutralize thrombin. Interestingly, although the primary role of PAI-1 is to block inappropriate fibrinolysis, the thrombin-neutralizing effect of PAI-1/vitronectin also makes PAI-1 an anticoagulant! Both PAI-1 and vitronectin are heparin-binding proteins, and their complexation does not appear to involve the heparin-binding domain of the

PAI-1, but may block the heparin-binding domain of the vitronectin. Thus, it is possible that endothelial cell heparan sulfate proteoglycan may play some role in the activation or the catabolism of PAI-1 (see Section II.H.2., p. 271). It is clear that exogenous heparin can alter the activities of both PAI-1 and vitronectin.

I. Vitronectin

1. Structure and heparin binding

The structure and biological activities of vitronectin have been reviewed [906, 1029-1031]. Vitronectin, also referred to as complement S protein and serum spreading factor, circulates in the blood in 65 and 75 kDa forms at a concentration of 250-450 µg/ml (0.1-0.5% of the plasma protein). In addition, a peptide, somatomedin B, which is identical to the first 44 amino acids of vitronectin, is found in the circulation at a concentration of 11-14 µg/ml, *although* its relationship to vitronectin has not been established [1032]. The human and rabbit vitronectins have been cloned and sequenced [1033-1035], and their domain structures have been characterized [reviewed in 1029]. As shown in Figure 8.10, vitronectin has the N-terminal somatomedin domain, followed by a connecting segment that contains an RGD cell attachment site and a collagen-binding site. In addition, there are two homoplexin-type repeats and a heparin-binding sequence in the C-terminal domain. As shown in Figure 8.11, vitronectin exists in a native, or "latent," form (form 1) that cannot bind heparin. In the latent form, the heparin-binding domain, indicated by a "+," is paired with the acidic N-terminal domain of the protein (arrowhead). Conversion of the latent form to a partially unfolded form (2) occurs in the presence of heparin. Alternatively, the latent form is induced by serpin-protease complexes, or by other agents (chaotropes, a basic peptide representing the heparin-binding sequences of vitronectin) to unfold completely. These changes in conformation yield the heparin-binding forms (forms 3 and 4, Figure 8.11) which undergo multimerization to aggregate forms which still binds heparin with high affinity ($K_d = 10^{-8}$ $\underline{M}$). The affinity of the heparin-binding forms of vitronectin for

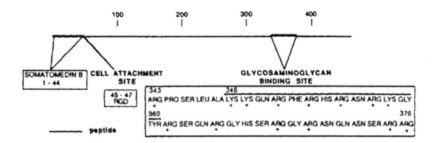

Figure 8.10. Structure of vitronectin. Reprinted from [1037], with permission.

heparin is comparable to the antithrombin affinity for heparin. Both antithrombin-binding and antithrombin-nonbinding heparin fractions (HA-heparin and LA-heparin—see Chapter 7) bind to vitronectin, as does heparan sulfate. Heparin binding to vitronectin is not dependent on the molecular weight of the heparin. Oligosaccharides as small as the antithrombin-binding pentasaccharide bind as well as higher molecular weight heparins. The heparin-binding domain has been localized to an Arg- and Lys-rich sequence near the C-terminal of the protein [1037-1039], namely amino acids 348-361 (see Figure 8.10):

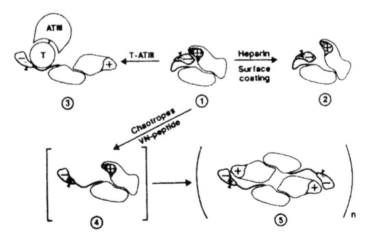

Figure 8.11. Structure of vitronectin. Reprinted from [1036], with permission.

$$\underline{K}\ \underline{K}\ Q\ \underline{R}\ F\ \underline{R}\ H\ \underline{R}\ N\ \underline{R}\ \underline{K}\ G\ Y\ \underline{R}$$
348   361

This peptide can form an amphipathic α-helix in which the arginine and lysine residues are on one surface of the helix and the hydrophobic residues are on the other.

2.   Biological activity

Vitronectin is synthesized exclusively by hepatocytes. It is secreted as a mixture of monomeric and oligomeric forms, and co-purifies with PAI-1, which stimulates its multimerization [1040]. Vitronectin appears in a variety of tissues, only some of which are involved in hemostasis. For example, it is found in α-granules of platelets, but also in the matrix of subendothelial tissues. Thus, it is transported to these tissues *via* the blood by mechanisms that are not completely understood. Vitronectin appears to play at least two roles in hemostasis. First, it plays a critical role in the stabilization of PAI-1, a protein that is the primary regulator of fibrinolysis (Section II.H., p. 270). Second, it participates in the clearance of inactive serpin/protease complexes from the circulation. An understanding of this latter activity requires some reiteration of our prior discussion of the heparin-mediated reaction of antithrombin with thrombin. Antithrombin has a high affinity for the heparan sulfate chains of endothelial cell heparan sulfate proteoglycan to which it binds in an activated form. When the heparan sulfate/antithrombin complex reacts with thrombin, the antithrombin is cleaved, and an inactive thrombin-antithrombin complex is formed. The affinity of the thrombin-antithrombin complex for heparan sulfate is much reduced, and the covalent complex dissociates from the endothelial cell surface heparan sulfate and appears in the circulation. When this complex encounters the latent form of vitronectin, the vitronectin binds to the complex, yielding a thrombin-antithrombin/vitronectin ternary complex. As a result, vitronectin undergoes the conformation change/multimerization reaction that yields the heparin-binding form of vitronectin [1041, 1042]. Although the thrombin-antithrombin portion of this tertiary complex no longer has a high affinity for heparinoids, the heparin-binding properties of the *vitronectin* are retained by the complex, and the vitronectin component of the ternary complex mediates the binding of the complex to endothelial cell heparan sulfate proteoglycan [1043]. The complex is then internalized and catabolized.[5] The uptake of the complex is

---

[5] It has been shown by Völker *et al.* [1044] that labeled vitronectin, in complex with thrombin-antithrombin, is internalized by endothelial cells and delivered in part to the lysosomes and in part to the basolateral face of the endothelial cells, apparently by a

markedly inhibited by treating the cells with heparin, heparinase, or β-xylosides (which prime heparan sulfate synthesis and secretion but release the β-xyloside-linked product from the endothelial cell surface—see Chapter 5), but not by RGD peptides [1043]. Similar results are found with other serpin/protease complexes. Thus, vitronectin plays a major role in the *clearance* of thrombin from the circulation following its neutralization by antithrombin (or by HCII). In fact, all of the thrombin-antithrombin complex that can be detected in the plasma is found in the ternary complex with vitronectin. This system for clearance of thrombin is distinct from the α_2-macroglobulin and the liver clearance systems for serpin/protease complexes discussed in Section I.D.3., p. 248. Vitronectin may also be involved in the clearance of thrombin-antithrombin complexes via gC1qR, a vitronectin-binding protein originally recognized as a binding protein for the globular head domains of complement component C1q [1045].

To reiterate, the pathway for metabolism of vitronectin may be described as follows. Vitronectin is synthesized by hepatocytes and secreted into the plasma in its native form, which cannot bind heparin or heparan sulfate. When inactive thrombin-serpin complexes encounter this latent vitronectin, they induce a conversion of the vitronectin into the heparin-binding form which is converted immediately to multimers that bind to the thrombin-serpin complexes. The tertiary complexes thus formed are bound to endothelial cell surface heparan sulfate proteoglycan and are internalized and catabolized by the endothelial cells. A further aspect of vitronectin metabolism that relates to its complexation with PAI-1 is described in Section II.H.2., p. 271.

One additional activity of vitronectin that may have physiological relevance is its ability to bind to plasminogen, thus inhibiting the t-PA-catalyzed conversion of plasminogen to plasmin [1019]. This suggests (a) that vitronectin released from activated platelets may block the initial plasminogen activation until the coagulation cascade can generate an effective procoagulant response, and (b) that vitronectin, which appears in platelet secretions and in the extracellular matrix, may retard or prevent fibrinolysis at these sites.

3.   Effect of exogenous heparin

The direct effects of exogenous heparin on vitronectin-mediated processes have not been characterized in *in vivo* studies, but can be inferred, at least in part, from the above discussion. Heparin will displace the complexes of

---

transcytosis process. This transcytosis mechanism may be involved in the distribution of vitronectin to extravascular systems.

vitronectin/serpin-protease complexes from the endothelial cells, thus preventing this aspect of the final catabolism of the neutralized proteases. It is not clear whether this would have any effect on hemostasis. However, these tertiary vitronectin/serpin-protease complexes would be expected to bind a portion of the exogenous heparin and to prevent a portion of the exogenous heparin from participation in antithrombin-mediated anticoagulant activity. Perhaps a more serious concern would be the possible competition between PAI-1 and heparin for vitronectin. The latent form of vitronectin is converted to the active form when it encounters either PAI-1 or heparin, and heparin may cause the dissociation of the vitronectin/PAI-1 complex (Section II.H.3., p. 274). Since vitronectin is required for the stabilization of PAI-1, it is possible that exogenous heparin will cause a destabilization of the PAI-1 and thus interfere with the normal regulation of t-PA and u-PA activities.

Finally, it seems clear that vitronectin, a major protein in the circulation, may neutralize the anticoagulant activity of exogenous heparin even when it is not in a complex with serpin-proteases, and that effective anticoagulant doses of heparin must be in excess of those required to satisfy the heparin-binding capacity of vitronectin.

J. Protein C Inhibitor

1. Structure and heparin binding

Protein C inhibitor (PCI) [6] is a 57-kDa heparin-dependent serpin present in human blood at a concentration of 5 µg/ml. It is identical to a previously identified plasma protein, plasminogen activator inhibitor-3 (PAI-3) [1046]. Its role in hemostasis is to block the action of activated protein C (Section I.D.4., p. 250), acting as a typical serpin suicide substrate [1047]. Like PAI-1, HCII, and antithrombin, its three-dimensional structure has been modeled on that of α_1-protease inhibitor [785]. Of the residues of APC inhibitor modeled on α_1-protease inhibitor, 44% are identical and 72% are identical or similar. Interestingly, unlike antithrombin, the D-helix of protein C inhibitor presents no significant positive surface for heparin binding. However, two amphipathic α-

---

[6] As described in Section I.D.4., p. 250, protein C exists in a precursor form, which is referred to simply as protein C, and an activated form, which is referred to as activated protein C (APC). Although the inhibitor that is discussed here is commonly referred to as "protein C inhibitor" (PCI), its inhibitory activity is directed against *activated* protein C and not against the precursor form.

helices, the A+ and the H, lie opposite each other, forming a groove that is 10 Å wide.

A+ helix:

$$\text{H } \underline{\text{R}} \text{ H H P } \underline{\text{R}} \text{ E M } \underline{\text{K}} \text{ } \underline{\text{K}} \text{ } \underline{\text{R}} \text{ V E D L H}$$

1   5   10   15

H helix:

$$\text{S E } \underline{\text{K}} \text{ T L } \underline{\text{R}} \text{ } \underline{\text{K}} \text{ W L } \underline{\text{K}} \text{ M F } \underline{\text{K}} \text{ } \underline{\text{K}} \text{ } \underline{\text{R}} \text{ Q L E L Y}$$

265   270   275   280

These helices direct their positively charged amino acids toward the groove, while their opposite sides are buried in the hydrophobic interior of the molecule. The positively charged amino acid side chains in the groove can accommodate the negatively charged sulfate and carboxyl groups of a heparin octasaccharide. The apparent differences between heparin binding by PCI and antithrombin and HCII are shown in Figure 8.4. The novel heparin-binding site of PCI is implicated by several observations. A monoclonal antibody directed against the A+ helix blocks heparin binding to APC inhibitor, whereas another antibody that recognizes neither the A+ nor the H helix has no effect on heparin binding. However, in a separate study, Pratt and Church [1048] compared several synthetic peptides derived from different domains of APC inhibitor for their abilities to block the binding of heparin by APC inhibitor. Interestingly, peptide 1-16, representing the A+ helix, had no effect on heparin binding, whereas peptide 264-283, representing the H helix was a potent inhibitor. In any event, the heparin-binding region of this serpin is clearly different from that of antithrombin, HCII, and PAI-1. In fact, the positive helix pair suggested by Heeb *et al.* [1046], which includes both helix A+ and helix H, is similar to that involved in heparin binding by platelet factor 4 (Section II.L., p. 285) [1049].

2. Biological activity

Activated protein C (APC) exerts its effect on hemostasis by proteolysis of factors Va and VIIa in a reaction that is mediated by protein S (Section I.D.4., p. 250). This effectively shuts down the coagulation pathway. Thus, APC is an anticoagulant. Genetic defects in APC result in thrombotic deficiencies [1050-1053]. Protein C inhibitor is a heparin-dependent serpin that neutralizes APC, thereby permitting the coagulation pathway to proceed. Thus, protein C inhibitor exerts a pro-coagulant activity. The stimulation of the reaction between protein C inhibitor and APC by heparin increases with the molecular weight of the heparin, requiring a heparin fragment with a molecular weight greater than 2500 for maximum activity [1048]. Protein C inhibitor also inhibits thrombin, plasma kallikrein, urokinase (u-PA), and tissue plasminogen activator (t-PA). Inactive complexes of protein C inhibitor with all of these proteases have been demonstrated both *in vivo* and *in vitro*. Thus, the neutralization of all of these proteases by APC inhibitor is physiologically relevant.

The APC-neutralizing activity of protein C inhibitor is markedly stimulated by heparin. However, an *in vivo* heparinoid-dependent mechanism of activation of protein C inhibitor in plasma has not been described. It has been shown, however, that protein C inhibitor activity of the kidneys is activated by cell surface proteoglycans [1054].

α_1-Protease inhibitor is also a potent serpin that neutralizes APC (in the absence of heparin), and, in fact, when APC is added to plasma, the amounts of the APC/protein C inhibitor complex that are formed are small when compared to the amounts of the APC/α_1-protease inhibitor complex [see reference 1055]. Thus, the physiological role of protein C inhibitor in the control of APC activity is not clear.

3. Effect of exogenous heparin

Exogenous heparin activates protein C inhibitor which thus gains the capacity to neutralize (a) APC (procoagulant effect), (b) t-PA and u-PA (pro-coagulant, antifibrinolytic effects), and (c) thrombin and other serine proteases of the coagulation pathway (anticoagulant effect). Stated another way, exogenous heparin effects exerted via protein C inhibitor include (a) prevention of the APC-mediated inactivation of factors Va and VIIIa so that coagulation can proceed, (b) neutralization of the activities of t-PA and u-PA so that clot lysis is blocked, and (c) neutralization of the serine proteases of the coagulation pathway so that coagulation cannot occur. The relative contributions of protein C inhibitor to each of these effects are not clear.

K. Tissue Factor Pathway Inhibitor

1. Structure and heparin binding

The structure and activity of tissue factor pathway inhibitor (TFPI), which has also been called anticonvertin, extrinsic pathway inhibitor (EPI), and lipoprotein associated coagulation inhibitor (LACI), have been reviewed [881, 1056]. Human TFPI has been cloned [1057]. The transmembrane glycoprotein has 276 amino acids (after removal of the signal sequence), with a molecular mass of 32 kDa. The primary sequence (Figure 8.12) has an acidic N-terminal domain, three Kunitz-type domains,[7] and a basic C-terminal domain, which has an unusually high concentration of basic amino acids and which appears to be the heparin-binding domain.

R I S K G G L I K H K T K R K R K K O R V K I A Y
246   250   255   260   265   268

The activity of TFPI, which can be purified by heparin affinity chromatography, is stimulated two- to three-fold by heparin [1060]. Low molecular weight heparin and unfractionated heparin appear to have similar activities in the stimulation of TFPI, suggesting that the heparin effect is independent of the molecular weight of the heparin [1061], but fractions with molecular weights

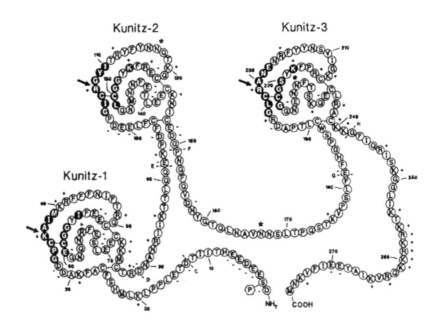

Figure 8.12. Primary structure of TFPI. Reprinted from [875], with permission.

[7] A Kunitz type domain contains about 60 amino acid residues, with six specifically spaced cysteines that are present in disulfide bonds. They are slow, tight-binding, reversible inhibitors of serine proteases that bind to the active site and inhibit the enzyme activity. In contrast to serpins, cleavage between the P1 and P1' residues of the Kunitz domains occurs very slowly, if at all [1058, 1059].

above 10,000 give the strongest binding [1062]. Antithrombin-binding and antithrombin-nonbinding heparin fractions bind equally well to TFPI [1062].

2. Biological activity

The exposure of tissue factor by vascular injury is the event that initiates the coagulation pathway. As described in Section I.B., p. 240, the coagulation pathway is "turned" by tissue factor, which is found as a complex with factor VIIa. Coagulation is then "turned off" by TFPI, with the result that there is a pulsed initiation event. This initiation event is then sustained and amplified by the other reactions of the coagulation pathway. TFPI plays two roles in the control of the coagulation pathway: (a) it inhibits factor Xa by forming a 1:1 complex with factor Xa, and (b) it blocks the activity of tissue factor. The latter effect results when the TFPI/factor Xa complex combines with the tissue factor/factor VIIa complex in the presence of Ca^{2+} to give a quaternary complex. In this complex the capacity of the tissue factor/factor VIIa complex to convert factors X and XI to factors Xa and XIa is neutralized, as shown.[8]

(1)   TFPI + factor Xa → TFPI/factor Xa

> (factor Xa inactive)

(2)   TF + factor VIIa → TF/factor VIIa

> (factor VIIa active)

(3) TFPI/factor Xa + TF/factor → VIIaTFPI/factor XA/TF-factor VIIa

> (both factors VIIa and Xa inactive)

Thus, TFPI exerts an anticoagulant activity [1063].

It has been established that, in the quaternary complex, the Kunitz 1 domain of TFPI binds to factor VIIa and the Kunitz 2 domain binds to factor Xa [1064]. Factor Xa is required for the TFPI-mediated inactivation of the TF/factor VIIa complex. These combined effects of TFPI result in an inhibition of coagulation due to (a) the inhibition of factor Xa, and (b) the prevention of further initiation of coagulation by tissue factor.

TFPI is synthesized by endothelial cells and circulates in the plasma at a level of ~100 ng/ml, where it is found in complexes with lipoproteins. A small amount of TFPI (~8 ng/ml) is found in platelets and is released on platelet

---

[8] In an alternative reaction sequence, TFPI could combine with a preformed TF/factor VIIa/factor Xa complex.

activation. However, there is a much larger pool of TFPI bound to the endothelium, apparently via its interaction with endothelial cell heparan sulfate proteoglycan. This pool, which is released following heparin administration, represents 50-90% of the total vascular TFPI [1065-1067]. This endothelium-bound TFPI is found in capillaries and venules, but not in large vessels [1068].

During the early purifications of TFPI, it was found that activity was lost due to the proteolytic loss of the C-terminal domain. Concurrent with the activity loss, the fragmented TFPI lost the ability to bind to heparin affinity matrices. Thus, the activity of TFPI, as well its ability to bind heparinoids, depends on the intact C-terminal domain [1069]. The complexation of TFPI with heparan sulfate proteoglycan on the endothelial cell surface and with lipoproteins in the blood may protect these two major pools of TFPI from the proteases to which it is so susceptible. The significance of the endothelial and the circulating pools of TFPI are not clear, but it may be that, like antithrombin and vitronectin, the primary activity of TFPI occurs on the surface of the endothelium, and the circulating pools represent a reservoir of TFPI that replaces the TFPI that is internalized by the endothelial cells or consumed in the neutralization of factor Xa and tissue factor/factor VIIa. It is known that heparin stimulates the activity of TFPI, but it has not been shown that endothelial cell surface heparan sulfate proteoglycan can stimulate TFPI *in vivo*. Both TFPI and antithrombin represent endothelial surface anticoagulants. Perhaps the bound TFPI supplements the anticoagulant activity of the bound antithrombin on the endothelial surface.

3.   Effect of exogenous heparin

Administration of heparin results in a two- to four-fold increase in plasma TFPI as a result of its displacement from the endothelium by the heparin [1065-1067]. This may lower the anticoagulant effect of TFPI on the endothelium, but it raises the anticoagulant activity of the blood as a result of the elevation of the plasma TFPI and the stimulation of the TFPI activity by heparin [1070, 1071]. The exogenous heparin apparently exerts its anticoagulant effect through both antithrombin and TFPI. The TFPI effect is mediated through the tissue factor-activated arm of the coagulation pathway (the extrinsic pathway), whereas the antithrombin effect is mediated through the neutralization of both factor Xa and thrombin [1061] (see Figure 8.1). The dynamics of this process in the presence of heparin are complicated by the fact that, whereas factor Xa is *required* for neutralization of the tissue factor/factor VIIa complex, the heparin-mediated antithrombin inactivation of factor Xa reduces the levels of factor Xa below those that are required for the TFPI effect. Jesty *et al.* [1072] have shown that

heparin stimulates the reaction of
TFPI with factor Xa, but inhibits the
reaction of TFPI/Xa with TF/VIIa.
However, a great deal more must be
learned in order to resolve the
respective antithrombin- and TFPI-
mediated effects of exogenous
heparin [see, for instance, 1073,
1074]. As the reader should, by now,
be aware, the effects of exogenous
heparin on the hemostatic
mechanisms are quite complex.

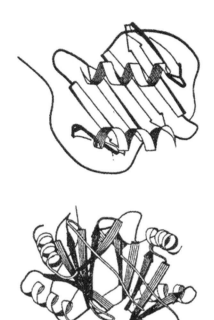

L.   Platelet Factor 4

1.   Structure and heparin binding

Under physiological conditions,
human platelet factor 4 exists as a
tetrameric protein containing four
identical subunits, each with 70
amino acids and a molecular weight
of 7772, as shown in Figure 8.13

Figure 8.13. Platelet factor 4 dimer (top)
and tetramer structures. Reprinted from
[1075], with permission.

[1076, 1077]. It binds to heparin with high affinity ($K_d = 3 \times 10^{-8}$ $\underline{M}$), eluting
from heparin-Sepharose with 1.4 $\underline{M}$ NaCl [1078]. It also binds to heparan sulfate
with similar affinities [906, and references therein]. In fact, because platelet
factor 4 has such a high affinity for heparin, it has been used as a scavenger for
binding heparin to reverse its anticoagulant effect [684].[9] The C-terminal
domains, which are essential for the biological activities of platelet factor 4
[1079], contain the heparin-binding sequences:

N G $\underline{R}$ $\underline{R}$ I C L D L Q A P L Y $\underline{K}$ $\underline{K}$ I I $\underline{K}$ $\underline{K}$ L L E S
50   55   60   65   70

These sequences are present in amphipathic α-helix regions of the polypeptide
chain [see 906]. The high heparin affinity apparently requires the native
tetrameric protein [1080] so that the heparin can wrap around the protein and
bind to two C-terminal domains [1081, 1082]. The capacity of platelet factor 4

---

[9] Note, however, that platelet factor 4 does not bind to the specific sequence in heparin
that is recognized by antithrombin.

to bind heparin drops as the molecular weight of the heparin drops below 5400 (approximately an 8- to 10-mer). The binding of platelet factor 4 to heparin occurs by relatively nonspecific electrostatic interactions [1083].

2.   Biological activity

Platelet factor 4 is synthesized by megakarocytes and is stored as a complex with chondroitin sulfate in the α-granules of platelets [1076, 1084, 1085]. It is released from platelets when they are activated and is cleared from the circulation by the liver [1086]. Its concentration in plasma is quite low. Although it has a number of interesting biological activities [reviewed in 1087], its physiological role is uncertain. In normal hemostasis, platelet factor 4 might be expected to bind to the endothelial cell surface heparan sulfate proteoglycan at the immediate site of platelet aggregation. In fact, platelet factor 4 is highly concentrated in platelet-rich thrombi [1088]. Thus, when platelets are activated, platelet factor 4 would bind to the endothelial cell heparan sulfate proteoglycan in the immediate area of the platelet plug, displacing antithrombin from the cell surface, thus disabling the primary thrombin-neutralizing effect of endogenous heparan sulfate proteoglycan [906]. This would allow coagulation to occur at the local site of injury.

3.   Effect of exogenous heparin

When exogenous heparin is administered, all of the platelet factor 4 that is not protected in the clot becomes bound to the heparin. In fact, platelet factor 4, even though it does not bind specifically to the antithrombin-binding heparin pentasaccharide, can be used clinically in place of protamine to neutralize excess heparin [1089, 1090]. Since the platelet factor 4 binds to and sequesters heparin so that the heparin cannot inhibit antithrombin, heparin must be in excess of the platelet factor 4 in order to obtain the desired anticoagulant effect of the heparin/antithrombin complexation [872, 1091]. Apparently there is no unique *in vivo* activity of platelet factor 4 that results from its binding to heparin.

M. Histidine-Rich Glycoprotein

1. Structure and heparin binding

Histidine-rich glycoprotein is a plasma protein with a molecular weight of 75,000 with 14% carbohydrate [see 906, 1092]. It binds to heparin in a cation-dependent fashion with a K_d of 7 x 10^{-9} M [1093]. As the molecular weight of the heparin drops below 5400 (approximately an 8- to 10-mer), the heparin-binding capacity of histidine-rich glycoprotein drops [872, 1093]. It is much less effective in the binding of heparan sulfate than in the binding of heparin [872]. The protein from human plasma has been cloned and sequenced [1094]. It has 507 amino acids, with four potential carbohydrate-binding sites. Of its total amino acids, Pro and His each represent more than 12%. The histidines are concentrated in the C-terminal domain of the protein. As a general rule, histidine residues do not seem to play an important role in heparin binding, and this may be the case for the His-rich regions in histidine-rich glycoprotein. However, the N-terminal region of His-rich glycoprotein exhibits sequence similarity to antithrombin in the region corresponding to the heparin-binding sequence of antithrombin and is a potential site of heparin binding [1095]:

$$E \underline{K} A L D L I N \underline{K} \underline{R} \underline{R} \underline{R}$$
$$\phantom{E \underline{K} A L D} 14 20 \phantom{\underline{R}} 25$$

Lane [906] has suggested that the following are additional potential sites of heparin binding:

$$\underline{R} Y \underline{R} \underline{K} Q A N \underline{K} A L E \underline{K} Y \underline{K}$$
$$135 \phantom{Y \underline{R} \underline{K} Q A N \underline{K} A L E \underline{K} Y} 148$$

$$\underline{R} V D \underline{R} I E \underline{R} V A \underline{R} V \underline{R}$$
$$158 \phantom{V D \underline{R} I E \underline{R} V A \underline{R} V} 169$$

$$\underline{R} \underline{R} \underline{R} G P G \underline{K} G P \underline{R}$$
$$421 \phantom{\underline{R} \underline{R} G P G \underline{K} G P} 430$$

2. Biological activity and effect of exogenous heparin

Histidine-rich glycoprotein is present in the plasma at a concentration of ~100 µg/ml [1096], with a much lower amount in platelets [1097]. It is synthesized in parenchymal cells of the liver [1098, 1099]. In addition to heparin, histidine-rich glycoprotein binds plasminogen [1100], thrombospondin

[1101], fibrinogen and fibrin [1102], and activated platelets [1103]. About 50% of the circulating plasminogen is found in a complex with histidine-rich glycoprotein, a complex from which the plasminogen must dissociate before it can bind to fibrinogen [1100]. When histidine-rich glycoprotein levels are elevated, the amount of free plasminogen is reduced, an effect that would be expected to result in antifibrinolytic activity. In fact, histidine-rich glycoprotein elevation is persistent in patients with histories of thromboembolic events [see 1104]. However, a causal role for histidine-rich glycoprotein in thrombosis has not been established.

The physiological role of histidine-rich glycoprotein is uncertain. However, its high affinity for heparin results in a neutralization of exogenous heparin. Thus it is one of the proteins that is titrated in the dosing of heparin [872, 1093].

N.   Thrombospondin

1.   Structure and heparin binding

Thrombospondins are a family of adhesive glycoproteins [1105, 1106]. The first recognized member of this family was thrombospondin-1 (TSP-1), which is found in the α-granules of platelets where it represents 3% of the platelet protein and 25% of the total protein secreted by activated platelets [1107]. Other members of the thrombospondin family do not appear to be involved in hemostasis and will not be discussed here [1108, 1109]. Human TSP-1 has been cloned and sequenced [1110-1112]. It is a homotrimer made up of 145-kDa subunits (1152 amino acids, after loss of an 18 amino acid signal sequence) that are cross-linked by disulfide bonds. As shown in Figure 8.14, the glycoprotein subunits are made up of a complex domain structure, beginning with a globular N-terminal domain which contains the major heparin-binding region, followed by a domain that may form an amphipathic α-helix and that contains the two cysteine residues (Cys 252 and Cys 256) that are involved in the disulfide cross-linking of the subunits. There follows A Cys-rich domain which is homologous with the N-terminal propeptide of the α1 chain of type I procollagen; three type I repeats containing 50-54 amino acids; three type II repeats, which are EGF-like sequences; and seven type III repeats, which are Ca^{2+}-binding domains. The last of the type III repeats contains an RGDA sequence, suggesting a possible role for integrin $\alpha_v \beta_3$ in the binding of TSP-1 in the extracellular matrix. Finally, there is a C-terminal domain that has no homologies with other proteins, which may be involved in cell adhesion.

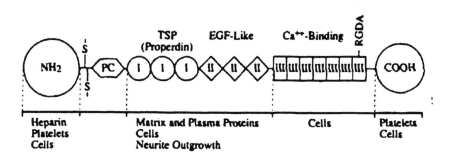

Figure 8.14. Domain structure of thrombospondin.. Reprinted from [1105], with permission.

Thrombospondin binds heparin with modest affinity, eluting from heparin-Sepharose with 0.55 $\underline{M}$ NaCl [1113]. Two sequences in the N-terminal domain, amino acids 23-32 and amino acids 77-83, each containing several basic amino acids, are involved in heparin binding [1110, 1114, 1115]. These are:

$$\underline{R}\ \underline{K}\ G\ S\ G\ \underline{R}\ \underline{R}\ L\ V\ \underline{K}$$
$$\ \ \ 23 \qquad\qquad\qquad\ \ 32$$

and

$$\underline{R}\ Q\ M\ \underline{K}\ \underline{K}\ T\ \underline{R}$$
$$\ \ 77 \qquad\quad\ \ 83$$

Both of these sequences are predicted to form α-helices [1115, 1116]. Site-directed mutagenesis of the basic residues in these sequences show that heparin binding by TSP-1 is significantly reduced when R23 and K24 are converted to Q and N, respectively; when R28 and R29 are converted to N and Q, respectively; or when K80 and K81 are converted to Q and N, respectively [1113]. This suggests that all of these basic amino acids are important for heparin binding and that the two sequences may interact with heparin cooperatively for high-affinity binding in a manner similar to that described for antithrombin.

In addition to the N-terminal globular domains, it has been reported that the type I repeats contain heparin-binding consensus sequences [1110, 1111, 1117]. Also, a second type of heparin-binding sequence has been identified by Roberts et al. [1118-1120]. These peptides were initially identified as sequences that blocked the binding of thrombospondin to heparin-BSA immobilized on plastic dishes. The peptides also blocked the heparin-dependent interactions of laminin and TSP-1 with human melanoma cells. The minimal sequence required for inhibition of these heparin-dependent activities is:

W S P W

but larger peptides containing this sequence are required for maximal activity. All four amino acids are required in the minimal peptide, but the proline can be substituted in some of the longer peptides. The same peptides that block the binding of TSP-1 to heparin-BSA strongly promote melanoma cell adhesion when immobilized on plastic. For direct binding to heparin-agarose, larger peptides containing this sequence, plus a flanking sequence containing basic amino acids are required. For example,

$$K \underline{R} F \underline{K} Q D G G W S H W S P W S S$$

is retained quantitatively by heparin agarose and eluted with 130-200 m$\underline{M}$ NaCl. The fact that these peptides appear to be heparin-binding peptides is an extremely interesting observation. However, as judged by the requirement for the basic amino acids in the extension for direct binding to heparin and the low salt concentrations required to elute the peptides from heparin-agarose, the binding is very weak and may be marginal at physiological salt concentrations. The authors suggest that these peptides interfere with heparin-protein interactions as a result of their binding to the heparin (rather than to the heparin-binding proteins). If this is the case, then the peptides may simply coat the heparin so that it is not available for binding to *any* peptide sequences, including those in the N-terminal domain of the TSP-1 (see Chapter 6). Thus, the fact that these peptides block binding of TSP-1 or laminin to heparin-BSA may not reflect a binding of heparin to these type I repeats in the intact protein. In fact, it is not known whether these sequences in the type I repeats are actually exposed on the surface of the intact TSP-1 so that they are available for interaction with heparin. This story will continue to be an interesting one as it unfolds.

2. Biological activity

Thrombospondin is a protein that binds to a number of prominent extracellular matrix components, including heparan sulfate proteoglycan, glycolipids, collagens, fibronectin, fibrinogen, fibrin, plasminogen, plasmin, t-PA, u-PA, and laminin [1121-1127]. As noted earlier, the C-terminal Ca^{2+}-binding domain contains an RGD sequence that may be involved in integrin binding [1110, 1128]. Although TSP-1 was first discovered in the α-granules of platelets, it is clear that it is synthesized and secreted by a variety of other cells, including endothelial cells, smooth muscle cells, fibroblasts, epithelial cells, and macrophages/monocytes [1106, 1123, 1124, 1129].

As might be anticipated from the variety of thrombospondin-binding ligands and the wide occurrence of thrombospondin, a number of effects of thrombospondin on cell adhesion, motility, and growth have been observed [1130]. Thrombospondin inhibits proliferation and spontaneous tube formation by endothelial cells and inhibits angiogenesis *in vivo* under some conditions [1131, 1132], but promotes angiogenesis under others [1133]. It promotes the migration of endothelial cells, but inhibits chemotaxis stimulated by basic fibroblast growth factor [1134]. In contrast, smooth muscle cell growth is stimulated by thrombospondin [530]. It also blocks focal adhesion formation in endothelial cells and fibroblasts [1135, 1136].

It is of interest here to note that many of the activities of thrombospondin require the heparin-binding N-terminal domain. For example, the inhibition of focal adhesion formation by thrombospondin depends on the heparin-binding domain of the TSP-1 and is reversed by heparin [1135, 1136]. Similarly, the N-terminal fragment of thrombospondin alone inhibits the mitogenic and proliferative responses of endothelial cells [1120], and the heparin-binding peptides from thrombospondin block focal adhesion formation [1136]. Furthermore, adsorption of CHO cells to thrombospondin-coated plates is markedly inhibited by heparin and by the thrombospondin-binding domain of thrombospondin, whereas CHO cell mutants that do not synthesize heparan sulfate do not adhere [1137].

Studies of the metabolism of thrombospondin in endothelial cells also show a prominent role for the heparin-binding capacities of thrombospondin. Thrombospondin is synthesized as a glycoprotein that undergoes oligosaccharide processing prior to secretion [1138]. The secreted thrombospondin is bound to endothelial cell surfaces and is internalized and catabolized in the lysosomes. The uptake of TSP-1 is mediated by both heparan sulfate proteoglycan and low density lipoprotein receptor-related protein (LRP), which are co-receptors for the TSP-1 [1139]. LRP is discussed in detail in Chapter 11. The cell surface binding to heparan sulfate proteoglycan is mediated through the N-terminal heparin-binding domain of the thrombospondin [1140, 1141]. The thrombospondin binds to porcine endothelial cells with $K_d=1.9 \times 10^{-7}$ $\underline{M}$, with 7.7×10^6 sites/cell. Both the intact thrombospondin and the N-terminal domain alone are internalized, whereas thrombospondin fragments lacking the heparin-binding domain are not bound or taken up. Heparin displaces the thrombospondin from the cell surface [633, 1142] and thus inhibits uptake, as does pretreatment of the cells with chlorate to prevent sulfation of the cell surface heparan sulfate chains (Chapter 5). In addition, it has been shown that CHO cell mutants with markedly reduced capacities to synthesize heparan sulfate chains show much reduced binding and internalization of thrombospondin [538]. Thus, endothelial cell surface heparan sulfate proteoglycan and thrombospondin cycle together from the cell surface back into the cell where they are delivered to the lysosomes for catabolism (see Chapter 5).

Despite the extensive literature on thrombospondin, the role of this protein in hemostasis is not well defined. Two circumstances must be considered: (a) the role of TSP-1 in the hemostatic responses following vessel injury, and (b) the role of TSP-1 in normal hemostasis. The literature has addressed the former role much more extensively than the latter. Thrombospondin is a major fraction of the protein that is secreted by platelets when they are activated. Normally, the circulating concentration of thrombospondin in the blood is in the p$\underline{M}$ range, but on platelet aggregation in response to injury, the levels rise to the n$\underline{M}$ range [1120, 1123] and then return to the lower levels as the healing progresses [1143, 1144]. Platelet activation following vascular injury releases a large quantity of TSP-1, most of which remains tightly associated with the platelet plug, a result of its cross-linking and stabilization of the platelets in the plug. It is clear that TSP-1 promotes platelet aggregation and stabilization of the platelet plug. However, a number of platelet surface proteins have been proposed as TSP-1 receptors, including GPIa/IIa, GPIa/IIIa, GPIIb, GPIV, and the vitronectin receptor, but the actual TSP-1 platelet receptor in not clearly established [1145-1147].

Several recent observations may implicate thrombospondin in fibrinolysis. It has been shown that thrombospondin binds to both scu-PA and tcu-PA with K_d's of ~40-50 n$\underline{M}$ and that the scu-PA complex with TSP-1 can be converted to tcu-PA by plasmin [1148]. In the TSP-1/u-PA complex, the u-PA is protected from inactivation by PAI-1. Also, it has been shown that the TSP-1 trimer forms a 1:1 molar complex with plasmin, blocking the plasmin activity [1127]. Thus, thrombospondin may play some role in the control of the orderly lysis of the fibrin clot. None of these proposed roles for TSP-1 suggest an involvement of its heparan sulfate-binding capacity.

The role of thrombospondin in normal hemostasis is less obvious. Thrombospondin is secreted into the extracellular matrix of endothelial cells and apparently appears as a constant and dynamic pool on the surface of these cells. The thrombospondin is bound to a portion of the cell surface heparan sulfate proteoglycan and is available for binding to its other ligands. It is not clear whether there are any specific sequences in heparan sulfate that are recognized by the thrombospondin or whether the binding of thrombospondin to the endothelial cells will interfere with the binding of other heparan sulfate-binding proteins that are involved in hemostasis. It has been reported that thrombospondin forms complexes with thrombin-serpin complexes [1149], raising the possibility that thrombospondin may be involved in clearing these complexes from the circulation. However, the effects of thrombospondin on normal hemostasis have not been reported.

3. Effect of exogenous heparin

The failure to demonstrate any direct role of TSP-1 in hemostasis suggests that exogenous heparin would displace TSP-1 from its heparan sulfate proteoglycan anchorage on the endothelium, but that this would have little effect on hemostasis. As for other heparin-binding proteins in the vasculature, TSP-1 will bind to some of the exogenous heparin, thus reducing its availability for interaction with antithrombin. Consequently, TSP-1 is one of several proteins that must be titrated in obtaining an effective dosage of anticoagulant heparin [908].

III. SUMMARY

A large number of proteins that are key players in the hemostatic pathways have been identified as "heparin-binding proteins." Table 8.1 lists these proteins and gives their normal roles and the effects of exogenous heparin on their activities. There are several general conclusions from this summary.

- First, for many, but not all, of these proteins, the heparin-binding properties play important roles in normal hemostasis. The primary heparinoid involved in these roles is the endothelial cell heparan sulfate proteoglycan.

- The binding constants for the interactions of these proteins with heparin or heparan sulfate are quite variable. Although the binding constants are very dependent on the ionic strength, the values reported in this literature have not always been measured at physiological ionic strengths. For example, despite a significant literature on the interactions of heparin with t-PA and u-PA, the more recent demonstrations that these interactions are minimal under physiological conditions virtually eliminates them from classification as true heparin-binding proteins.

- Few of the studies of these protein interactions with heparin have been extended to heparan sulfate (and especially to endothelial cell heparan sulfate). This is a particularly important point for proteins that require high molecular weight fractions of heparin for binding and activity. Heparins contain extended sequences of highly sulfated disaccharides, but heparan sulfates have relatively short blocks of highly sulfated sequences that are separated by blocks of unsulfated disaccharides (Chapter 2). It is important to determine whether the interaction of high molecular weight heparin with each protein is mimicked by heparan sulfates that have quite different (and variable) sequences.

- Some of the heparin-binding proteins that apparently have no significant interactions with heparan sulfate in normal hemostasis become very important players when exogenous heparin is present. Two notable examples are HCII and vWF. HCII, in contrast to antithrombin, apparently is not activated by endothelial cell heparan sulfate, but it is activated by *exogenous* heparin and specifically inhibits thrombin. Similarly, vWF has no obvious interactions with endothelial cell heparan sulfate in normal hemostasis, but exogenous heparin blocks the binding of vWF to platelets and thus prevents platelet plug formation, thus inhibiting the formation of arterial thrombi.

- When endothelial cell heparan sulfate *does* participate in normal hemostasis, the types of roles that it plays are quite diverse. For example, (a) heparan sulfate activates antithrombin, contributing to the anticoagulant properties of the endothelial cell surface; (b) it serves as a receptor for complexes of inactive serpin-proteases with vitronectin and participates in the endocytosis/catabolism of these ternary complexes; (c) it binds thrombospondin for endocytosis/catabolism; (d) it may activate TFPI for inhibition of factor Xa and/or tissue factor; and (e) it may bind platelet factor 4 at sites of platelet plug formation with the displacement of antithrombin, thus rendering this localized site pro-coagulant.

- A major unanswered question concerning the participation of endothelial cell heparan sulfate proteoglycan in the hemostatic activities of a variety of proteins is whether these proteins must compete for the available endothelial cell heparan sulfate for binding. It seems that antithrombin is quite "polite" in occupying only a small percent of the available cell surface heparan sulfate proteoglycan, but the other heparin-binding proteins in the circulation may be more greedy. If that is the case, then the competition among the heparin-binding proteins for the available heparan sulfate may introduce a new dynamic into the hemostatic process.

- For none of these heparin-binding proteins have the structure-activity relationships, the hemostatic roles, and the metabolism been so extensively addressed as for antithrombin. Antithrombin is the prototype to which all of the other heparin-binding proteins will be compared. The descriptions of antithrombin in Chapter 7 and in the present chapter suggest approaches for studies of all of the other proteins in this group.

- One final point that emerges from the studies of the hemostatic heparin-binding proteins is that the heparin-binding properties of these proteins, or of any other protein that binds to heparin, can no longer be considered random, biologically insignificant properties. In any given case, the heparin-binding property may have little biological significance, but certainly the biological significance of heparin binding has been established

Table 8.1. Heparin-binding Proteins in Hemostasis [a]

Protein [b]	Plasma conc. [c]	Role of protein in normal hemostasis	Role of heparan sulfate in normal hemostasis	Effect of exogenous heparin	Requires high MW heparin for activity?
Antithrombin	2.5 μM	Neutralization of thrombin and factor Xa	Activates antithrombin	Activates all circulating antithrombin; displaces antithrombin from endothelial cell surface; inhibits new clot formation	Yes-thrombin No-factor Xa
Thrombin		Catalysis of clot formation; regulation of rates of coagulation and fibrinolysis	Binds thrombin and antithrombin simultaneously (approximation)	Inhibits thrombin activity by direct binding to thrombin	No(?)
HCII	1 μM	Neutralization of thrombin in extravascular tissues	None	Activates all circulating HCII; inhibits new clot formation	Yes
vWF	40 nM	Binding of platelets to subendothelial surfaces that are exposed by vessel injury; initiation of platelet plug formation; stabilization of factor VIII and VIIIa	None	Blocks vWF/GPIb interaction; inhibits vWF-dependent platelet plug formation	Yes
t-PA	70 pM	Initiates fibrinolysis by converting plasminogen to plasmin	None	Probably no effect at physiological salt concentrations	---
Urokinase	40 pM	Probably minimally involved in hemostasis	None	Probably no effect at physiological salt concentrations	---

PAI-1	500 p$\underline{M}$	In complex with vitronectin, neutralizes t-PA and u-PA to block fibrinolysis	Probably none, except when in complex with vitronectin (see below)	May displace PAI-1 from its complex with vitronectin, thus labilizing PAI-1 and causing an elevation of t-PA and u-PA levels	Not reported
Vitronectin	250-450 µg/ml	Stabilization PAI-1; mediation of endocytosis/catabolism of serpin-proteases complexes	Receptor for endocytosis/catabolism of serpin-protease complexes	May displace PAI-1 from its complex with vitronectin, thus labilizing PAI-1 and raising t-PA and u-PA levels; inhibition of catabolism of serpin-protease complexes; reduction of available heparin for anticoagulant activity	No
Protein C inhibitor	90 n$\underline{M}$	Prevention of the shut-down of the coagulation pathway by inhibition of APC; also inhibits thrombin, t-PA, and u-PA	Although APC inhibitor activities are heparin or heparan sulfate dependent, a role for heparan sulfate in vivo has not been demonstrated	Activates APC inhibitor, thus blocking the destruction of factors V and VII by APC, thereby exerting a pro-coagulant activity	Yes
Tissue factor pathway inhibitor	2-3 n$\underline{M}$	Inhibition of initiation of coagulation by tissue factor; inhibition of factor Xa	Possible endothelial cell anchor for TFPI	Activates all circulating TFPI; TFPI is displaced from endothelial cell surface; stimulates anticoagulant activity of TFPI, but blocks tissue factor inhibition of the coagulation activities of tissue factor CHECK	No

296

Platelet factor 4		Blockage of anticoagulant activity of endothelial cell surfaces at the site of tissue injury by displacement of antithrombin from heparan sulfate proteoglycan	May bind platelet factor 4 at platelet plug with displacement of antithrombin	No direct action; binds heparin so that it cannot exert its anticoagulant activities	Yes
Histidine-rich glycoprotein		Unknown	None	No direct action; binds heparin so that it cannot exert its anticoagulant activities	Yes
Thrombospondin	pM range	Stabilization of platelet plug by cross-linking activated platelets; possible role in regulation of fibrinolysis	Binding of TSP-1 to endothelial cells for endocytosis/catabolism	Probably none; may block stabilization of platelet plug	Yes
Fibronectin	700 nM	May cross-link platelets to solidify platelet plug	None		

[a] Since thrombin, plasmin, t-PA, and u-PA are rapidly inactivated by their respective serpins and removed from the circulation, the levels of these proteirs in the plasma are very low. With the exception of vWF and TFPI, which are synthesized by endothelial cells, most of these factors are synthesized by hepatocytes.

[b] HCII, heparin cofactor II; t-PA, tissue plasminogen activator, PAI-1, plasminogen activator inhibitor I; TF, tissue factor.

[c] Data taken from Halkier [873]

- for a number of proteins, and when new heparin-binding proteins are identified, the potential importance of heparan sulfate interactions must be assessed. Heparin-binding proteins probably do not encounter heparin *in vivo*, but there is a reasonable likelihood that they interact with heparan sulfate in a physiologically significant way.

ADDITIONAL REFERENCES

Biessen, E. A. L., van Teijlingen, M., Vietsch, H., Barrett-Bergshoeff, M. M., Bijsterbosch, M. K., Rijken, D. C., van Berkel, T. J. C., and J., K. (1997) Antagonists of the mannose receptor and the LDL receptor-related protein dramatically delay the clearance of tissue plasminogen activator, *Circulation* 95, 46-52.

Björquuist, P., and Boström, S. (1997) Determination of the kinetic constants of tissue factor/factor VII/factor VIIa and antithrombin/heparin using surface plasmon resonance, *Thromb. Res.* 85, 225-236.

Bombeli, T., Mueller, M., and Haeberli, A. (1997) Anticoagulant properties of the vascular endothelium, *Thromb. Haemostas.* 77, 408-423.

Ciaccia, A. V., Willemze, A. J., and Church, F. C. (1997) Heparin promotes proteolytic inactivation by thrombin of a reactive site mutant (L444R) of recombinant heparin cofactor II, *J. Biol. Chem.* 272, 888-893.

Clezardin, P., Lawler, J., Amiral, J., Quentin, G., and Delmas, P. (1997) Identification of cell adhesive active sites in the N-terminal domain of thrombospondin-1, *Biochem. J.* 321, 819-827.

Debrock, S., and Declerck, P. J. (1997) Neutralization of plasminogen activator inhibitor-1 inhibitory properties: identification of two different mechanisms, *Biochim. Biophys. Acta* 1337, 257-266.

Higazi, A. A.-R., Mazar, A., Wang, J., Quan, N., Griffin, R., Reilly, R., Henkin, J., and Cines, D. B. (1997) Soluble human urokinase receptor is composed of two active units, *J. Biol. Chem.* 272, 5348-5353.

Han, J.-H., Van Deerlin, V. M. D., and Tollefsen, D. M. (1997) Heparin facilitates dissociation of complexes between thrombin and a reactive site mutant (L444R) of heparin cofactor II, *J. Biol. Chem.* 272, 8243-8249.

Hardy, M. M., Feder, J., Wolfe, R. A., and Bu, G. (1997) Low density lipoprotein receptor-related protein modulates the expression of tissue-type plasminogen activator in human colon fibroblasts, *J. Biol. Chem.* 272, 6812-6817.

Hosseini, G., Liu, J., and de Agostini, A. J. (1996) Characterization and hormonal modulation of anticoagulant heparan sulfate proteoglycans synthesized by rat ovarian granulosa cells, *J. Biol. Chem.* 272, 22090-22099.

Iversen, N., Sandset, P. M., Abildgaard, U., and Torjesen, P. A. (1996) Binding of tissue factor pathway inhibitor to cultured endothelial cells. Influence of glycosaminoglycans, *Thromb. Res.* 84, 267-278.

Kluszynski, B. A., Kim, C. P., and Faulk, W. P. (1997) Zinc as a cofactor for heparin neutralization by histidine-rich glycoprotein, *J. Biol. Chem.* 272, 13541-13547.

Kojima, T., Katsumi, A., Yamazaki, T., Muramatsu, T., Nagasaka, T., Ohsumi, K., and Saito, H. (1996) Human ryudocan from endothelium-like cells binds basic fibroblast growth factor, midkine, and tissue factor pathway inhibitor, *J. Biol. Chem.* 271, 5914-5920.

Lawrence, D. A., Palaniappan, S., Stefansson, S., Olson, S. T., Francis-Chmura, A. M., Shore, J. D., and Ginsburg, D. (1997) Characterization of the binding of different conformational forms of plasminogen activator inhibitor-1 to vitronectin. Implications for the regulation of pericellular proteolysis, *J. Biol. Chem.* **272**, 7676-7680.

Lijnen, H. R., van Hoef, B., and Collen, D. (1996) Characterization of the murine plasminogen/urokinase-type plasminogen-activator system, *Eur. J. Biochem.* **241**, 840-848.

Liu, L., Freedman, J., Hornstein, A., Fenton II, J. W., Song, Y., and Ofosu, F. A. (1997) Binding of thrombin to the G-protein-linked receptor, and not to glycoprotein Ib, precedes thrombin-mediated platelet activation, *J. Biol. Chem.* **272**, 1997-2004.

Matsui, T., Kunishima, S., Hamako, J., Katayama, M., Kamiya, T., Naoe, T., Ozeki, Y., Fujimura, Y., and Titani, K. (1997) Interaction of von Willebrand factor with the extracellular matrix and glycocalicin under static conditions, *J. Biochem. (Tokyo)* **121**, 376-381.

Mikhailenko, I., Krylov, D., Argraves, M. K., Roberts, D. D., Liau, G., and Strickland, D. K. (1997) Cellular internalization and degradation of thrombospondin-1 is mediated by the amino-terminal heparin binding domain (HBD). High affinity interaction of dimeric HBD with the low density lipoprotein receptor-related protein, *J. Biol. Chem.* **272**, 6784-6791.

Molino, M., Bainton, D. F., Hoxie, J. A., Coughlin, S. R., and Brass, L. F. (1997) Thrombin receptors on human platelets. Initial localization and subsequent redistribution during platelet activation, *J. Biol. Chem.* **272**, 6011-6017.

Narita, M., Bu, G., Olins, G. M., Higuchi, D. A., Herz, J., Broze Jr., G. J., and Schwartz, A. L. (1995) Two receptor systems are involved in the plasma clearance of tissue factor pathway inhibitor *in vivo*, *J. Biol. Chem.* **270**, 24800-24804.

Pijuan-Thompson, V., and Gladson, C. L. (1997) Ligation of integrin $\alpha_5\beta_1$ is required for internalization of vitronectin by integrin $\alpha_v\beta_3$, *J. Biol. Chem.* **272**, 2736-2743.

Poletti, L. F., Bird, K. E., Marques, D., Harris, R. B., Suda, Y., and Sobel, M. (1997) Structural aspects of heparin responsible for interactions with von Willebrand factor, *Arterioscler. Thromb. Vasc. Biol.* **17**, 925-931.

Preissner, K. T. (1995) Vitronectin as a link between protease cascades and cell adhesion in hemostasis, in *Vascular Control of Hemostasis*, ed. van Hinsbergh, V. W. M., Harwood Academic Publishers, United Kingdom, pp. 169-186.

Rao, C. N., Reddy, P., Liu, Y., O'Toole, E., Reeder, D., Foster, D. C., Kisiel, W., and Woodley, D. T. (1996) Extracellular matrix-associated serine protease inhibitors (M_r 33,000, 31,000, and 27,000) are single-gene products with differential glycosylation: cDNA cloning of the 33-kDa inhibitor reveals its identy to tissue factor pathway inhibitor-2, *Arch. Biochem. Biophys.* **335**, 82-92.

Scott, P. E., Purkall, D. B., and Ruddy, S. (1997) Effects of thrombospondin purified by heparin affinity or ion-exchange chromatography on the alternative complement pathway, *J. Immunol. Methods* **202**, 213-216.

Seiffert, D. (1997) The glycosaminoglycan binding site governs ligand binding to the somatomedin B domain of vitronectin, *J. Biol. Chem.* **272**, 9971-9978.

Seiffert, D., and Smith, J. W. (1997) The cell adhesion domain in plasma vitronectin is cryptic, *J. Biol. Chem.* **272**, 13705-13710.

Shirk, R. A., Church, F. C., and Wagner, W. D. (1996) Arterial smooth muscle cell heparan sulfate proteoglycans accelerate thrombin inhibition by heparin cofactor II, *Arterioscler. Thromb. Vasc. Biol.* **16**, 1138-1146.

Urano, T., Ihara, H., Takada, Y., Nagai, N., and Takada, A. (1996) The inhibition of human factor Xa by plasminogen activator inhibitor type 1 in the presence of calcium ion, and its enhancement by heparin and vitronectin, *Biochim. Biophys. Acta* **1298**, 199-208.

van't Veer, C., and Mann, K. G. (1997) Regulation of tissue factor initiated thrombin generation by the stoichiometric inhibitors tissue factor pathway inhibitor, antithrombin-III, and heparin cofactor-II, *J. Biol. Chem.* **272**, 4367-4377.

van't Veer, C., Golden, N. J., Kalafatis, M., and Mann, K. G. (1997) Inhibitory mechanism of the protein C pathway on tissue factor-induced thrombin generation. Synergistic effect in combination with tissue factor pathway inhibitor, *J. Biol. Chem.* **272**, 7983-7994.

Wang, Z., Mottonen, J., and Goldsmith, E. J. (1996) Kinetically controlled folding of the serpin plasminogen activator inhibitor 1, *Biochemistry* **35**, 16443-16448.

Weisel, J. W., Nagaswami, C., Young, T. A., and Light, D. R. (1996) The shape of thrombomodulin and interactions with thrombin as determined by electron microscopy, *J. Biol. Chem.* **271**, 31485-31490.

Yamanobe, F., Mochida, S., Ohno, A., Ishikawa, K., and Fujiwara, K. (1997) Recombinant human tissue factor pathway inhibitor as a possible anticoagulant targeting hepatic sinusoidal walls, *Thromb. Res.* **85**, 493-501.

Yoneda, A., Kojima, K., Matsumoto, I., Yamamoto, K., and Ogawa, H. (1996) Porcine vitronectin, the most compact form of single-chain vitronectin: the smallest molecular mass among vitronectins was ascribed to deletion and substitution of base pairs, and proteolytic trimming of the peptide, *J. Biochem. (Tokyo)* **120**, 954-960.

Zhuang, P., Chen, A. I., and Peterson, C. B. (1997) Native and multimeric vitronectin exhibit similar affinity for heparin. Differences in heparin binding properties induced upon denaturation are due to self-association into a multivalent form, *J. Biol. Chem.* **272**, 6858-6867.

Chapter 9. Fibroblast Growth Factors

I. INTRODUCTION

A. The FGF Family

The fibroblast growth factors are members of a family of proteins that affect the growth, migration, differentiation, and survival of mesoderm- and neuroectoderm-derived cells. There are nine members of this family that are related by amino acid sequence similarities and activities. These include the first discovered and most studied members of the family, namely, acidic fibroblast growth factor (aFGF) and basic fibroblast growth factor (bFGF) as well as several oncogene products (int-2, hst/K-FGF, FGF-5, and FGF-6), keratinocyte growth factor (KGF), and the more recently reported members, FGF-8

301

(androgen-induced growth factor—AIGF) [1150, 1151] and FGF-9 (glial activating factor—GAF) [1152, 1153]. In addition to their structural similarities and activities, these growth factors all adhere to heparin, a property that greatly facilitates their purification. Heparin and heparan sulfate modulate the activities of FGFs [534, 1154, 1155], sometimes stimulating and sometimes inhibiting their activities [1156-1160]. A number of reviews have described these proteins [1161-1164]. The biological activities of these growth factors require that they bind to specific cell surface receptors. It has been established that the receptors themselves also bind heparin and heparan sulfate. Thus, cell surface heparan sulfate proteoglycans mediate the biological activities of the FGFs. This chapter will review the literature in this area with primary focus on the heparin-binding properties of these growth factors and their receptors, and the influence of heparinoid binding on the growth factor activities.

B.   Nomenclature

The 1989 review by Burgess and Maciag details the early discoveries of acidic and basic fibroblast growth factors and describes the nomenclature of these proteins [1161]. As is common in the early development of an area of biology, several names appeared in the literature for growth factors that were eventually found to be identical to aFGF and bFGF. However, the most easily identified feature of this group of proteins is their ability to bind to heparin. Consequently, a more systematic nomenclature was adopted in which members of this family were called "heparin-binding growth factors" (HBGFs), with aFGF being designated HBGF-1 and bFGF being designated HBGF-2 [1154, 1161]. As additional oncogene members of this family were discovered, they were designated HBGF-3, -4, -5, etc. However, many growth factors that are not related to the fibroblast growth factors also bind to heparin. Thus, the most recent stage in the evolution of the nomenclature of this group of proteins came from the recommendations of a committee of workers in this field [1163] who proposed that the terminology FGF-1, FGF-2, FGF-3, etc. be used. This terminology provides that the FGFs are related by their sequence similarities alone and that they may thus be distinguished from the many other heparin-binding growth factors that lack this sequence homology. Furthermore, the committee pointed out that FGF activity is not restricted to cells of mesodermal or neuroectodermal origin and that the fibroblast growth factors belong to a common family because they are structurally related, although not necessarily biologically related. Thus, acidic fibroblast growth factor now becomes FGF-1; basic fibroblast growth factor becomes FGF-2; etc. The names that have been used for various members of the family are given in Table 9.1.

Table 9.1. Nomenclature of Fibroblast Growth Factors
and Their Receptors

Name	Names found in literature
FGF-1	aFGF, HBGF-1, ECGF (endothelial cell growth factor)
FGF-2	bFGF, HBGF-2
FGF-3	int-2
FGF-4	hst-1, ks, k-fgf hst/ks
FGF-5	
FGF-6	
FGF-7	KGF (keratinocyte growth factor); amphiregulin
FGF-8	AIGF (androgen-induced growth factor)
FGF-9	GAF (glial-activating factor)
FGFR-1	mammalian flg, chicken cek1
FGFR-2	human or murine bek
	human k-Sam, chicken cek3
FGFR-3	murine flg-2, chicken cek2
FGFR-4	(no other names)

The literature still contains aFGF and HBGF-1, bFGF and HBGF-2, etc., although the most common usages for the two most studied members of this family remain aFGF and bFGF.

II. FGF STRUCTURES AND PROPERTIES

A. Isolation of FGFs by Heparin Affinity Chromatography

One of the prominent physical properties of the fibroblast growth factors is their capacity to bind to heparin. When FGFs bind to heparin, they become more resistant to denaturation by acid or heat [1165, 1166] and to attack by proteases such as thrombin, trypsin, plasmin, subtilisin, papain, and thermolysin [622, 1167, 1168]. Advantage has been taken of the heparin-binding properties of this group of proteins in the purification of FGF-1 and FGF-2 by chromatography on heparin affinity matrices [528, 1154, 1169, 1170]. Perhaps the most sophisticated procedure is that of Shing [665], which exploits the fact that fibroblast growth factors bind both to copper and to heparin. Thus, an affinity column with both copper and heparin attached allows the removal of contaminating proteins by displacing all proteins that bind to one of these ligands from the column, and then all proteins that bind to the other. Finally, elution of the column with a salt gradient containing imidazole, a copper chelator, results in extensive purification not only of FGF-1 and FGF-2, but of subspecies of each as well (below). FGF-1 has an acidic isoelectric point and elutes from heparin affinity columns with 0.8-1 $\underline{M}$ NaCl, whereas FGF-2 has a basic isoelectric point and elutes with 1.4-1.6 $\underline{M}$ NaCl. FGF-1 has a greater affinity for copper than FGF-2.

B. Primary Structures

The properties of FGFs are shown in Table 9.2. These proteins are synthesized by a variety of cell types. FGF-2 was originally identified as a protein containing 146 amino acids. However, improved methods for extraction of FGF-2 from cells and for detecting FGF-2 showed that the 146 amino acid sequence is a protease-truncated form of a primary translation product that contains 154 amino acids (18 kDa) [1171-1174]. In addition, it became clear that several other molecular size species of FGF-2 are synthesized in most cell types [1175-1179]. The genes for FGFs from several species have been cloned [reviewed in 1161, 1180].

FGFs are related by their structural similarities, which range from 35 to 80% identities among the nine members of the family. Homologies for a given

Table 9.2. Properties of Fibroblast Growth Factors

Name	Primary structure (AUG initiation)	Other initiation sites	Signal sequence?
FGF-1	154 amino acids, 16.5 kDa	None; however, species containing amino acids 15-154 and amino acids 21-154 are formed during isolation	None
FGF-2	154 amino acids	Yes; upstream CUGs	None
FGF-3	231 amino acids (ORF); 28.5 kDa	Yes, upstream CUG	Yes
FGF-4	206 amino acids; 22 kDa		Yes
FGF-5	267, 268 ? amino acids	Yes; yields 32.5- and 38-kDa forms; also upstream AUG that suppresses expression	Yes
FGF-6	208 amino acids	Three in-frame AUGs; only first one used	Yes
FGF-7	194 amino acids; 22.5 kDa		Yes
FGF-8			
FGF-9	209 amino acids, 23.5 kDa	None	None

FGF in different species are high (90% identities). All FGFs have conserved Cys residues at positions 16 and 83 (FGF-1 numbering), and there are one to two other Cys residues that are not conserved. FGF-1, FGF-2, and FGF-9 lack a signal sequence, but all other members of the family have typical sequences that direct them into the cell secretory pathway. Despite similar activities and metabolism, FGF-1 and FGF-2 are readily distinguished by their isoelectric points of 5 to 6 and 9.6, respectively [1154]. They are 55% identical [1181]. The primary structures obtained from the cDNA sequences show that the members of the family range in size from 140 to ~300 amino acids. Some confusion arose in earlier studies when multiple species of several of the FGFs were obtained in cell or tissue extracts. For example, FGF-1 occurs as a full-length 154 amino acid species, but is also isolated as two smaller species, a species containing amino acids 15-154 and a species containing amino acids 21-154. Both of these are formed as artifacts due to protease cleavages during the isolation procedures [1171, 1182]. However, FGF-2 has multiple translation initiation sites. For example, the original 18-kDa species of human FGF-2 is formed by typical translation initiation at an AUG codon. Additional species of FGF-2 are all larger than the 18-kDa species and have been identified as forms with additional N-terminal amino acid sequences that result from translation initiation at several different CUG codons upstream from the AUG codon [1178]. Species that have been identified include 24.0-, 23.0-, 22.5-, and 18.0-kDa forms. The proportion of each species varies with the developmental stage of the organism and the tissue or cell type examined. Also, at least in some cells, the 18-kDa form is found in the cytoplasm, whereas the larger forms appear in the nucleus (see Section III.C.4., p. 314).

C. Crystal Structures

The crystal structures of FGF-2 [1183, 1184] and of analogs of FGF-2 and FGF-1 [1] [1185, 1186] have been reported. The three-dimensional structure of FGF-1 is essentially the same as that of FGF-2 [1185]. FGF-2, shown in Figure 9.1, consists of 12 antiparallel β-strands, which, for description, can be numbered in sequence from the N-terminal. Six of the strands (1, 4, 5, 8, 9, and 12) form a β-barrel that is closed at one end by the remaining β-strands (2, 3, 6, 7, 10, and 11). The core of the barrel is packed with hydrophobic amino acid side chains, while the surface is rich in charged amino acids.

---

[1] The FGF-1 analog was the bovine form with Cys 47 and His 93 changed to Ala and Gly, respectively; the FGF-2 analog was the human form with cysteines 70 and 80 changed to serines.

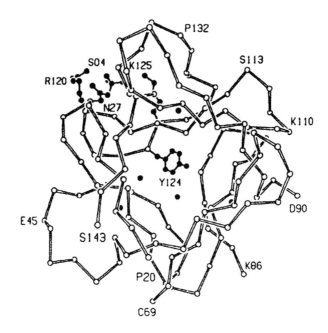

Figure 9.1. Crystal structure of FGF-2. Reprinted from [1184], with permission.

There are four cysteines in FGF-2 at residues 25, 69, 87, and 92, with cysteines 25 and 92 being the highly conserved cysteine residues in the FGFs [1161, 1187, 1188]. None of the four cysteines in FGF-2 is found in a disulfide linkage, although Cys 87 and Cys 92 are only 7.3 Å apart [1183]. All four cysteines can be replaced with serines without loss of activity, although replacement of Cys 25 and Cys 92 together reduces the affinity for heparin [1189, 1190]. A high resolution NMR study of FGF-2 yields a structure consistent with that of the crystal structures [1191].

III. METABOLISM OF FGFs

A. Introduction

1. Overall pathway

The pathway of FGF metabolism is not completely understood. FGF-1 and FGF-2, the two major forms of FGF, are synthesized without a signal sequence, and yet appear in abundance in the extracellular matrix of cells and in basement membranes. FGFs also appear inside cells—most notably in the nucleus. FGFs apparently can be delivered to the nucleus, either from their sites of synthesis in the cytoplasm or from their extracellular sites. The most apparent pathway of metabolic flow of FGF-1 and FGF-2 is (a) release from the cell, (b) storage in the extracellular matrix or in basement membranes, (c) internalization, and (d) delivery, at least in part, to the nucleus.

From the standpoint of a role for heparinoids in FGF metabolism, the receptor-mediated endocytosis is of particular significance. Cell surface receptors for FGFs are of two kinds: (a) high affinity FGF receptors (FGFRs), which are transmembrane proteins having intracellular tyrosine kinase domains, and (b) low affinity receptors, which are heparan sulfate proteoglycans. The heparan sulfate chains of the heparan sulfate proteoglycan can bind simultaneously to the FGFs and the high affinity receptors for receptor function. Apparently the FGFs can be internalized on either their high affinity receptors or on the heparan sulfate proteoglycan. This is discussed in more detail in Section IV.B., p. 318.

The effects of FGFs on cells take several forms. FGFs cause phosphorylation of specific proteins by activating the tyrosine kinases of their receptors when they bind at the cell surface. They are internalized on their receptors, on heparan sulfate proteoglycan, or on both. They induce the synthesis of *c-fos* and u-PA. They stimulate DNA synthesis and induce cell division. The mitogenic effect requires several hr of exposure of cells to FGF ("long-lasting effect") [1192]. A particularly pertinent question with respect to the metabolic flow of FGFs from their sites of synthesis to the outside of the cell and to the nucleus is "what role do the FGFRs and the heparan sulfate proteoglycans play in directing the metabolic flow and the induction of the various activities observed when cells respond to FGFs?" These questions are addressed below.

2.   High affinity receptors

High affinity receptors for the FGFs are widely distributed on the surfaces of many different cell types [1193]. The literature on the high affinity FGF receptors has been reviewed [1194-1198]. Like the FGFs themselves, the FGFRs represent a family of proteins, with four such receptors currently recognized. And like the FGFs, the receptors have been given a variety of names, which are presented in Table 9.1. FGFR-1 [661, 1199-1204], FGFR-2 [1201, 1204-1207], FGFR-3 [1204, 1206, 1208], and FGFR-4 [1204] from several species have been cloned and sequenced.

The structural features of the FGFRs are depicted in Figure 9.2. The FGFRs have (a) N-terminal extracellular regions that contain two or three immunoglobulin-like (Ig-like) domains with typical conserved disulfide-linked cysteine residues that define the beginning and end of each Ig-like loop, (b) a single transmembrane domain, and (c) an intracellular tyrosine kinase domain that is activated by FGF binding to the receptor [1209] and that has a 14 amino acid insert between two portions of the kinase (a "split kinase"). Binding of FGFs to FGFRs results in autophosphorylation of the tyrosine kinase domain of the receptor and phosphorylation of target substrates, including phospholipase C-γ [1210] and other proteins [1211-1213]. It has also been reported that there is a serine kinase associated with FGFR-4 [1214]. In the extracellular domains, several segments between the Ig-like domains are of particular interest. There is an acidic box between domains I and II that contains 4-8 consecutive acidic amino acids (as well as several additional acidic amino acids in close proximity) and that may be involved in binding of copper and calcium [1215] and a heparin-binding region—described in greater detail in Section IV.C., p. 322— that terminates with the amino acid that is linked to the N-terminal cysteine of Ig-like loop II. The first residue of the heparin-binding domain is 25-34 amino acids downstream from the first residue of the acidic box.

In addition to the four types of FGFRs, there are a variety of expressed

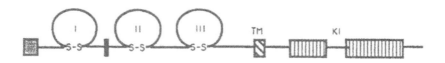

Figure 9.2. The domain structure of FGFR. I, II, and III represent Ig-like domains; the solid bar between Ig-like domains I and II represents the acid box; TM is the transmembrane domain; and the intracellular cross-hatched bars represent the split tyrosine kinase (K1) domain. Reprinted from [1208], with permission.

forms of the FGFRs that result from alternate splicing of the mRNA [reviewed in 1164, 1194, 1197, 1198]. These include (a) membrane-bound forms that contain full structures with three Ig-like loops, and the transmembrane and tyrosine kinase domains, (b) forms that lack Ig-like domain I or domain I and the acidic box, and (c) forms with truncated intracellular regions [1216]. Also, there are expressed forms that lack the transmembrane domain and that contain loop I, loops I and II, or loops II and III, with or without the tyrosine kinase domain. There are also mutations in human FGFR genes associated with dystrophic syndromes of bone growth and development [see 1217, and references therein]. Finally, and most significantly, the mRNA splicing of the Ig-like domain III of FGFR-1, -2, and -3 yields three alternative versions of domain III, referred to as IIIa, IIIb, and IIIc. The IIIa splice forms terminate within domain III and yield secreted FGF-binding proteins with no apparent signaling role, but the IIIb and IIIc isoforms are functional forms of the receptors. Thus, within each FGFR type, multiple isoforms exist [1153, 1196, 1201, 1204, 1207, 1208, 1218-1220]. Sequence requirements for regulation of RNA splicing of FGFR-1 have been reported [1221].

The specificities for binding the different FGFs lie primarily in Ig-like domains II and III. The Ig-like domain I, which shows much less homology among FGFRs than domains II and III [see for example, 1196, 1201], is sometimes missing in functional forms of the receptor [1203], but the two-loop forms retain FGF binding capacity very similar to that of the three-loop forms [1196, 1222-1224]. The role of loop I in binding is not clear, although it may interact with loops II and III to modulate the binding activity [1223, 1225, 1226]. In fact, it appears that the major contribution to the FGF-binding specificity is derived from the variable C-terminal half of the Ig-like domain III , with contributions from loop II [see 1225, and references therein]. Several groups have reported the receptor specificities for individual FGFs [1153, 1227]. A comprehensive study of the receptor activities of the variant forms has been reported by Ornitz et al. [1198] who have transfected BaF3 cells with FGFR-4 or with the IIIb and IIIc forms of FGFRs-1,-2, and -3 and compared the mitogenic capacities of all nine FGFs with each of the FGFR variants. FGF-1 is the only member of the FGF family that stimulates DNA synthesis with all of these FGFR variants. Consequently, FGF-1 is used as an internal standard to express the relative capacities of all FGFs with all FGFRs. The variations in the responses are quite striking and suggest that the responses of cells to various FGFs must depend on the developmental expression of both the FGFs and the receptors. Also, since the responses of cells to FGFs depend on receptor dimerization (below), these workers point out that the responses to *heterodimeric* receptor complexes may extend the repertoire of responses that are elicited by the FGFs.

In these specificity studies, the mitogenic response was dependent on the exogenous heparin that was present in the assays. However, the primary emphasis in assigning the FGFR-binding specificities was on the interactions between the FGFRs and the FGFs. In HSPG-producing cells, it is clear that the cell surface heparan sulfate proteoglycan also plays a role in binding specificity, and the question of heparan sulfate contributions to these FGF/FGFR interactions has not yet been addressed. It has been shown that the amounts of HSPG on the cell surface and the structures of the heparan sulfate chains are altered by the growth state of the cells and by responses of the cells to growth factors such as glucocorticoids [1228], interleukin 1 [1229], and the FGFs themselves [619, 1230] (see Section IV.G., p. 334). The binding of FGFs to heparan sulfate does not seem to depend on the core proteins of the proteoglycans, since syndecan, glypican, and perlecan proteoglycans have been shown to bind to FGF [416, 1231-1234]. It is also clear that effects of heparin differ for the different FGFs [1219]. Heparin is not an appropriate mimic of these heparan sulfate structures. Consequently, a final understanding of FGF specificities of the FGFRs must consider these heparan sulfate structures as well.

B.   Biosynthesis and Secretion of FGFs

Newly synthesized FGF-1 and FGF-2 lack signal sequences but appear in significant quantities outside of the cell. It has been demonstrated that they do not progress through the endoplasmic reticulum and the Golgi to the outside of the plasma membrane *via* the normal secretory pathway [1235]. Interestingly, cells transfected with an FGF-2 that has a signal sequence fused onto its N-terminal become transformed [1236]. For FGFs that lack a signal sequence, it has been suggested that the FGF that is found outside the cell is derived from damaged cells [1237]. However, several workers have reported that this cannot account for the levels of the FGFs found in the extracellular matrix [1238]. It has been suggested that bFGF is released via normally occurring plasma membrane disruptions [1239]. The secreted FGF is stored in association with heparan sulfate proteoglycans in the extracellular matrix of cells [1240-1242] or in basement membranes [1172, 1238]. In fact, FGFs can be displaced from the cell matrix by heparin and heparin fragments [1243]. Also, isotonic trypsinization of cells, which cleaves the extracellular portion of the core protein of the proteoglycan, releases FGF in complexes with heparan sulfate chains. Similar results have demonstrated that when GPI anchored heparan sulfate proteoglycans are released from cell surfaces by phosphatidylinositol-specific phospholipase C or when heparan sulfate is released with heparin lyase III [1244], FGFs are released. It may be that some of these release mechanisms represent natural mechanisms for release of FGFs from their extracellular matrix stores so that they can manifest their paracrine activities. In fact, it has been shown that lysosecithin-treated cells release a heparanase that cleaves cell surface HSPG [477]. Nevertheless, the immobilization of FGF on the heparan sulfate chains apparently plays some role in the regulation of its mitogenic activity [1158].

C.   Endocytosis and Intracellular Translocation of FGFs

1.   Introduction

Exogenous FGF-1 and FGF-2 bind to cell surface receptors, are endocytosed, and are translocated to the lysosomes [1245] or to the cell nucleus. The endocytosis requires binding of the growth factors to FGFRs, to heparan sulfate proteoglycan, or to both. FGFRs are high affinity, low capacity receptors for FGFs, with K_d's in the range of 10^{-11} $\underline{M}$ [1196, 1246, 1247]. The number of FGFRs per cell varies from undetectable amounts to quite high levels. For example, in CHO cells and Chinese hamster lung cells, FGFR levels have been reported to be 20,000/cell and 4300/cell, respectively [1248, 1249].

Heparan sulfate proteoglycans are low affinity, high capacity receptors for the FGFs, with K_d's in the range of 10^{-9} $\underline{M}$ [1250]. Because of the abundance of heparan sulfate proteoglycan on the surfaces of most cells, it is difficult to saturate the heparan sulfate receptors with FGFs [1251]. The number of heparan sulfate proteoglycans receptor sites is from one to three orders of magnitude greater than the number of high affinity sites [1252] and may be $>10^6$/cell [1249].

The receptor binding of FGFs is complicated by the fact that FGFs may bind *simultaneously* to FGFRs and heparan sulfate chains of heparan sulfate proteoglycans, *separately* to heparan sulfate chains, and *separately* to FGFRs. A further complication is that FGFs elicit multiple cellular responses, each of which depends on the mode of FGF binding and internalization [see, for example, 1160]. Some of these responses occur on receptor dimerization before the entry of FGF into the cells (e.g., activation of tyrosine kinase) whereas others occur after FGF entry (e.g., effects that may result from the transport of FGF into the nucleus). Furthermore, some responses may be elicited through binding to heparan sulfate proteoglycan alone, whereas others may be elicited only when FGF binds to *both* FGFR and heparan sulfate chains. Although there has been some debate about whether binding of FGFs to FGFRs alone elicits responses, it has been shown that FGF-2 can induce DNA synthesis in the absence of heparan sulfate [1253]. The resolution of some of these complexities has involved the use of (a) cells that express FGFRs but not heparan sulfate proteoglycans (e.g., FGFR-producing cells that cannot synthesize heparan sulfate proteoglycans) (b) FGFR-producing cells that normally make heparan sulfate proteoglycan but are prevented from making *sulfated* heparan sulfate chains by treatment with chlorate—see Chapter 5), (c) cells that produce heparan sulfate proteoglycans but not FGFRs, (d) cells that express both FGFRs and heparan sulfate proteoglycans, or (e) cells that produce neither FGFRs nor heparan sulfate proteoglycans, e.g., BaF3 cells [1198] or myeloid cell lines [1254].

2.   Endocytosis on heparan sulfate proteoglycans

Apparently uptake of FGF *via* heparan sulfate proteoglycan can result from the direct binding of the growth factor to the heparan sulfate chains without simultaneous binding to an FGFR. Most cell types secrete heparan sulfate proteoglycan into the extracellular matrix constantly and then internalize it rapidly, giving it a relatively short extracellular half-life (Chapter 5). Thus, FGF appears to "take a ride" to the inside of the cell on the heparan sulfate proteoglycan. This results in a large pool of intracellular FGF that is catabolized much more slowly than most endocytosed proteins [1255, 1256]. The stabilization of the internalized FGF has been attributed to its demonstrated resistance to denaturation [1165, 1166] and proteolytic breakdown when the FGF is complexed to heparinoids [535, 622, 1167, 1168, 1257]. Such FGF/heparan sulfate complexes have not been isolated from the intracellular contents of cells. However, direct studies of heparan sulfate proteoglycan metabolism have shown that, following internalization of heparan sulfate proteoglycan, the heparan sulfate chains are cleaved from the proteoglycan (perhaps by an endoglucuronidase in the plasma membrane) and themselves persist inside the cell as relatively large oligo- or polysaccharides for extended time periods (Chapter 5). In fact, it is likely that the formation of an FGF/HS complex protects *both* polymers from catabolism by their respective catabolic enzymes. In any event, the notion that the internalized FGF exists inside the cells in a complex with heparan sulfate chains is certainly consistent with studies on the metabolism of both FGF and heparan sulfate proteoglycan. FGF that is internalized on heparan sulfate proteoglycan alone (i.e., without FGFR) does not appear in the cytosol or in the nucleus [1251, 1252, 1258, 1259]. As expected, FGF internalized on heparan sulfate chains does not activate the tyrosine kinase of the FGFRs [1252], but it does induce u-PA synthesis to maximal levels [1252]. Also, internalized FGF does not stimulate DNA synthesis or cell proliferation [1252]. These observations suggest that the stimulation of cell growth by FGF depends on its endocytosis on FGFR, a pathway that *does* result in transport of FGF into the nucleus (below).

3. Endocytosis on FGFR/heparinoid receptor complexes

FGF is taken up more rapidly by its high affinity receptors than by heparan sulfate proteoglycan alone [1249, 1260]. When cells are exposed to FGF, their cell surface FGFRs are rapidly downregulated to about 20% of their original values [1192, 1248, 1255, 1260] as FGF and FGFR enter the cell together. The binding of an FGF to one of the high affinity receptors activates the tyrosine kinase on the intracellular domain of the receptor [1261, 1262]. It also leads to the endocytosis of the FGF and translocation of a portion of the internalized FGF to the cytoplasm and into the nucleus [1263, 1264]. Binding and internalization of FGF on an FGFR is necessary in order to obtain the FGF-mediated stimulation of DNA synthesis and cell proliferation [1252, 1264]. Thus, endocytosis of FGF on an FGFR leads coincidentally to the appearance of FGF in the nucleus and DNA synthesis, suggesting that nuclear FGF elicits cell division [1252, 1259, 1263].

A complete understanding of the endocytosis of FGF on an FGFR is complicated by the requirement for cell surface heparan sulfate proteoglycan for the maximal binding and internalization of the FGF/FGFR complex. The proteoglycan requirement has been confirmed in many studies [1259, 1260, 1263, 1265], but may not be found in all cases [1251, 1253]. FGF stimulates cell growth in cells that produce both FGFR and heparan sulfate proteoglycan, but loses this activity when these cells are treated with chlorate to prevent sulfation of the heparan sulfate chains (Chapter 5). Similarly, cell mutants that have high FGFR levels but lack the capacity for heparan sulfate proteoglycan synthesis are ineffective in FGF endocytosis or FGF stimulation of cell growth [1252, 1259]. Thus, FGF-mediated activities that depend on FGFR also require heparan sulfate proteoglycan, and apparently the complex that is actually taken up by the cells contains FGF, FGFR, and heparan sulfate proteoglycan. This observation complicates the picture that we have of FGF endocytosis. For example, some heparan sulfate proteoglycans have multiple heparan sulfate chairs attached to their core proteins (Chapter 2), and one could imagine that one such chain might participate in the formation of a ternary FGF/FGFR/heparan sulfate complex while another may form a binary FGF/HS complex. Thus, both complexes might enter the cell together on the same proteoglycan. However, as the proteoglycan carrying these complexes enters the cell, the heparan sulfate chains will be cleaved from the core protein (Chapter 5), allowing the ternary FGF/FGFR/heparinoid and binary FGF/heparan sulfate complexes to travel separate routes and elicit separate responses. Furthermore, it is not known whether the heparan sulfate chains of the internalized FGF/FGFR/heparan sulfate proteoglycan complex play any role in the targeting of the FGF or in mediating any of the activities of the FGF. Many workers have suggested that the long half-life of the FGF after it has been taken up by cells is due to its complexation with heparan sulfate chains and that it is actually this complex that is translocated to the nucleus [451, 503, 1237, 1259],; yet it is clearly necessary that FGFR also participate in the nuclear translocation, and, with our present knowledge, it is difficult to determine whether the FGFR and the heparan sulfate chains play separate roles in the translocation and whether FGF is translocated to the nucleus in a complex with heparan sulfate and/or FGFR. It is interesting to speculate that if an heparan sulfate/FGF complex enters the nucleus, both members of this complex may play some role in the nuclear activities usually attributed to FGF; i.e., by forming a complex with FGF, heparan sulfate may be protected from catabolism so that it may be delivered to the nucleus where it plays a role in the control of cell division.

4.   Translocation of FGF to the nucleus

The translocation of FGF to the nucleus appears to be important in the stimulation of cell growth [1266, 1267]. Extracellular FGF that is targeted to the nucleus (a) must enter the cell in an FGF/FGFR/heparan sulfate complex, (b) must avoid the typical shunting of endocytosed proteins to the lysosomes, (c) must be transferred first to the cytosol, and (d) must be transferred from the cytosol to the nucleus. In these steps the heparan sulfate facilitates the binding of FGF to the FGFR on the cell surface and protects the endocytosed FGF from proteolytic degradation, while FGFR plays a role in the targeting of the FGF to the cytosol. Furthermore, it has been shown that FGFR-1 is also translocated to the nucleus when cells are treated with FGF-1 or FGF-2 [641, 642]. The flow of FGF from the cytosol to the nucleus appears to occur by free diffusion [1268]. Generally, proteins that are targeted to the nucleus contain nuclear translocation sequences (NTSs), also referred to as nuclear localization sequences (NLSs) [1269]. Such sequences, found in a variety of proteins that target to the nucleus, are relatively short peptide domains that contain multiple basic amino acids. Some of these are listed in Table 9.3. The positions of nuclear localization sequences in FGF sequences are not entirely clear. For FGF-2, which occurs (a) as an 18-kDa protein formed by translation initiation at an AUG and (b) as 22-, 22.5-, and 24-kDa proteins with N-terminal extensions formed by translation initiation at CUGs upstream from the AUG (Section II.B., p. 304), there appear to be NTSs in the N-terminal extensions (Table 9.3). It has been found in several

Table 9.3. Nuclear Translation Sequences (NTSs) in Proteins Targeted to the Nucleus

NTS	Protein	Ref.
NTSs on non-FGF proteins		
P^{126} K K K C K W E	SV40 large T	[1269]
P^{279} P K K A R E V	Polyoma large T	"
A^1 P T K R K G S	SV40 VP1	"
P^{316} N K K K R K L	SV40 VP2	"
P^{120} A A K R V K L D	Human c-Myc	"
G^{160} Q A K K K K L	Nucleoplasmin	"
G^{29} K K R S K A	Yeast histone 2B	"
G^{296} N K A K R Q R	v-Rel	"
S^{190} R K R P R P	Polymoma large T	"
K^{281} R P R P	Adenovirus E1A	"
K^3 I P I K	Yeast mat α_2	"
G^{22} K R K R K S	c-erb-A	"
S^{127} K R V A K R K L		
S^{181} H W K Q K R K F	c-myb	"
P^{521} L L K K I K Q		
P^{337} P Q K K I K S	N-myc	"
P^{316} Q P K K K P	p53	"
F^{250} K R K H K K D I S Q N K R A V R R	HSP70	"
G R K K R R Q V E R A P	HIV tat I	"
Q R R R A P	HIV tat II	"
P^{277} G K R K K E M T K Q K E V P	Nucleolin	"
NTSs on FGFs		
G^{20} N Y K K P K L	FGF-1, near N-terminal	[1276]
R Q R G R	FGF-2, found in several N-terminal extension sequences	[1273]
R^7 L R R, and R^{28} R R K	FGF-3	[1278]

studies that the species of FGF-2 that are translocated to the nucleus are primarily the larger species that contain N-terminal extensions having potential NTS sequences [1270-1272]. In fact, when the N-terminal sequences are fused to β-galactosidase, a cytosolic protein, the β-galactosidase is translocated to the nucleus [1273]. But, to confuse the story, it has been shown that removal of these N-terminal extensions does not impair the transport of FGF to the nucleus [640].

In contrast to FGF-2, FGF-1 does not occur as multiple species with N-terminal extensions because there is a translation termination codon just upstream from the normal AUG initiation codon [1274]. Thus, it appears that the NTS for FGF-1 must lie within the main sequence. Such a sequence—residues 21-27, shown in Table 9.3—has been identified near the N-terminal of FGF-1 [1275]. A deletion mutant lacking amino acids 21-27 does not appear in the nucleus [1276] and does not induce DNA synthesis or cell proliferation [1275]. It has also been shown that when the NTS (residues 21-27) of FGF-1 is incorporated into a membrane-permeable peptide, this peptide is translocated to the nucleus and elicits a mitogenic response [1277]. However, it has been reported that when FGF-1 is modified by the removal of amino acids 1-27, a sequence that includes the putative NTS, it still targets to the nucleus and stimulates cell growth [1276]!

One interesting observation concerning the NTSs shown in Table 9.3 is that they look a lot like amino acid sequences that bind to heparin (Chapter 6), raising the possibility that these sequences may bind to heparan sulfate chains. If this can occur, then heparan sulfate chains might play a role in nuclear targeting. However, it has been shown that although the removal of the NTS from FGF-1 prevents nuclear targeting, it does not eliminate the capacity of FGF-1 to bind to heparin [1275]. Clearly, further studies are required to resolve these uncertainties.

IV. INTERACTION OF HEPARINOIDS WITH FGFs AND THEIR RECEPTORS

A. FGF Interaction with the FGFR/HS Receptor Complex

We now have the background to address the question of how heparinoids interact with the FGFs and the FGFRs and to appreciate the physiological significance of such interactions. However, before addressing the specific protein and heparinoid domains that participate in these interactions, it may be helpful to put the reactions into the context of a model for cell surface binding of FGF to its receptors.

A model of the FGF/FGFR/heparinoid interactions that is consistent with much of the literature is illustrated in Figure 9.3. According to this model, the FGF/FGFR interaction is mediated by a strand of heparan sulfate that serves as a template for the formation of the FGF/FGFR/heparinoid complex. This is reminiscent of the thrombin/antithrombin/heparinoid complex in which heparinoids act as a template for bringing thrombin and antithrombin together (Chapter 7). In the present case, the FGFR and the heparan sulfate proteoglycan are anchored in the cell membrane with the heparan sulfate chain already bound to

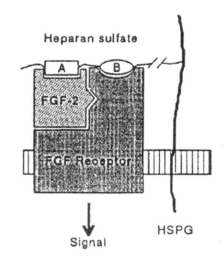

Figure 9.3. Model for FGF/FGFR/HS complex. FGFR and the HSPG are bound in the cell membrane with FGF bound to both. A "B" segment of the heparan sulfate chain of the proteoglycan is bound to the FGFR prior to arrival of the FGF. FGF completes the complex formation, becoming bound to both the FGFR and the "A" segment of the same heparan sulfate chain. Reprinted from [624], with permission.

the FGFR. This receptor complex is available to interact with incoming FGF. FGF then binds to both the FGFR and the heparan sulfate chains of the proteoglycan. One might say that the FGF receptor is, in fact, the heterodimeric FGFR/heparan sulfate complex that is present on the cell surface. The model suggests that each member of the FGF/FGFR/heparinoid complex has separate binding sites for the other two members and, as a corollary, that subtle differences in structures of the different forms of FGF, FGFR, and heparan sulfate might affect the cell binding and the resulting cellular responses. It also suggests that, in cell culture, exogenous heparin could (a) fulfill the role of the endogenous heparan sulfate or (b) compete with the endogenous heparan sulfate chains for binding to both the FGF and the FGFR. The model implies that there are separate binding sequences on the heparinoid for the FGF and the FGFR, and that both proteins must be bound to the heparan sulfate in order to obtain a cellular response to FGF, consistent with the many reports showing that both heparan sulfate and the high affinity receptor are required for the mitogenic effects of FGFs. Recent modifications of this model propose that *two* molecules of FGFR must be bound up in this complex in order to obtain a mitogenic response and that two molecules of FGF may also occur in the complex. These modified models are described in Section IV.H., p. 337.

We will now turn our attention to the experimental background and bases for this model. First we will describe the *peptide* sequences in the FGFs and FGFRs that bind to each other and to heparinoids, and we will then turn our attention to the *heparinoid* sequences that bind to these peptide sequences on FGFs and FGFRs. Finally, we will discuss these results in the context of the current models for the complex formed from FGF, FGFR, and the heparinoid.

B. FGF Domains that Bind Heparinoids

1. Overall findings

The amino acid residues of FGFs that bind to heparin have been deduced from several types of studies, including (a) identification of heparin-binding peptide fragments of FGF, (b) observations of the effects of structure modifications of FGFs on heparin binding, (c) interpretations of the crystallographic studies of the three-dimensional structures of FGFs, and (d) projections from molecular modeling of the heparin/FGF complex. The general conclusion from these studies is that hexa- and octasaccharide fragments of heparinoids bind to clusters of basic amino acids on the surfaces of the FGFs. Most of the amino acids in these clusters lie close together in the primary structure, but several more distant amino acids also appear to be involved in heparin binding. For FGF-2, the putative heparin-binding domain suggested by the crystal studies is represented by a cluster of basic amino acids on the surface of the protein, including lysines 26, 119, 125, 129, and 135 and arginines 44 and 120. A similar binding domain is inferred for FGF-1. Although these residues are derived from both the N- and C-terminal regions of these FGFs, many of these occur in the following sequences near the C-terminal:

For FGF-2:

$$\underline{K}\ \underline{R}\ T\ G\ Q\ Y\ \underline{K}\ L\ G\ S\ \underline{K}\ T\ G\ P\ G\ Q\ \underline{K}$$
$$119\qquad\quad 125\qquad\quad 130\qquad\qquad 135$$

The corresponding peptide in FGF-1 is:

$$\underline{K}\ \underline{K}\ N\ G\ S\ C\ \underline{K}\ \underline{R}\ G\ P\ \underline{R}\ T\ H\ Y\ G\ Q\ \underline{K}$$
$$112\qquad\quad 118\qquad\quad 123\qquad\quad 128$$

The FGF-1 sequence contains one more basic amino acid residue than the FGF-2 sequence. All studies point to these primary sequences as the most important, but not the only, portions of the FGFs that participate in heparin binding. Both electrostatic and nonelectrostatic interactions contribute to the binding. Results obtained for both FGF-1 and FGF-2 are consistent with these conclusions. The work leading to these conclusions is now described in more detail.

2.   Early work

In early FGF-1 studies, it was shown that reductive methylation of Lys 118 (equivalent to Lys 125 in FGF-2 [1279]) or replacement of Lys 118 in FGF-1 with Glu [1261] gives marked reduction in the affinity of the growth factor for heparin, but not in its binding to FGFR. Also, thrombin digestion studies implicated Lys 118 and Arg 122 in FGF-1 in heparin binding [1167]. In an extensive study, Baird *et al.* [535] prepared synthetic peptide segments of FGF-2 and tested them for binding both to heparin and to cell receptors, and for their ability to modulate FGF-2 activities. FGF-2 peptides 24-68 and 93-120 bound to heparin. Interestingly, these same peptides gave significant inhibition of binding of FGF-2 to cell surface receptors and inhibited mitogenic activity. These results suggested that these two sequences participate in both heparin binding and FGFR binding. However, peptide 106-115, a sequence within the larger peptide 93-120, is very potent in inhibition of FGF-2 binding to FGFR, but has much lower heparin-binding capacity than the larger peptide. This suggests that the receptor binding region lies in amino acids 106-115, which begins in the middle of β-strand 9 and forms a distorted antiparallel β-turn loop on the surface before terminating at the beginning of β-strand 10. Data also suggest that the heparin-binding region lies just downstream from the receptor-binding domain—i.e., in peptide 116-120, which lies in a β-bend-β-strand-loop region on the surface of FGF-2—and that peptide 24-68 contains segments that also participate in heparin and receptor binding. This is consistent with the conclusions about the main heparin-binding domain in FGF-2 deduced from the studies described below. Phosphorylation of Thr 112 of FGF-2 by protein kinase A increases the receptor binding by three- to eight-fold [1280].

3.   Crystal structures

The three-dimensional structures of FGFs-1 and -2 are described in Section II.C., p. 305 and are illustrated in Figure 9.1. In all of the crystal studies, the crystals were formed from solutions saturated with $(NH_4)_2SO_4$. As a consequence, sulfate radicals became bound to the proteins, forming counterions for several of the basic amino acid clusters on the surfaces of the FGFs [1183, 1184]. This suggests that the basic amino acid clusters represent sites to which the sulfate moieties of heparinoids bind. Of the basic amino acids conserved on the surfaces of FGF-1 and FGF-2, five in the C-terminal (K119, R120, K125, K129, and K135 [2]) form a prominent cluster. These sequences occur in a β-turn (K119-G122), a short β-strand (Q123-L126), and a loop (S128-I137). One of the sulfates found in the crystallized FGF-2 has its four oxygen atoms H-bonded to side chains of Asn 27, Arg 120, and Lys 125 and to the main chain imino group of Arg 120. A sulfate radical is found in the same relative position in FGF-1, where the residue that corresponds to Asn 27 in FGF-2 is Asn 18 in FGF-1. The second sulfates in FGF-1 and FGF-2 are bound about 8 Å from the first and are not so clearly defined. In FGF-2, this sulfate may H-bond to the main chain amide of Leu 126 and to side chains of Lys 119 and Lys 129 [1184]. Crystal structure results are consistent with the peptide sequences shown in Section IV.B.1., p. 318.

These basic amino acid clusters are located ~25 Å from the presumed receptor binding loop, and the two sites are on different faces of the molecule (see below), consistent with the proposition that the heparinoid may contact both FGF and FGFR in the FGF/FGFR/heparinoid complex.

4.   Site-directed mutagenesis

Several modifications of FGF-2 by recombinant DNA techniques have also yielded results consistent with the implicated heparin-binding sequences. For example, removal of a 40 amino acid segment from the N-terminal does not reduce heparin affinity, but removal of 40-42 residues from the C-terminal of both FGF-1 and FGF-2 markedly lowers heparin binding [1281]. The heparin-binding sequences shown in Section IV.B.1., p. 318 are present in the 40 residue C-terminal segments.

A more extensive site-directed mutagenesis study has been reported by Thompson *et al.* [662], who have presented the model for heparin docking to FGF-2 shown in Figure 9.4 (see color insert). As described above, X-ray data suggest that heparin binding to FGF-2 involves Asn 27 and the basic amino acid residues in the 119-135 peptide. When a heparin pentasaccharide is docked to FGF-2, several sequences appear to come into contact with the heparinoid segment. These include K119, R120, K125, K129, and K135, K26 and R81, as well as several uncharged amino acids in the 119-135 sequence, namely T121,

---

[2] FGF-2 numbering

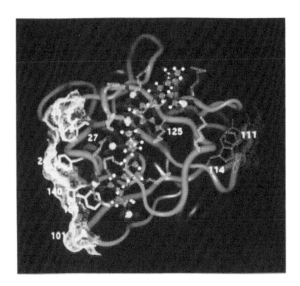

Figure 9.4. Model of the FGF-2/heparin complex. The docked heparin pentasaccharide lies from top to bottom across the middle of the FGF, with the basic amino acids that bind the heparinoid shown in orange. The primary and secondary FGFR-binding sites are shown in yellow and purple, respectively. Reprinted from [1282], with permission.

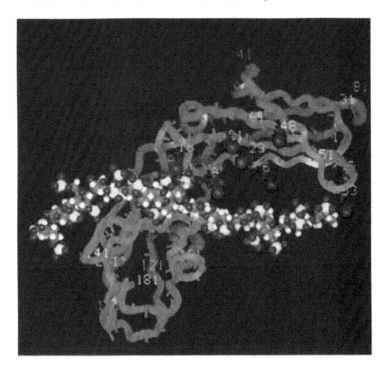

Figure 9.5. Model of FGFR-1 with a strand of heparin bound to its K18K segment. Reprinted from [661], with permission.

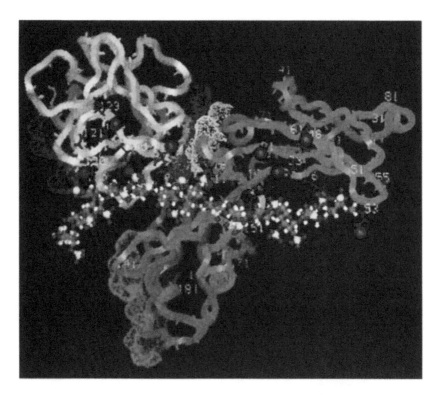

Figure 9.6. Model of the complex formed when FGF-2 is added to the heparin/FGFR-1 complex shown in Figure 9.5. Reprinted from [661], with permission.

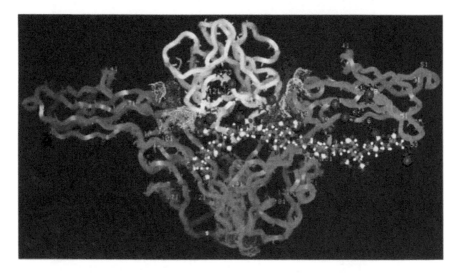

Figure 9.7. Model of the complex formed when a second molecule of FGFR-1 is added to the heparin/FGF-2/FGFR-1 complex shown in Figure 9.6. Reprinted from [661], with permission.

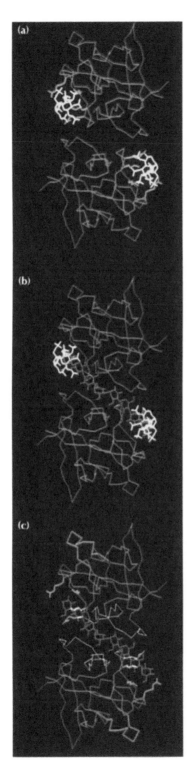

Figure 9.8. Models of binding of heparinoids to an FGF-1 dimer: (a) binding of sucrose octasulfate; (b) binding of a heparin octasaccharide, the ends of which are shown to overlap with the sucrose octasulfate; and (c) binding of octasaccharide alone, but with the putative FGFR-1 binding sites shown in green at the two opposite ends of the complex. Reprinted from [1305], with permission.

Q123, and Q134. Each of these amino acids was converted individually to alanine by site-directed mutagenesis, and the resulting FGF-2 mutants were tested for heparin binding by (a) measurement of the NaCl concentration required for elution from heparin-Sepharose, and (b) isothermal titration calorimetry. The results show that K125 (equivalent to K118 in FGF-1) contributes about 20% of the total binding energy and that the total peptide, K119-K135, contributes 70% of the total binding energy. However, more than half of the binding affinity of the 119-135 peptide is derived from heparinoid interactions with T121, Q123, and Q134. In fact, only 30% of the total binding energy is derived from electrostatic interactions. Interestingly, N27, near the N-terminal, and Q134, in the 119-135 peptide, contribute 25% of the binding energy. It may be recalled that FGF-2 binds more strongly to heparin than does FGF-1. Despite this, Q123 of FGF-2 appears as an arginine in FGF-1, suggesting that the FGF-1 peptide should bind heparin more strongly. However, the replacement of Q123 in FGF-2 with arginine gives no change in heparin binding. Thus, other sources for the difference in FGF-1 and FGF-2 affinities for heparin must be sought. There is almost absolute conservation of amino acids in the heparin-binding sites of these two FGFs, the main differences appearing to lie in the replacements of K26 and R81 in FGF-2 with serine and leucine in FGF-1. Based on FGF-2 binding data, these differences may lower the heparin binding energy by ~1 kcal/mole, accounting, at least in part, for the weaker heparin binding by FGF-1.

The identity of the FGF-1 sequence shown in Section IV.B.1, p. 318 as the primary heparin-binding sequence is confirmed by a site-directed mutagenesis by Wong *et al.* [647] who selected three peptide sequences in FGF-1 that conformed to the Cardin and Weintraub consensus sequences for heparin binding (see Chapter 6). Mutations only in the sequence shown above gave significant reduction of heparin binding.

5. Conclusions

FGF-1 and FGF-2 present very similar three-dimensional structures, and they bind heparinoids in virtually identical manners, using the same relative amino acid positions on the surface of the proteins for heparin binding [662]. Most of the binding energy in each case derives from interactions of the heparinoid with the peptide sequences shown in Section IV.B.1, p. 318, but it is surprising to find that so little of the total binding energy appears to come from electrostatic interactions. One qualification with respect to the modeling of heparinoid binding to FGF-2 is that the antithrombin-binding heparin pentasaccharide was chosen as the heparinoid for the modeling in this work. This pentasaccharide may not be an appropriate sequence for specific recognition by the growth factor. One further problem is that the structure given for the antithrombin-binding pentasaccharide in the paper is incorrect. Hopefully, this does not alter the conclusions derived from the modeling work.

C. FGFR Domains that Bind Heparinoids

1. FGFR-1

The structural features of the fibroblast growth factor receptors (FGFRs) have been described in Section III.A.2., p. 308 (see Figure 9.2). FGFR-1 binds to heparin affinity columns and is eluted with 0.5 $\underline{M}$ NaCl [1283]. From the standpoint of the interactions of FGFRs with FGFs and heparinoids, the most notable features of the FGFR structures are found in Ig-like domains II and III and in the connecting region between domains I and II that just precedes Ig-like domain II. This latter region contains a sequence of acidic amino acids, referred to as the "acidic box," as well as the putative heparin-binding domain. The identity of the heparin-binding domain is based on a report by Fan *et al.* [1283], who isolated a single, specific 18 amino acid segment of the receptor, referred to as K18K from a mixture of peptide fragments of FGFR-1:

$$\underline{K} \text{ M E } \underline{K} \text{ } \underline{K} \text{ L H A V P A A } \underline{K} \text{ T V } \underline{K} \text{ F } \underline{K} \text{ } (\rightarrow \text{Cys})$$
$$71 \qquad\qquad\qquad 80 \qquad\qquad\qquad 88 \quad 89$$

This is the only heparin-binding peptide found among FGFR-1 fragments. The peptide is rich in lysine, but has no arginine, and it has structural features like those of other heparin-binding proteins (Chapter 6). In FGFR-1, this peptide lies just upstream from the cysteine residue that forms the N-terminal residue of Ig-like domain II, i.e., the amino acid residue in FGFR-1 that follows this 18 amino acid heparin-binding fragment is the domain II N-terminal cysteine (Cys 89). Also, the peptide begins 33 amino acid residues from the first acidic residue in the acidic box (see Figure 9.2).

The peptide binds to FGF, but only in the presence of heparin. It also inhibits FGF binding to intact cells, apparently by binding to the cell surface HSPG, thus preventing the FGFR-1 from binding (see Figure 9.3). The effects of mutations in this peptide domain of FGFR-1 on the binding of FGF were determined by measuring the binding of FGFs to (HSPG-producing) cells transfected with altered FGFRs. Substitution of residues 71, 74, or 75 eliminated the binding of FGFs, whereas substitution of residues 83 and 86 together increased the K_d for FGF binding by the cells (i.e., lowered the affinity) 13-fold. These changes in FGF binding correlated well with changes in heparin binding of the mutated FGFR-1's. The results show that the heparin-binding domain and the cell surface heparan sulfate are essential for binding of FGF to cells and for manifestation of the biological activities of FGFs.

Another study has implicated additional amino acid residues in heparin binding [1284]. Modeling and site-directed mutagenesis suggest that, in addition to the residues in the K18K sequence, residues within the Ig-like domain II loop participate in the binding of heparan sulfate proteoglycan to the receptor.

Pantoliano *et al.* have cloned an extracellular portion of FGFR-1 containing the Ig-like domains II and III and the pre-domain II segment containing the above heparin-binding peptide [661] and have modeled the three-dimensional structure of FGFR-1, matching its primary, secondary, and tertiary structure based on the homologous CH_2 domain of the IgFc structure [1285]. The segment of FGFR-1 with a strand of heparin bound to it is shown in Figure 9.6 (see color insert described in Section IV.H.2., p. 337). The three-dimensional model of this soluble form of FGFR-1 shows a surface cluster containing all of the lysine and arginine residues in the K18K peptide above. In addition, R113, K118, and R120, which lie downstream from the K18K peptide and within the disulfide closed loop of the Ig-like domain II, are aligned on the surface with the K18K peptide. The soluble form of FGFR-1 binds heparin with a K_d of 1.01 x 10^{-4} $\underline{M}$ in an incubation mixture containing 50 mM Hepes and 100 nm NaCl—i.e., under conditions approximating physiological salt concentrations [661]. This is not particulary strong binding, but, as the authors point out, most cells have 10^5-10^6 copies of heparan sulfate proteoglycan/cell [476], a concentration that would tend to drive the formation of these FGFR/heparinoid complexes on cell surfaces. Although almost nothing is known about the sequence of the *heparinoids* that bind to FGFRs, these workers have modeled the docking of a 13-mer heparinoid sequence, approximately 60 Å long, to FGFR-1 (Figure 9.6). Of the 13 monosaccharide residues in this hypothetical sequence, 6-7 interact with the binding cleft while the remaining residues are protected by the FGFR and sterically precluded from binding to another FGFR-1 molecule.

It has been reported that Ca^{2+} or Mg^{2+} is required for the binding of heparin to FGFR, but not for FGF/heparin binding [659]. It has been suggested that the FGFR/heparan sulfate receptor complex is restrained by the metal ion and heparan sulfate in a conformation that restricts transphosphorylation of the receptor and that this restraint is removed when FGF binds to the receptor.

2.   Other FGFRs

Although similar experiments have not been carried out for FGFR-1 from other species or for FGFRs-2, -3, or -4, there is a remarkable conservation of the

K18K sequence in all FGFRs, especially among the vertebrates. This is illustrated in Table 9.4, which lists the 18 residue sequences found just prior to the N-terminal cysteine of Ig-like domain II for a variety of FGFRs from several species. It also shows a 17-amino acid peptide found in the same position in a sea urchin FGFR [1286]. An examination of the vertebrate FGFRs in Table 9.4 leads to the following conclusions:

- The peptide is always 18 amino acid residues long.

- It always terminates with a basic amino acid that is linked to the first Cys of Ig-like domain II.

- Although Lys and Arg residues appear to be interchangeable in the positions of the basic amino acids, these residues are always in identical positions in every FGFR.

- There is always a methionine in position 72 and an acidic amino acid in

Table 9.4 K18K Sequences from Different FGFRs

FGFR	Species	K18K sequence	Reference
FGFR-1	Human	K M E K K L H A V P A A K T V K F K	[1201, 1287, 1288]
	Murine	K M E K K L H R V P A A K T V K F K	[1203, 1204, 1289]
	Rat	K M E K K L D R V P A A K T V K F K	[1290]
	Chick	K M E K K L H A V P A A K T V K F K	[1199, 1200]
	Xenopus	K M E K K L H A V P A A K T V K F R	[1202]
FGFR-2	Human	K M E K R L H A V P A A N T V K F R	[1201, 1205, 1291]
	Murine	K M E K R L H A C P A A N T V K F R	[1248]
	Rat	K M E K R L H A V P A A N T V K F R	[1292]
	Chick	K M E K R L H A V P A A N T V K F R	[1206]
FGFR-3	Human	R M D K K L L A V P A A N T V R F R	[1208]
	Murine	R M D K K L L A V P A A N T V R F R	[1204]
	Chick	K M E K K L L A V P A A N T V R F R	[1206]
FGFR-4	Human	R M E K K L H A V P A G N T V K F R	[1247]
FGFR	Sea urchin	R M E P E K P L P S N T K V R L E	[1286]

position 73; other amino acid residues in these sequences are also identical or highly conserved.

- The number of amino acids between the first amino acid of the acidic box and the first residue of the heparin-binding domain (Figure 9.2) is 25-34; i.e., the heparin-binding domain in any FGFR sequence can be easily located by searching for this sequence about 30 amino acid residues downstream from the beginning of the acidic box, or just preceding the first cysteine of the Ig-like domain II.

- FGFR-1 contains six basic amino acids in this sequence, whereas the remaining vertebrate FGFRs contain only five.

- The features of these K18K sequences are remarkably similar to that of the sequence identified as the heparin-binding domain of FGFR-1. Thus, it is hard to escape the conclusion that these are heparin-binding domains in all of the FGFRs. Furthermore, the similarities of these sequences suggest that all FGFRs must recognize similar heparinoid sequences.

The sea urchin FGFR shows some similarities in its 17-amino acid sequence, which contains only four basic amino acids [1286]. It also contains an acid box ~30 amino acids upstream from this sequence.

D. Heparinoid Sequences that Bind FGFs

1. Introduction

Heparin and heparan sulfate, as well as oligosaccharide fragments of these glycosaminoglycans, bind to FGFs. The relative affinities of different FGFs for heparinoids vary widely. Also, the binding of different heparinoid species to any single FGF shows variability. These simple observations are indications enough that the structure of the heparinoid in these interactions is important in establishing the strength of the binding and, in turn, the extent of cellular responses to FGFs. Given the structural variabilities of heparinoids, several general comments, which are documented in Sections IV.D.2., p. 327 and IV.F.2., p. 331, may be in order. Most notably, heparins generally bind more strongly to FGFs than do heparan sulfates. The latter polymers, of course, exhibit greater structural diversity than heparins (Chapter 2) and are the heparinoids that are physiologically involved in FGF responses. It is also generally true that *polymeric* heparinoids bind more tightly to FGFs than the oligosaccharide fragments prepared from them. Furthermore, polymeric heparinoids may bind several mols of FGF per mole of heparinoid. These facts should be kept in mind as we proceed through the following discussion for several reasons: (a) there is a tendency to place undue emphasis on results from the heparin binding studies, while the more stringent structural requirements for the heparan sulfate binding to FGFs and their receptors are the physiologically significant parameters, (b) FGF binding to heparinoids alone may not be sufficient for biological responses to the growth factor; the heparinoid may bind to both the FGF and the FGFR; and (c) the (relatively) weak binding of *heparan sulfates* to FGFs and FGFRs is probably the more physiologically important parameter.

The model for the FGF/FGFR/heparinoid complex for FGF binding to cell surfaces in Figure 9.3 proposes that the biological activities of FGF depend on the binding of both the FGF and the FGFR to either cell surface heparan sulfate or exogenous heparin. However, heparinoids will bind to FGFs in the absence of the receptors. Furthermore, as will be described, smaller fragments of heparin are required for binding to FGF alone than for binding to both FGF and FGFR. Therefore, the assays that are required for measurement of the binding of heparinoids and their fragments to FGFs are different from those needed to observe the role of heparinoids in productive growth factor/receptor complex formation.

For the measurement of the binding of any FGF to heparinoids, the most common assay involves determination of the NaCl concentration that is required for elution of the the heparinoid from immobilized FGF (or the elution of FGF from immobilized heparin) [see for example, 561, 627, 662]. The more tightly the FGF binds to the heparinoid, the higher the salt concentration required. In fact, it has been shown that for FGF-2 and other heparin-binding proteins, the K_d's for the protein/heparin dissociations are directly correlated with the concentration of NaCl required to elute the heparinoid from the immobilized protein [662]. Other assays have also been used to study these interactions. These include (a) binding of syndecan-transfected cells to FGF immobilized on a multiwell plate and the measurement of the heparinoid concentrations required for inhibition of the binding [561]; (b) mixing FGF and the heparinoid, either of which may be labeled, and then measuring the amount of the complex recovered on a nitrocellulose filter as a function of the heparinoid concentration [627]; (c) using the above systems to study the effects of various heparinoids on the inhibition of the binding of some standard heparin to the FGF [561, 627]; or (d) measurement of the interaction of the FGF and the heparinoid by calorimetry [661]. In these assays, the incubation conditions can be varied as desired, and various heparins and heparan sulfates, their fragments, and their structurally modified forms can be tested. It is important to remember in evaluating these or any similar studies, that only binding observed at 0.15 $\underline{M}$ salt concentrations may be of physiological significance.

2. FGF-2

The binding of heparinoids to FGF-2 has been much more extensively studied than the binding to FGF-1 and so will be addressed first. The initial studies on heparinoid sequences that bind to basic FGF were reported by Habuchi *et al.* [625] and by Turnbull *et al.* [626]. Both groups measured the binding of heparan sulfate and its fragments to FGF-2 affinity matrices. Habuchi *et al.* concluded that the minimum sequence that binds to FGF-2 is an octasaccharide that contains three contiguous IdoA-2-SO$_4$→GlcNSO$_3$ disaccharide units (plus one additional disaccharide). The FGF-2 binding unit reported by Turnbull *et al.* is a 14-mer containing seven contiguous IdoA-2-SO$_4$ →GlcNSO$_3$ disaccharide units. This group later reported that IdoA-2-SO$_4$→ GlcNSO$_3$-containing heparan sulfate sequences as small as octa- and decasaccharides binds to FGF-2 [628]. Both groups noted that the GlcN residues must be N-sulfated. Thus, this work suggests that the following structure is required for FGF-2 binding:

$$IdoA-2-SO_4 \rightarrow GlcNSO_3 \rightarrow [IdoA-2-SO_4 \rightarrow GlcNSO_3]_{2-3} \rightarrow IdoA-2-SO_4 \rightarrow GlcNSO_3$$

In more recent work, Maccarana *et al.* [627] set out to find the smallest and least sulfated oligosaccharide structure that would bind to FGF-2. This work led to the demonstration that the pentasaccharide

$$HexA \rightarrow GlcNSO_3 \rightarrow HexA \rightarrow GlcNSO_3 \rightarrow IdoA-2-SO_4$$

represents the minimal heparinoid fragment required for FGF-2 binding, albeit a structure that does not bind as tightly as heparin and some of its fragments. Thus, according to the latter study, only a single IdoA-2-SO$_4$ residue is required in the FGF-2 binding sequence, and this IdoA-2-SO$_4$ must be present on the downstream end of an oligosaccharide sequence in which all of the GlcN residues are N-sulfated. The uronic acid residues in the internal and nonreducing terminal positions may be either GlcA or IdoA.

In addition to the work with oligosaccharide fragments of heparinoids, several interesting observations have been made using polymeric heparin derivatives. First, it is confirmed that 2-O-sulfation but not 6-O-sulfation is required for FGF-2 binding since 6-O-desulfated heparin (which retains most of the 2-O-sulfated uronic acid residues) binds well to FGF-2 affinity matrices whereas 2-O-desulfated heparin does not [367, 627]. Second, when heparan sulfate is subjected to affinity chromatography on immobilized FGF-2, only a portion of the total heparan sulfate pool from various tissues adheres to FGF-2. The heparan sulfate fraction that binds to FGF-2 protects the FGF from cleavage by trypsin, whereas the portion that does not bind lacks this protective effect [625]. This result indicates that FGF-2 exhibits specificity in binding to heparan sulfate chains; i.e., *not all heparan sulfate chains from a given cell or tissue type contain sequences that will bind to FGF-2.* This interesting result is discussed further in Section V.A., p. 342. In contrast to heparan sulfate, it has been shown that heparin has multiple binding sites for FGF-2. For example, heparin with an average molecular weight of 15,000 can bind six to seven molecules of FGF-2 [1293].

Although the minimal structure for FGF-2 binding may require only one IdoA-2-SO$_4$ unit on the reducing terminal disaccharide unit of a pentasaccharide, additional 2-O-sulfate substituents, and even 6-O-sulfate substituents, do not interfere with binding. After all, heparin, which is enriched in IdoA-2-SO$_4$→ GlcNSO$_3$-6-SO$_4$ disaccharides, binds more tightly to FGF-2 than most heparan sulfates. Also, it has been demonstrated that the hexasaccharide

$$[IdoA-2-SO_4 \rightarrow GlcNSO_3-6-SO_4]_2 \rightarrow IdoA-2-SO_4 \rightarrow AMan_R-6-SO_4$$

binds strongly to FGF-2 [305, 367, 627], and that, in modeling the oligosaccharide binding to FGF-2, 6-O-sulfate substituents do not interfere [627]. Furthermore, the original work of Habuchi *et al.* and Turnbull *et al.* demonstrated that oligosaccharide sequences in which *all* of the IdoA residues are 2-O-sulfated bind tightly to FGF-2. Thus, the minimal pentasaccharide structure that contains only one 2-O-sulfated residue can be further 2-O-sulfated and 6-O-sulfated without interfering with FGF-2 binding.

3.   FGF-1

Although there have been quite extensive studies on the heparinoid sequences that bind to FGF-2, there have been very few reports of similar work on FGF-1. In fact, the effects of structurally varied heparinoids on the FGF-1/FGFR complex have been much more extensively studied than the direct binding of heparinoids to FGF-1 alone (see Section IV.F.2., p. 331). Bârzu *et al.* [1294] have used FGF-1-Sepharose affinity chromatography to study the FGF-1-binding of oligosaccharides formed by nitrous acid cleavage of heparin. These workers find that a hexasaccharide,

$$[IdoA-2-SO_4 \rightarrow GlcNSO_3-6-SO_4]_2 \rightarrow IdoA-2-SO_4 \rightarrow AMan_R-6-SO_4,$$

is the smallest heparin fragment that binds to FGF-1. We have already noted that this same heparin hexasaccharide fragment binds to FGF-2. Higher oligosaccharides also bind FGF-1, but these have not been fully characterized. All of these fragments were tested for their ability to induce cell growth in the presence of FGF-1 (see Section IV.F.2., p. 331).

The binding of heparin and fragments generated by heparin lyase cleavage of heparin has been studied by light scattering and ultracentrifugal analyses [1295]. This work shows that a 16 kDa heparin can bind 14-15 molecules of FGF-1, with K_d's in the range of 50-140 nM. Digestion of such an FGF-1/heparin complex with a mixture of heparin lyases I and II yields a product containing oligosaccharide/FGF-1 complexes. Dissociation of the protected oligosaccharides from these complexes by denaturation of the FGF-1 yields an oligosaccharide mixture, the most prominent component of which migrates on gradient PAGE gels at the position of the standard tetrasaccharide,

$$\Delta^{4,5}\text{HexA-2-SO}_4 \rightarrow \text{GlcNSO}_3\text{-6-SO}_4 \rightarrow \text{IdoA-2-SO}_4 \rightarrow \text{GlcNSO}_3\text{-6-SO}_4$$

This suggests that this is the minimum sized heparin segment protected by FGF-1. The same group reported on the effects of heparin lyase-generated oligosaccharides on the stabilization of FGF-1, measured by changes in CD and fluorescence spectra of FGF-1 in the presence of heparinoids [1296]. Although the disaccharide

$$\Delta^{4,5}\text{HexA-2-SO}_4 \rightarrow \text{GlcNSO}_3\text{-6-SO}_4$$

is inactive, the following tetra- and hexasaccharides do stabilize FGF-1:

$$\Delta^{4,5}\text{HexA-2-SO}_4 \rightarrow \text{GlcNSO}_3\text{-6-SO}_4 \rightarrow \text{IdoA-2-SO}_4 \rightarrow \text{GlcNSO}_3\text{-6-SO}_4$$

$$\Delta^{4,5}\text{-HexA-2-SO}_4 \rightarrow \text{GlcNSO}_3\text{-6-SO}_4 \rightarrow [\text{IdoA-2-SO}_4 \rightarrow \text{GlcNSO}_3\text{-6-SO}_4]_2$$

Also, mixtures containing uncharacterized octa- and decasaccharide fragments stabilize FGF-1.

E.   Heparinoid Sequences that Bind to FGFR

There have been no direct studies on the specific heparinoid structures that bind to FGFRs. However, there is some indirect evidence discussed in Section IV.F.2., p. 331 that shows that 6-O-sulfated GlcN residues are required for FGF-2/FGFR-1 interaction. Also, it has been shown that a mouse neuroepithelial cell line produces HSPG-containing heparan sulfate chains that bind specifically to FGFR-1, even in cells producing both FGFR-1 and FGFR-3 [1284].

F.   Heparinoid Domains that Stabilize FGF/FGFR Complexes

1.   Assays for measuring heparinoid effects on FGF/FGFR/heparinoid complex formation

To this point, we have discussed the surface structures of heparinoids, FGFs, and FGFRs that participate in each binary interaction. However, there is more to be learned when we examine the heparinoid structures that bind to *both* FGF and FGFR. At the outset, we must consider the ways in which one can identify the specific heparinoid structures that stabilize the FGF/FGFR complexes. The most obvious assay involves incubation of individual FGFs and FGFRs with various heparinoids under physiological conditions and observation of the effects of variations in heparinoid structure and concentrations on complex formation [561, 661, 1297]. When FGFR is not available for such assays, one may measure the effect of heparinoids on binding of FGFs to cells that express FGFRs but not heparan sulfate proteoglycans [624]. Both of these assays show the effects of heparinoids on direct binding of FGFs to FGFRs without regard to the physiological events that follow FGF binding.

Another common assay for FGF/FGFR complex formation utilizes FGFR-expressing cells and measures the effects of heparinoids on the biological responses of exogenous FGF when various heparinoids are added. In such assays, one must decide what activity to measure and must contend with the fact that most cells express relatively high concentrations of cell surface heparan sulfate proteoglycan. Thus, exogenous heparinoids would compete with the endogenous heparan sulfate in stimulating the FGF/FGFR interaction [366, 367, 561]. To circumvent this problem, Rapraeger *et al.* [586] described an assay in which the sulfation of the endogenous heparan sulfate by cultured cells is inhibited by chlorate (Chapter 3), effectively eliminating the participation of the endogenous heparan sulfate proteoglycan in FGF responses. Under these conditions, the response of cells to FGFs is lost, but exogenous heparinoids of known structures can be tested for their ability to restore FGF-mediated responses. This assay, which has allowed the examination of the size and structure of heparinoid sequences that are required for simultaneous binding of FGF-2 and its receptor, has been applied in the study of the effects of FGFs on Swiss 3T3 cells and MM14 mouse myoblasts [624], on adrenocortical endothelial (ACE) cells [366, 367, 561], on F32 cells, a lymphoid cell line [560, 1298], and on CHO cells [1297], and has become the primary assay for assessing the effects of structurally characterized exogenous heparinoids and heparinoid fragments on FGF/FGFR interactions. In most cases the effect of exogenous heparinoids plus FGF on DNA synthesis or cell proliferation has been measured. Of course, these responses are multistep events, and the effects of replacement of endogenous heparan sulfate proteoglycan with exogenous heparinoids on endocytosis and intracellular translocation are not known. One thing that is clear, however, is that the responses of cells to exogenous heparinoids are substantially different in the absence of chlorate than in its presence [366, 367, 561], as will be discussed in Section IV.F.3., p. 333.

2. Effects of heparinoids on FGF/FGFR complex formation

Table 9.5 summarizes the results obtained in the above assays when heparinoids are incubated with chlorate-treated cells in the presence of FGF-1, FGF-2, or FGF-4. The heparinoids tested have included heparin and several modified heparins, including 6-O-desulfated heparin, 2-O-desulfated heparin, carboxyl-reduced heparin (CR-heparin), and amidomethylsulfonated heparin (AMS-heparin).[3] AMS-heparin has a negatively charged sulfonate group attached to its carboxyl groups via an amide linkage so that a negative charge remains at this position, albeit in a different chemical form. When oligosaccharide fragments of these heparinoids are tested, it is observed that a minimum sized fragment is necessary to obtain a response, and this size is given in parentheses below the response level. Generally, when the active oligosaccharides identified in these assays are further fractionated on either FGF affinity matrices or on anion exchange columns, the most highly sulfated fragments, which elute late from both types of columns, are found to be the most active. Results have been obtained with several cell types. These results may be summarized as follows:

- Heparin, and oligosaccharides prepared from heparin, promote the FGF/FGFR interactions. The minimal size of the oligosaccharides required for this activity seems to vary in different reports, but is in the deca- to dodecasaccharide range. These minimal sized oligosaccharides are two disaccharide units longer than the shortest oligosaccharides found to bind to to the FGFs alone.

- Heparan sulfates are active, but generally less active than heparin.

- Since carboxyl-reduced heparin fails to induce the activity of any of the FGFs, the carboxyl groups of the heparinoids are required for activity .

---

[3] All of the heparin derivatives are described in Chapter 4.

Table 9.5. Effect of Heparinoid Structure on FGF/FGFR Complex Formation [1]

FGF	Cell type	Heparin		HS	6-O-DS-heparin		2-O-DS-heparin		AMS-heparin		CR-heparin		Ref.
		Poly	Oligo	Poly	Poly	Oligo	Poly	Olig	Poly	Oligo	Poly	Olig	
1	3T3	+4			—		—						[624]
1	ACE	+4	+4 (10)	+	+	+ (10)	—	—	+2	+3 (10)	—	—	[366]
2	3T3	+4	+4 (12)		—		—						[624]
2	ACE	+4	+4 (10)	+3	+3	+4 (10)	—		+3	+4 (10)	—	—	[367]
2	F32 [2]	+4	+2 (6)				+4						[560]
4	3T3	+4	+4 (10)	+6	+6		—		+4	+4 (10)	—		[624]
4	ACE	+4	+4 (10)	+2	+2	+ (10)	—		+4	+4 (10)	—	—	[366]

[1] If cell is blank, the heparinoid was not tested. A dash indicates that the heparinoid was inactive. The effects of both polysaccharides (poly) and oligosaccharides (oligo) are shown. Numbers in parentheses indicate the minimum sized oligosaccharide yielding the level of activity indicated.

[2] HS= heparan sulfate; 6-O-DS-heparin=6-O-desulfated heparin; 2-O-DS-heparin=2-O-desulfated heparin; AMS-heparin=amidomethylsulfonate-heparin; CR-heparin=carboxyl-reduced heparin. These derivatives are described in Chapter 4.

[3] F32 is a lymphoid cell line that is dependent on heparinoids for mitogenic activity

332

- The results obtained with the 2-O-desulfated and the 6-O-desulfated heparins vary with both cell type and FGF type. The most surprising results are those with FGF-4, where, in contrast to FGF-1 and FGF-2, both of the desulfated heparinoid derivatives are active.

- The negative charges of the carboxyl groups can be replaced by the negative charges of the amidomethylsulfonate substituents. However, AMS-heparin is somewhat less active than heparin in FGF-1 and FGF-2 assays.

For each cell type, the differences for the different FGFs cannot be attributed to the FGFR, which does not change. Also, it should be noted that no studies have been carried out on the structural features of heparinoids that are necessary for binding to FGF-4 alone. This fact and the different behaviors of Swiss 3T3 cells and ACE cells indicate that there is a great deal more to be learned about heparin interactions with FGFs and their receptors.

3.   Effects of heparinoids and heparinoid oligosaccharides on normal FGF-mediated cell growth (i.e., in the presence of FGF but not chlorate)

The results described above utilize cells whose heparan sulfate metabolism is altered by treatment with chlorate. Although this work helps in the understanding of heparinoid effects in FGF/FGFR recognition, it is also instructive to consider the effect of heparinoids on cells that are expressing normal levels of both heparan sulfate proteoglycan and FGFR. ACE cells, with their endogenous heparan sulfate proteoglycan and FGFR, are induced to grow by FGF-1, FGF-2, or FGF-4 alone; i.e., no exogenous heparinoid is required. When these FGFs are added to ACE cells at levels that induce maximal DNA synthesis or cell growth, quite distinct results are obtained, depending on which FGF is supplied [366]. When FGF-2 is used, maximal responses are obtained without added heparinoid. This indicates that the heparan sulfate proteoglycan expressed by these cells is quite adequate to support FGF-2 binding to the FGFR. However, at levels of heparin greater than 8 µg/ml (much higher than the 0.25 µg/ml required to get maximal response in chlorate-treated ACE cells), cell growth is inhibited. Presumably these levels of heparin are so high that the FGF-2 and the FGFR molecules are bound to different chains of the heparin or the endogenous heparan sulfate. This prevents the template action of the heparinoid.

When the same experiments are carried out with cells that are treated with FGF-1- and FGF-4 in the absence of exogenous heparin, the growth is poor and low levels of exogenous heparin actually stimulate cell growth. Identical results are obtained using deca- and dodecasaccharides prepared from heparin. However, higher levels of heparin once again inhibit ACE cell growth. These are interesting results because they suggest that the structures of the heparan sulfate chains on the surface of the ACE cells are adequate to support maximal rates of growth in the presence of FGF-2 but are not adequate to support growth with FGF-1 and FGF-4. Exogenous heparin supplies the necessary structures for FGFR-1 and FGFR-4. This adds support to the suggestion that the structure of the cell surface heparan sulfate chains can play a role in controlling the responses of the cell to different heparin-binding growth factors.

Such results are helpful in trying to understand the nature of the interactions among FGFs, FGFRs, and heparinoids, but they could have physiological significance as well, since the high concentrations of heparan sulfate proteoglycans on the surface of cells may, in fact, block the binding of some growth factors to receptors [1297]. It is apparent that such inhibitory activity would depend on the structure of the heparan sulfate chains.

G.   The FGF/FGFR Interaction

1.   The effect of heparin on the binding of FGF-2 to FGFR-1

Several groups have studied the effects of heparinoids on the interactions of FGF and FGFR [661, 1299-1302]. It has been shown that FGF-2 binds to cells at both high affinity sites (FGFRs) and low affinity sites (heparan sulfate proteoglycans). The kinetic on-rates for the formation of FGF/heparinoid complexes and FGF/FGFR complexes are similar, but the off-rate for the FGF/FGFR complex dissociation is much lower than the off-rate for the FGF/heparinoid complex dissociation. Thus, the higher affinity of FGF-2 for FGFR-1 appears to be due largely to a low kinetic off-rate for the FGF-2/FGFR-1 dissociation [1300, 1301]. However, the high affinity binding of FGF-2 to FGFR-1 depends on the presence of heparinoid, i.e., the high binding affinity is observed only in the presence of the heparinoid. The heparinoid is not *required* for binding of FGF-2 to receptor; it simply increases the affinity of the binding [1302]. Similar results have been reported for binding of FGF-1 to its receptor [1299].

Pantoliano *et al.* have used calorimetry to study the binding interactions of the components of the FGF/FGFR/heparin complex and have reported that the complex that forms contains two mols of FGFR-1 and one each of FGF-2 and heparin. As discussed below, when heparin is present, FGF-2 has two binding sites for FGFR-1—a primary and a secondary binding site. The binding constants for the formation of this complex are shown in Table 9.6. In the absence of heparinoid, there is strong binding of FGF-2 to its receptor, FGFR-1 (reaction 1), and the K_d for this binding is eight-fold lower in the presence of low molecular weight heparin (reaction 2). The binding of the second mole of FGFR-1 (reaction 3) is much weaker than the binding of the first. Also shown

Table 9.6. Binding Interactions in the Formation of
FGFR-1/FGF-2/Heparinoid Complexes [1]

Rxn	Reaction	K_d (nM)
1	FGFR-1/FGF-2 → FGFR-1 + FGF-2	41
2	FGFR-1/FGF-2/LMWH → FGFR-1 + FGF-2/LMWH [2]	4.9
3	(FGFR-1)₂/FGF-2/LMWH → FGFR-1 + FGFR-1/FGF-2/LMWH	1900
4	FGF-2/LMWH → FGF-2 + LMWH	470
5	FGFR-1/LMWH → FGFR-1 + LMWH	85,000

[1] from reference [661].

[2] low molecular weight heparin.

are the constants for the binding of FGF-2 and FGFR-1 to low molecular weight heparin (reactions 4 and 5). The binding of FGF-2 to heparin is quite strong, but the binding of FGFR-1 is much weaker. In the absence of heparin (reaction 1), FGFR-1 binds to its primary binding site on FGF-2 (below). In the presence of heparinoid, apparently the binding at the secondary site (low affinity site) is enhanced. It has been reported that heparin induces a small conformation change in FGF-2 [1303]. This conformation change may correlate with the increase in binding affinity at the secondary binding site.

2. Binding domains

Just as the FGFs and the FGFRs contain specific domains for their interactions with heparinoids, they must also present specific surface domains for interacting with each other. As discussed in Section III.A.2, p. 308, the specificity for binding of different members of the FGF family by FGFRs appears to lie in Ig-like domains II and III, with the greatest contribution from the variable C-terminal half of domain III. Little is known about the actual amino acid sequences in FGFRs that make contact with FGFs.

In contrast, there is information on the amino acid sequences of FGFs that are involved in FGFR recognition. In fact, we have noted that there are *two* domains on FGF-2 that recognize FGFR. The early work of Baird *et al.* [535] demonstrated that the FGF-2 peptide fragment comprising amino acids 106-115,

Y R S R K Y S S W Y,
106   110   115

is the smallest peptide that blocks binding of FGF-2 to FGFR without much effect on FGF-2 binding of heparin. This sequence appears as a surface-exposed loop in FGF-2. Consistent with these results, Reich-Slotky *et al.* [1304] have prepared chimeras of FGF-2 and FGF-5 (KGF) and have shown that the N-terminal portions of these two FGFs play important roles in receptor recognition and that residues 91-110 of FGF-5 also contribute to receptor recognition of the KGFR.

A different approach to the identification of FGFR-binding domains of FGF-2 has been taken by Springer *et al.* [1282] who examined the three-dimensional structure of FGF-2 and noted two regions that might be involved in FGFR binding. One is the region containing the above 106-115 sequence on one side of the FGF-2 molecule, and the second is an unusual cluster of hydrophobic amino acids on the opposite surface of the protein. These residues were replaced individually with alanine, and the mutant FGF-2s were tested for FGFR-1 binding, heparinoid binding, and mitogenic activity in Swiss 3T3 fibroblasts. As anticipated, none of these substitutions affect heparinoid binding. Curiously, in the absence of heparin, changes in which residues 109-114 are replaced individually with alanine give no detectable change in binding of FGF-2 to FGFR-1. Nevertheless, these latter changes *do* decrease the mitogenic activity of FGF-2, indicating that these residues are important for cell proliferation, even though they did not appear to participate in FGFR-1 binding in the absence of a heparinoid.

The second potential FGFR-1 binding site on FGF-2, namely the cluster of hydrophobic amino acids, was also examined. This cluster contains Y24, Y103, L140, and M142 and is surrounded by several basic amino acids, including R22, K26, R44, and K46. Single replacements of the hydrophobic amino acids with alanine have no effect on heparinoid binding, but give a 10- to 400-fold reduction in FGF-2/FGFR-1 binding. A double replacement, L140A/M142A, gives a 1500-fold decrease in FGFR-1 binding. When the surrounding basic amino acids are changed, only the R44A change affects FGFR binding, giving a 3- to 4-fold decrease in binding. These substitutions that reduce FGF-2/FGFR-1 binding also result in a marked reduction in the mitogenic activities of FGF-2. The authors conclude that FGF-2 contains a *primary* FGFR binding site, which includes Y24, Y103, L140, and M142, and a *secondary* FGFR binding site, which includes, at least in part, K110, Y111, and W114. But why is the secondary site called a binding site at all? The answer to this question lies in the fact that, *in the presence of a heparinoid*, FGF-2 binds two mols of FGFR-1; i.e., the heparinoid induces the FGFR-1 occupation of the secondary binding site. Neither FGF-2 alone nor heparin alone induce FGFR-1 dimerization. The results obtained when heparin is added to a mixture of FGF-2 and FGFR-1 will be discussed in greater detail in Section IV.H.2., p. 337. However, at this point it may be useful to recall that site-directed mutations in both the primary and the secondary FGFR-1 binding sites on FGF-2 destroy the mitogenic activity of the growth factor, which was measured in cells that express both FGFR and heparan sulfate proteoglycan. Therefore, the authors conclude that formation of this quaternary complex containing FGF-2, HSPG, and two mols of FGFR-1 is necessary for mitogenesis.

H.   Proposed Mechanisms of Interactions in FGF/FGFR/Heparinoid Complexes

1.   Introduction

The initial observations that led to the concept that heparinoids stabilize the binding of FGFs to their receptors were the findings that the mitogenic activity of FGF-2 is lost in cells that do not have typical levels of normally-sulfated heparan sulfate proteoglycan. Cell surface heparan sulfate is further implicated in the FGF activity since mutants that exhibit diminished levels of cell surface heparan sulfate show a reduced sensitivity to FGF-2 [15, 559, 560, 582, 586]. Furthermore, as we have discussed, both the FGFs and the FGFRs have heparin-binding domains. Finally, the mitogenic activity of FGF-2 requires oligosaccharide sequences that are longer than the shortest oligosaccharides that bind to FGF-2. This suggests that the interaction of FGF-2 with its receptors requires a heparinoid sequence long enough to bind to both the FGF-2 and the receptor. Thus, a model like the one shown in Figure 9.3 represents an early view of the role played by heparinoids. More recent work has led to several additional models. All of these invoke the generally accepted view that FGFR must be induced by FGF to form dimers in order to transmit the growth factor signal into the cell. One of these models suggests that FGFR dimers are formed as a result of the binding of two molecules of FGFR to one molecule of FGF in the presence of a heparinoid. Two additional models suggest that the heparinoid induces FGF dimerization, or stabilizes FGF dimers, and that the FGF dimers can then bind two molecules of FGFR to yield the FGFR dimers.

2.   Dimerization of FGFR by a single molecule of FGF

Based on the results described in Section IV.G., p. 334, Pantoliano *et al.* [661] have proposed that FGF-2 forms a *quaternary* complex with heparan sulfate and *two* mols of FGFR-1. This model is shown in Figures 9.5, 9.6, and 9.7 (see color inserts). Before FGF binds to the cells, the HSPG and the FGFR are present on the cell surface in a complex like that illustrated in Figure 9.5 (see color insert). Although the K_d for the FGFR/heparan sulfate interaction is relatively high (Table 9.6), the high concentration of HSPG on the cell surface would normally drive the FGFR/heparan sulfate association. According to a recent report, this binary complex is restrained by Ca^{2+} or Mg^{2+}, which stabilizes the FGFR/HSPG complex and prevents the transphosphorylase activity of the

FGFR [659].[4] FGF binds to the FGFR/HSPG complex to give the ternary FGF/FGFR/HSPG complex shown in Figure 9.6 (see color insert). In this complex, the FGFR is bound to the secondary FGFR-binding site on the FGF, and the heparan sulfate now is bound to both FGFR and FGF. Finally, a second mole of FGFR binds to the primary FGFR-binding site of FGF to give the quartenary complex shown in Figure 9.7 (see color insert). In this final complex, the original restraints on the tyrosine kinase activity are released, and transphosphorylation can occur [659].

In examining the part of the complex formed with the FGF-2 and the FGFR-1 that is bound to the secondary binding domain, the elongated cluster of nine basic amino acids in FGFR-1 (Section IV.C., p. 322) becomes aligned with the heparin-binding site on FGF-2 (Section IV.B., p. 318) such that the two proteins can be docked to a single strand of heparinoid that is 60 Å long (13 monosaccharides), as shown in Figure 9.6. The heparinoid spans the two proteins, contacting both. Interestingly, the symmetry of the model with respect to the *two* FGFR-1s in Figure 9.7 suggests that the heparin-binding sites on the two FGFRs and the FGF-2 are aligned. However, the model does not show the heparinoid strand extending across the second FGFR-1. This is because the two FGFR-1 molecules are in a mirror image orientation. Thus, the N- to C-terminal sequences of the heparin-binding peptides on the two FGFR-1 molecules would run in opposite directions. However, a single heparinoid strand can run in only one direction; i.e., from its nonreducing terminal to its reducing terminal direction (or vice versa!). Thus, if the heparin strand can bind to FGFR-1 in only one orientation, it would be impossible for a single heparinoid to bind to both FGFR-1s. Furthermore, we have already shown that the segment of heparin required to form a mitogenically active complex containing FGF-2, FGFR-1, and heparin need be only two disaccharides longer than the minimum size required for binding to FGF alone. This small increment in size may allow simultaneous binding to FGF and one FGFR, but it is not of sufficient length to accommodate the FGF plus *two* FGFRs.

3.   Sucrose octasulfate induction of dimer formation

The above model shows only a single FGF-2 molecule in the complex with the heparinoid and two FGFR-1s. Although it seems to fit a great deal of data in the literature, Ornitz *et al.* [560] have reported that in the presence of heparin but not FGFR, FGF-2 forms dimers. FGF-2 dimer formation occurs over a range of FGF-2 and heparin concentrations, but requires that the molar ratio be 1 heparin (average molecular weight) to 2 FGF-2s. The equilibrium lies far in

---

[4] The requirement for the metal ion for stalibilation of the FGFR/HSPG complex had not been reported at the time that this original model was proposed.

the direction of the FGF-2 monomers. Whether this finding represents a significant departure from the above model is unclear. Given the high number of FGF-2 binding sequences in a single heparin chain, it is reasonable to anticipate that two or more mols of FGF-2 could bind side by side to one chain of heparin, and, in fact, this is observed in the binding studies (Section IV.D., p. 325). The probability that two molecules of FGF-2 would bind to heparan sulfate, the physiological heparinoid, would seem to be less (see Section IV.B.1., p. 318)

The idea that heparinoids can induce FGF dimer formation has been extended by Zhu *et al.* [1305] who have described a 2.7-Å crystal structure for a 1:1 complex of FGF-1 and sucrose octasulfate, a heparin analog. It had been shown previously that sucrose octasulfate binds tightly to FGFs and stabilizes these growth factors [1303, 1306]. The crystal structure in Figure 9.8 (see color insert) shows that sucrose octasulfate binds to the 112-127 amino acid sequence of FGF-1, with contacts at K112, R116, Lys118, and R122, a region described above as part of the heparin-binding domain of FGF-1 (Section IV.B.1., p. 318). Two molecules of sucrose octasulfate appear to induce FGF-1 dimer formation in the crystals, with the two FGF-1 molecules related by a two fold rotation (not seen in FGF-1 crystals lacking sucrose octasulfate), and two molecules of sucrose octasulfate cross-linking the two FGF-1s (Figure 9.8a). The authors suggest that a single heparinoid octasaccharide could replace the two sucrose octasulfate molecules, accommodating FGF-1 dimerization with a single strand of heparinoid, as shown with such an octasaccharide overlapping the two sucrose octasulfates in Figure 9.8b. The modeling suggests that a heparin disaccharide would accommodate most of the binding contacts found in each of the 112-128 sequences of the dimer, but the octasaccharide would be the minimal size that would span the distance required to induce dimerization. Finally, in Figure 9.8c, the sucrose octasulfate is completely substituted by the octasaccharide, and the FGFR- binding domains of the two FGF-1 molecules are shown in green at positions opposite the positions of heparin binding. Thus, according to this model, a short segment of a heparinoid would induce FGF dimerization through interactions with the heparin-binding domains of both molecules of FGF-1, leaving the FGFR-binding domains of the FGF molecules available for unhindered binding of two molecules of FGFR, thus forming an FGFR dimer. There is one caveat, however. Although the model shown in Figures 9.8b and 9.8c appears to be symmetrical, it is not. For the single strand of heparin that runs from its nonreducing terminal to its reducing terminal, the two FGFs in the model are bound in opposite directions, i.e., one FGF-1 runs from its N- to its C-terminal along the heparin, while the other runs from its C- to its N-terminal.

4. Induction of dimer formation by unsulfated di- and trisaccharides

In the discussion to this point, the results for all of the FGFs that have been tested have indicated that heparin oligosaccharides that induce mitogenic activities of the FGFs must be (a) several disaccharides longer than the smallest oligosaccharide required to bind to the growth factor alone, and (b) highly sulfated. In this light, it is interesting to consider results by Ornitz et al. [560, 1298] who have reported the effects of various heparinoids on the FGF-2-mediated DNA synthesis in a lymphoid cell line, F32. These cells express neither FGFRs nor HSPG. Thus, when they are transfected with FGFR-1, they are dependent on exogenous heparinoids for DNA synthesis in the presence of FGF-2. As found in the studies described above, highly sulfated heparin-derived octasaccharides and higher oligosaccharides induce DNA synthesis in these cells, whereas the corresponding hexasaccharides do not. A more recent report on F32 cells now shows that several synthetic *unsulfated heparinoid di- and trisaccharides* are able to stimulate FGF-2-mediated DNA synthesis in F32 cells [1298]. These results, recorded in Table 9.7, show that the methyl glycosides of two GlcNAc-containing trisaccharides are almost as effective as heparin in inducing DNA synthesis in F32 cells! In fact, sucrose octasulfate, which behaves as a heparin mimic, is much less active than these unsulfated trisaccharides. Several other di- and trisaccharides shown in Table 9.7 have lower but significant activities, whereas rather modest changes in the structures of the active trisaccharides result in a complete loss of activities. In fact, some of the changes that result in the reduction of activities are the addition of N- or O-sulfates. Preliminary crystallographic studies show that the disaccharide GlcA→GlcNAc-OMe binds to FGF-2 at two sites, one of which is

Table 9.7. Stimulation of FGF/FGFR Complex Formation by Synthetic Oligosaccharides

Heparinoid [a]	Relative mitogenic activity
Heparin	1.0
GlcA→GlcNAc→GlcA-OMe	1.2
IdoA→GlcNAc→GlcA-OMe	3.2
IdoA→GlcNSO$_3$→GlcA-OMe	471
IdoA→GlcNSO$_3$-OMe	70
GlcA→GlcNAc-OMe	26
GlcA→GlcNSO$_3$-OMe	25
GlcNAc→GlcA-OMe	464
IdoA→GlcNAc→GlcA→Gal-OMe	Inactive
IdoA→GlcNAc-6-SO$_4$→GlcA→AMan$_R$	Inactive
GlcA → GlcNAc-6-SO$_4$ → GlcA AMan$_R$	Inactive
IdoA → GlcNAc-OMe	Inactive
Sucrose octasulfate	1381

[a] All of these structures have the same linkages as found in heparin and heparan sulfate. Data from [1298].

similar to the heparin-binding site, and that both di- and trisaccharides can induce dimerization of FGF-2 [1298].

More extensive crystallography, as well as modeling studies of FGF-2 and co-crystals of FGF-2 and IdoA→GlcNAc→GlcA-OMe has shed further light on the effects of these small unsulfated oligosaccharides [1307]. FGF-2 by itself associates to form dimers and perhaps oligomers. Such FGF-2 dimers are stabilized by binding of the unsulfated trisaccharide at an interface between the two FGF-2 molecules that is distinct from the heparin-binding site; i.e., these FGF-2 dimers are different from those described in Figure 9.8. Furthermore, the true heparin-binding sites of the FGF-2 molecules in the dimer are readily accessible, and the binding of heparin to these sites further stabilizes the dimer. Also, FGFR-binding sites on the FGF-2 dimer are accessible so that one molecule of the receptor can bind to each FGF-2 molecule in the dimer, resulting in the dimerization of the FGFR that is necessary for signal transduction. Thus, according to this model, the minimal FGF/FGFR/heparinoid complex required for signal transduction would contain two molecules each of FGF and FGFR and one strand of a heparinoid octa- or decasaccharide. In contrast to the other models, this model would account for FGF signaling at high FGF concentrations in the absence of heparinoid. Furthermore, since these FGF dimers are aligned head to tail, they can bind the heparinoid with the same alignment of the heparinoid along each FGF molecule (nonreducing terminal to reducing terminal, and N- to C-terminal). However, both this model and the previous one seem to discard the idea that heparinoid binding to the FGFR plays a role in FGF-mediated activities.

5.   FGF-2 signaling at high heparin concentrations

The three models described above are not mutually exclusive, and it is possible that signal transduction by FGFs can take place by several mechanisms. However, no single model seems to accommodate all of the data in the literature concerning the interacting domains of FGF, FGFR, and heparan sulfate and the effects of modifications of these domains on signal transduction. Clearly further explorations of the signaling question are needed.

The problem is further complicated by a report by Krufka et al. on the effects of high heparin concentrations on signal transduction [1160]. In this study, Swiss 3T3 and F32 cells that produce FGFR and HSPG were treated with chlorate, resulting in dramatic reductions in (a) FGF-2 binding to the cells and (b) the response of the cells to FGF-2. At optimal concentrations of exogenous heparin, the FGF-2 binding to receptor and the mitogenic responses of the chlorate-treated cells to FGF-2 were restored. At higher heparin concentrations, where separate heparin chains can bind to FGF-2 and to FGFR, binding of FGF-

2 to the receptor and the receptor transphosphorylation were both diminished as expected. However, *the mitogenic signaling was not altered.* The authors conclude that the high-affinity signaling complex that is necessary for receptor transphosphorylation differs from the signaling complex that triggers mitogenesis. They suggest that heparan sulfate participates in two hierarchies of receptor activation *in vivo*: (a) heparan sulfate binds FGF-2 to FGFR and activates mitogenic signaling, perhaps without receptor dimerization; and (b) heparan sulfate stabilizes an FGF/FGFR complex that contains FGFR dimers or multimers. As the authors note, this work adds weight to the suggestion that more than one model may operate in the transduction of FGF signals.

V.   INVOLVEMENT OF HEPARINOIDS IN FGF METABOLISM: CAVEATS

A.   FGF Signaling *in Vivo*

Under normal circumstances, cells expressing both FGFR and heparan sulfate proteoglycan may exhibit the full array of responses to FGFs. This is what one would expect to occur *in vivo*, where FGF responses are controlled by the mechanisms that regulate the expression of (a) the types, amounts, and variant forms of the receptor, (b) the structures and amounts of the heparan sulfate chains, and (c) the amounts and types of FGF that are accessible.

Heparan sulfate proteoglycans are generally available on the surfaces of most cells. However, if the heparan sulfate *structure* does not support complex formation for the FGFs and/or the FGFRs that are available, then the cells cannot respond to the FGFs. Heparan sulfate synthesis is dependent on the cellular expression of one of the proteoglycan core proteins described in Chapter 2, although it appears that the cell type and not the specific core protein determines the *structure* of the heparan sulfate that is synthesized. The structures of the heparan sulfate chains vary from one cell type to the next. Heparan sulfate species that have been tested for mitogenic activity have shown relatively low activity when compared to the activities with heparin, yet it is clear that the endogenous cell surface heparan sulfate elicits functional responses to FGFs. How can we resolve this apparent dilemma? There are several considerations. First, not all heparan sulfate chains from a given source contain the sequences necessary to bind to the growth factor. This is reflected in the findings that only a fraction of the total heparan sulfate that is applied to an FGF-2 affinity matrix binds to the column and that the proportions that bind vary significantly with the source of the heparan sulfate, ranging from 90% of porcine aorta heparan sulfate to 54% of EHS tumor heparan sulfate [625]. Thus, if heparan sulfate activity is compared at the same concentrations used for optimal heparin activity, and only a fraction the heparan sulfate sample is active, then the heparan sulfate is being tested at a suboptimal level.

A second point is that it may not be necessary to obtain a maximal response to FGF in order for FGF to fulfill its physiological roles. In this connection, it has been shown that different heparan sulfate proteoglycan species have quite different activities in the stimulation of FGF-2 binding to cells [1297]. In addition, it has been shown that the structures of the heparan sulfate chains on the core protein produced by neuroendothelial cells of developing mouse embryos exhibit binding affinities that are selective for FGF-2 when this growth factor is being expressed on embryonic day 9. However, on embryonic day 11, when the cells switch to FGF-1 expression, they start to produce, *on the same core protein*, heparan sulfate chains that are selective for FGF-1 [1308]. Thus, the mechanisms that control the responses of cells to FGFs are dynamic processes that regulate the changes in the types of the FGFs, the types and structures of the FGFRs, and the structures of heparan sulfate proteoglycans. Furthermore, these results indicate that significant progress in understanding the interactions of heparinoids with FGFs and FGFRs should focus on endogenous heparan sulfates rather than on exogenous heparin derivatives.

It is these *in vivo* responses that one seeks to duplicate in cell culture. The experimental approaches that have been used for the *in vitro* work have contributed to an understanding of FGF activities, but there are several caveats about these approaches that researchers should recognize.

B.   Heparan Sulfate vs Heparin in FGF Signaling

1.   Differences in structures of endogenous HSPG and exogenous heparin

We have a good picture of the *minimal* heparinoid structure that is required to bind to FGF-2, but almost nothing is known about the heparinoid structures that bind to the other FGFs, or to any of the FGFRs. A minimal structure will allow docking of a heparinoid to an FGF so that actual contacts can be deduced. Such modeling, for FGF-2 at least, will also be facilitated by the knowledge that addition of any further sulfate groups to the minimal structure does not significantly reduce the FGF-2/heparinoid binding. Of the four types of FGFRs, the putative heparin-binding domains are highly conserved (Table 9.4), suggesting that all FGFRs may bind to the same, or to very similar, heparinoid sequences. However, because the heparinoid sequences that bind to FGFRs have not been determined, the possible similarities of these sequences need to be tested experimentally.

When the heparinoid structures required as a template for binding to *both* FGF and FGFR are considered, the stringency of the heparinoid structural requirements will be greater than that for either FGF or FGFR alone. Significant differences in the structurally modified variants of *heparin* that support mitogenesis with the different FGFs have already been reported (Table 9.5). However, it is heparan sulfate—not heparin—that mediates the actual physiological roles of the heparinoid in these activities. Heparin and heparan sulfate are quite different structures. Heparin and the FGF-stimulating oligosaccharides prepared from it are highly sulfated along their entire lengths. Heparan sulfates, however, are block structures with blocks of unsulfated disaccharides interspersed with blocks of sulfated disaccharides. The sulfated blocks are not as highly sulfated as those of heparin and, generally, they appear to extend for only 6-8 monosaccharide residues. Thus, if highly sulfated deca- or dodecasaccharide sequences are required for the FGF- and FGFR-mediated responses, then one must question whether the sulfated blocks that are found in cell surface heparan sulfates actually extend for 10 to 12 monosaccharide units. This is the same question that we asked in regard to the heparan sulfate-mediated thrombin/antithrombin complexation in Chapter 7, and it will be a recurrent theme as further heparin-binding proteins are investigated at this level.

The specific involvement of heparan sulfate in the responses of cells to FGFs notwithstanding, there has been a tendency for workers in this area to try to determine the heparinoid structures required for FGF responses using heparin and its derivatives rather than heparan sulfates. As noted in Chapter 6, it is impossible to mimic heparan sulfate structures with heparin or any derivatives that can be prepared from it. However, heparin is readily available, and heparan sulfates are less available (and more expensive). Nevertheless, it is clear that questions concerning the actual functional heparinoid structures in FGF responses would be better addressed using the heparan sulfates produced by the cells that are being studied. A viable alternative would be to use "heparin by-products," which are true heparan sulfate mimics, with a range of block structures (Chapter 2), and which can be obtained in larger amounts. Approaches for preparing fragments of these block structures are described in Chapters 3 and 4.

2. Differences in cellular metabolism of FGF in the presence of endogenous HSPG and exogenous heparin

There is another problem in using heparin to study the activities of FGFs and FGFRs. In cells that have no cell surface heparan sulfate chains (mutant cells, or chlorate-treated cells), exogenous heparin will serve to replace the cell surface heparan sulfate chains in potentiating FGFR-mediated responses to FGF. This yields useful information, but also raises several concerns. First, *in vivo,* most cells would never encounter heparin, which is (a) produced by mast cells, (b) much more extensively sulfated than a typical heparan sulfate, and (c) more active with FGFs than heparan sulfate. When heparin is added to cells that produce their own HSPG, a nonphysiological condition is established, since both heparan sulfate proteoglycan, which is anchored in the cell membrane, and heparin, which is not, are available to interact with (and compete for) FGF and FGFR.

Heparin may also be a concern during the endocytosis of FGF. It may be that FGF can be internalized in the presence of heparin as long as both FGF and FGFR are present, but this is uncertain. Certainly, in the absence of FGFR, exogenous heparin will not be very effective in mediating FGF endocytosis. In any event, if the core protein of the HSPG plays any role in the internalization of FGF, exogenous heparin will not be able to mimic this effect. Even though exogenous heparin mediates a number of cellular responses to FGFs, it is not clear how much understanding may be lost by substituting heparin for the endogenous HSPG. For example, a critical step in the induction of a cellular response to FGF appears to be the dimerization of FGFR. Several mechanisms for FGF induction of dimerization have been discussed, but an additional mechanism may occur in the presence of heparin. High heparin levels can cause FGF dimerization (and oligomerization) as a result of binding several FGF molecules *side by side*, on a single strand of heparin, and these side-by-side FGFs may induce FGFR dimerization or oligomerization. This mechanism, which was not considered in Section IV.G., p. 334, may result in atypical cellular responses.

Also, since heparin generally binds FGFs more strongly than the endogenous heparan sulfates, exogenous heparin may inhibit any FGF responses that depend on internalization on heparan sulfate proteoglycan in the absence of FGFRs. The effects of exogenous heparin on those responses depend on the heparin concentrations; i.e., at high heparin concentrations, where some heparin chains bind to the FGF and different chains bind to the FGFR, FGF/FGFR/heparinoid complex formation is blocked, as is binding of FGF to cell surface FGFR or to heparan sulfate proteoglycan. Thus, high concentrations of exogenous heparin would be expected to reverse the FGF effects on cells, although this effect is not always observed (see Section IV.H.5., p. 341). The effects of heparin that are observed *in vitro* may offer useful insights, but they do not reflect physiological conditions.

The two approaches used to avoid this competition—use of mutant cells that do not produce endogenous HSPG, and use of chlorate to suppress sulfation of the endogenous HSPG—also generate non-physiological conditions. These may or may not, cause problems.

3.   Heparan sulfate structures formed in the presence of chlorate

Heparinoid structures that are synthesized by cells in the presence of chlorate are quite different from any structures that are normally synthesized either *in vivo* or *in vitro*. chlorate-treated cells produce an unsulfated heparan proteoglycan (Chapter 5). The polysaccharide chains that accumulate in the proteoglycan are [GlcA→GlcNAc]$_n$ chains in which a significant proportion of the GlcNAc residues are de-N-acetylated and ready to receive a sulfate group (Chapter 2). Such products have been identified in chlorate-treated cells in culture. These products, which have anionic carboxyl groups on the GlcA residues, cationic amino groups on many of the GlcN residues, and very few IdoA residues, apparently appear on the cell surface [5] but lack the ability to bind FGFs and FGFRs. Thus, although chlorate "disarms" the cell so that it cannot respond to the growth factors, it also results in the synthesis of an unnatural heparan sulfate structure that appears on the cell surface. Since heparan sulfate binds to many structures on and near the surface of cells (collagen, fibronectin, laminin, thrombospondin, etc.), the structure of the cell surface and the extracellular matrix may be altered significantly. Furthermore, all cellular responses that depend on heparan sulfate (e.g., other growth factors) will be altered, even though one observes only those changes that are measured.

4.   Proteoglycan structures of mutant cells that do not synthesize heparan sulfate chains

Chlorate-treated cells are different from the CHO cell mutants that produce modified heparan sulfate structures or no heparan sulfate at all. The most commonly used CHO cell mutants have defects in the enzymes that add the necessary linkage region xylose and galactose residues needed to initiate glycosaminoglycan chain synthesis. Consequently, the core proteins without any heparinoid chains attached appear on the surfaces of these cells (see Chapter 5). As in chlorate-treated cells, the cell surface and extracellular matrix would be significantly altered. It is likely that such mutations would be lethal *in vivo*. Despite these differences between chlorate-treated cells and CHO cell mutants, both cell systems seem to respond similarly to FGFs in the presence of exogenous heparinoids.

---

[5] The fact that ClO_4^--treated cells make normal levels of heparan proteoglycan has been demonstrated by showing that [^{3}H]glucosamine, which is incorporated into these polymers, appears in a proteoglycan with attached heparan chains.

VI. SUMMARY

The heparan sulfate-mediated antithrombin/thrombin complexation represents the prototype for both understanding of heparinoid-protein interactions and the methodologies that are required to investigate such interactions. Although the heparinoid/FGF/FGFR interactions represent the second most extensively studied examples of heparin-binding proteins, there is a great deal more to be learned about the latter interactions before we reach the same level of understanding as has been developed in the antithrombin case. The FGF studies are in a relatively early stage compared to the antithrombin studies that were initiated in the early 1970s. There are also some clear differences between the two cases. Perhaps most notable is the fact that the effects of antithrombin reactions are observed either on the cell surface or in the plasma, whereas the FGF reactions are initiated on the cell surface but are manifest largely inside the cell, with the important mitogenic effects apparently requiring transport, by poorly understood mechanisms, to the nucleus. Consequently, much of the development of the heparin-antithrombin field could be conducted with (relatively) abundant proteins in the test tube, whereas the FGF studies have required much more emphasis on cell culture approaches. Also, the multiplicity of FGFs and FGFRs presents a greater challenge in the FGF case. Despite these differences, the two stories come to similar conclusions, namely, that the initial interactions of the proteins that are involved occur on the surface of cells and that two proteins that have some modest affinity for each other are stabilized by their binding side by side on the cell surface heparan sulfate template.

In the FGF case, the receptor for FGF is represented by a complex between an FGFR and heparan sulfate proteoglycan on the cell surface. What advantage might such a heterodimeric complex offer over the simple high affinity receptor, which, itself, appears in four forms, each of which can undergo alternate splicing to yield variants that modulate FGF responses (Section III.A.2., p. 308). If one assumes that there are specific and relatively rare binding sequences in the heparan sulfate for each protein, the most obvious advantage would be that any alteration of heparan sulfate structure that may occur during its rapid turnover (see Chapter 5) would allow a high degree of control of the cellular response to FGFs that could not be obtained with the high affinity receptor alone. Thus, the model allows for a finely tuned cellular response to FGFs, since the FGFRs as well as the heparan sulfate structures can be rapidly modified in the cellular responses to the environment. Finally, perhaps the advantage in having a variety of forms of FGF is the distinct responses to each FGF that can be obtained with different cell types, thus allowing tissues to use the FGFR and heparan sulfate variabilities to generate precise and discriminating responses to FGFs.

ADDITIONAL REFERENCES

Aviezer, D., Iozzo, R. V., Noonan, D. M., and Yayon, A. (1997) Suppression of autocrine and
    paracrine function of basic fibroblast growth factor by stable expression of perlecan
    antisense cDNA, *Mol. Cell. Biol.* **17**, 1938-1946.
Blunt, A. G., Lawshé, A., Cunningham, M. L., Seto, M. L., Ornitz, D. M., and MacArthur, C. A.
    (1997) Overlapping expression and redundant activation of mesenchymal fibroblast
    growth factor (FGF) receptors by alternatively spliced FGF-8 ligands, *J. Biol. Chem.* **272**,
    3733-3738.
Bonneh-Barkay, D., Shlissel, M., Berman, B., E., S., Admon, A., Vlodavsky, I., Carey, D. J.,
    Asundi, V. K., Reich-Slotky, R., and Ron, D. (1997) Identification of glypican as a dual
    modulator of the biological activity of fibroblast growth factors, *J. Biol. Chem.* **272**,
    12485-12421.
Cizmeci-Smith, G., Langan, E., Youkey, J., Showalter, L. J., and Carey, D. J. (1997) Syndecan-4 is a
    primary-response gene induced by basic fibroblast growth factor and arterial injury in
    vascular smooth muscle cells, *Arterioscler. Thromb. Vasc. Biol.* **17**, 172-180.
Delehedde, M., Deudon, E., Boilly, B., and Hondermarck, H. (1996) Heparan sulfate proteoglycans
    play a dual role in regulating fibroblast growth factor-2 mitogenic activity in human
    breast cancer cells, *Exptl. Cell Res.* **229**, 398-406.
Faham, S., Hileman, R. E., Fromm, J. R., Linhardt, R. J., and Rees, D. C. (1996) Heparin structure
    and interactions with basic fibroblast growth factor, *Science* **271**, 1116-1120.
Friedl, A., Chang, Z., Tierney, A., and Rapraeger, A. C. (1997) Differential binding of fibroblast
    growth factor-2 and -7 to basement membrane heparan sulfate. Comparison of normal
    and abnormal human tissues, *Am. J. Pathol.* **150**, 1443-1455.
Hata, J.-I., Takeo, J., Segawa, C., and Yamashita, S. (1997) A cDNA encoding fish fibroblast growth
    factor-2, which lacks alternative translation initiation, *J. Biol. Chem.* **272**, 7285-7289.
Ishihara, M., Kariya, Y., Kikuchi, H., Minamisawa, T., and Yoshiba, K. (1997) Importance of 2-O-
    sulfate groups of uronate residues in heparin for activation of FGF-1 and FGF-2, *J.
    Biochem.* **121**, 345-349.
Jayson, G. C., and Gallagher, J. T. (1997) Heparin oligosaccharides: inhibitors of the biological
    activity of bFGF on Caco-2 cells, *Brit. J. Cancer* **75**, 9-16.
Mohammadi, M., McMahon, G., Sun, L., Tang, C., Hirth, P., Yeh, B. K., Hubbard, S. R., and
    Schlessinger, J. (1997) Structures of the tyrosine kinase domain of fibroblast growth
    factor receptor in complex with inhibitors, *Science* **276**, 955-960.
Perez-Castro, A. V., Wilson, J., and Altherr, M. R. (1997) Genomic organization of the human
    fibroblast growth factor receptor 3 (FGFR3) gene and comparative sequence analysis
    with the mouse fgfr3 gene, *Genomics* **41**, 10-16.
Pineda-Lucena, A., Jiménez, Á., Lozano, R., Nieto, J. L., Santoro, J., Rico, M., and Giménez-
    Gallego, G. (1996) Three-dimensional structure of acidic fibroblast growth factor in
    solution: effects of binding to a heparin functional analog, *J. Mol. Biol.* **264**, 162-478.
Romero, A., Pineda-Lucena, A., and Giménez-Gallego, G. (1996) X-ray structure of native full-
    length human fibroblast-growth factor at 0.25 nm resolution, *Eur. J. Biochem.* **241**, 453-
    461.
Shi, J., Friedman, S., and Maciag, T. (1997) A carboxyl-terminal domain in fibroblast growth factor
    (FGF)-2 inhibits FGF-1 release in response to heat shock *in vitro*, *J. Biol. Chem.* **272**,
    1142-1147.
Song, H. H., Shi, W., and Filmus, J. (1997) OCI-5/rat glypican-3 binds to fibroblast growth factor-2
    but not to insulin-like growth factor-2, *J. Biol. Chem.* **272**, 7574-7577.
Suzu, S., Kimura, F., Matsumoto, H., Yamada, M., Hashimoto, K., Shimamura, S., and Motoyoshi,
    K. (1997) Identification of binding domains for basic fibroblast growth factor in
    proteoglycan macrophage colony-stimulating factor, *Biochem. Biophys. Res. Commun.*
    230, 392-397.
Taipale, J., and Keski-Oja, K. (1997) Growth factors in the extracellular matrix, *FASEB J.* **11**, 51-59.

Tumova, S., and Bame, K. J. (1997) The interaction between basic fibroblast growth factor and heparan sulfate can prevent the *in vitro* degradation of the glycosaminoglycan by Chinese hamster ovary cell heparanases, *J. Biol. Chem.* **272**, 9078-9085.

Chapter 10. Extracellular Superoxide Dismutase

I.   SUPEROXIDE GENERATION DURING THE INFLAMMATION PROCESS

Inflammation is a complex process by which tissues respond to injury or infection. The process involves the migration of leukocytes from the circulation through the endothelium to the site of injury and the destruction of invading organisms or damaged tissue [1309, 1310]. Neutrophils, which normally do not produce superoxide ($\cdot O_2^-$), become activated when they reach the site of infection and express a "respiratory burst" in which they consume oxygen at a markedly increased rate, converting it to $\cdot O_2^-$. The pathways for the formation and metabolism of $\cdot O_2^-$ are shown in Figure 10.1. The enzyme that converts O_2 to $\cdot O_2^-$ catalyzes an NADPH-coupled one-electron reduction of O_2 at the cell surface [1311]. The enzyme is dormant prior to the activation of the neutrophils. Actually, $\cdot O_2^-$, which is quite toxic, does not persist in the tissues but is converted to H_2O_2, $\cdot OH$, and HOCl, all of which are very reactive species that play a significant role in the microbicidal action of the neutrophils. As shown in Figure 10.1, two moles of $\cdot O_2^-$ undergo a spontaneous dismutation reaction that leads to the formation of H_2O_2 and O_2. The same reaction is *catalyzed* by

351

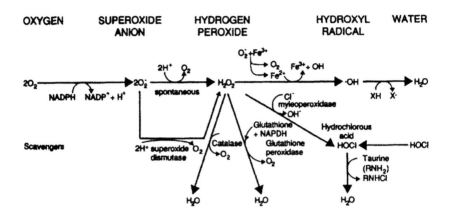

Figure 10.1. Pathway for the production of various activated oxygen species. Reprinted from [1310], with permission.

superoxide dismutases. H_2O_2 can be reduced to ·OH by ferrous ions or to HOCl by myeloperoxidase. When these active oxygen species are no longer required, the H_2O_2 may be destroyed by catalase or glutathione peroxidase. Thus, superoxide dismutases plays a significant role in the microbicidal activity of neutrophils.

II.   SUPEROXIDE DISMUTASE ISOZYMES

There are three superoxide dismutases in mammalian tissues [1312]. CuZn-superoxide dismutase (CuZn-SOD), the first described superoxide dismutase, occurs in all eukaryotes [1313] and in bacteria [1314, and references therein], and is found in the cytosol and nucleus of all cell types [1315]. CuZn-SOD is composed of two identical, 16 kDa subunits, each of which contains one Cu^{2+} and one Zn^{2+} atom [1312, 1313]. Its activity is sensitive to cyanide.

Mn-superoxide dismutase (Mn-SOD) is found in bacteria and in the mitochondrial matrix [1316]. It contains four identical, 22-kDa subunits, each with one Mn^{2+} atom [1312, 1313]. It shows no sequence homologies with CuZn-SOD [1317]. Its activity is insensitive to cyanide.

The isozyme of primary interest here is extracellular superoxide dismutase (EC-SOD). It was originally described by Marklund as a cyanide-sensitive superoxide dismutase activity in extracellular fluids that does not respond to antibodies against CuZn-SOD [1318]. It is found in extracellular matrices, associated with heparan sulfate proteoglycan [1319-1321], with much smaller amounts in the plasma, lymph, synovial fluid, and cerebrospinal fluid [1322-1324]. EC-SOD, in contrast to CuZn-SOD and Mn-SOD, is a glycoprotein that binds to concanavalin A and to wheat germ agglutinin. Treatment of EC-SOD with endoglycosidase F, an enzyme that releases N-linked oligosaccharides, reduces the apparent molecular weight of the enzyme [1325]. The enzyme can be separated from the other superoxide dismutases by lectin affinity chromatography [1318, 1325].

EC-SOD, CuZn-SOD, and Mn-SOD are not immunogenically cross-reactive. The tissue distribution of the three isozymes has been compared in a number of mammalian species [1326]. Of the three isozymes, only EC-SOD is secreted. The latter is produced by only a few cell types, notably fibroblasts and glial cells [1327]. The location of these three isozymes suggests that each plays a role in its own compartment, since $\cdot O_2^-$ is not readily transported across membranes. The physiological role of these enzymes is discussed in Section V., p. 361.

III. THE STRUCTURE OF EC-SOD

A. General Structural Features

EC-SOD occurs in several variant forms, one of which binds tightly to heparin [1318, 1325]. The heparin-bound form retains its activity. EC-SOD is composed of four identical 30-kDa subunits and has a pI of 4.5 [1328]. The enzyme contains one mole of Cu^{2+} and one mole of Zn^{2+} per subunit. It is much more sensitive (approx. three-fold) to cyanide and azide than CuZn-SOD. It is inactivated by H_2O_2, diethyldithiocarbamate (a Cu^{2+} chelator) and sodium dodecyl sulfate [1329].

The human placental EC-SOD has been cloned and sequenced [1330]. Each subunit is composed of 222 amino acids, with 18 additional amino acids that represent the putative signal sequence that is removed before secretion. The N-terminal half of the molecule contains one N-glycosylation site at Asn 89, but shows no homology with any other proteins. The C-terminal half (residues 96-193) shows about 50% sequence similarity with the C-terminal two-thirds of eukaryotic CuZn-SODs. Thus, some inferences concerning the three-dimensional structure of EC-SOD may be drawn from by examining the three-dimensional structure of CuZn-SOD [1331]. The crystal structure of CuZn-SOD has allowed the identification of the amino acid side chain ligands for Cu^{2+} and Zn^{2+}, the cysteines that form an intrasubunit disulfide bridge, and an arginine at the entrance to the active site. These critical residues are all conserved in EC-SOD. Thus, the active site of EC-SOD is presumed to be very similar to that of CuZn-SOD.

Interestingly, rat C_6 glioma cells produce an EC-SOD that is somewhat distinct from the human EC-SOD [1324, 1332]. This enzyme is a homo*dimer* of an M_r 34- to 36-kDa subunit that is highly processed as it passes through the secretory pathway. The subunits contain 224 amino acids plus a 20 amino acid signal sequence. The enzyme has a pI of 6.1-7.2—significantly higher that the pI 4.5 of human EC-SOD. In addition, the rat enzyme exhibits a lower heparin affinity than the EC-SOD from other mammalian species. Despite these differences, the rat enzyme has an overall homology with the human enzyme of 68% and a higher homology with the catalytic active site and the heparin-binding site (below) of the human enzyme.

B.   Heparin Binding by EC-SOD

1.   Introduction

In the initial report of the purification and properties of EC-SOD, it was noted that highly purified superoxide dismutase that is isolated by lectin affinity chromatography, and that gives a single band on polyacrylamide gel electrophoretograms, can be further resolved into three fractions by heparin affinity chromatography [1318]. The literature refers to these three fractions as types A, B, and C of the enzyme. Fraction A does not bind to heparin-Sepharose, fraction B binds but is eluted with 0.2 $\underline{M}$ NaCl, and fraction C binds and is eluted with 0.5 $\underline{M}$ NaCl. A variety of other glycosaminoglycans have been tested for their binding to EC-SOD by measuring the concentration of NaCl required for elution of EC-SOD from glycosaminoglycan-substituted Sepharose columns [1320]. Approximate NaCl concentrations required for elution from chondroitin sulfate, dermatan sulfate, heparan sulfate (0.7 SO₄/disaccharide), heparan sulfate (1.5 SO₄/disaccharide) and heparin (2.3 SO₄/disaccharide) are 0.1, 0.2, 0.28, 0.4, and 0.47 $\underline{M}$. Thus, binding to non-heparinoid glycosaminoglycans is minimal under physiological salt conditions, whereas the concentration of NaCl required for elution of EC-SOD from the various heparinoids increases with the increasing degree of sulfation. There is no difference in binding affinity of EC-SOD for the antithrombin-binding and antithrombin nonbinding fractions of heparin. The displacement of EC-SOD from these glycosaminoglycans with NaCl indicates that the binding is electrostatic in nature, probably involving the sulfate and carboxyl groups of the heparinoids. Attempts to fractionate *heparin* on EC-SOD-Sepharose (to determine whether there are specific structural motifs in the heparin that are involved in the binding) have not been carried out.

The binding of heparin by EC-SOD is reflected in other ways as well. For example, the activity of EC-SOD is inhibited approximately 15% by heparin and, to a lesser degree, by heparan sulfate and other glycosaminoglycans. The heparin inhibition is reversed by protamine [1333]. In addition, the apparent molecular weight of EC-SOD is markedly increased in the presence of heparin, as indicated by gel filtration studies [1333]; i.e., the heparin and EC-SOD chromatograph together as a complex.

2.   The heparin-binding domain of EC-SOD

Even though it has a pI of 4.5, the enzyme interacts with heparin primarily through electrostatic interactions,. Thus, one might expect to find a region of localized positive charge in EC-SOD. Such a region has been identified in the C-terminal region of the enzyme, which contains 10 basic amino acid side chains with a cluster of six of these at positions 210-215 [1334, 1335]. The C-terminal sequence, with the basic amino acids underlined, is shown:

G P G L W E R Q A R E H S E R K K R R R E S E C K A A
196   200   210   220

Computer analysis of the sequence suggests that it will form an α-helix [S. L. Marklund, cited in 1336]. Note that the Gly-Pro-Gly sequence at 196-198 would tend to form a β-bend in the polypeptide sequence. Such a bend is presumed to turn the heparin-binding domain of the human protein out into the solvent so that it is much more available for heparinoid binding.

Several observations implicate this sequence as the heparin-binding domain of EC-SOD. When a few lysine or arginine residues are derivatized with trinitrobenzene sulfonic acid or phenylglyoxal, respectively, heparin binding is lost [1333]. This sequence is further implicated in heparin binding by the finding that treatment of recombinant EC-SOD C, which has a high affinity for heparin, with trypsin or endoproteinase Lys-C results in cleavage at K211 with loss of heparin binding [1335]. Furthermore, the transfection of a variety of site-directed mutant cDNAs—some with stop codons in the C-terminal region and some with amino acid substitutions—into CHO cells leads to the production of modified EC-SODs with reduced or lost affinity for heparin [1334]. These studies suggest that residues beyond Glu 216 play little role in heparin binding and that the six-amino acid cluster of contiguous Lys and Arg residues plays a major role. It may be noted that this sequence does not conform to the consensus heparin-binding sequences proposed by Cardin and Weintraub (Chapter 6). Furthermore, since EC-SOD is a homotetramer, the heparin binding may result from cooperative interactions of two or more of the subunits with the heparin, as described below.

Further evidence for the importance of this sequence in heparin binding comes from the finding that a human phenotypic variant form of the enzyme, which has a reduced affinity for heparin, contains a mutation in which Arg 213 is replaced with Gly [1337]. The lowered heparin affinity apparently results in a reduced binding of the EC-SOD to the endothelial heparan sulfate proteoglycan [1338]. This variant form of EC-SOD also is more resistant to proteolytic cleavage than the normal EC-SOD [1339].

3. The multiple forms of EC-SOD

Fractions A, B, and C of EC-SOD exhibit increasing affinities for heparin. However, the human genome appears to contain a single gene for EC-SOD that codes for the full 222 amino acid C type EC-SOD subunits. Furthermore, only the C type EC-SOD is produced by a variety of human cell lines [1327, 1340]. This raises the question of the origin of the A and B types of the enzyme. An examination of the subunit structure of these forms shows that all subunits were of either the A type (no heparin binding) or the C type (high heparin affinity). This suggests that the B type of EC-SOD is a heterotetramer containing both A and C type subunits. Indeed, when the cells were cotransfected with *both* A and C type cDNAs, the B type isomer was formed by the cells.

The C type enzyme can be converted to A and B type enzymes by treatment with trypsin, plasmin, or endoproteinase Lys-C [1335, 1340]. These enzymes give a stepwise loss of the heparin-binding domain from the subunits of the tetrameric protein and a corresponding progressive loss of heparin affinity. However, the full enzymatic activity is retained. The cleavage is blocked by heparin or dextran sulfate, which apparently bind to the heparin-binding domain and block the access of the proteases. The ready conversion of the heparin-binding form of EC-SOD to heparin-nonbinding forms suggests that such proteolytic conversions of the C form occur *in vivo* to give rise to the A and B forms of the enzyme. Indeed, a study of the structural features of EC-SOD from human umbilical cords has shown that the forms with reduced heparin affinity are composed of subunits that are truncated at the C-terminal ends; they lack Lys212-Ala222 [1325]. Thus, the present view of the A, B, and C forms is that the full-length C form of the enzyme is secreted by cells and bound to cell surface heparan sulfate proteoglycan. By regulatory mechanisms that are not fully understood, the C form is proteolytically cleaved to yield enzymes containing the C-terminal truncated subunit. Thus, one might expect to find tetrameric EC-SOD forms containing two types of subunits: one with the full-length sequence and a second with the cleaved C-terminal. This should result in five tetrameric forms of the enzyme with a range of heparin affinities. Indeed, Adachi *et al.* have demonstrated that five such forms can be resolved by heparin-HPLC [1341].

An interesting case is presented by the rat glioma cells (see Section III.A., p. 353) which produce only one form of EC-SOD [1324, 1332]. The amino acid sequence of the heparin-binding domain of the rat enzyme shows high sequence similarity with that of the human C form, but lacks the Gly-Pro-Gly sequence that flanks the heparin-binding site in the human EC-SOD and allows the bending of the heparin-binding domain out into the solvent. The rat enzyme does not bind heparin as tightly as the human C form. The authors suggest that the rat enzyme shows reduced heparin affinity because the heparin-binding domain in the rat EC-SOD is not as highly exposed to the solvent [1332]. The lack of the β-bend in the rat enzyme may also make the heparin-binding domain less accessible to proteolytic loss, thus preventing the formation of multiple forms of the enzyme.

The lack of a β-bend in the rat enzyme may be only part of the reason for its reduced heparin affinity. In contrast to the homotetrameric form of EC-SOD found in most mammalian species, the rat enzyme is a homodimer. Carlsson *et al.* [1342] have compared the amino acid sequences of the rat and mouse EC-SODs and have shown that a single amino acid in the N-terminal region of the enzyme, corresponding to Asp24 in the human enzyme, is essential for the formation of tetramers. When Asp24 in the mouse enzyme is converted to Val, the amino acid found in this position in the rat enzyme, the mouse enzyme forms a dimer that exhibits reduced heparin affinity. Conversely, when the Val in the rat enzyme is replaced with an Asp, the altered enzyme forms a tetramer with high heparin affinity. Thus, the four heparin-binding domains of the tetrameric form of the enzyme appear to act in a cooperative manner to enhance the heparin binding.

IV. METABOLISM OF EC-SOD

A. Metabolism *in Vivo*

1. Tissue distribution

EC-SOD occurs widely in all tissues of mammalian species [1324, 1326, 1343]. EC-SOD is the major isozyme of superoxide dismutase found in the extracellular fluids [1322]. Nevertheless, 90-99% of the total EC-SOD in mammals is found in the extravascular space of the tissues [1319, 1324, 1326, 1328, 1329]. Furthermore, EC-SOD is found in *all* mammalian tissues [1319, 1326], where it appears to be bound to cell surface heparan sulfate proteoglycans. The enzyme form in the tissues is the homotetrameric form, composed almost exclusively of the high heparin affinity type C subunit. However, the plasma EC-SOD is a mixture of heterotetramers made up of variable numbers of the full size C type subunits and subunits that have lost the heparin-binding domain. Only about 1% of the total EC-SOD is present in the vasculature, and this enzyme is apparently equilibrated between the surface heparan sulfate proteoglycan of the endothelium and the plasma.

2. Release of bound EC-SOD by intravenous heparinoids

The suggestion that the vascular EC-SOD is bound largely to the endothelium is supported primarily by the finding that intravenous injection of heparins or heparin-like polysaccharides causes an immediate increase of EC-SOD in the plasma of humans, dogs, pigs, rabbits, guinea pig, and mice [1324, 1328, 1341, 1344-1346]. The release occurs without a change in the total EC-SOD. Antithrombin-binding and antithrombin-nonbinding heparin fractions are equally effective in releasing the enzyme. The rat, which lacks the more potent heparin-binding species of EC-SOD (see Sections III.A. and III.B.3., pp. 353 and 356), does not exhibit such a heparin-induced rise. The degree of increase varies from 2- to 11-fold, depending on the species. The form of the enzyme that accounts for this release is the high heparin affinity form. The effect of heparin is reversed by protamine, which is used clinically to neutralize heparin [1344]. Also, protamine alone causes the release of the enzyme, probably by competing with EC-SOD for the endothelial cell surface heparan sulfate proteoglycan.

The half-life of the heparin-released enzyme in the plasma is 90 min, an interval that is much shorter than the 10-20 h $t_{1/2}$ for the normal turnover of the bound EC-SOD that results from its slow release into the plasma and its uptake and catabolism in the liver [1347, 1348]. The short half-life of the plasma EC-SOD after heparin administration probably results from its rapid re-association with the endothelial heparan sulfate proteoglycan as the heparin is rapidly catabolized (Chapter 5). Consistent with this conclusion, repeated injection of heparin results in a repeated elevation of EC-SOD [1341]. Interestingly, when type C (high heparin affinity) EC-SOD is injected subcutaneously or intramuscularly, it is retained locally in the tissue where it first appears and has a $t_{1/2}$ 10-fold longer than type A (nonheparin binding) EC-SOD [1340]. Thus, binding to the endothelium appears to sustain the levels of total EC-SOD in the vasculature. As suggested by these authors, it is likely that enhanced local proteolytic activity in tissues, e.g., as a result of inflammation, could lead to "a significant loss of EC-SOD from the interstitium."

3.   Variations in serum levels of EC-SOD

Significant differences in serum EC-SOD levels have been observed in different individuals [1349]. As noted earlier, alterations in the heparin-binding domain of EC-SOD result in diminished binding of the enzyme to the endothelial cell surface and a corresponding increase in the level of serum enzyme. With respect to serum levels of EC-SOD, individuals fall into two groups. Group I has less than 120 ng of enzyme/ml serum, with EC-SOD types A, B, and C present. Group II has greater than 400 ng enzyme/ml serum, and the enzyme is primarily of the C type. These variations run in families, suggesting that they are genetically transmitted. The most notable pathological condition in which EC-SOD levels were altered is renal disease [1350].

4.   Glycation and turnover

The loss of the heparin-binding domain clearly contributes to the heterogeneity of circulating EC-SOD, but one other modification of the protein adds further heterogeneity, namely non-enzymatic glycosylation of the enzyme ("glycation") [1336]. Glycosylation can be observed *in vitro* in the presence of varying concentrations of glucose. The reaction is blocked when the Lys residues are derivatized with trinitrobenzenesulfonic acid and proceeds more slowly in the presence of heparin. Furthermore, the derivatized enzyme retains its activity, but shows a diminished affinity for heparin. These results suggest that the nonenzymatic glycosylation sites are on Lys residues in the heparin-binding domain of the enzyme and not at the active site. The glycosylated enzyme can also be demonstrated in the serum of normal subjects and is markedly elevated in diabetic patients, even though serum levels of the enzyme are not elevated in the latter individuals.

B.   Metabolism in Cell Culture

The broad tissue distribution of EC-SOD raises the question of whether the enzyme is synthesized by all cell types or whether it is secreted by only a few cell types and taken up by the heparan sulfate proteoglycan on the surface of other tissues. This question has been addressed by Marklund [1327], who has measured the production of EC-SOD by a wide range of cell lines in culture. The results show that EC-SOD is expressed in only a few cell types—fibroblasts and glial cells, but not in endothelial cell lines, epithelial cell lines, or amnion-derived cells. The enzyme is also expressed in smooth muscle cells [cited in 1337]. Among fibroblast cell lines, the rates of synthesis vary over 100-fold. All of the cells that produce EC-SOD express the high heparin affinity C type of the enzyme. Particularly interesting is the finding that endothelial cells from human umbilical cords, a rich source of the enzyme, do not express the enzyme. Taken together, the findings suggest that EC-SOD, in contrast to CuZn-SOD and Mn-SOD, is produced in a limited number of cell types and that the secreted enzyme then passes through the circulation and is bound by cell surface heparan sulfate proteoglycan of a wide variety of cells.

EC-SOD has also been expressed in transfected CHO cells [1351]. Again, the product, a glycoprotein, is the C type enzyme and is secreted into the culture medium.

The uptake of EC-SOD by cells has been demonstrated by incubation of ^{125}I-labeled EC-SOD with cells in the presence and absence of heparin [1320]. Results show that all anchorage-dependent cells bind similar amounts of the enzyme. The binding is saturable and the bound enzyme is displaced by heparin, by highly sulfated heparan sulfate, or, to lesser degrees, by other glycosaminoglycans. In contrast to anchorage-dependent cells, *E. coli* and blood cells bind very little EC-SOD. These findings suggest that blood cells have no cell surface heparan sulfate proteoglycan and are consistent with the suggestion that EC-SOD, secreted by a limited number of cell types, can be taken up from the circulation and bound to cell surface heparan sulfate proteoglycan. Since EC-SOD retains most of its activity when bound to heparinoids, the cell-bound EC-SOD can protect host cells from the toxic effects of $\cdot O_2^-$ while bacteria, which lack this protective effect, can be destroyed.

Curiously, virtually all of the EC-SOD in cell cultures is recovered in the culture medium. This seems to be in contradiction to the finding that EC-SOD is readily taken up by cells. Interestingly, a significant portion (an average of 11%) of the total EC-SOD synthesized in these cultures is found *inside* the cells. This may mean that an unusually large pool of enzyme accumulates in the secretory organelles during passage to the culture medium. It could also mean that most of the cell lines that express EC-SOD make very little cell surface heparan sulfate proteoglycan, which would be expected to bind at least part of the secreted enzyme, but this does not seem very likely. A more likely scenario is that the intracellular pool results from the *endocytosis* of the secreted enzyme on heparan sulfate proteoglycan (see Chapter 5) and stabilization of the enzyme internally as an EC-SOD/heparan sulfate complex. This is what is found for fibroblast growth factors and other growth factors (Chapters 8 and 9).

Since fibroblasts, one of the primary cell types that produces EC-SOD, generally respond to cytokines, Marklund has tested the effects of various factors on the production of EC-SOD by human dermal fibroblast cells [1352]. Expression of the enzyme is stimulated by interferon-γ, but no effects are seen with interferon-α, granulocyte-macrophage colony stimulating factor, leukotriene B_4, prostaglandin E_2, formylmethionylleucylphenylalanine, platelet-activating factor, indomethacin, or interleukins-2, -3, -4, -6, or -8. Interleukin-1-α, tumor necrosis factor-α, and transforming growth factors-α and -β variably enhance or depress the interferon-γ effect. Generally, the responses are not immediate but take place over several days in culture. There is no correlation between the effects of these cytokines on EC-SOD production and the production of the other SOD isozymes.

V.   HSPG IN THE REGULATION OF EC-SOD ACTIVITY

A.   Introduction

It is interesting to ask what role heparan sulfate proteoglycan may play in the physiological activities of EC-SOD; i.e., does the binding of EC-SOD to heparan sulfate proteoglycan really make some important contribution to the specific functions of EC-SOD? In fact, the significance of the EC-SOD/heparan sulfate interaction remains a matter of speculation, but there are some general aspects of heparan sulfate proteoglycan behavior that have received little attention in the literature. For purposes of this discussion, it useful first to re-state the role of EC-SOD and then to consider how heparan sulfate proteoglycan may modulate the EC-SOD activities.

O_2^- is a very reactive species, capable of destroying both pathogens and host cells. It participates in the inflammatory reactions, which are designed to respond locally and specifically at sites of tissue injury. $\cdot O_2^-$ plays a role as a precursor of H_2O_2, $HOCl$, and $\cdot OH$. It also reacts with nitric oxide to form another potent oxidant, peroxynitrite [1353]. These reactive oxygen agents must be produced only at the site of injury and must be destroyed as soon as they are no longer useful. Otherwise, damage to normal tissue can result. Consequently, elaborate mechanisms operate to control inflammation and minimize tissue injury, but these mechanisms are imperfect and inflammatory responses can persist beyond their useful lifetimes.

The roles of the superoxide dismutases must be considered in terms of the overall flow of oxygen metabolites at the tissue injury site (Figure 10.1). The respiratory burst of neutrophils at the site of inflammation yields high levels of potentially harmful $\cdot O_2^-$ in the tissues. The spontaneous dismutation of $\cdot O_2^-$ to H_2O_2 and O_2 destroys the O_2^- and generates a reactive O_2 species, namely H_2O_2, for which efficient enzymatic breakdown mechanisms exist (catalase, myeloperoxidase). Thus, excess H_2O_2 does not persist in the tissues. However, the dismutation of $\cdot O_2^-$ is the major pathway for the destruction of $\cdot O_2^-$, and apparently the spontaneous dismutation, although a very rapid reaction, is not fast enough to shuffle the $\cdot O_2^-$ on to H_2O_2 and reduce the tissue levels of $\cdot O_2^-$ to nontoxic levels. This is especially true in pathologic conditions, where the levels of these toxic oxygen species are greatly elevated. Hence, the SOD-catalyzed dismutation appears to protect tissues from toxic $\cdot O_2^-$ levels. Presumably CuZn-SOD protects against intracellular $\cdot O_2^-$, Mn-SOD protects against mitochondrial $\cdot O_2^-$, and EC-SOD protects against extracellular $\cdot O_2^-$.

B. Is EC-SOD Really Necessary?

The spontaneous dismutation of $\cdot O_2^-$ is a very rapid reaction. Thus, one might question the need for SODs. Two types of demonstrations of the critical roles of the SODs have been reported: (a) those using animals that express reduced or elevated levels of EC-SOD, and (b) clinical studies with ischemia/reperfusion patients. In both types of studies, high O_2 tensions lead to the production of $\cdot O_2^-$ levels that are beyond the capacity of the body to respond. Animals tolerate normal O_2 tensions, but high O_2 tension leads to tissue destruction due to elevated levels of $\cdot O_2^-$. Even EC-SOD null mutant mice develop normally, with no blood abnormalities and no induction of other SOD isoenzymes or other antioxidant enzymes [1354]. However, when either normal animals or the EC-SOD null mutants are subjected to high O_2, the deleterious effects of $\cdot O_2^-$ are observed. The possible mechanisms of these effects have been discussed [1354].

It has been shown that overexpression of superoxide dismutases in transgenic mice enhances the survival of mice at high O_2 tension [1355, 1356]. Transgenic expression of both CuZn-SOD [1357] and Mn-SOD [1358] also prolongs the survival of mice in high oxygen tension. However, EC-SOD null mutant mice are significantly more sensitive to lung damage by high O_2 concentrations than wild-type mice [1354]. Thus, the SOD-mediated reduction of the concentration of $\cdot O_2^-$ appears to reduce the toxic effects that result from conversion of O_2 to $\cdot O_2^-$.

Reperfusion following ischemia also leads to tissue damage. During reperfusion, which re-introduces oxygen-rich blood into a circulatory system that has been partially damaged by the ischemia, oxygen free radicals are formed. These are key mediators of postischemic adhesion of leukocytes to postcapillary venules and the ensuing local tissue damage. The protective effect of SOD in ischemia/reperfusion has been demonstrated. [1345, 1359]. Intravenous injection of SOD exerts protective effects in a variety of pathophysiological models [1360]. It has also been shown that intravenous injection of smaller doses of recombinant CuZn-SOD at the time of reperfusion prevents the leukocyte adhesion and the resulting tissue damage [1345]. Thus, there are clearly opportunities for clinical usage of EC-SOD. In this connection, it is of interest that, by recombinant techniques, the cloned human EC-SOD has been used to produce EC-SOD in the mammary glands of transgenic mice [1346]. This suggests that the same technique could be used to produce large quantities of the enzyme in farm animals for therapeutic purposes.

C.   Effects of Heparan Sulfate Proteoglycan on EC-SOD

The only obvious role of heparan sulfate proteoglycan in the modulation of EC-SOD activity is to anchor the C type of the enzyme (up to 99% of the total) to cell surfaces. Does this offer any advantages or disadvantages over the circulating forms of the enzyme? It should be recalled that EC-SOD retains most of its catalytic activity even when attached to heparan sulfate (see Section III.B., p. 354); thus, both bound and circulating forms of the enzyme are active. There appear to be two advantages of the binding of EC-SOD to cell surfaces. One is the localized protection of cell surfaces from the effects of $\cdot O_2^-$. A second appears to be the extended $t_{1/2}$ of the enzyme. For example, when the type C (high heparan sulfate affinity) EC-SOD is injected subcutaneously or intramuscularly, it is retained locally in the tissue with $t_{1/2}$ 10-fold greater than that of the type A (nonheparan sulfate binding) EC-SOD [cited in 1340]). In fact, advantage has been taken of the extended $t_{1/2}$. Since CuZn-SOD is cleared rapidly, the human recombinant EC-SOD (rEC-SOD) offers advantages for clinical use since its $t_{1/2}$ seems to be extended as a result of its binding to the heparan sulfate proteoglycan on the endothelial cell surface [1361]. Similarly, a chimeric protein in which CuZn-SOD is coupled to the heparin-binding domain of protein C inhibitor through the Gly-Pro-Gly motif has been used to improve the therapeutic efficacy of CuZn-SOD [1362]. Also, substitution of CuZn-SOD with polylysine to facilitate association with negatively charged cell membranes highly potentiates the ability of the enzyme to protect activated polymorphonulear leukocytes against self-inactivation [1363]. Thus, the regulation of the toxic effects of $\cdot O_2^-$ by EC-SOD is potentiated by its binding to heparan sulfate proteoglycan, largely as a result of the extension of the $t_{1/2}$ of the enzyme.

Despite the perceived advantages of an EC-SOD with an extended half-life, it appears that a long $t_{1/2}$ may not be necessary for the well-being of the individual. As noted in Section IV.A.3., p. 359, a significant number of healthy persons have markedly elevated plasma EC-SOD, a result of an R213G mutation in the heparin-binding domain of the enzyme that occurs frequently (e.g., in 6% of the people in Japan) [1337, 1349]. All such individuals are heterozygous for the mutated subunit. The fact that alterations affecting the EC-SOD heparin-binding domain are apparently common in diverse populations suggests that they are not associated with any severe negative effects, at least not in individuals that are heterozygous for the mutation.

The appearance of types A and B EC-SOD in the circulation apparently results from local proteolytic activity in tissues. Circulating EC-SOD can also be elevated by injection of heparin (see Section IV.A.2., p. 358). In fact, intravenous injection of heparin gives protection against the $\cdot O_2^-$ effect in ischemia/reperfusion [1345]. Although this heparin effect apparently results from the release of EC-SOD from cell surface heparan sulfate proteoglycan *in vivo*, it has also been shown that heparin can inhibit the formation of $\cdot O_2^-$ by cultured neutrophils [1364]. Thus, heparin may have an effect both in preventing $\cdot O_2^-$ formation and in destruction of $\cdot O_2^-$ after it is formed. Furthermore, many other changes occur in the vasculature when heparin is injected (Chapter 8). Thus, it remains speculative as to whether proteolytic or heparin-induced release of EC-SOD from its cell anchorage plays some role in the scavenging of $\cdot O_2^-$. It may be that EC-SODs A and B have protective functions in the plasma, whereas C mainly acts as a protector on cellular surfaces [1324]. The importance of the putative cell surface association of EC-SOD C remains unclear. Whether it is necessary to displace the EC-SOD from the cell surface in order to realize its effects in the circulation is uncertain.

D.   Is EC-SOD Turned Over as a Result of HSPG Metabolism?

We have characterized the cell surface pool of EC-SOD as a reservoir of enzyme that equilibrates with the EC-SOD in the fluids that contact the cells. However, the cell surface heparan sulfate proteoglycan is in a constant state of metabolic flux, a result of its continuous secretion and endocytosis (Chapter 5). When heparan sulfate proteoglycan is endocytosed, proteins that are bound to its heparan sulfate chains (e.g., thrombospondin, the fibroblast growth factors, lipoprotein lipase, etc.) are endocytosed with it (Chapters 8 and 9). In fact, Ohta et al. [1365] have shown that endothelial cells degrade EC-SOD by internalization of the enzyme. Does the internalization occur with the enzyme complexed to heparan sulfate and, if so, does such complexation affect its activities? These questions have not been addressed. However, they do raise a number of questions:

- Does rapid endocytosis of EC-SOD account for the observed intracellular pools of this isozyme that are found in cultured cells?

- What percentage of the EC-SOD that is extracted from tissues is actually recovered from inside the cells?

- Does the amount of EC-SOD attached to cell surface heparan sulfate proteoglycan change as the structure of the heparan sulfate chains is altered by the cells? For example., what effect would increasing levels of ClO_4^- have on the secretion/endocytosis of EC-SOD?

- If EC-SOD does enter the cell, does it retain its activity—and thus play a protective role inside the cell?

- If EC-SOD enters the cells, is it catabolized in the lysosomes?

If the EC-SOD on cell surfaces is internalized and catabolized by cells, then the cell surface pool is a more dynamic pool than has been suggested. For example, although the endothelial cell itself can generate its own pool of heparan sulfate proteoglycan, it must rely on cells that produce EC-SOD for its supply of the enzyme. Consequently, as EC-SOD disappears into cells that cannot produce the enzyme, it must be replaced by EC-SOD synthesized by cells that can—primarily fibroblasts. According to this scenario, the surface pool of EC-SOD on cells that do not produce the enzyme would be constantly renewed by enzyme derived from source cells. The question then arises as to how such enzyme can be delivered continuously to the cells.

ADDITIONAL REFERENCES

Abdel-Aziz, A. F., and El-Naggar, M. M. (1997) Superoxide dismutase activities in serum and white blood cells of patients with some malignancies, *Cancer Lett.* **113**, 61-64.

O'Donnell, V. B., and Azzi, A. (1996) High rates of extracellular superoxide generation by cultured human fibroblasts: involvement of a lipid-metabolizing enzyme, *Biochem. J.* **318**, 805-812.

Tsan, M.-F. (1997) Superoxide dismutase and pulmonary oxygen toxicity, *Proc. Soc. Exp. Biol. Med.* **214**, 107-113.

Chapter 11. Heparin-Binding Proteins in Lipoprotein Metabolism

---

I. OVERVIEW

Triglycerides and cholesterol are transported to and from tissues by a number of lipoproteins—circulating particles composed of neutral lipids, phospholipids, and one or more "apolipoproteins" [1366-1370]. The apolipoproteins fall into four groups—A, B, C, and E—each of which includes subgroups.[1] The compositions and properties of the lipoproteins are shown in Table 11.1.

The lipoproteins are classified into groups based on their particle densities. As a general description, lipoproteins are composed of characteristic neutral lipids (triacylglycerols, cholesterol esters) which form an internal hydrophobic core that is coated with (a) phospholipids and (b) characteristic apolipoproteins.

---

[1] The protein designated "apolipoprotein D" was originally believed to be involved in lipoprotein metabolism, but was later found to be retinol-binding protein, a member of the lipocalin family [1371].

367

Table 11.1. Compositions and Properties of the Major Classes of
Lipoproteins [a]

Lipoprotein group	Diameter (nm)	Density (g/cm³)	Protein (% of wt.)	Major apolipoproteins	Major lipid
Chylo-microns	500-1000	<0.95	1-2	apoA-I, apoA-II, apoB-48, apoC's-II and III, apoE	TAG
VLDLs	30-70	0.95-1.006	7-10	apoB-100, apoC's-I, II, & III, apoE	TAG
IDLs	25-50	1.006-1.019		apoB-100, apoE	CE
LDLs	15	1.019-1.063	75	apoB-100	CE
HDLs	10-15	1.063-1.21	50	apoA, apoE, apoC-II	CE

[a] Abbreviations: VLDL, very low density lipoprotein; IDL, intermediate density lipoprotein; LDL, low density lipoprotein; HDL, high density lipoprotein; TAG, triacylglycerol; CE, cholesterol ester.

In performance of their metabolic roles, these lipoproteins circulate in the plasma and deliver both structural and metabolic lipids to, or remove them from, those tissues for which such lipid exchange is of physiological benefit. Our interest here stems from the fact that a number of the proteins that participate in the metabolism of lipoproteins are heparin-binding proteins. Furthermore, it is clear that cell surface heparan sulfate proteoglycan plays significant, though not yet fully defined, roles in the metabolic flow of triacylglycerols and cholesterol. These roles involve the interaction of cell surface heparan sulfate proteoglycans with (a) lipases, (b) several heparin-binding apolipoproteins that are exposed on lipoprotein surfaces, and (c) lipoprotein receptors.

Although the pathways involved in lipoprotein metabolism are quite complex, they may be presented in a general outline as shown in Figure 11.1. This figure focuses on the liver as the primary source and/or repository of the lipoproteins. It is clear from the precursor-product relationships of the

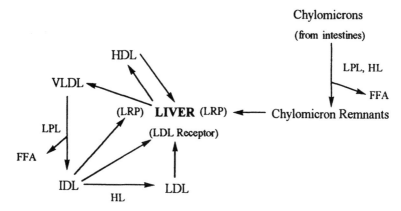

Figure 11.1. Metabolic pathways of lipoproteins

Abbreviations: VLDL, very low density lipoprotein; IDL, intermediate density lipoprotein; LDL, low density lipoprotein; HDL, high density lipoprotein;LPL, lipoprotein lipase; HL, hepatic lipase; LDLR, LDL recptor; LRP, LDL receptor-like protein.

lipoproteins, and from the lipoprotein compositions shown in Table 11.1, that the lipid and apolipoprotein compositions of many of the precursors are different from those of their products. This is because, as the lipid compositions of the lipoproteins are altered by the primary metabolic conversions, the apolipoproteins are transferred to other lipoproteins. Lipid transfers also occur. Lipid exchanges take place by two different mechanisms. One involves binding of the lipoproteins to cell surface receptors. The bound lipoproteins are metabolized on the cell surface by enzymes (that are also bound to cell surface receptors) or internalized by the cells and metabolized further in the lysosomes. The second type of lipid exchange is mediated by carrier proteins that facilitate transfer of triacylglycerols and cholesterol from one lipoprotein type to another [reviewed in 1372]. Thus, the metabolism shown in Figure 11.1 describes a continuum of changes in which the lipoproteins undergo alterations in their lipid and apolipoprotein compositions that are accompanied by changes in their sizes and their densities.

  Dietary (exogenous) lipids enter the circulation through the thoracic duct in the form of chylomicrons which are synthesized in the small intestine. Triacylglycerols make up 99% of the content of chylomicrons, which also contain the heparin-binding apolipoprotein E (apoE), as well as apolipoproteins B-48 (apoB-48) and C (apoC). Chylomicrons are involved in the delivery of dietary fatty acids to tissues. This occurs as a result of the binding of the chylomicrons to endothelial cell surfaces where their triacylglycerols are hydrolyzed by lipoprotein lipase, a heparin-binding protein on the surface of endothelial cells. The released fatty acids are taken up by the tissues. The resulting *chylomicron remnants* are taken up by the liver by a receptor-mediated endocytotic mechanism that involves hepatocyte cell surface heparan sulfate proteoglycan.

Very low density lipoproteins (VLDLs), also triacylglycerol-rich lipoproteins, contain apoB-100 and apoE, both of which bind to heparin, as well as several species of apoC. VLDLs, which are synthesized and secreted by the liver, deliver *endogenous* fatty acids to the tissues, again by binding to the endothelium where they are attacked by lipoprotein lipase. Special note should be taken of the fact that chylomicrons and VLDLs, the primary substrates for lipoprotein lipase, are the only lipoproteins that are rich in apoC, a cofactor for lipoprotein lipase. Lipoprotein lipase converts VLDL to *VLDL remnants* and intermediate-density lipoproteins (IDL). The latter conversion is accompanied by loss of apoC, which does not appear in IDL. The VLDL remnants are taken up by the liver; the IDL is converted by lipoprotein lipase and hepatic lipase (both are heparin-binding proteins) to low-density lipoproteins (LDL), which are rich in cholesterol esters. During this conversion, apoE is lost but apoB-100 is retained by the LDL. LDL is taken up by the liver, and by peripheral tissues, distributing cholesterol for use as a membrane component. The involvement of lipoprotein lipase in the metabolism of chylomicrons and VLDLs has been reviewed by Goldberg [1373].

High-density lipoproteins (HDLs) are secreted by the liver and the intestine. They may also be assembled from their components in the plasma or on cell surfaces. They (a) assimilate cholesterol transferred from other lipoproteins or released as a result of membrane turnover or cell lysis, (b) serve as a site for fatty acylation of the cholesterol by lecithin:cholesterol acyl transferase (LCAT), and (c) transport cholesterol esters to peripheral tissues or to the liver for final excretion. HDL contains mainly apolipoprotein A (apoA),[2] but it also carries small amounts of apoC and apoE. HDL appears to be a carrier of apolipoproteins that are synthesized (primarily) in the liver and transferred to other lipoprotein classes. Thus, chylomicrons and VLDL enter the circulation without apoC or apoE, but they acquire these apolipoproteins from HDL. The apoC-II residing on the surface of these two classes of lipoprotein is required as a cofactor for the lipoprotein lipase which converts them to chylomicron remnants and VLDL remnants, respectively.

As Figure 11.1 shows, chylomicron remnants, IDL, LDL, and HDL, at various stages in their metabolism in the plasma, are taken up and catabolized by the liver. The apolipoproteins that characterize each of these lipoprotein classes play roles in their uptake. Lipoprotein lipase and hepatic lipase are also endocytosed and catabolized by the liver [1374]. Normally, lipoprotein lipase

---

[2] ApoA is not to be confused with apo(a)—"apo-little-a"—described in Chapter 8.

(LPL) circulates in the picomolar range in plasma, almost exclusively in a form that is bound to lipoproteins [1375, 1376]. The amount of LPL in plasma increases after fat feeding or after infusion of triacylglycerol-rich emulsions [1377, 1378]. In fact, in normal lipoprotein metabolism, the free fatty acids released by the actions of the lipases on the lipoproteins cause the release of LPL from the endothelium [1379, 1380]. It is this released LPL that becomes associated with the cholesterol-rich lipoproteins in the circulation [1375, 1381] and is cleared by the liver [1382, 1383]. Not only is LPL itself cleared by the liver, but the LPL that is bound to circulating lipoproteins may promote the interaction of the associated lipoproteins with heparan sulfate and receptors on cell surfaces; i.e., LPL stimulates the uptake of the lipoproteins to which it binds [1384].

Consistent with the concept that these multicomponent lipoprotein complexes are intermediates in an elaborate series of metabolic conversions, there is heterogeneity with respect to size and composition within each lipoprotein class. This heterogeneity is a result of the dynamic processes in which the metabolic changes in the lipid components are accompanied by transfer of both lipids and apolipoproteins to other lipoproteins.

An emerging picture is one in which many of the lipoprotein-cell interactions involve the receptor-mediated binding of the lipoproteins to cell surface receptors. In some cases, these receptors require the presence of heparan sulfate proteoglycan as a receptor, or a co-receptor, a phenomenon that may be similar to that described in Chapter 9 for FGF/FGFR interactions. Thus, in addition to the typical protein receptors that are involved in endocytosis, heparan sulfate proteoglycan plays a co-receptor role in recognition and binding of both lipoproteins and lipases.

We will now turn our attention to a more detailed description of the heparin-binding proteins that participate in lipoprotein metabolism.

II. LIPASES

A. Introduction

We will concern ourselves here with three lipases that are involved in supplying lipids to tissues. The first is pancreatic lipase (PL) which processes dietary lipids in the intestine. The others are lipoprotein lipase (LPL) and hepatic lipase (HL), both of which are involved in the metabolism of circulating lipoproteins. LPL converts chylomicrons to chylomicron remnants and VLDLs to VLDL remnants and IDLs. In each case, the *primary* action of LPL is to release fatty acids from the tracylglycerols and the phospholipids of these lipoproteins. HL plays a primary role in these same activities in the liver, but plays a larger role in the hydrolysis of phospholipids during the conversion of IDL to LDL. A fourth member of this family in mammals has now been described, phosphatidylserine-phospholipase A$_1$ (PS-PLA1) [1385]. This enzyme, a heparin-binding protein that is secreted from activated platelets, is the first phospholipase that hydrolyzes fatty acids from the *sn*-1 position and has a strict head group specificity of the substrate. Its biological function is not known.

These four lipases are members of a family of proteins that are very similar in their primary and three dimensional structures [1370, 1386]. PL, LPL, HL, and PS-PLA1 exhibit 30% amino acid identities and 75-80% similarities [1387-1390]. Among species, each of the lipases is even more highly conserved [1391]. For example, bovine and human HLs have 93% identity and 99% similarity [1389]. In all of these lipases, there is a remarkable conservation of 10 cysteine residues, all of which are disulfide linked, suggesting a similar overall tertiary folding [1387]. Also, there is conservation of eight serine residues in LPL and HL across all species [1392]. There is maximum conservation in the internal regions of these proteins, especially in the hydrophobic residues of the central β-pleated sheets and the regions around the catalytic site [1388]. There is also a great deal of similarity in their cDNAs and their gene structures [1393-1395]. Thus, it is clear that much can be learned about each of these enzymes by examining the differences in their interactions and their activities in light of the similarities and differences in various domains of their structures.

B.   Comparative Properties of the Lipases

Because studies of each of these lipases have contributed to an understanding of the activities and interactions of the others, it is of interest to compare and contrast their properties at the outset of our discussion. Table 11.2 gives a compilation of these properties for the three well-characterized members of the family. The molecular sizes of the lipases are essentially the same, not only when compared to each other, but also when each of the lipases is compared among species. However, the terminal sequences may not line up. For example, when LPL is compared to PL, it is seen that LPL is missing the first 25 amino acids found in the N-terminal of PL, but has a 10 amino acid extension on the C-termimal [1390]. Since these enzymes are secreted glycoproteins, they are synthesized with typical signal sequences that are removed in the secretory pathway. The enzymes have two to four potential N-glycosylation sites.

Lipases are esterases that have catalytic triads composed of serine, histidine, and aspartic acid residues. Thus, they are related to other esterases and to the serine proteases (see Chapters 7 and 8). The positioning of the catalytic triad residues is very similar, and the regions in the area of the catalytic site are highly conserved. While PL is active in its monomeric form, LPL is active only as a homodimer. When this dimer dissociates to monomers or forms higher aggregates, enzyme activity is lost [1396]. HL has been reported to be active as a monomer or as a homodimer, depending on the tissue source [1397, 1398]. The "esterase" activities of these enzymes are functionally distinguished from the "lipase" activities by measurement of the *esterase* activities using water-soluble esters of small fatty acids (p-nitrophenyl butyrate, tributyrin) as substrates. *Lipase* activities are assayed using water-insoluble triacylglycerols, such as triolein [1399-1402]. These enzymes are inhibited by typical inhibitors of serine hydrolases, such as diisopropyl fluorophosphate [1385] and phenylboronates [1403].

LPL and HL differ in their substrate specificities and in their roles in lipoprotein metabolism. We have already mentioned that LPL, which is found on the surface of endothelial cells, acts on the triacylglycerols of triacylglycerol-rich lipoproteins, chylomicrons, and VLDLs. It also acts on phospholipids. HL, which is largely confined to the surface of hepatocytes, also acts on triacylglycerols and phospholipids, but is more active on phospholipids than on triacylglycerols [1390, 1404, 1405]. Both enzymes remove fatty acids from the 1 and 3 positions of *sn*-glycerol, and do not distinguish between these positions [1406]. Since heparin releases both LPL and HL from their sites on endothelial cells and hepatocytes, respectively, they appear in plasma together following heparin administration. Their separate activities can be distinguished, however, because the activity of LPL is almost completely inhibited by 1 $\underline{M}$ NaCl, whereas salt has only a minimal effect on HL activity [1405]. Also, HL can be selectively denatured by sodium dodecyl sulfate [1407].

Each of these lipases requires a protein cofactor. LPL requires *apoC-II*, HL uses *apoE*, and PL requires *colipase*. These protein cofactors are distinct and unrelated proteins, and they play different roles in the activity of their respective enzymes. The activity of PL, which is secreted by the pancreas into the small intestine, is inhibited by bile salts, which block the binding of the enzyme to micellar substrates. Colipase, an 11-kDa protein, restores this binding and is resistant to the effects of bile salts [1390, 1405]. However, colipase does not *stimulate* PL activity. In contrast, LPL activity is virtually dependent on activation by either free or lipoprotein-bound apoC-II [1400, 1405, 1408-1414]. Thus, LPL acts primarily on lipoproteins that have exposed apoC-II on their surfaces. Finally, although HL exhibits activity in the absence of a cofactor, its activity is stimulated three-fold by apoE [1415].

Table 11.2. Properties of Members of the Lipase Family

Property	Lipoprotein lipase	Hepatic lipase	Pancreatic lipase
No. of amino acids	Human: 448 Bovine: 450 Mouse: 447	Human: 476 Rat: 472	Human: 449
Monomer MW (without CHO)	55,000	53,200	
% Carbohydrate	8.3	~21	
Active form	Homodimer	Monomer/ dimer [a]	Monomer
Activity on TAGs	High	Modest	High
Activity on P-lipids	modest	high	none
Cofactor required	apoC-II	apoE	Colipase (11 kDa)
pI	4.0	4.1	
Effect of 1 $\underline{M}$ NaCl	Complete inhibition	Minimal	
Heparin binding? [b]	Yes (1.0-1.2 $\underline{M}$)	Yes (0.5-0.7 $\underline{M}$)	No
Crystal structure	No	No	Yes

[a] Depends on the tissue source.
[b] Numbers in ( ) refer to the concentration of NaCl required to elute the enzyme from immobilized heparin.

LPL and HL are heparin-binding proteins, but PL is not. As with other heparin-binding proteins that we have discussed, heparin affinity chromatography has been of great value in the purification of LPL and HL [1416-1418]. Although PL does not bind to heparin and plays no role in the metabolism of circulating lipoproteins, it has been of great value in the development of an understanding of the other two lipases. Most importantly, PL is the only member of the family whose crystal structure has been solved. Consequently, it serves as a model for the three-dimensional structures of the other two. Since the structures of heparin-binding lipases can be modeled on that of PL, it is possible to deduce many aspects of the interactions of the heparin-binding lipases from their similarities to, and differences from, PL. We will address these interactions below.

III. HEPARIN BINDING LIPASES

A. Introduction

The most extensively studied enzyme in these metabolic pathways is lipoprotein lipase. The structure, synthesis, and metabolic role of LPL have been reviewed [1370, 1373, 1419, 1420]. LPL is an enzyme that was first described as "clearing factor," a name reflecting its marked ability to reduce lipemia following intravenous heparin administration [1421]. LPL is a glycoprotein that is normally bound via HSPG to the surface of the capillary endothelium. It performs at least two critical roles in lipoprotein metabolism [1373]. First, it acts on chylomicrons and very low density lipoproteins (VLDL), releasing free fatty acids for uptake by the tissues. In performing this role, it remains bound to the endothelial wall, and its lipoprotein substrates bind to the enzyme. In effect, it tethers its substrates to the endothelial cell surface HSPG. The cleavage of triacylglycerols in these lipoproteins thus occurs on the cell surface, allowing the release of free fatty acids, which can then be taken up by cells that carry LPL. Alternatively, LPL facilitates the uptake of its substrate lipoproteins and their catabolism in the lysosomes. Thus, LPL plays a primary role in the delivery of fatty acids to cells in proportion to the levels of LPL on their surfaces. When exogenous heparin is administered, the heparin displaces the enzyme from the endothelium into the plasma—which normally contains only small amounts of enzyme. The released enzyme then binds to *circulating* lipoproteins and acts on them, yielding free fatty acids that no longer are directed specifically to the tissues that may need them most [1422]. The free fatty acids that are formed in the clearing action of LPL are taken up primarily by the liver.

A recently recognized role of LPL is its stimulation of the cellular binding and endocytosis of lipoproteins [1384, 1423-1427]. This activity is discussed in more detail below.

B. *In Vivo* Metabolism of LPL

1. General features of LPL metabolism

There is a single LPL gene that is not linked to the gene for HL [1394, 1395]. One, two, or three different mRNAs are found, the number varying with the species and the tissue [1387]. The mRNA for LPL is found in many tissues, but its levels are quite different and are high in cardiomyocytes and adipocytes. LPL is synthesized, not by endothelial cells, but by the parenchymal cells of muscle and adipose tissue. From its sites of synthesis, it is transported, by mechanisms that seem to involve heparan sulfate proteoglycan, across the extracellular matrix of the parenchymal cells and across basement membranes to the endothelial cells where it becomes associated with the heparan sulfate proteoglycan on the luminal surface of the cells [1387, 1389, 1428].

In contrast to LPL, HL is synthesized and secreted by hepatocytes and is captured by the heparan sulfate proteoglycan on the surface of these cells. HL is largely localized to the hepatocyte surfaces. When ^{125}I-labeled HL is injected into rats, it is taken up primarily by the liver [1382, 1383, 1429]. The failure of HL to bind to the endothelium suggests that it recognizes heparan sulfate structures that are unique to the hepatocyte cell surface.

HL and LPL are bound to lipoproteins in the plasma where they are found in only small amounts [1375, 1376]. It appears that these enzymes normally dissociate from their endothelium-bound complexes with heparan sulfate proteoglycan and form complexes with circulating lipoproteins that are taken up rapidly by the liver. Thus, the metabolic flow of LPL is as follows:

- The enzyme is synthesized by myocytes or adipocytes and transported to endothelial cell surfaces.

- The cell surface enzyme is endocytosed by the endothelial cells (below) and/or released slowly into the plasma where it is bound to circulating lipoproteins.

- Enzyme in the plasma is rapidly cleared by the liver, where it is catabolized.

HL, however, remains bound to the hepatocyte heparan sulfate proteoglycan, and any of this enzyme that dissociates into the plasma is rapidly recaptured, endocytosed, and catabolized. As a corollary, one would expect that the surface-bound HL is in a constant state of flux on the hepatocyte cell surface. Thus, it seems that both of these enzymes appear at steady state levels on the surfaces of their respective cells.

LPL is subject to developmental and dietary regulation. For example, the LPL level in the heart increases for several weeks after birth [1430]. LPL synthesis rates differ in fed and fasting animals [1431, 1432]. Also, the critical roles played by LPL in lipoprotein metabolism are reflected in the altered lipoprotein metabolism that is found in humans and other species that carry genetically defective forms of the enzyme [1433-1436]. Mutations that give rise to inactive forms of LPL result in hyperlipoproteinemia and high triacylglycerol levels in the plasma [1387, 1400, 1437].

2. The effect of heparin on LPL metabolism

When humans or other animals are treated with heparin, two lipases—HL and LPL—are displaced from the vascular endothelial and hepatocyte cell surfaces and released into the circulation [1429, 1438-1440]. Following their release, both lipases are ultimately taken up by the hepatocytes and catabolized. However, circulating heparin prevents the re-binding of the lipases to the hepatocyte surface heparan sulfate proteoglycan and slows the clearance of the enzymes. As the heparin disappears from the circulation (Chapter 5), the lipases are taken up and catabolized by the hepatocytes. Both unfractionated heparin and low molecular weight heparin release equal amounts of LPL, but the low molecular weight heparin is not as effective as the unfractionated heparin in preventing uptake in the liver [1409]. Thus, the lipase disappears from the circulation more rapidly when low molecular weight heparin is administered [1429]. This has been further demonstrated by injecting rats with heparin oligosaccharides of increasing sizes. A mixture of hexasaccharides and octasaccharides is the smallest fraction that will release LPL [1440, 1441], but quantitative release requires 10-times more of this fraction than of a heparin fraction with an average molecular mass of 12 kDa [1429]. Also, when the oligosaccharide fraction is injected, the elevated level of LPL in the plasma falls to normal levels after 1 hr, whereas the LPL level remains high after an hour when the high molecular weight heparin fraction is injected. In fact, when ^{125}I-LPL is perfused into liver, high molecular weight heparin blocks its binding and catabolism whereas the oligosaccharide fraction has no effect. Interestingly, there is no fraction that gives *selective* release of either of the lipases [1429]. Following the release of LPL by heparinoids, the rate of chylomicron clearance from the blood is markedly decreased [1440].

These results correlate well with the fact that the affinity of LPL for heparinoids parallels the increase in their effectiveness in releasing lipases *in vivo* [33, 1409, 1441]. Also, low molecular weight heparin and unfractionated heparin both bind quantitatively to immobilized LPL and are eluted in broad peaks with a NaCl gradient. Unfractionated heparin elutes at 0.4-0.7 $\underline{M}$ salt whereas low molecular weight heparin elutes at 0.3-0.6 $\underline{M}$ salt [1409].

3.   The effect of lipids on metabolism

Interestingly, lipid emulsions cause release of endothelial LPL, which rises gradually in the plasma after the emulsion is administered [1380]. In fact, in normal lipoprotein metabolism, the free fatty acids released by the actions of the lipases on the lipoproteins cause the release of LPL from the endothelium [1379, 1380]. The released LPL binds to circulating lipoproteins and is cleared by the liver [1375, 1381-1383]. It has also been reported that lysolecithin induces endothelial cells to secrete a heparanase-like activity that causes the release of depolymerized heparan sulfate from the cells, an effect that may contribute to the release of LPL [477].

C.   Metabolism of LPL in Cell Culture

1.   Introduction

The metabolism of LPL has been studied in a number of different cell lines, including cardiomyocytes and various adipocyte cell lines, or cells that differentiate into adipocytes in culture [634, 1442-1453]. Endothelial cells have also been useful in these studies, even though they themselves do not produce LPL [635, 1454-1456]. The results of these *in vitro* studies may be summarized as follows:

- LPL, synthesized in the endoplasmic reticulum as a monomer, becomes a homodimer as it passes to the Golgi and, more particularly, as the high-mannose oligosaccharides that are added in the endoplasmic reticulum are enzymatically modified.

- The secreted LPL binds to cell surface heparan sulfate proteoglycan. A portion of this is released into the culture medium pool, while the remainder is endocytosed and either recycled to the surface or catabolized in the lysosomes.

- Heparan sulfate proteoglycan plays a critical role in the routing of the secreted enzyme.

- Insulin and other hormones regulate the metabolism of LPL.

    We will now describe these observations in greater detail.

2. Cells that synthesize LPL

Several cell types differentiate into LPL-producing cells in culture. For example, 10T1/2 mouse embryo fibroblasts are induced to differentiate into adipocytes by 5-azacytidine [1443], 3T3-L1 fibroblasts differentiate into adipocytes when they grow to confluence [1444]; primary cultures of cells from rat epididymal fat pads differentiate into adipocytes in culture [1457]; and monocytes differentiate into macrophages that show a marked increase in LPL synthesis [1458]. Adipocytes formed in culture accumulate lipid droplets [1443] and show an elevated level of glycerol phosphate dehydrogenase, a marker enzyme for adipocytes [1443, 1445]. Lipoprotein lipase levels are quite low before differentiation, but increase markedly for several days after the adipocyte phenotype begins to appear [1444, 1445].

The rates of LPL synthesis in cultured cells are regulated by insulin, cAMP, thyroxin, catacholamines, tumor necrosis factor-α, tumor necrosis factor-α, and other hormones [634, 1387, 1444, 1450, 1457, 1459-1461]. In particular, insulin doubles the synthesis rate [1450]. In addition, glucose levels influence the rate of LPL secretion. In the presence of glucose, rat adipocytes produce LPL as a 55-kDa monomer. In glucose-free medium, however, a 49-kDa monomer with only 10% of the activity accumulates [1432, 1450]. This is probably due to the effects of glucose levels on N-linked oligosaccharide synthesis (below). However, it must be noted that metabolism of heparan sulfate proteoglycan is also markedly altered in glucose-limiting medium (Chapter 5), and this might have an effect on LPL metabolism.

In adipocytes and cardiac myocytes, newly synthesized LPL enters the secretory pathway in the endoplasmic reticulum and is transported in approximately 1 hr to the cell surface as a homodimeric, active enzyme. Although heparan sulfate does not appear to serve as a chaperone for the secretion of LPL [1462], the secreted enzyme is bound to heparan sulfate proteoglycan on the surface of the cells and is either released into the culture medium or endocytosed. The percentage that is released to the medium varies with the cell type and with the culture conditions. The enzyme that is released into the culture medium loses activity rapidly, apparently as a result of the dissociation of the dimeric enzyme into monomers. The enzyme that re-enters the cell is apparently internalized together with heparan sulfate proteoglycan [1463]. This material is either recycled to the cell surface or carried to the lysosomes where it is catabolized. Again, the relative amounts that are recycled and catabolized vary with cell type. The cell surface half-life of the newly secreted LPL is approximately 1 hr [1445].

It is a curious finding in these studies that adipocytes and cardiac myocytes produce cell surface heparan sulfate proteoglycan that binds LPL tightly. Yet, *in vivo*, the enzyme produced by these cells must be released from their cell surface heparan sulfate proteoglycan in order to be transported to the endothelium. It is possible that the affinity of LPL for the HSPG can be controlled metabolically by regulation of the HSPG structure, or that the enzyme is released together with heparan sulfate as a result of heparanase activity [477].

3. Effects of heparin on LPL metabolism in culture

Just as heparin releases LPL and HL from their heparan sulfate proteoglycan complexes *in vivo*, it also releases these enzymes from the surfaces of cultured cells. Heparin causes the release of LPL from the cell surface with biphasic kinetics [1446, 1452, 1464, 1465]. An initial rapid release occurs as the heparin displaces the LPL that is bound to the cell surface. The slower release rate is actually a measure of the rate of arrival of newly synthesized enzyme on the cell surface. Because heparin prevents the binding of LPL to heparan sulfate proteoglycan, it also prevents the endocytosis and catabolism of the enzyme. In cells that synthesize their own LPL, a significant fraction of the internalized enzyme is delivered to the lysosomes where it is catabolized. Thus, heparin blocks endocytosis/catabolism and causes a marked increase of enzyme levels in the culture medium pool [1466, 1467].

LPL binds to avian adipocytes with a K_a of 0.6 x 10^8 $\underline{M}^{-1}$ and is internalized and catabolized [634]. The bound enzyme can be released from the cell surface, not only by heparin, but also by treatment of the cells with heparin lyases or endoglucuronidase, both of which cleave heparan sulfate chains (Chapter 3). One further indication that heparan sulfate chains bind LPL is seen in cells that are grown in the presence of chlorate. As described in Chapter 5, chlorate blocks sulfation reactions, and thus causes the cells to synthesize an undersulfated heparan sulfate proteoglycan to which the enzyme binds poorly [558]. Thus, in chlorate-treated cells, there is a direct secretion of newly synthesized enzyme into the culture medium In a similar manner, CHO cell mutants that are unable to synthesize proteoglycans (Chapter 5) synthesize both LPL and HL, but these enzymes are released into the culture medium [1468, 1469].

4. LPL metabolism in endothelial cells

Although LPL is bound to endothelial cells *in vivo*, these cells do not synthesize it. Nevertheless, endothelial cells exhibit an active LPL metabolism in their *in vitro* processing of the enzyme. Calf aorta endothelial cells bind LPL with a K_a of 0.7 x 10^7 M^{-1} [1455]. Treatment of these cells with proteases or with endoglucuronidase removes the heparan sulfate and markedly reduces the binding. Similarly, porcine aorta endothelial cells bind ^{125}I-LPL at 4 ^{0}C but the bound enzyme can be completely released by heparin treatment. When the temperature is shifted to 37 ^{0}C, the amount of LPL that is released by heparin is reduced as endocytosis proceeds [635]. The internalized LPL is recycled to the surface, where it is bound once again to heparan sulfate proteoglycan. The recycled LPL can now be released to the culture medium pool by heparin. The *catabolism* of the enzyme in endothelial cells is very slow compared to that in adipocytes and myocytes [635, 1445, 1470]. It is suggested that the recycling measured in monolayer cultures may be mechanistically similar to the transcytosis that may occur *in vivo* as the enzyme migrates from its site of synthesis in subendothelial cells to the surface of the endothelium. In fact, direct evidence for such transcytosis has been obtained by showing that bovine aortic endothelial cell monolayers cultured in a two-chamber culture dish transport ^{125}I-LPL from the lower chamber (basal side of the cells) to the upper chamber (apical side), with very little release of enzyme into the culture medium pool [1456]. Heparinase treatment of the basolateral side of the cells reduces the amount of LPL appearing on the apical side. An apparent transcytosis role for heparan sulfate proteoglycan has also been suggested for vitronectin transport (Chapter 8).

5.   N-linked oligosaccharides and dimerization

Newly synthesized LPL is a secreted glycoprotein. It enters the secretory pathway in the endoplasmic reticulum where it becomes N-glycosylated. The N-linked oligosaccharides are modified in both the endoplasmic reticulum and the Golgi prior to secretion. Depending on the species, the LPL may carry one, two, or three N-linked oligosaccharides [1393, 1410, 1471]. The enzyme is initially formed as a monomer in the endoplasmic reticulum but appears as a dimer in the Golgi. Failure to dimerize in the secretory pathway results in the formation of an inactive enzyme that is not efficiently secreted from the cell. The dimerization of the monomers is controlled by the extent of maturation of the N-linked oligosaccharides. Thus, failure to generate the appropriately structured N-linked oligosaccharides results in the accumulation of an inactive, monomeric form of the enzyme in the cell [1472]. The following observations support this conclusion:

- Replacement of Asn 43 in human LPL with Ala prevents dimerization, but replacement of Asn 359 (in a second potential N-glycosylation sequence) or an Asn that is not in a potential N-glycosylation site with Ala does not [1410, 1473].

- Treatment of cells with tunicamycin, which blocks the synthesis of the initial high mannose oligosaccharide, yields a non-glycosylated form of LPL which does not dimerize [1418, 1445, 1474-1476].

- Treatment of cells with N-methyl castanospermine or castanospermine, which block trimming of the glucose residues from the initial high mannose oligosaccharide attached to the protein (the initial maturation step), prevents dimerization [1449, 1476, 1477]].

- Carbonyl cyanide m-chlorophenylhydrazone, an ionophore, causes the accumulation of an inactive monomeric form of the enzyme in the endoplasmic reticulum. However, when the ionophore is removed, the accumulated enzyme is secreted as an active enzyme [1445, 1446].

- In glucose-free medium, a 49-kDa monomer with only 10% of the activity accumulates [1450]. Glucose deprivation prevents normal synthesis and maturation of N-linked oligosaccharides [620, 621].

However, the following treatments, which also alter the maturation of the N-linked oligosaccharides, *do not* prevent formation of an active enzyme:

- An Asn291Ser mutant of LPL, expressed in COS cells, is secreted from the cells and has the same heparin-binding affinity as the normal protein [540]. However, the mutant enzyme has reduced activity, for reasons that are not caused by its failure to dimerize.

- Monensin, an ionophore, causes the accumulation of *active* enzyme in the Golgi [1446].

- Brefeldin A, which inhibits movement of proteins from the endoplasmic reticulum and induces disassembly of the Golgi, causes the intracellular accumulation of active LPL that is not secreted from the cell [1449].

- When maturation of N-linked oligosaccharides is blocked in *cld/cld* mouse adipocytes or in castanospermine-treated adipocytes, inactive LPL accumulates in the endoplasmic reticulum. Treatment of these cells with brefeldin A, which allows translocation of Golgi enzymes into the endoplasmic reticulum, yields active enzyme [1478].

- Attachment of an endoplasmic reticulum retention signal, KDEL, to the C-terminal of LPL results in accumulation of *active* enzyme in the endoplasmic reticulum [1449].

- Deoxymannojirimycin and methyl-deoxynojirimycin, which block mannose trimming, allow the formation of an active LPL that is secreted from the cell [1449, 1451, 1476]. Thus, trimming of the mannoses is not critical.

Based on these observations, it is clear that only the oligosaccharide linked to Asn 43 is involved in the formation of an active dimer and that an early step in the glycosylation modification reactions is required for dimerization of the monomers and formation of the active enzyme. It appears that the step that is critical for dimerization is the removal of glucose residues from the initial high mannose oligosaccharide [1475].

6.   Other cell surface receptors for LPL

Although it has long been felt that cell surface heparan sulfate proteoglycans are the sole receptors for LPL, recent results have suggested that there are several LPL-binding proteins on cell surfaces in addition to heparan sulfate proteoglycan [1463, 1479, 1480]. It has been shown that there are at least three components that bind to LPL on the surface of human aortic endothelial cells: (a) a 220-kDa heparan sulfate proteoglycan [1463], (b) a 116-kDa protein that is released from the surfaces of these cells by heparin, referred to as "hrp-116" (heparin-releasable protein-116) [1479], and (c) an 85-kDa glycoprotein that binds to LPL [1480]. Interestingly, hrp-116 shows sequence similarity to the N-terminal region of apoB and is apparently formed as a normal proteolytic cleavage product of apoB-100 (NTAB—N-terminal peptide of apoB) [1481]. It is suggested that the heparan sulfate proteoglycan is a high-density, low-affinity receptor for lipoprotein lipase, whereas hrp-116 is a low-density, high-affinity receptor, and that these may act cooperatively in LPL binding and uptake in a manner similar to the cooperative interaction between FGFRs and the heparan sulfate proteoglycan in FGF uptake (Chapter 9). However, the precise roles of these additional LPL-binding proteins remain to be established. As discussed below, LDL receptor-related protein (LRP) may also be a receptor for LPL in some cells.

D.   Binding Domains of LPL

1.   Introduction

It is clear from the above discussion that the three-dimensional structure of LPL must present a number of specific domains. These include (a) its substrate binding domain, which may be appropriately subdivided into domains for small (ester) and large (triacylglycerol) substrates, (b) its catalytic site, (c) its heparin-binding domain, (d) its subunit interaction domains, i.e., the domains that allow the monomers to associate, (e) its apoC-II recognition domain, and (f) domains for interaction with other cell surface proteins or receptors, including low density lipoprotein receptor-like protein (LRP) (see Section IV.B.1.b., p. 397). In addition, it is recognized that lipases must act at a lipid-water interface; thus, an interface recognition domain must also be present. Identification of these domains has involved (a) studies with peptide fragments, (b) examination of mutagenized LPL, (c) determination of the properties of chimeras of LPL and HL, and (d) deductions based on modeling of LPL against the crystal structure of PL. Obviously, it takes surface recognition domains on both LPL and each of its ligands for each of these interactions. Thus, there must be LPL recognition domains on the lipoprotein substrates, the heparin or heparan sulfate, the apoC-II, and the other proteins that bind to LPL and participate in its actions. Most of the domains on these LPL "ligands," including the subunit/subunit interaction domain, have not been studied extensively.

2.   General structural features of LPL

LPLs from man, bovine, guinea pig, rat, and chicken have been cloned and characterized [see 1420, 1471 for reviews]. These enzymes show 73-77% amino acid identities. All are glycoproteins, but there is some variation in the number of potential N-glycosylation sites [1393, 1410, 1471], with the most N-terminally located glycosylation site in each sequence (e.g., Asn 43 in human, Asn 45 in chicken, etc.) being essential for dimerization and secretion of the active enzyme (see Section III.C.5., p. 381) [1410]. Interestingly, it has been shown for the chicken enzyme that the N-linked oligosaccharide at Asn 45 is sulfated on one of the core GlcNAc residues [1428]. Lipoprotein lipase is active only as a dimer. When it dissociates spontaneously or as a result of treatment with guanidinium chloride, it loses all activity [1396, 1482].

3.   Pancreatic lipase as a model for the LPL structure

In order to obtain the clearest possible appreciation of the studies that have addressed the identification of these domains, it is helpful to have a picture of the three-dimensional structure of LPL. Since the only view of the three-dimensional structure of LPL that is available is that modeled on the crystal structure of PL, we must first examine the PL structure.

The sequences of PLs from several species [1391, 1483, 1484], as well as the crystal structure of the human enzyme [1391] and the porcine enzyme [1485], have been reported. The PL structure, illustrated in Figure 11.2, is divided into two regions: an N-terminal half (residues 1-336) and a C-terminal half (residues 337-449). The N-terminal region at the top of Figure 11.2 is a typical α/β structure with a central parallel β-sheet. The C-terminal region is formed by two layers of β-sheet, each with four antiparallel strands. The two sheets of the C-terminal unit are folded upon each other, forming a β-sandwich type structure. In the N-terminal unit, the active site, formed by serine 132, aspartic acid 156, and histidine 241, is covered by a "lid," a disulfide-closed loop formed by residues 216-239, and indicated by the heavy lines in Figure 11.2. The lid contains an α-helical structure connected to the disulfide-linked cysteines by two elongated polypeptide stretches.

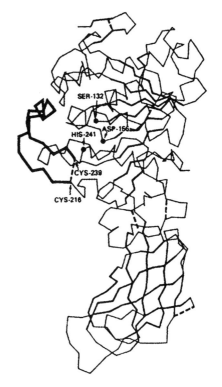

Figure 11.2. The crystal structure of human pancreatic lipase. Reprinted from [1400], with permission.

Even if this loop were simply removed, it would be difficult for bulky triacylglycerol substrates to access the active site serine. Thus, formation of the enzyme-substrate complex must require significant conformational changes in the area of the active site in the enzyme in order to open the lid and permit access of the substrate to the catalytic site. This structure with the lid covering the catalytic site is referred to as the "closed structure."

Crystal structures of human PL have also been reported in which the lipase is cocrystallized with procolipase [1486] and in which the lipase is co-crystallized with procolipase *and* phospholipid [1487]. Procolipase, which contains an N-terminal pentapeptide that is lost when it is converted to colipase, exhibits the same activity as the colipase in the presence of bile salts [see 1488, and references therein]. In its complex with PL, procolipase binds almost completely to the C-terminal region of PL, with no significant change in the conformation of either protein [1486, 1487]. However, when crystals are formed in the presence of both procolipase and phospholipids, the lid is folded back to expose the catalytic site to substrate [1487, 1489]. This structure, referred to as the "open structure," shows the lid folded back with its secondary structure altered and with the phospholipid bound in the active site. The opening of the lid not only renders the catalytic site accessible to solvent, but also exposes a number of hydrophobic residues over the entrance to the active site [1400]. In addition, it creates an oxyanion hole at the entrance to the active site that stabilizes the transition state of the substrate as the ester bond is being cleaved. Although the catalytic serine is surrounded by hydrophobic amino acid side chains, it is totally exposed to solvent. In the open structure, the N-terminal domain of the lipase interacts with the procolipase (which binds almost completely to the C-terminal domain in the closed structure) via the open lid, a result of a significant bending of the C-terminal region of the enzyme toward the N-terminal region as the substrate binds. The open lid and the bound procolipase form an extensive hydrophobic region capable of interacting strongly with lipid/water interfaces that may be covered with bile acids, a property gained by the presence of the procolipase. The details of these changes have been described [1486, 1487]. Several studies have shown that the substrate specificities of the different members of the lipase family are determined by the lid structure [1385, 1490-1492].

4.   The three-dimensional structure of LPL

The three-dimensional closed and open structures of human LPL have been modeled on the corresponding PL structures [1390]. One view of this model is presented in Figure 11.3 (see color insert), which shows the C-terminal half of the molecule on the left and the lid covering the active site on the lower right. The clusters above represent the heparin-binding domains (see Section III.E.2., p. 389). The N-terminal half of LPL, residues 1-335, contains the catalytic triad made up of Ser 132, Asp 156, and His 241 lying in a groove covered by a lid loop, residues 237-261, that blocks the active site serine in the closed form. Again, this loop swings away in the open form. The N-terminal half has the same α/β structure as PL, with a central β-sheet. The C-terminal domain exhibits the same β-sandwich type structure seen in PL. The two enzymes have about 30% amino acid identity and 40% similarity. They are about the same size. Additions and deletions of amino acids occur mostly in the loops. This three-dimensional structure is consistent with many aspects of the LPL structure that had been probed earlier by a variety of techniques. These latter results will be discussed and integrated into the structure shown in Figure 11.3.

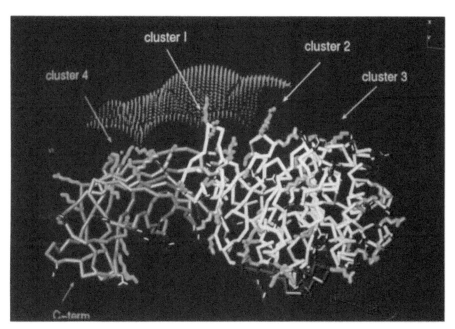

Figure 11.3. The modeled structure of lipoprotein lipase. Reprinted from [1390], with permission.

5.   Binding domains of LPL

Because the members of the lipase family exhibit two-domain structures, they have lent themselves to a particularly interesting study in which Schotz and co-workers have compared the activities of native human LPL and rat HL with chimeras of these two lipases in which the N- and C-terminal halves of each enzyme are joined to the C- and N-terminal halves, respectively, of the other to obtain rat HL→human LPL and human LPL→rat HL (N-→C-terminal) chimeras [1405, 1493]. Both chimeric forms of these lipases are enzymatically active. Since the native LPL and HL exhibit significant differences in their cofactor requirements, heparin affinities, and substrate specificities (Table 11.2), it is possible to judge which of the two major domains of each enzyme contributes to each type of property. For example, the results of these studies, which are summarized in Table 11.3, show that the LPL→HL chimera is inhibited by NaCl and is stimulated by apoC-II—both behaviors that are exhibited by native LPL but not by native HL. This suggests that these two responses must be mediated through the N-terminal domain of LPL, which is present in the LPL→HL chimera but missing in the HL→LPL chimera. In addition, while both enzymes act on both triolein and phospholipids, the HL→LPL chimera shows the specificity of the HL whereas the LPL→HL chimera shows the specificity of the LPL, indicating that the enzyme specificity is controlled in the N-terminal portion of these enzymes. Finally, the salt concentrations required to elute the chimeric proteins from heparin-Sepharose suggest that the heparin-binding domain is in the C-terminal domain of these proteins. The latter result is contrary to some other reports, and, indeed, the

Table 11.3. Properties of Chimeric Forms of HL and LPL

	Enzyme		Hybrid (N-domain/C-domain)	
	LPL	HL	LPL/HL	HL/LPL
V_{max} on triolein (nmole/hr)	53	16	51	13
V_{max} on phospholipid (nmole/hr)	4	35	5	76
apoC-II stimulation?	Yes	No	Yes	No
NaCl sensitive?	Yes	No	Yes	No
$\underline{M}$ NaCl for elution from heparin affinity matrix	1.1	0.75	0.75	1.1

From references [1405, 1493].

heparin-binding site appears to be more complex than can be determined by any single study, as we will now discuss.

E.   The Heparin/LPL Interaction

1.   Oligosaccharides sequences that bind to LPL

The active homodimer form of LPL elutes from heparin affinity columns with 1.0-1.2 $\underline{M}$ NaCl, depending on the species from which it is derived [1389, 1408, 1448, 1494-1496]. However, when the dimer dissociates into monomers, these elute from the column with 0.5 $\underline{M}$ NaCl and are detected using antibodies raised against the active enzyme [1389]. This suggests that both monomer units in the dimer participate in the heparin binding [1497, 1498].

Only a few studies have addressed the structures of the heparin sequences that bind to these lipases. It is reported that heparan sulfate binds LPL less strongly than heparin and that N-desulfated heparin is unable to bind the enzyme [1499]. Also, it has been shown that both antithrombin-binding and antithrombin-nonbinding fractions of heparin exhibit equal affinity for LPL [1499, 1500]. In addition, heparin can be fractionated on a LPL affinity chromatography column into fractions that bind to the LPL and fractions that do not [1501]. With regard to the length of the heparinoid segment that is required for LPL binding, an early report showed that heparins with molecular masses of 10 to 18.4 kDa form 1:1 molar complexes with LPL dimers, whereas low molecular weight heparins, with molecular masses of 6.6 to 8 kDa form 2:1 (heparin:LPL) molar complexes [1489]. This would indicate that a sequence longer than low molecular weight heparin, which, on the average, contains 5-7 disaccharides, is required to span both monomers in the dimer.

A second study reported that when oligosaccharides prepared from bovine aortic endothelial cell-derived *heparan sulfate* are chromatographed on a bovine LPL affinity column, the smallest oligosaccharide that binds to LPL is a decasaccharide composed of five IdoA-2-$SO_4 \rightarrow$ GlcNSO$_3$-6-SO_4 disaccharide units, i.e., the most highly sulfated disaccharide that occurs commonly in heparinoids [1502]. Interestingly, such sequences represent only about 4% of the total heparan sulfate chains, indicating that only a limited number of LPL binding sites are found in this endothelial cell heparan sulfate. When *heparin*-derived oligosaccharides are chromatographed on the bovine LPL affinity column, heparin tetrasaccharides do not bind, but hexa-, octa-, and decasaccharides exhibit K_d's of 188-, 18.7-, and 5.3 x 10^{-8} $\underline{M}$ [33]. The stoichiometry of binding, i.e., moles of oligosaccharide per mole of enzyme, was not reported. Again, the highest affinities are found for the fully sulfated decasaccharide, although a decasaccharide having only a slightly lower affinity was found to have the same structure with one of the internal IdoA-2-SO$_4$ residues replaced with a GlcA. The yields of LPL-binding oligosaccharides from heparin are higher than those from heparan sulfate, as expected.

Finally, the question of the binding of heparin, heparan sulfate, and heparin oligosaccharides to dimeric LPL has been revisited by Lookene *et al.* who have determined both the affinities and the stoichiometries of heparinoid/LPL binding [1503]. A variety of techniques were used in this work, including stabilization of enzyme activity by heparinoids, sucrose density gradient sedimentation of the LPL/heparinoid complexes, surface plasmon resonance (with immobilized heparinoids) to determine association and dissociation rate constants, and fluorescence and circular dichroism to observe complex formation. These results have led to the following conclusions:

- The K_d values for binding of heparinoid oligosaccharides are about 10-fold lower in these studies than those previously reported (above).

- In 0.15 $\underline{M}$ NaCl, heparan sulfate binds to dimeric LPL with a K_d of 0.3 n$\underline{M}$, i.e., with quite high affinity.

- Monomeric LPL has a 6000-fold lower affinity for heparin than dimeric LPL.

- Heparin decasaccharides are the shortest oligosaccharides that will bind with maximum affinity to LPL. The decasaccharides form a 1:1 complex with dimeric LPL.

- Heparinoid binding to dimeric LPL causes no major changes in conformation of the protein.

- Electrostatic interactions account for 44% of the interaction energy for heparan sulfate, 49% for heparin, and 60% for affinity fractionated decasaccharides.

These results have implications with respect to the mode of interaction of the LPL dimers with heparinoids, discussed in Section III.E.4., p. 392.

2. LPL peptide sequences that bind to heparinoids

Several approaches have been taken to identify the heparin-binding domains of LPL. For example, the more classical approach in which the enzyme is cleaved with proteases and heparin-binding peptides are studied has identified a fragment containing amino acids 229-370 as a heparin-binding sequence. The binding of this large peptide was attributed to an arginine- and lysine-rich sequence included in amino acids 229-312 [1504]. In addition, several heparin-binding consensus sequences (Chapter 6) have been identified in both LPL and HL.

Finally, the modeling of the three-dimensional structure of LPL on the PL structure has allowed the identification of clusters of basic amino acids along the surface of the enzyme, opposite the lid, where heparin or heparan sulfate might bind [1390]. We will first describe the heparin-binding domains in the modeled LPL structure and will then discuss evidence based on the heparin-binding activities of naturally occurring and site-directed mutants of the LPL structure.

Calculations of the positive electrostatic potential on the surface of LPL show four clusters of positive charges that lie opposite the surface that contains the lid over the catalytic site. These clusters are illustrated in Figure 11.3 (see color insert). The sequences that contain these clusters are shown below. In these sequences, the basic amino acids *that appear in the clusters* are underlined. Also, in Cluster 4, "X" is used to indicate *any* amino acid in long gaps between the basic amino acids that form the clusters:

Cluster 1:   R A K R S S K
                        294   300

Cluster 2:   R C S S K E A F K G L C L S C R K N R
                        263   282

Cluster 3:   K K V N R
                            147   151

Cluster 4.:   K X₈₅ R X K X₆ K
                            319   414

Several points are of interest.

- Clusters 2 and 4 contain some basic amino acids in the *linear* sequences that do not appear in the surface clusters. This is not illustrated for cluster 4, which has a long intervening gap between K319 and R405, and a shorter gap between K407 and K414.

- According to this analysis, there are 15 basic amino acids in these four clusters, suggesting that it is unlikely that a single amino acid plays a uniquely critical role in heparin binding.

- These clusters lack acidic aspartic acid and glutamic acid residues in their immediate vicinities.

- In PL, the regions of the clusters contain fewer charged amino acids and these are more symmetrically distributed so that there is no concentration of positive electrostatic potential on the surface.

- In HL, the number of Arg and Lys residues in clusters 1-4 is only 12, compared to 15 for LPL [see references 1505, 1506-1511].

- The binding of heparinoids to the face of LPL that lies opposite the entry to the catalytic site allows the interaction of the enzyme with its substrates while it is bound to cell surface heparan sulfate proteoglycan, as has been observed.

- Chymotryptic cleavage of LPL yields an N-terminal domain of the enzyme (residues 1-392) that retains its heparin-binding capacity. Therefore, cluster 4 may not play a significant role in heparin binding [1512]. However, the studies of the chimeric constructs of HL and LPL (see, Section III.D.5., p. 387) support the importance of the C-terminal domain in heparin binding.

- The total length across clusters 1-4 is sufficient to accommodate an octasaccharide. However, since one group of these four clusters *occurs on each monomer* of the homodimer, we must consider the monomer-monomer interactions in the dimer that will allow us to rationalize the binding of heparin to the *dimer*. This is discussed further below.

3.   Effects of mutations in LPL

The interactions of LPL with a variety of ligands (e.g., apoC-II, heparin, substrate, etc.) have prompted attempts to identify specific amino acid moieties in the enzyme that participate in these interactions. There are a number of naturally occurring missense, nonsense, and splice site mutations in LPL [reviewed in reference 1496]. In addition, amino acid replacements have been made in (a) the catalytic triad amino acids and their surrounding residues [1388, 1392, 1400, 1401, 1410], (b) the potential N-glycosylation sites [1388, 1428], (c) the lid [1388, 1490-1492], (d) residues that may control sorting during the cellular processing of the enzyme [1437], (e) residues that may stabilize the dimeric form of the enzyme [1496], (f) residues that may be involved in binding of the enzyme at a lipid-water interface [1389, 1401, 1513], (g) residues that may stabilize the tertiary or quaternary structures [540, 1388, 1389, 1401, 1435, 1436, 1514], and (h) residues that are in the heparin-binding domain [1389, 1515, 1516]. The work on the heparin-binding interactions will be discussed separately. Other studies have

- confirmed the assignment of the residues forming the catalytic triad and shown that other conserved serine residues can be replaced without total loss of activity,

- demonstrated that N-glycosylation at the most N-terminal glycosylation site, but not at other potential glycosylation sites, is essential for the formation of an active enzyme,

- suggested that residues 125-142 may form a lipid-binding site,

- shown that reducing the amphipathicity of the α-helices in the lid, or removing the two C-terminal amino acids, results in loss of activity on triolein but not activity on tributyrin,

- demonstrated that any structural alteration that prevents dimerization of the monomers in the secretory pathway results in the accumulation of inactive enzyme in the endoplasmic reticulum or the Golgi, and shown that LPL is not a GPI anchored protein [1389, 1517].

A number of natural and site-directed mutations in lipoprotein lipases have been investigated to try to identify the positions of the heparin-binding sequences. A list of these mutagenesis studies is given in Table 11.4. Most of these reports give results consistent with the involvement of clusters 1, 2, and 3. There have been no reports of mutagenesis studies on the residues in cluster 4. Several mutant forms listed in Table 11.4 show that amino acid alterations in the most highly conserved regions of the enzyme, not in the putative heparin-binding regions, also reduce the affinity of the enzyme for heparin. Presumably these changes result in an altered conformation in the regions for heparin binding. In addition, one of the mutations is in a sequence *between* clusters 1 and 2, and this change appears not to perturb the conformation of the heparin-binding domains sufficiently to reduce the heparin binding.

4.   Monomer-dimer considerations in the binding of heparinoids to LPL

We have noted that the heparin-binding affinity of LPL, as judged by the NaCl concentrations required to elute the LPL from a heparin affinity column, is much higher for the active dimer form than for the inactive monomer. An early report showed the ratio of heparin to LPL dimer in the heparin/LPL complex to be 1:1 for unfractionated heparin (MW ~12,000) and 2:1 for low molecular weight heparin (MW ~6-8000, containing a mixture of octa- to tetradecasaccharides) [33, 1489]. However, two reports show that the smallest oligosaccharide that binds with high affinity to LPL is a highly sulfated decasaccharide [33, 1503], and Lookene *et al.* [1503] have presented convincing evidence that a heparin

Table 11.4. The effects of mutations in putative heparin-binding sequences on the binding of LPL to heparin

LPL source	Mutation	Comment	Result of mutation	Ref.
Human	R279A, K280A, R282A, K296A, or R297A	From putative heparin-binding sequences in clusters 1 and 2	Significant reduction in heparin binding	[1518]
Human	K147A, K148A, K292A, R294A or K304A	From putative heparin-binding sequences in clusters 1 and 3	No effect on heparin binding	[1518]
Human	K280A, K292A, R294A, R297A, K300A, or E289A	From putative heparin-binding sequences in clusters 1 and 2	Significant reduction in heparin binding	[1516]
Human	N291S	Residue located between putative heparin-binding clusters 1 and 2	Normal synthesis, processing, secretion, and heparin binding, but activity reduced to about half	[540]
Human	G188E	From region of central homology; not in putative heparin-binding domain	Secretion of inactive enzyme with reduced heparin binding, presumably due to conformational alteration	[1519]
Human	A176T	From region of central homology; not in putative heparin-binding domain	Secretion of inactive enzyme with reduced heparin binding, presumably due to conformational alteration	[1520]
Avian	R^{281}KNR^{284} → various mono- & disubstituted sequences	From putative heparin-binding sequence in avian enzyme cluster 2	Modest reductions in heparin-binding	[1515]

decasaccharide will form a 1:1 complex with dimeric LPL. The latter authors conclude, therefore, that *the decasaccharide can bind to both subunits of the active LPL.*

These observations must be rationalized in terms of the interfaces between the two monomers of the dimer and between the dimer and a single, relatively short segment of heparin or heparan sulfate. Several authors have speculated on the appearance of the interface between the two monomers in dimeric LPL in the absence of heparin [1390, 1493]. In the structures that have been proposed, the monomers are aligned side by side, and the primary consideration has been whether the monomers are aligned in a head-to-head or a head-to-tail orientation.

van Tilbeurgh *et al.* [1390] suggest that the most favored view of the dimer is one in which the monomers are aligned head to tail with the heparin-binding clusters facing a heparin *dodecamer* that is bound between the two subunits, as shown in Figure 11.4. Such an arrangement allows substrate access to the active site, but presents several problems. First, the contacts between the two monomers are largely through the dodecasaccharide, so that the model does not suggest a dimer structure that exists in the absence of heparin. A second potential problem concerns the lack of symmetry in the heparinoid interactions with the two subunits. In this head-to-tail orientation, the heparin-binding clusters on one subunit are running in the N→C direction, while those on the other are running in the C→N direction. However, a single strand of heparin can run in

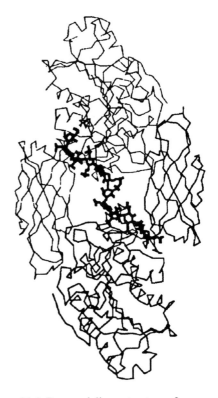

Figure 11.4. Proposed dimer structure of lipoprotein lipase. Here, two monomers of LPL are aligned head to tail, with the open lid shown on the sides opposite the monomer-monomer interaction sites. A dodecamer strand of heparinoid is shown in the thick lines between the two monomers. Reprinted from [1390], with permission.

only one direction (nonreducing terminal to reducing terminal). Thus, on one subunit, the heparinoid and the protein are running parallel to each other whereas on the other they are running antiparallel to each other. Consequently, one LPL monomer must bind to one face of the heparinoid while the other must bind to the opposite face. There is precedence in the case of antithrombin (Chapter 7) for a single strand of heparin binding to one helix of the protein via sulfate and carboxyl groups on one face of the heparin strand and to a second helix using a different array of sulfate and carboxyl groups on the opposite face, but in the antithrombin case, the two helical segments of the protein are quite different. Nevertheless, if symmetry is not required for these interactions, the model shown in Figure 11.4 could be an accurate representation of the interactions of heparin and LPL However, a crystal structure showing the heparin decasaccharide bound to LPL would resolve this dilemma, but, to date, there has been no success in obtaining such crystals for *any* heparin-binding protein.

IV. THE ROLE OF HSPG IN THE ENDOCYTOSIS OF LIPOPROTEINS

A. Introduction

Clearly HSPG plays an important role in localizing the action of LPL and HL to the surfaces of endothelial cells and hepatocytes, respectively. It also plays a role in the endocytosis and catabolism of lipoproteins illustrated in Figure 11.1.

The uptake of lipoproteins and LPL, both separately and together, is mediated by a number of cell receptors, the understanding of which has developed over more than 30 years. In many recent reports, a role of heparan sulfate proteoglycan in the endocytosis of lipoproteins has been recognized. It is our goal here to identify the various receptors and to describe the interactive roles played by heparan sulfate proteoglycan and these receptors in the endocytosis and catabolism of LPL, HL, and the various lipoproteins.

Several receptors are involved in lipoprotein uptake and catabolism. These include the LDL receptor, the low density lipoprotein receptor-like protein (LRP), gp330, the VLDL receptor, three recently reported LDL receptor homologs expressed in brain tissue, and perhaps another receptor, known as the remnant receptor (for uptake of chylomicron and VLDL remnants). These proteins represent a family of lipoprotein receptors that have been described in several reviews [1521-1529]. The common structural features of this family, illustrated in Figure 11.5, are (a) a C-terminal cytoplasmic domain containing

one or more NPXY sequences which are markers for clustering to coated pits
and rapid endocytosis, (b) a single transmembrane domain, (c) multiple copies
of cysteine-rich complement-type repeats, (d) multiple copies of cysteine-rich
epidermal growth factor-type repeats, and (e) one or more copies of a "YWTD"-
containing spacer region similar to one found in epidermal growth factor
precursor.

The term "receptor" is generally reserved for *proteins* that are attached to
cell membranes and that exhibit some specificity for binding one or a few
structurally related ligands. Yet, we have already seen that heparan sulfate
proteoglycan serves as a "receptor" for LPL and HL, and that these two enzymes
are bound to cell surface HSPG as they act on their substrates and during their
own endocytosis and catabolism. Furthermore, both apoB and apoE are
"heparin-binding proteins"; therefore, lipoproteins that contain these two
apolipoproteins may potentially bind to HSPG on the cell surface. Since HSPG

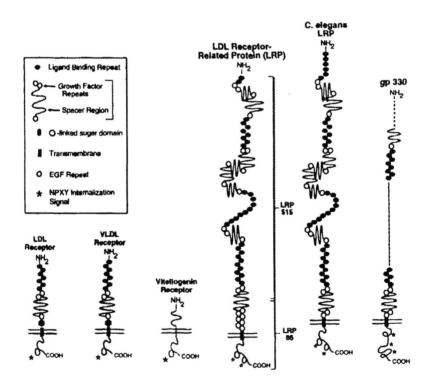

Figure 11.5. Structures of lipoprotein receptors. Reprinted from [1528], with permission
from Annual Review of Biochemistry, vol. 63, © 1994, by Annuals Reviews, Inc.

is continuously internalized by cells (Chapter 5), such HSPG-bound lipoproteins may be internalized in association with HSPG and catabolized. Thus, we might suggest that HSPG, although not a typical receptor, should be considered an honorary member of the group of receptors that participate in lipoprotein metabolism.

B. Lipoprotein Receptors

1. Receptors that act primarily in the liver

a. The LDL receptor

The first receptor that was recognized, and the one that has been most extensively studied, is the LDL receptor (LDLR). The LDL receptor has been characterized by Brown and Goldstein and by others, and this work has been reviewed [1523]. The LDL receptor is the prototype for the family. This receptor, 70% of which is located in the liver [1530], clusters in clathrin-coated pits where it binds LDL. Following endocytosis, the endosomes that contain the LDL/LDLR complex become acidified, resulting in the release of LDL from the receptor and the recycling of the receptor to the cell surface. LDL is delivered to the lysosomes and catabolized. The LDL receptor is a prime player in the clearance of cholesterol from the circulation. It also plays a role in the uptake of a portion of IDL, the LDL precursor. The apoB-100 in LDL is a recognition marker for the LDL receptor. In contrast, the apoB-48 that is found in chylomicrons and chylomicron remnants lacks the C-terminal domain of apoB-100 that is required for receptor binding. Thus chylomicron remnants are not recognized by the LDL receptor, and these are endocytosed by another receptor, LRP (see below) [1531].

In addition to apoB-100 (4536 amino acids), the LDL receptor recognizes the structurally unrelated ligand, apoE (299 amino acids). In lipoproteins containing both apoE and apoC, the apoC seems to mask the apoE and excess apoE may be needed for binding to the LDL receptor. [1532-1534].

b. The LDL receptor-like protein

i. LRP structure

The metabolism of chylomicrons and VLDL by LPL and HL results in the formation of chylomicron remnants and VLDL remnants, lipoproteins that are rich in apoE. These particles are normally taken up and catabolized by the liver. Since the LDL receptor recognizes both apoB and apoE, it was initially presumed that the apoE-rich chylomicron and VLDL remnants would be taken up by the LDL receptor. However, patients with homozygous familial hypercholesterolemia (FH), as well as Watanabe rabbits, both of which have defective LDL receptors, fail to accumulate remnants in their plasma, leading to the suggestion that there is a distinct "remnant receptor." The endocytosis of remnants requires the presence of apoE in the particles since defective apoE results in type III hyperlipoproteinemia with accumulation of remnants in the plasma [1535-1537]. Thus, a second receptor that recognizes apoE seems to play a central role in the clearance of these particles. The cloning of a new receptor by a homology screening approach resulted in the identification of a 100-kDa receptor with apoE specificity that is similar to the LDL receptor [1538]. This receptor, which has structural as well as biological similarity to the LDL receptor, is referred to as "LDL receptor-related protein, or LRP. It is illustrated in Figure 11.5. Although LRP has properties expected for the remnant receptor [1530, 1539], repeated attempts to obtain conclusive biochemical evidence for the identity of LRP and the remnant receptor have been equivocal (see, for example, reference [1540]). As we shall see, in contrast to LDL receptor-mediated endocytosis, heparan sulfate proteoglycan plays a role in LRP-mediated endocytosis. Thus, we will describe LRP in greater detail.

Most of the work on the receptor role of LRP has been carried out using β-migrating VLDL (β-VLDL) as the lipoprotein ligand. β-VLDL is a mixture of abnormal cholesterol ester- and apoE-rich remnant-like particles that accumulate in animals fed diets high in saturated fat and cholesterol or in the plasma of patients with type III (apoE-defective) hyperlipoproteinemia. β-VLDL, especially in the presence of added apoE, has a very high affinity for LRP [1426, 1539, 1541-1543].

Studies on LRP have been reviewed [1527-1529]. The predicted amino acid sequence of the cloned LRP shows that it is composed of several distinctive domains that are similar to those found in the LDL receptor. LRP is present on the surfaces of a number of cell types, and the mRNA for LRP is found in many tissues, but its primary function is observed in the liver [1544]. By comparison with the LDL receptor (839 amino acids), LRP is very large (4525 amino acids) and it has multiple copies of several of the types of extracellular domains found in the LDL receptor (Figure 11.5). The protein undergoes processing during secretion, the most unusual feature of which is the cleavage into two chains which remain tightly associated, but not covalently linked, as the protein takes up its position in the cell membrane. The resulting β-chain (85 kDa), from the original C-terminal of the protein, contains the intracellular and transmembrane domains, and a short extracellular domain, whereas the α-chain contains a large (515 kDa) extracellular domain.

LRP was found by Strickland *et al.* and by Kristensen *et al.* to be identical to the α_2-macroglobulin receptor (α_2MR) [1545, 1546], which binds and internalizes "activated" α_2-macroglobulin (α_2M*). α_2-Macroglobulin is a 720-

kDa homotetramer that circulates in the plasma and interacts with all known classes of proteases [reviewed in 898, 1547]. Activation occurs when α_2-macroglobulin encounters a protease and is cleaved, causing a conformation change in the α_2-macroglobulin that results in the entrapment of the protease (like a Venus flytrap), and the exposure of its receptor binding domain [1547]. Alternatively, activation can be obtained in the laboratory by treatment of α_2-macroglobulin with methylamine, which causes the cleavage of an intramolecular thiol ester bond followed by the same conformation changes. The receptor-binding domain of $\alpha_2 M^*$ is recognized by the α_2-macroglobulin receptor ($\alpha_2 MR$) [1374] and is rapidly cleared from the circulation. The $\alpha_2 M^*/\alpha_2 MR$ complex enters cells via clathrin-coated pits and is degraded in lysosomes [1548, 1549]. The binding region for $\alpha_2 M^*$ has been identified within residues 776-1399 [1550, 1551]. *Because of the identity of LRP and $\alpha_2 MR$, we will refer to this receptor henceforth as LRP/$\alpha_2 MR$.*

ii.   LRP/$\alpha_2 MR$ ligands

Most receptors exhibit high ligand specificity. LRP/$\alpha_2 MR$, however, serves as a receptor for a number of quite diverse ligands. In fact, it shows the least specificity of all known receptors. Since LRP and $\alpha_2 MR$ are the same protein, it is clear that LRP must bind and internalize all of the ligands of $\alpha_2 MR$ that have been identified. This includes all species of $\alpha_2 M^*$ [1526]. Also, we have already implied that LRP/$\alpha_2 MR$ does not bind LDL, but does bind the apoE-rich β-VLDL [1530]. Among other ligands for LRP/$\alpha_2 MR$ are LPL [1423], HL [1552], plasminogen activator/plasminogen activator inhibitor complexes, but not uncomplexed plasminogen inhibitors [1526, 1553-1555], lactoferrin, [1526, 1556], and thrombospondin [1139, 1557]. LRP/$\alpha_2 MR$ also binds Ca^{2+} and it has been reported that this binding is obligatory for β-VLDL binding [1530]. For some of these cases, it has been shown that binding of ligands to LRP/$\alpha_2 MR$ requires HSPG [1552].

An additional LRP/$\alpha_2 MR$ ligand, one that remains tightly associated with LRP/$\alpha_2 MR$ during its isolation, is referred to as α_2-macroglobulin receptor-associated protein ($\alpha_2 MRAP$, or simply RAP—also often referred to as the 39 kDa LRP-associated protein, and now as LRP-associated protein-1, or LRPRAP1) [1558-1561]. $\alpha_2 MRAP$, a heparin-binding protein [1558, 1559], has been cloned [1561]. $\alpha_2 MRAP$ shares a common epitope with members of the LDL receptor family [1562]. It inhibits the binding of β-VLDL/apoE, chylomicrons and all other ligands of LRP/$\alpha_2 MR$ [1526, 1563]. It also binds to calmodulin and undergoes phosphorylation/dephosphorylation, although the significance of this is unclear [1564]. Finally, it has been shown that malaria sporozoites utilize a specific interaction of their circumsporoszoite protein with LRPRAP and HSPG as a major mechanism of host invasion [1565]. Although its role is not fully established, it has been proposed that α_2MRAP is a modulator of ligand binding to LRP/α_2MR [1374, 1559, 1566-1568] and that α_2MRAP serves as a *chaperone* that controls LRP/α_2MR folding during its biosynthesis and secretion [1569].

iii. LRP/α_2MR vs the LDL receptor

As shown in Figure 11.5, LRP/α_2MR is a large protein with many more binding domains than found in the LDL receptor. However, the types of domains are quite similar. How then does one distinguish LRP/α_2MR from the LDL receptor? A number of approaches have been used in addressing this question. Most notable is the use of cells that lack the LDL receptor [1374] or cells that lack LRP/α_2MR [1552]. In these cases, comparison of the binding/uptake responses of mutant and wild-type cells helps to identify which receptor interacts with the ligand in question.

Another common approach to this problem is to observe the blocking of the binding and/or cellular uptake of one type of ligand (e.g., α_2-macroglobulin) by another (e.g., β-VLDL). However, with the size of LRP/α_2MR, it may be possible for one ligand to bind to one site on the receptor without steric inhibition of binding of a second ligand to another site. The following are examples of results that have been obtained using such competition approaches:

- Lactoferrin blocks the uptake of β-VLDL and chylomicron remnants by liver parenchymal cells [1543].

- Lactoferrin does not block uptake of α_2M* [1543].

- α_2MRAP blocks binding of LPL to cells [1424].

- The binding of the u-PA/PAI-1 complex (Chapter 8) to cells is blocked by α_2MRAP and by the t-PA/PAI-1 complex [1553].

- LPL binding to LRP/α_2MR is blocked by anti-LRP antibody. Also, α_2M* displaces LPL from LRP/α_2MR [1374].

- HL binding to LRP/α_2MR is inhibited by α_2MRAP and by anti-LRP [1552].

Although there are some results of such competition studies that do not fully support the binding of certain ligands to LRP/α_2MR, it seems to be generally true that α_2MRAP inhibits the binding of *all* ligands to the receptor [1526] and that inhibition of ligand binding and uptake by α_2MRAP is diagonostic for the involvement of LRP/α_2MR in uptake of the ligand.

One problem that complicates the assignment of an LRP/α_2MR role in the binding of certain ligands is that although LRP/α_2MR can internalize ligands directly, there are a number of cases in which heparan sulfate proteoglycan plays an obligatory role in the binding/endocytosis activity of LRP/α_2MR and its complexes with various ligands. We will address this role in Sections IV.D., IV.E., and IV.F, on pp. 405, 406, and 409.

2. Receptors that act primarily in the peripheral tissues

a. GP330

The LDL receptor and LRP/α_2MR play their primary roles in lipoprotein metabolism in central pathways that occur in the liver. Several additional lipoprotein receptors have been identified in the peripheral tissues. One of these is gp330, a member of the lipoprotein receptor family shown in Figure 11.5. Gp330 is a large glycoprotein expressed by several types of absorptive epithelial cells, including those of kidney proximal tubules. It is not found in liver [1570, 1571]. It has been localized in clathrin-coated pits of epithelial cells of the kidney [1570, 1572] and in the embryo [1573]. The rat [1574] and human [1575] forms of gp330 have been cloned. The latter contains 4655 amino acids with a molecular mass of 520 kDa. It has a 25 amino acid signal sequence, a 23-amino acid transmembrane region, and an extracellular region containing 4397 amino acids. The extracellular domain contains the typical functional motifs of the lipoprotein receptor family [1575]. Gp330 has a receptor profile similar to that of LRP/α_2MR It binds α_2MRAP [1576], plasminogen activator/ plasminogen activator inhibitor complexes [1526], activated α_2-macroglobulin [1577], LPL [1423, 1578], and lactoferrin. The biological function of gp330 is not known.

b. The VLDL receptor

The VLDL receptor is a recent member of the LDL receptor family [1579]. It has been cloned from a rabbit heart cDNA library. It is expresssed in heart, muscle, and adipose tissue, but its mRNA is barely detectable in liver. The protein has the five domains typical of other members of the family. It specifically recognizes and internalizes apoE-containing lipoproteins, including VLDL, β-VLDL, and IDL, but does not bind LDL. The VLDL receptor binds α_2MRAP and mediates the uptake and catabolism of LPL and u-PA/PAI-1 complexes [1580]. Its role may be to mediate the uptake of triacylglycerol-rich, apoE-containing lipoproteins in nonhepatic tissues.

c. The apoER2 receptor

A late member of the LDL receptor family is human apoE receptor 2 [1581]. This receptor contains five functional domains that resemble those of the LDL receptor family. The receptor binds β-VLDL with high affinity but has only low affinity for LDL. It is expressed in multiple mRNA species in the brain of humans and rabbits. Interestingly, the receptor mRNA is induced by nerve growth factor.

d.   The LR8B receptor

Another late member of the LDL receptor family is referred to as LR8B to designate that is is an LDL receptor analogue with eight ligand binding repeats [1582]. It is an 894 residue glycoprotein found in the brain of chicken and mouse.

e.   The sorLA-1 receptor

A human receptor designated sorLA-1 (sorting protein-related receptor containing LDL receptor class A repeats), which may be involved in the uptake of lipoproteins, has been isolated from human brain and cloned. The protein has domains similar to those of the lipoprotein receptor family, as well as a Vps10p/sortilin domain and six fibronectin type III repeats [1583].

C.   HSPG as a Receptor

1.   Introduction

In addition to LPL and HL, several other proteins that play roles in lipoprotein metabolism are heparin-binding proteins. These include apoB, apoE, and α_2MRAP. We have already noted that α_2MRAP is a heparin-binding protein, and there is little more that is known about the physiological significance of this fact. However, the heparin-binding apolipoproteins are more extensively characterized. Properties of the apolipoproteins have been reviewed [643, 1584, 1585].

2.   Apoliprotein B

The cloning of apoB-100 has been reported [1586]. ApoB is a glycoprotein that occurs in several forms: apoB-100, the full-length protein of 550 kDa; apoB-48, a truncated version limited to the N-terminal 48% of apoB-100; and apoB-apo(a), a covalent complex consisting of apoB-100 linked by a disulfide bridge to apo(a), a polymorphic protein homologous to plasminogen (Chapter 8) [1587]. ApoB-100 (4536 amino acids), which is synthesized in the liver, and apoB-48 (2152 amino acids), which is synthesized in the intestine, are products of a single gene [1588]. The truncation that leads to apoB-48 synthesis results from apoB mRNA editing that converts apoB codon 2153 to a stop codon [1589]. Thus, apoB-48 lacks the C-terminal half of the protein, which contains the LDL receptor-binding domain [1590, 1591]. No function has been attributed to this region, although it has been shown that LPL binds to the N-terminal half of the molecule [1592, 1593]. It has also been shown that a 116 kDa heparin-releasable protein, which appears on the surface of endothelial cells and which binds LPL, is a metabolic product that results from the proteolytic cleavage of apoB-100 (see Section III.C.6., p. 383) [1481, 1592].

The heparin-binding domains of apoB have been examined by isolation of heparin-binding peptides formed from apoB. Weisgraber and Rall isolated a series of such peptides from CNBr/V-8 protease cleavage of apoB-100 and showed that sequences that correspond to residues 3134-3209 and residues 3356-3489 bound most tightly [1594]. Both of these have regions containing clusters of arginine and lysine residues. Thus, these authors conclude that the primary heparin-binding sequences are in the C-terminal region of apoB-100 and therefore would not be present in apoB-48. A similar approach by Hirose *et al.* [1595] identified four CNBr heparin-binding peptides that correspond to residues 2016-2151, 3109-3240, 3308-3394, and 3570-3719. These sequences are essentially the same as those identified in the Weisgraber and Rall study. Hirose *et al.* suggest that the following clusters that fall within the isolated peptides are involved in heparin binding:

$$V V R K Y R A A L$$
$$2081 \qquad\qquad 2088$$

$$T T K K Y R I T E$$
$$2119 \qquad\qquad 2126$$

$$Y K K N K H R H$$
$$3150 \qquad\qquad 3157$$

$$T R K R G L K L$$
$$3361 \qquad\qquad 3368$$

$$G R R Q H L R V$$
$$3670 \qquad\qquad 3677$$

In fact, these sequences were among those that led Cardin and Weintraub to suggest their consensus binding sequences for heparin (Chapter 6). It should be noted that neither apoB-100 nor the peptides that are released from it, bind with very high affinity to heparin. For example, 96% of unfractionated heparin does not bind to apoB-100 in 10 mM Tris, pH 8, 10 mM CaCl$_2$, 0.01% NaN$_3$, and the remaining 4% is eluted with 0.5 M NaCl in the same buffer (this protein might have eluted at lower concentrations of NaCl, which were not tested) [1596]. Thus, in contrast to frequent statements in the literature, apoB binds heparin only weakly.

Heparin releases LDL from cell surfaces [1597], an observation that suggests that the LDL receptor binding domain lies near the heparin-binding domain. Chemical modification of lysine and arginine residues in apoB decreases the binding to heparin [1439, 1598].

3.   Apolipoprotein E

ApoE is a component of chylomicrons, VLDL, remnants, and of some subclasses of HDL. It is synthesized primarily in the liver and is observed in several isoforms, including three that occur with specific single amino acid substitutions and three that result from variations in the sialylation of O-linked oligosaccharides [1599]. Interestingly, the high level of sialic acid in newly synthesized apoE is, in large part, lost in the circulation. Human apoE, which has been both sequenced [1600] and cloned [1601, 1602], contains 299 amino acids (M_r=34.2 kDa), with a highly ordered secondary structure [[1600, 1603]. Heparinoids exhibit several effects on apoE and the lipoproteins that contain it. *In vivo*, heparan sulfate proteoglycan modulates the synthesis and secretion of apoE [1604]. *In vitro*, heparin causes the precipitation of apoE-containing lipoproteins [1598]. Also, both apoE [1605, 1606] and the lipoproteins that contain it bind to heparin affinity columns and are eluted with buffers containing NaCl or NH_4HCO_3 [1607]. The affinity of apoE and its lipoproteins for heparin, however, is relatively low. For example, precipitation of apoE-containing lipoproteins occurs at physiological salt concentrations when heparin and $MnCl_2$ are added in increasing concentrations [1598]. These apoE-containing lipoproteins can be purified by binding to heparin-Sepharose in 0.025 $\underline{M}$ NaCl and elution with 0.095 NaCl in 5 m$\underline{M}$ Tris buffer (in the absence of $MnCl_2$), i.e., at salt concentrations *below* physiological levels [1607]. Finally, apoE itself has been isolated by heparin affinity chromatography by binding the apoE to heparin-Sepharose in 0.25 $\underline{M}$ NH_4HCO_3 and eluting with 0.75 $\underline{M}$ NH_4HCO_3, without attempting to use intermediate salt levels [1605]. Also, apoE has been purified by heparin affinity chromatography by elution from heparin-Sepharose with NaCl [1606]. In this case the apoE was eluted in a salt gradient with 0.4 $\underline{M}$ NaCl. It is not clear whether appropriate species of heparan sulfate proteoglycan (e.g., heparan sulfate proteoglycan from the liver parenchymal cells) bind with similar low affinity.

Several approaches have been taken to identify the heparin-binding region(s) of apoE. In one, monoclonal antibodies with known epitopes on apoE were used to inhibit heparin binding to apoE [1608]. This work leads to the conclusion that residues 142-147 and residues 243-272 are involved in heparin binding when apoE is presented in the lipid-free state. When apoE is complexed to phospholipids, only the first site is exposed. This is a sequence enriched in basic amino acids:

$$\underline{R}\,\underline{K}\,L\,\underline{R}\,\underline{K}\,\underline{R}$$
$$142 \qquad\qquad 147$$

In a second approach, apoE was treated with thrombin and the resulting heparin-binding peptides were identified [1609]. This work identified two sequences,

$$L\,\underline{R}\,\underline{K}\,\underline{R}$$
$$144 \qquad 147$$

and

$$G\,E\,\underline{R}\,L\,\underline{R}\,A\,\underline{R}\,M$$
$$211 \qquad\qquad\qquad 218$$

one of which is contained in the sequence identified in the MAb study. Thus, both studies identify similar regions of apoE as heparin-binding domains. The binding of heparin to an extended version of the 142-147 sequence induces significant changes in secondary structure of the peptide [1610]. It is interesting to note that the *receptor*-binding sequence of apoE has been shown to reside within residues 140-160, i.e., in the same region that seems to contain one of the heparin-binding sequences [1584, 1611-1614]. It has been suggested that the interaction of apoE-containing lipoproteins with the LDL receptor is through ionic interactions and that this explains why heparin and other polyanions inhibit this binding [1615].

D. The Metabolism of β-VLDL

It was noted earlier that the receptor activity of LRP/α_2MR has been studied using β-VLDL as the ligand (see Section IV.B.1.b., p. 397) and that heparan sulfate proteoglycan seems to play a role in LRP/α_2MR-mediated endocytosis of beta-VLDL (see Section IV.C., p. 402). The binding and internalization of β-VLDL are markedly stimulated by the addition of apoE [reviewed in 1539, 1616]. In fact, when rat hepatoma cells are transfected with apoE cDNA, the amount of β-VLDL that they bind is increased 2- to 3.5-fold, whereas transfection with apoE-Leiden, a defective apoE, gives the same β-VLDL binding as found in the untransformed cells [1617], see also [1618]. Consequently, uptake of β-VLDL by cells is usually measured in the presence of added apoE. The internalized β-VLDL/apoE accumulates in the endosomes and the lysosomes. In the lysosomes, the β-VLDL is catabolized [1544] and the released cholesterol is esterified by acyl-CoA:cholesterol acyltransferase [1566]. Thus, the uptake of β-VLDL can be assayed by measuring the incorporation of [^{14}C]oleate into cholesterol esters. Uptake of triacylglycerol-rich lipoproteins by the VLDL receptor is also stimulated by apoE [1619].

A significant body of data indicates that LRP/α_2MR is involved in β-VLDL uptake [1530, 1533, 1544, 1620, 1621]. However, the kinetics of the initial phase of the clearance suggested that some other component may be responsible for the initial sequestration of remnants. This led to the demonstration that heparan sulfate proteoglycan is involved. Treatment of HepG2 cells, normal human fibroblasts, and familial hyperlipoproteinemia fibroblasts with heparin lyases I and III removes 80, 80, and 95 %, respectively, of β-VLDL/ApoE-binding sites [1617, 1622]. CHO cell mutants that produce no proteoglycans (Chapter 5) do not bind β-VLDL/ApoE, even though they produce LRP/α_2MR [1622]. Interestingly, the latter cells bind α_2-macroglobulin, indicating (a) that the LRP/α_2MR is functional, and (b) that HSPG is not required for α_2-macroglobulin binding. β-VLDL that is bound to apoE-transfected rat hepatoma cells can be displaced by heparin or by the heparin analog, suramin (Chapter 8), again implying that the β-VLDL binds to HSPG [1617, 1623]. Finally, when heparin lyase I is infused into mice, there is a marked reduction in the rate of clearance of chylomicron remnants and β-VLDL, whether or not apoE is co-administered [1624].

If binding of β-VLDL requires excess apoE *in vitro*, where does it obtain this apoE *in vivo*? It has been postulated that as β-VLDL (and normal chylomicron and VLDL remnants) are taken up by the liver, they enter the space of Disse where they are sequestered in an environment surrounding the parenchymal cells, which are secreting apoE. The remnants bind the excess apoE, thus facilitating their internalization by the LRP/α_2MR; i.e., the secreted apoE is recaptured by the parenchymal cells along with the lipoprotein remnants. This is referred to as the "secretion-recapture" mechanism [1539, 1616] and is illustrated in Figure 11.6. In fact, when β-VLDL is re-isolated from apoE-transfected rat hepatoma cells, it is found to contain significantly elevated levels of apoE [1617]. According to the secretion-recapture model, the binding of remnants to the cell surface requires either (a) binding of the lipoproteins first to HSPG followed by transfer to LRP/α_2MR, as illustrated in Figure 11.7, or (b) binding simultaneously to both HSPG and LRP/α_2MR, in a manner similar to one of the mechanisms proposed for HSPG-mediated FGF/FGFR binding.

E.   The Effect of LPL and HL on Lipoprotein Metabolism

It was indicated in the earlier discussion that the action of LPL bound to the endothelium results in the release of free fatty acids, and that one effect of the free fatty acids is to cause the release of LPL into the plasma where it circulates in association with lipoproteins until it is cleared by the liver. Thus, the remnants that enter the space of Disse are, at least in part, already complexed

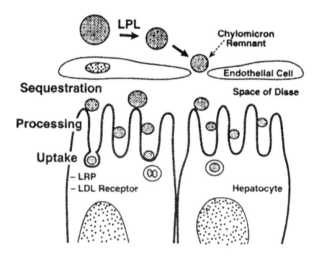

Figure 11.6. The secretion-recapture model for lipoprotein uptake. Remnants from the plasma enter the space of Disse and become bound to HSPG. They then become enriched in apoE that is secreted by the hepatocytes and are taken up via the LRP/α_2MR in a process that may be mediated by the LPL that is bound to the remnants or by the HL that is present on the surface of the hepatocytes (see text). Reprinted from [1539], with permission.

to LPL. In 1975,Felts *et al.* [1625] proposed that the LPL attached to chylomicron remnants might be a signal that allows specific recognition of these particles by the liver. It has now been demonstrated in a number of cell types that the binding and internalization of β-VLDL, and of other types of lipoproteins, are markedly stimulated by LPL [1384, 1423 82, 1426, 1427, 1619, 1626-1630, but see 1631]. Similar stimulatory effects have been observed when LPL is replaced with HL [1617, 1632, 1633]. The stimulation of binding by LPL ranges from 30-to 40-fold up to about 80-fold, depending on the cell type [1423, 1427, 1627]. The stimulation is observed when LPL is added to cells together with various lipoproteins, when cells are preincubated with LPL before the addition of lipoproteins, or when cells are transfected with cDNAs for lipases [1634]. The binding of LPL to lipoproteins requires the lipid-binding (C-terminal) domain of the LPL and does not depend on any apolipoprotein components of the lipoproteins [1384, 1424, 1635-1637]. In fact, it has been shown that apoE can inhibit LPL-mediated lipoprotein catabolism and might play a role in modulating this process [1638]. It has been demonstrated that sequences between amino acid residues 313 and 348 in the bovine LPL, and between 378 and 423 in the human LPL, are required for binding of LPL to

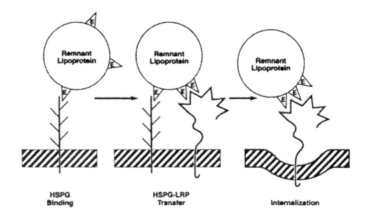

Figure 11.7. Possible mechanism for participation of both heparan sulfate proteoglycan and LRP in lipoprotein uptake. Reprinted from [1539], with permission.

LRP/α_2MR [1636, 1637] and that LPL binds to the α-subunit of LRP/α_2MR [1424].

Interestingly, both LRP/α_2MR and HSPG are involved in the LPL- and HL-mediated uptake of lipoproteins [1531]. Chondroitin sulfate proteoglycan and dermatan sulfate proteoglycan, both of which are more abundant in the artery wall, also may play a role in LPL-mediated lipoprotein uptake, but their effects are much less pronounced than those of HSPG [1425].

We have already discussed the findings that implicate HSPG in the uptake of LPL alone. A number of similar observations demonstrate the absolute requirement of HSPG for the LPL-mediated uptake of lipoproteins:

- In cultured fibroblasts and HepG2 cells, heparinase treatment of the cells reduces the LPL-enhanced binding of chylomicrons 90-95% [1374, 1384, 1426, 1427].

- The effect of LPL on chylomicron binding to cells is minimal in CHO cell mutants that do not synthesize proteoglycans [1384].

- Heparin prevents the stimulation of β-VLDL binding to rat hepatoma cells that are transfected with HL cDNA [1426, 1639].

- Chlorate treatment of rat hepatoma cells prevents the stimulation of β-VLDL binding to rat hepatoma cells transfected wih HL cDNA [1639].

The requirement for LRP/α_2MR in the LPL-mediated binding and uptake of lipoproteins has also been established by a similar set of results:

- LRP/α_2MR binds LPL and may mediate its metabolism [1374, 1423, 1424, 1426, 1636, 1637, 1640].

- The LPL-stimulated catabolism of lipoproteins is partially or completely blocked by antibodies that bind LRP/α_2MR [1374, 1526].

- The LPL-stimulated catabolism of lipoproteins is partially or completely blocked by α_2M* or α_2MRAP, both of which bind to LRP/α_2MR [1374, 1526].

- Cross-linking experiments show that LPL can be linked to a protein of the size of LRP/α_2MR [1423].

- LPL binds directly to LRP/α_2MR in the absence of proteoglycans [1374].

- LPL stimulates the binding of lipoproteins to purified LRP/α_2MR without a requirement for proteoglycans [1374].

F. Models of Lipoprotein Endocytosis

The cell surface interactions that may play a role in lipoprotein clearance from the plasma have become numerous indeed. Receptors in the liver include the LDL receptor, LRP/α_2MR, and HSPG. Lipoproteins can bind directly to their specific receptors; they may bind to HSPG via apoB or apoE; or they may bind to LPL or HL, both of which can bind to HSPG, to LRP/α_2MR, or to both. Although lipoprotein uptake may have seemed (relatively) clear several years ago, it has now become quite complex. This complexity arises in part from the recently documented requirement for HSPG in several of the binding/endocytosis mechanisms. The following scenarios for lipoprotein binding to the cell surface prior to uptake and catabolism in the liver must be considered:

- Lipoproteins may bind directly to their specific receptors—the LDL receptor for LDL and IDL; and the LRP/α_2MR for the remnants of VLDL and chylomicrons.

- Lipoproteins may bind directly to HSPG via apoB or apoE

- Lipoproteins may bind cooperatively to *both* HSPG and their specific receptors (Figure 11.8).

- Lipoproteins may be concentrated on the cell surface by binding to HSPG, and then transferred to their receptors (Figure 11.7).

- Lipoproteins may bind to LPL or HL, which in turn may bind to LRP/α_2MR. In this case, the lipases *tether* the lipoproteins in the circulation to LRP/α_2MR on the cell surface.

- Some lipoproteins may bind to LPL or HL, which in turn are bound to HSPG. Again, the lipases become *tethers* that bind the lipoproteins in the circulation to HSPG on the cell surface.

- Lipoproteins may bind cooperatively to *both* LPL (or HL) and their receptors. It is also possible that HSPG may be a transient member of such complexes.

- Lipoproteins may be concentrated on the cell surface by binding to LPL or HL and then transferred to their receptors.

What actually happens is not clear. Many of these mechanisms may operate, and the relative contributions of each may vary with the cell type. For example, Fernández-Borja *et al.* have shown that in the absence of LPL (or HL), human fibroblasts internalize triacylglycerol-rich lipoproteins into endosomes and lysosomes. However, LPL itself localizes on the cell surface, not to LRP/α_2MR, but to HSPG. When lipoproteins are added together with LPL, they also localize in a manner identical to that of HSPG-bound LPL [1629]. The lipoproteins that bind in the presence of LPL are internalized slowly (typical of HSPG uptake) and appear in small, widely scattered vesicles, i.e., not in endosomes. Thus, two modes of uptake may occur, depending on the presence of LPL. However, a mechanism that will explain the absolute dependence on HSPG for LRP/α_2MR-mediated uptake of lipoprotein lipase remains to be defined.

The different behaviors of various cell types in lipoprotein internalization are highlighted by a report of Obunike *et al.* [1628]. These workers found that the $t_{1/2}$ of cell surface-bound LDL and LPL at 37 ^{0}C were 1 hr in THP-1 cells (a macrophage cell line) and 6 hr in fibroblasts. The authors showed that LRP/α_2MR and not the LDL receptor is involved in LPL-mediated degradation of LDL in fibroblasts and suggested that internalization of LPL/LDL/HSPG complexes account for a significant proportion of the LDL uptake in fibroblasts [1626]. Perhaps the most significant conclusion from this work is that the relative importance of LPL-mediated LDL uptake may differ among cell types.

ADDITIONAL REFERENCES

Bergö, M., Olivecrona, G., and Olivecrona, T. (1996) Diurnal rhythms and effects of fasting and refeeding on rat adipose tissue lipoprotein lipase, *Am. J. Physiol.* **271**, E1092-E1097.

Breedveld, B., Schoonderwoerd, K., Verhoeven, A. J. M., Willemsen, R., and Jansen, H. (1997) Hepatic lipase is localized at the parenchymal cell microvilli in rat liver, *Biochem. J.* **321**, 425-430.

Hardy, M. M., Feder, J., Wolfe, R. A., and Bu, G. (1997) Low density lipoprotein receptor-related protein modulates the expression of tissue-type plasminogen activator in human colon fibroblasts, *J. Biol. Chem.* **272**, 6812-6817.

Küchenhoff, A., Harrach-Ruprecht, B., and Robenek, H. (1997) Interaction of apo E-containing lipoproteins with the LDL receptor-related protein LRP, *Am. J. Physiol.* **272**.

Lookene, A., Savonen, R., and Olivecrona, G. (1997) Interaction of lipoproteins with heparan sulfate proteoglycans and with lipoprotein lipase. Studies by surface plasmon resonance technique, *Biochemistry* **36**.

Macri, J., and Adeli, K. (1997) Studies on intracellular translocation of apolipoprotein B in a permeabilized HepG2 system, *J. Biol. Chem.* **272**, 7328-7337.

Mehta, K. D., Chang, R., Underwood, J., Wise, J., and Kumar, A. (1996) Identification of a novel *cis*-acting element participating in maximal induction of the human low density lipoprotein receptor gene transcription in response to low cellular cholesterol levels, *J. Biol. Chem.* **271**, 33616-33622.

Mikhailenko, I., Krylov, D., Argraves, M. K., Roberts, D. D., Liau, G., and Strickland, D. K. (1997) Cellular internalization and degradation of thrombospondin-1 is mediated by the amino-terminal heparin binding domain (HBD). High affinity interaction of dimeric HBD with the low density lipoprotein receptor-related protein, *J. Biol. Chem.* **272**, 6784-6791.

Narita, M., Bu, G., Olins, G. M., Higuchi, D. A., Herz, J., Broze Jr., G. J., and Schwartz, A. L. (1995) Two receptor systems are involved in the plasma clearance of tissue factor pathway inhibitor *in vivo*, *J. Biol. Chem.* **270**, 24800-24804.

Nielsen, M. S., Brejning, J., García, R., Zhang, H., Hayden, M. R., Vilaró, S., and Gliemann, J. (1997) Segments in the C-terminal folding domain of lipoprotein lipase important for binding to the low density lipoprotein receptor-related protein and to heparan sulfate proteoglycans, *J. Biol. Chem.* **272**, 5821-5827.

Olsson, U., Camejo, G., Hurt-Camejo, E., Elfsber, K., Wiklund, O., and Bondjers, G. (1997) Possible functional interactions of apolipoprotein B-100 segments that associate with cell proteoglycans and the apoB/E receptor, *Arterioscler. Thromb. Vasc. Biol.* **17**, 149-155.

Pillarisetti, S., Paka, S., Sasaki, A., Vanni-Reyes, T., Yin, B., Parthasarathy, N., Wagner, W. D., and Goldberg, I. J. (1977) Endothelial cell heparanase modulation of lipoprotein lipase activity: evidence that heparan sulfate oligosaccharide is an extracellular chaperone, *J. Biol. Chem.* **272**, XXXXXXXXXXXXX.

Ranganathan, G., Vu, D., and Kern, P. A. (1997) Translational regulation of lipoprotein lipase by epinephrine involves a trans-acting binding protein interacting with the 3' untranslated region, *J. Biol. Chem.* **272**, 2515-2519.

Sehayek, E., Wang, X.-X., Vlodavsky, I., Avner, R., Levkovitz, H., Olivecrona, T., Olivecrona, G., Willnow, T. E., Herz, J., and Eisenberg, S. (1996) Heparan sulfate-dependent and low density lipoprotein receptor-related protein-dependent catabolic pathways for lipoprotein lipase in mouse embryonic fibroblasts, *Isr. J. Med. Sci* **32**, 449-454.

Seo, T., and St. Clair, R. W. (1997) Heparan sulfate proteoglycans mediate internalization and degradation of β-VLDL and promote cholesterol accumulation by pigeon macrophages, *J. Lipid Res.* **38**, 765-779.

Wittrup, H. H., Tybjaerg-Hansen, A., Abildgaard, S., Steffensen, R., Schnohr, P., and Nordestgaard, B. G. (1997) A common substitution (Asn291Ser) in lipoprotein lipase is associated with increased risk of ischemic heart disease, *J. Clin. Invest.* **99**, 1606-1613.

Ziere, G. J., van der Kaaden, E., Vogelezang, C. J. M., Boers, W., Bihain, B. E., Kuiper, J., Kruijt, J. K., and van Berkel, T. J. C. (1996) Blockade of the α_2-macroglobulin receptor/low-density-lipoprotein-receptor-related protein on rat liver parenchymal cells by the 39-kDa receptor-associated protein leaves the interaction of β-migrating very-low-density lipoprotein with the lipoprotein remnant receptor unaffected, *Eur. J. Biochem.* **242,** 703-711.

Chapter 12. Epilog

I. INTRODUCTION

A question that arises early in considering heparin-binding proteins is whether the binding of heparin by proteins has any general physiological significance. In fact, it has become evident to workers in this field that it does. Furthermore, it is clear that the heparin-binding properties of proteins are, for the most part, a reflection of important interactions of these secreted proteins with the heparan sulfate chains of cell surface HSPG. This is documented in many cases in the preceding chapters. This chapter presents a brief recapitulation of the most important features of the metabolism and roles of HSPG, and the nature of the interactions between heparinoids and the proteins to which they bind.

II. ROLES OF HSPG

A. Protein Activation

The classical example of a heparin-binding protein is antithrombin, a serpin that inactivates thrombin and factor Xa. This activity is a critical one in

413

the maintenance of hemostasis. Antithrombin is not active unless it is bound to a heparinoid.   When heparin is administered as a drug, it activates antithrombin. However, it is the endothelial cell surface HSPG that normally mediates the activation of antithrombin. This is detailed in Chapter 7. Antithrombin is not the only serpin that is activated by heparinoids. Others, described in Chapter 8, include heparin cofactor II, plasminogen activator inhibitor-1, and protein C inhibitor. In addition, some proteins that are not serpins are also activated by heparinoids, including tissue factor pathway inhibitor. For some of the heparin-binding proteins in the circulation, a physiological role of HSPG in the binding or activation has not been established.

B.   Localization and Protection of Bound Proteins

Since HSPG is a cell surface structure that is either anchored in the membrane outer leaflet (glypicans) or embedded in the membrane via a transmembrane segment, proteins that bind to HSPG *in vivo* become localized to cell surfaces. In some cases, the bound protein exhibits its activity while bound to cell surface HSPG (for example, lipoprotein lipase, extracellular superoxide dismutase, and antithrombin). In other cases, the binding of proteins to HSPG is a prelude to endocytosis of the protein (for example, thrombospondin, lipoprotein lipase, various lipoproteins). HSPG-bound proteins are also stabilized and protected from proteolysis on the cell surface. For example, the $t_{1/2}$ of superoxide dismutase is markedly extended as a result of its attachment to heparan sulfate on the endothelial cell surface. Proteins are also protected from proteases by heparan sulfate chains *after* they are internalized (for example, the FGFs).

In some cases heparan sulfate proteoglycan binds proteins that bind to other non-heparan sulfate-binding proteins. In this manner, the HSPG-bound protein serves as a tether for a protein that does not itself bind to HSPG. Lipoprotein lipase, for example, binds to HSPG and to certain lipoproteins, thus tethering the lipoproteins to the HSPG. Also, vitronectin binds to HSPG and to serpin-proteases complexes, tethering these complexes to the cell surface as a prelude to endocytosis. As a consequence of this type of activity, the role of HSPG is extended beyond its direct adhesion to heparin-binding proteins. Thus, it is now evident that many proteins bind to the endothelial cell surface via heparan sulfate proteoglycan. What is not yet known is whether there are many more proteins that are tethered to the proteoglycan.

An interesting question with regard to the binding of multiple proteins to cell surface HSPG is whether there is competition among these proteins for the available heparan sulfate proteoglycan. It is established that FGF and antithrombin bind only to small fractions of the endothelial heparan sulfate proteoglycan, but it is not known whether the same heparan sulfate chains recognize these (and other) proteins.

C. Chaperones

Heparinoids cannot serve as chaperones for heparin-binding proteins during the secretion of such proteins. This follows from the fact that sulfated species of heparan sulfate chains are necessary for protein binding, and the sulfation of heparan sulfate chains occurs quite late in the secretory pathway, just prior to their secretion. Failure of HSPG to act as a chaperone during secretion has been demonstrated for lipoprotein lipase. It seems likely, however, that HSPGs can serve as chaperones for proteins to which they bind during transcytosis, i.e., proteins that bind to HSPG on the *outside* of cells can be internalized on the HSPG and re-secreted, apparently in complexes with the heparan sulfate chains. If the endocytosis occurs on one surface of the cells and the re-secretion occurs on another, the protein has utilized its heparin-binding capacity to gain passage across the cell. During transit, the heparan sulfate chains may protect the bound proteins from protease action. Transcytosis in this manner appears to occur for vitronectin and lipoprotein lipase, both of which are synthesized by subendothelial cells and transported to the luminal side of the endothelium.

D. Co-Receptors

HSPG serves as a co-receptor for certain proteins and protein complexes that are recognized by typical cell surface receptors. For example, FGFs bind to cell surface FGF receptors which transmit signals into the cell, both as a result of their protein kinase activities and as a result of their endocytosis on these receptors. In a manner that is not totally understood, HSPGs strengthen the binding of the FGFs to their receptors and enhance receptor dimerization. A similar phenomenon has been documented for the endocytosis of lipoproteins on LRP/α_2M receptors. It may be that HSPGs play similar roles for other proteins, but there have been few attempts to address these possibilities.

III. HSPG METABOLISM

A. Variability

Many cells express multiple heparan sulfate core proteins, often simultaneously and in a developmentally regulated fashion. The core protein may control the trafficking of the different HSPGs *after* they are secreted, but it seems that the structures of the heparan sulfate chains found on different core proteins within a single cell type are the same. Thus, the primary roles of the core proteins are (a) the regulation of trafficking of the secreted HSPG and (b) the control of the amount of HSPG produced by the cells. However, the structures of the heparan sulfate chains that appear in different cellular pools may be different, and this may result from poorly understood aspects of proteoglycan trafficking. In addition, the cell can respond to changing environments by changing the structures of its heparan sulfate chains, and these structures can control the responses of cells to various proteins, including growth factors, as described for FGF. Changes in the structures of the heparan sulfate chains may occur rapidly as a result of the rapid metabolic turnover of the proteoglycan

B. Binding and Release

Everything we know about the metabolic flow of endocytosed proteoglycans is in accord with the suggestion that both the protein and the heparan sulfate in protein/heparan sulfate complexes are trafficked together. Thus, both components of such complexes must be considered in attempts to understand the metabolism of either the heparan sulfate chains or the heparin-binding proteins. The problem is complicated by the possibility that heparan sulfate chains on several different core proteins may be internalized simultaneously and that several heparin-binding proteins may be complexed at the same time (a) on different segments of the same heparan sulfate chain, (b) on different heparan sulfate chains on the same core protein, or (c) on different proteoglycans. Attempts to understand the control of the metabolic flow of the heparan sulfate chains would be greatly facilitated (a) if cells that produce a single core protein were chosen, (b) if the number of heparin-binding proteins in each study were limited to one or two, and (c) if the trafficking and catabolism of both heparan sulfate and a heparin-binding protein were followed together.

Metabolism of heparan sulfate-binding proteins varies—some recycle without catabolism; some are routed to the nucleus; and some are simply catabolized. The mechanisms that control these effects are unknown. Also unknown are the mechanisms that control the release of heparin-binding proteins from heparan sulfate chains on the cell surface or the release of heparan sulfate chains into the extracellular milieu of the cells.

C. Tools

There are many unexploited opportunities for exploration various aspects of heparan sulfate proteoglycan metabolism. A few examples illustrate this point.

- Chlorate and selenate are agents that inhibit sulfation. Chlorate has been used extensively, but selenate may give different results.

- There are a number of mutants in the biosynthesis of GPI anchors, but the heparan sulfate metabolism in these cells, and their cells of origin, has not been studied.

- There are inhibitors that block GPI anchor synthesis, but glypican metabolism has not been studied in the presence of these inhibitors.

- 2-Fluoro-2-deoxy-D-glucose is incorporated by chondrocytes into keratan sulfate in place of D-galactose, and into chondroitin sulfate as the 2-fluoro-2-deoxy analog of D-glucuronic acid, but it has not been determined whether this analog will yield interesting modifications of heparan sulfate structures.

- Transition state analogs for UDPGlcA work quite well as inhibitors of the glucuronosyl transferases involved in detoxification, but have not been studied as inhibitors of glycosaminoglycan biosynthesis.

IV. HEPARIN-BINDING PROTEINS

A. Families

Heparin-binding proteins often occur as members of protein families. In some cases, all members of the family seem to respond to heparinoids (for example, the FGFs and the FGFRs), but in others (for example, the serpins), some members of a family do not respond to heparin. These differences may be related to the different roles played by the proteins.

B.   Heparin-Binding Domains

In Chapter 6 we pointed out that as we gain more knowledge of the heparin-binding domains of proteins, it becomes less easy to offer generalizations that allow one to identify these domains. The nature of these domains may be summarized by several generalizations, which are extensions of the earlier suggestions. It is evident that clusters of arginines and lysines on the surfaces of proteins participate in binding to heparinoids. Although many of the amino acids in such clusters are found in single short peptide segments of the protein, it is now obvious that the heparin-binding domains often are formed by several basic peptides that are brought together by the folding of the protein. Interestingly, several cases have been reported in which paired helices, separated by a groove that will accommodate the heparinoid, form the heparin-binding domain. Examples include antithrombin, heparin cofactor II, thrombin, protein C inhibitor, PAI-1, and perhaps thrombospondin. A particularly interesting observation is that one of the antithrombin helices that plays a major role in heparin-binding, when studied as an isolated peptide, appears to take primarily a β-strand conformation. It has become apparent that although it is useful to look for "consensus sequences" for heparin binding, such approaches are not totally satisfactory. The clusters of basic amino acids that form the heparin-binding domain can only be recognized by crystallography or, to some extent, by insightful use of site-directed mutagenesis. In the latter case, it is much more useful to observe the effects of replacements of a single amino acids rather than simultaneous replacement of two or three amino acids, or a whole peptide segment.

A problem that occurs frequently in the literature is the attempt to align two or more symmetrical subunits of a protein with a single strand of heparinoid. If symmetry is assumed in these heparinoid/protein alignments, then *both* subunits of the protein must take the same orientation (N- to C-terminal) with respect to the orientation of the heparinoid (nonreducing terminal to reducing terminal); i.e., like proteins, heparinoids have directionality, a point that is often overlooked.

A related problem is that heparinoids are relatively linear molecules that do not bend as readily as proteins. As described in Chapter 6, a 180° bend in a heparinoid chain may require 20 disaccharide units. Thus, heparin-binding domains of proteins must adapt to the linearity of the heparinoid chain, and models that violate this principle should be viewed with caution.

C. Specific Heparinoid Structures in Protein/Heparinoid Complexes

Remarkably, the first-studied, and the most intensively studied, heparin-binding protein, namely antithrombin, is the only heparin-binding protein found to date that exhibits rigid structural requirements in the heparin segment that gives functional binding. Of course, there are only a few other well-characterized examples of heparinoid sequences that are required for productive heparin binding to proteins. Perhaps the proteins that will exhibit the most exacting heparinoid structural requirements for productive binding are the enzymes that participate in heparinoid biosynthesis and catabolism. Although many of these proteins have been studied, determination of the ranges of heparinoid structures (at the hexa- and octasaccharide levels) that are recognized by these proteins is difficult and has not received as much attention as, for example, the FGFs.

A case for recognition of unique heparinoid structures by proteins other than antithrombin is emerging. Several examples illustrate this point.

- Heparan sulfates from different tissues bind to FGF affinity matrices incompletely. The proportions that bind differ for HSPGs from different tissue sources.

- Hepatic lipase is bound specifically to the HSPG produced by hepatocytes. It does not show high affinity for endothelial cell HSPG.

- Heparin cofactor II is a heparin-binding protein that does not bind to the HSPG that appears on the surface of endothelial cells.

However it seems that some heparin-binding proteins exhibit very little specificity in the heparinoid sequences to which they bind. An example is thrombin, which has been reported to bind any highly sulfated heparin hexasaccharide. Also, despite the finding that some heparan sulfate chains do not bind to FGFs, mentioned above, heparin binding to FGF-2 requires only that the heparinoid sequence contain at least one IdoA-2-SO$_4$ residue. This requirement is met by many sequences found in heparins and in many heparan sulfates.

Most of the studies of the binding of specific heparinoid structures to proteins have been carried out using heparins and heparin derivatives, but not heparan sulfates. Unfortunately, heparin cannot be converted to derivatives that will mimic the structures of heparan sulfates.

Appendix

Other Heparin-Binding Proteins

An Incomplete List

Presented here is a list of heparin-binding proteins. This list does not include proteins that have been discussed in detail in this book. Each protein has one or more references are given. These are not necessarily definitive references, but they represent a point of departure for those who may wish to learn more about these proteins.

This is a list that will become outdated rapidly. New heparin-binding proteins are described frequently in the literature.

Heparin-binding protein	Reference
α_1-Adrenoceptors	[1641]
Amphoterin (neurite growth-promoting protein, p30)	[1642]
Bone morphogenetic protein 2	[1643]
BSP-30K	[1644]
C4β-binding protein	[1645]
Cathepsin D	[1646]
Collagens	[1647, 1648]
Complement factor H	[1649, 1650]
Cytochrome c	[1651]
Diamine oxidase	[1652, 1653]
Elastase	[1654]
Fibrin	[905]
Glycoprotein Bb (bovine herpesvirus 1 gB)	[1655, 1656]
Granulocyte-macrophage colony stimulating factor	[1657]
Heparin biosynthesis, catabolism enzymes	
Heparin-binding-EGF like growth factor	[1658]
Heparin-binding protein-44	[1659]
Hepatocyte growth factor	[1660]
Hepatopoeitin A	[1661]
HIP (Heparan sulfate/heparin Interacting Protein)	[1662]
HIV (gp 120) and HSV surface glycoproteins	[1663, 1664]
Hyaluronidase	[1665]
Insulin-like growth factor binding protein 5	[1666]
Insulin-like growth factors	[1667]

Interferon-γ	[1668]
Interleukin 3	[1657]
Kinogen	[1669]
Lactoferrin	[1670]
Laminin	[1671, 1672]
Leuserpin-2	[922]
L-Selectin	[1673]
L-Type Ca^{2+} channels	[1674]
Malaria circumsporiozoite protein (CS protein)	[1675]
Mast cell proteases	[66, 1676, 1677]
Midkine (pleiotropin) (heparin binding growth-associated factor-HB-GAM) (p18/heparin-binding neurotrophic factor)	[1678]
Myosin ATPase	[1679]
Neural cell adhesion molecule	[1680]
Neurite promoting factor (amphoterin, p30)	[1681]
Neutrophil heparin-binding protein	[1682]
Osteogenin	[1683]
Phosphatidylinositol-4-PO$_4$ kinase	[1684]
Phospholipase A2	[1685, 1686]
Placental heparin-binding growth factor	[1687, 1688]
Platelet-derived growth factor	[1689]
Probasin	[1690]
Retinol-induced heparin-binding protein	[1691, 1692]
Spermadhesins	[1693]
Tenasin	[1694, 1695]
Transforming growth factor-β	[1696]
Trypsin	[1697]
Tumor necrosis factor	[1698]
Vascular endothelial growth factor	[1699]
wingless Protein	[1700]
Xanthine oxidase	[1701, 1702]

References

1. Rodén, L. (1989) Highlights in the history of heparin, in *Heparin: Chemical and Biological Properties, Clinical Applications*, eds., Lane, D. A. & Lindahl, U., CRC Press, Inc., Boca Raton, FL, pp. 1-23.
2. Marcum, J. A. (1996) The discovery of heparin revisited: the peptone connection, *Perspect. Biol. Med.* **39**, 610-625.
3. Nader, H. B., and Dietrich, C. P. (1989) Natural occurrence, and possible biological role of heparin, in *Heparin: Chemical and Biological Properties, Clinical Applications*, eds., Lane, D. A. & Lindahl, U., CRC Press, Inc., Boca Raton, FL, pp. 81-96.
4. Nader, H. B., Ferreira, T. M., Toma, L., Chavante, S. F., Dietrich, C. P., Casu, B., and Torri, G. (1988) Maintenance of heparan sulfate structure throughout evolution: chemical and enzymic degradation and ^{13}C-NMR--spectral evidence, *Carbohyd. Res.* **184**, 292-300.
5. Toledo, O. M. S., and Dietrich, C. P. (1977) Tissue specific distribution of sulfated mucopolysaccharides in mammals, *Biochim. Biophys. Acta* **497**, 114-122.
6. Gomes, P. B., and Dietrich, C. P. (1982) Distribution of heparin and other sulfated glycosaminoglycans in vertebrates, *Comp. Biochem. Physiol.* **73B, No. 4**, 857-863.
7. Cássaro, C. M. F., and Dietrich, C. P. (1977) Distribution of sulfated mucopolysaccharides in invertebrates, *J. Biol. Chem.* **252**, 2254-2261.
8. Murch, S. H., Winyard, P. J. D., Koletzko, S., Wehner, B., Cheema, H. A., Risdon, R. A., Philips, A. D., Meadows, N., Klein, N. J., and Walker-Smith, J. A. (1996) Congenital enterocyte heparan sulphate deficiency with massive albumin loss, secretory, diarrhoea, and malnutrition, *Lancet* **347**, 1299-1301.
9. Linker, A., and Hovingh, P. (1975) Structural studies of heparitin sulfates, *Biochim. Biophys. Acta.* **385**, 324-333.
10. Höök, M., Kjellén, L., Johansson, S., and Robinson, J. (1984) Cell-surface glycosaminoglycans, *Annu. Rev. Biochem.* **53**, 847-869.
11. Casu, B. (1987) Structure and biological activity of heparin, *Adv. Carbohyd. Chem. Biochem.* **43**, 51-134.
12. Casu, B. (1989) Methods of structural analysis, in *Heparin. Chemical and Biological Properties, Clinical Applications*, eds., Lane, D. A. & Lindahl, U., CRC Press, Inc., Boca Raton, FL, pp. 25-49.
13. Kjellén, L., and Lindahl, U. (1991) Proteoglycans: structures and interactions, *Annu. Rev. Biochem.* **60**, 443-475.
14. Gallagher, J. T., Turnbull, J. E., and Lyon, M. (1992) Heparan sulphate proteoglycans: molecular organisation of membrane-associated species and an approach to polysaccharide sequence analysis, in *Heparin and Related Polysaccharides. Advances in Experimental Medicine and Biology, Vol. 313*, eds., Lane, D. A., Björk, I. & Lindahl, U., Plenum Press, New York, pp. 49-57.
15. Lindahl, U., Lidholt, K., Spillman, D., and Kjellén, L. (1994) More to "heparin" than anticoagulation, *Thromb. Res.* **75**, 1-32.
16. Rodén, L. (1980) Structure and metabolism of connective tissue proteoglycans, in *The Biochemistry of Glycoproteins and Proteoglycans*, ed. Lennarz, W. J., Plenum Press, New York, pp. 267-371.
17. Lindahl, U., and Kjellén, L. (1987) Biosynthesis of heparin and heparan sulfate, in *Biology of Proteoglycans*, eds., Mecham, R. & Wight, T., Academic Press, Orlando, FL, pp. 59-103.
18. Lindahl, U., Kusche, M., Lidholt, K., and Oscarsson, L.-G. (1989) Biosynthesis of heparin and heparan sulfate, in *Heparin and Related Polysaccharides, Ann. New York Acad. Sci.*,

vol. 556, eds., Ofosu, F. A. & Danishefsky, I., New York Acad. Sci., New York, pp. 36-50.

19. Lindahl, U. (1989) Biosynthesis of heparin and related polysaccharides, in *Heparin. Chemical and Biological Properties, Clinical Applications*, eds., Lane, D. A. & Lindahl, U., CRC Press, Inc., Boca Raton, FL, pp. 159-189.

20. Salmivirta, M., Lidholt, K., and Lindahl, U. (1996) Heparan sulfate: a piece of information, *FASEB J.* **10**, 1270-1279.

21. van den Born, J., Gunnarsson, K., Bakker, M. A. H., Kjellén, L., Kusche-Gullberg, M., Maccarana, M., Berden, J. H. M., and Lindahl, U. (1995) Presence of N-unsubstituted glucosamine units in native heparan sulfate revealed by a monoclonal antibody, *J. Biol. Chem.* **270**, 31303-31309.

22. Lindahl, U., Larm, O., Wirén, E., and Von Figura, K. (1979) Identification of N-sulphate disaccharide units in heparin-like polysaccharides, *Biochem. J.* **179**, 77-87.

23. Guo, Y., and Conrad, H. E. (1989) The disaccharide composition of heparins and heparan sulfates, *Anal. Biochem.* **176**, 96-104.

24. Tekotte, H., Engel, M., Margolis, R. U., and Margolis, R. K. (1994) Disaccharide composition of heparan sulfates: brain, nervous tissue, organelles, kidney, and lung, *J. Neurochem.* **62**, 1126-1130.

25. Hovingh, P., and Linker, A. (1982) An unusual heparan sulfate isolated from lobsters (*Homarus americanus*), *J. Biol. Chem.* **257**, 9840-9844.

26. Pejler, G., Danielsson, Å., Björk, I., Lindahl, U., Nader, H. B., and Dietrich, C. P. (1987) Structure and antithrombin III-binding properties of heparin isolated from the clams *Anomalocardia brasiliana* and *Tivela mactroides*, *J. Biol. Chem.* **262**, 11413-11421.

27. Kosakai, M., and Yosizawa, Z. (1981) Sulfated oligosaccharides isolated from the deamination products of heparins, *J. Biochem. Tokyo* **89**, 1933-1944.

28. Bienkowski, M. J., and Conrad, H. E. (1985) Structural characterization of the oligosaccharides formed by depolymerization of heparin with nitrous acid, *J. Biol. Chem.* **260**, 356-365.

29. Linker, A. (1979) Structure of heparan sulphate oligosaccharides and their degradation by exo-enzymes, *Biochem. J.* **183**, 711-720.

30. Linhardt, R. J., Grant, A., Cooney, C. L., and Langer, R. (1982) Differential anticoagulant activity of heparin fragments prepared using microbial heparinase, *J. Biol. Chem.* **257**, 7310-7313.

31. Yamada, S., Yoshida, K., Sugiura, M., Sugahara, K., Khoo, K.-H., Morris, H. R., and Dell, A. (1993) Structural studies on the bacterial lyase-resistant tetrasaccharides derived from the antithrombin III-binding site of porcine intestinal heparin, *J. Biol. Chem.* **268**, 4780-4787.

32. Yamada, S., Murakami, Tsuda, H., Yoshida, K., and Sugahara, K. (1995) Isolation of the porcine heparin tetrasaccharides with glucuronate 2-O-sulfate: heparinase cleaves glucuronate 2-O-sulfate-containing disaccharides in highly sulfated blocks in heparin, *J. Biol. Chem.* **270**, 8696-8705.

33. Larnkjær, A., Nykjær, A., Olivecrona, G., Thøgersen, H., and Østergaard, P. B. (1995) Structure of heparin fragments with high affinity for lipoprotein lipase and inhibition of lipoprotein lipase binding to α_2-macroglobulin-receptor/low-density-lipoprotein-receptor-related protein by heparin fragments, *Biochem. J.* **307**, 205-214.

34. Loganathan, D., Wang, M. M., Mallis, L. M., and Linhardt, R. J. (1990) Structural variation in the antithrombin III binding site region and its occurrence in heparin from different sources, *Biochemistry* **29**, 4362-4368.

35. Casu, B., Oreste, P., Torri, G., Zoppetti, G., Choay, J., Lormeau, J.-C., and Petitou, M. (1981) The structure of heparin oligosaccharide fragments with high anti-(factor Xa) activity containing the minimal antithrombin III-binding sequence: chemical and ^{13}C NMR studies, *Biochem. J.* **197**, 599-609.

36. Tsuda, H., Yamada, S., Yamane, Y., Yoshida, K., Hopwood, J. J., and Sugahara, K. (1996) Structures of five sulfated hexasaccharides prepared from porcine intestinal

heparin using bacterial heparinase: structural variants with apparent biosynthetic precursor-product relationships for the antithrombin III-binding site, *J. Biol. Chem.* **271**, 10495-10502.

37. Desai, U. R., Wang, H.-M., and Linhardt, R. J. (1993) Specificity studies of the heparin lyases from *Flavobacterium heparinum, Biochemistry* **32**, 8140-8145.
38. Atha, D. H., Stephens, A. W., and Rosenberg, R. D. (1984) Evaluation of critical groups required for the binding of heparin to antithrombin, *Proc. Natl. Acad. Sci. U.S.A.* **81**, 1030-1034.
39. Thunberg, L., Bäckström, G., and Lindahl, U. (1982) Further characterization of the antithrombin-binding sequence in heparin, *Carbohyd. Res.* **100**, 393-410.
40. Ototani, N., Kikuchi, M., and Yosazawa, Z. (1982) Structure and biological activity of finback-whale (*Baleonoptera physalus L.*) heparin octasaccharides, *Biochem. J.* **205**, 23-30.
41. Ototani, N., and Yosizawa, Z. (1981) Antithrombin activity of heparin octasaccharide, *J. Biochem. (Tokyo)* **90**, 1553-1556.
42. Lane, D. A., and Lindahl, U. (1989) Preface, in *Heparin. Chemical and Biological Properties, Clinical Applications*, eds., Lane, D. A. & Lindahl, U., CRC Press, Inc., Boca Raton, FL, pp. ix-x.
43. Gallagher, J. T., and Walker, A. (1985) Molecular distinctions between heparan sulfate and heparin: analysis of sulphation patterns indicates that heparan sulfate and heparin are separate families of N-sulphated polysaccharides, *Biochem. J.* **230**, 665-674.
44. Nader, H. B., McDuffie, and Dietrich (1974) Heparin fractionation by electrofocussing: presence of 21 components of different molecular weights, *Biochemistry* **57**, 488-493.
45. McLean, J. (1916) The thromboplastic action of cephalin, *Am. J. Physiol.* **41**, 250-257.
46. Jorpes, J. E., and Gardell, N. (1948) On heparin monosulfuric acid, *J. Biol. Chem.* **176**, 267-276.
47. Cifonelli, J. A., and Dorfman, A. (1960) Properties of heparin monosulfate (heparitin monosulfate), *J. Biol. Chem.* **235**, 3283-3286.
48. Cifonelli, J. A. (1968) Reaction of heparitin sulfate with nitrous acid, *Carbohyd. Res.* **8**, 233-242.
49. Cifonelli, A. J., and King, J. (1977) Structural characteristics of heparan sulfates with varying sulfate contents, *Biochemistry* **16**, 2137-2141.
50. Fransson, L.-Å., L. Cöster, L., Carlstedt, I., and Malmström, A. (1985) Domain structure of proteoheparan sulphate from confluent cultures of human embryonic skin fibroblasts, *Biochem. J.* **231**, 683-687.
51. Dietrich, C. P., Nader, H. B., Britto, L. R. G., and Silva, M. E. (1971) Chemical composition of heparitin sulfate: fractionation and characterization of four acidic mucopolysaccharides in heparitin sulfate from beef lung tissue, *Biochim. Biophys. Acta.* **237**, 430-441.
52. Ayotte, L., Mushayakarara, E., and Perlin, A. S. (1980) Fractionation of heparin and heparan sulfate as barium salts: high-field, n.m.r.-spectral observations on heterogeneity, *Carbohyd. Res.* **87**, 297-301.
53. Delaney, S. R., and Conrad, H. E. (1983) Changes in disaccharide composition of heparan sulphate fractions with increasing degrees of sulphation, *Biochem. J.* **209**, 315-322.
54. Maccarana, M., Sakura, Y., Tawada, A., Yoshida, K., and Lindahl, U. (1996) Domain structure of heparan sulfates from bovine organs, *J. Biol. Chem.* **271**, 17804-17810.
55. Knecht, J., Cifonelli, J. A., and Dorfman, A. (1967) Structural studies on heparitin sulfate of normal and Hurler tissues, *J. Biol. Chem.* **242**, 4652-4661.
56. Höök, M., Lindahl, U., and Iverius, P.-H. (1974) Distribution of sulfate and IdoA residues in heparin and heparan sulphate, *Biochem. J.* **137**, 33-43.
57. Lindblom, A., Bengtsson-Olivecrona, G., and Fransson, L.-Å. (1991) Domain structure of endothelial heparan sulphate, *Biochem. J.* **279**, 821-829.

58. Hampson, I. N., Kumar, S., and Gallagher, J. T. (1984) Heterogeneity of cell-associated and secretory heparan sulfate proteoglycans produced by cultured human neuroblastoma cells, *Biochim. Biophys. Acta* **801,** 306-313.

59. Lyon, M., Steward, W. P., Hampson, I. N., and Gallagher, J. T. (1987) Identification of an extended N-acetylated sequence adjacent to the protein-linkage region of fibroblast heparan sulphate, *Biochem. J.* **242,** 493-498.

60. Nader, H. B., Dietrich, C. P., Buonassisi, V., and Colburn, P. (1987) Heparin sequences in the heparan sulfate chains of an endothelial cell proteoglycan, *Proc. Natl. Acad. Sci. U.S.A.* **84,** 3565-3569.

61. Turnbull, J. E., and Gallagher, J. T. (1991) Sequence analysis of heparan sulphate indicates defined location of N-sulphated glucosamine and iduronate 2-sulphate residues proximal to the protein-linkage region, *Biochem. J.* **277,** 297-303.

62. Ashikari, S., Habuchi, H., and Kimata, K. (1995) Characterization of heparan sulfate oligosaccharides that bind to hepatocyte growth factor, *J. Biol. Chem.* **270,** 29586-29593.

63. Ögren, S., and Lindahl, U. (1975) Cleavage of macromolecular heparin by an enzyme from mouse mastocytoma, *J. Biol. Chem.* **250,** 2690-2697.

64. Ögren, S., and Lindahl, U. (1976) Metabolism of macromolecular heparin in mouse neoplastic mast cells, *Biochem. J.* **154,** 605-611.

65. Robinson, H. C., Horner, A. A., Höök, M., Ögren, S., and Lindahl, U. (1978) A proteoglycan form of heparin and its degradation to single-chain molecules, *J. Biol. Chem.* **253,** 6687-6693 .

66. Matsumota, R., Šali, A., Ghildyal, N., Karplus, M., and Stevens, R. L. (1995) Packaging of proteases and proteoglycans in the granules of mast cells and other hematopoietic cells: a cluster of histidines on mouse mast cell protease 7 regulates its binding to heparin serglycin proteoglycans, *J. Biol. Chem.* **270,** 19524-19531.

67. Lindahl, U. (1964) The linkage of heparin to protein, *Biochem. Biophys. Res. Commun.* **17,** 254-259.

68. Lindahl, U., and Rodén, L. (1965) The role of galactose and xylose in the linkage of heparin to protein, *J. Biol. Chem.* **240,** 2821-2826.

69. Lindahl, U., Cifonelli, J. A., Lindahl, B., and Rodén, L. (1965) The role of serine in the linkage of heparin to protein, *J. Biol. Chem.* **240,** 2817-2820.

70. Esko, J. D. (1991) Genetic analysis of proteoglycan structure, function and metabolism, *Curr. Opin. Cell Biol.* **3,** 805-816.

71. Forsee, W. T., and Rodén, L. (1981) Biosynthesis of heparin: transfer of N-acetylglucosamine to heparan sulfate oligosaccharides, *J. Biol. Chem.* **256,** 7240-7247.

72. Navia, J. L., Riesenfeld, J., Vann, W. F., Lindahl, U., and Rodén, L. (1983) Assay of N-acetylheparosan deacetylase with a capsular polysaccharide from *E. coli* K5 as substrate, *Anal. Biochem.* **135,** 134-140.

73. Fritz, T. A., Gabb, M. M., Wei, G., and Esko, J. D. (1994) Two N-acetyl-glucosaminyltransferases catalyze the biosynthesis of heparan sulfate, *J. Biol. Chem.* **269,** 28809-28814.

74. Zhang, L., and Esko, J. D. (1995) Accumulation of a pentasaccharide terminating in α-N-acetylglucosamine in an animal cell mutant defective in heparan sulfate biosynthesis, *J. Biol. Chem.* **270,** 12557-12562.

75. Lindahl, U., Bäckström, G., Jansson, R., and Hallén, A. (1973) Biosynthesis of heparin. II. Formation of sulfamino groups, *J. Biol. Chem.* **248,** 7231-7241.

76. Riesenfeld, J., Höök, M., and Lindahl, U. (1980) Biosynthesis of heparin: assay and properties of the microsomal N-acetyl-D-glucosaminyl N-deacetylase, *J. Biol. Chem.* **255,** 922-928.

77. Riesenfeld, J., Höök, M., and Lindahl, U. (1982) Biosynthesis of heparin: concerted action of early polymer-modification reactions, *J. Biol. Chem.* **257,** 421-425.

78. Kusche, M., and Lindahl, U. (1991) Biosynthesis of heparin: use of *E. coli* K5 capsular polysaccharide as a model substrate in enzymatic polymer-modification reactions, *Biochem. J.* **275,** 151-158.

79. Bame, K. J., Reddy, R. V., and Esko, J. D. (1991) Coupling of N-deacetylation and N-sulfation in a Chinese hamster ovary cell mutant defective in heparan sulfate N-sulfotransferase, *J. Biol. Chem.* **266,** 12461-12468.

80. Lidholt, K., and Lindahl, U. (1992) Biosynthesis of heparin: the D-glucuronosyl and N-acetyl-D-glucosaminyl transferase reactions and their relation to polymer modification, *Biochem. J.* **287,** 21-29.

81. Lind, T., Lindahl, U., and Lidholt, K. (1993) Biosynthesis of heparin/heparan sulfate: identification of a 70 kDa protein catalyzing both the D-glucuronosyl- and the N-acetyl-D-glucosaminyl transferase reactions, *J. Biol. Chem.* **268,** 20705-20708.

82. Lindahl, U., Bäckström, G., Malmström, A., and Fransson, L.-Å. (1972) Biosynthesis of IdoA in heparin: epimerization of D-glucuronic acid on the polymer level, *Biochem. Biophys. Res. Commun.* **46,** 985-991.

83. Höök, M., Lindahl, U., Bäckström, G., Malström, A., and Fransson, L.-Å. (1974) Biosynthesis of heparin. III. Formation of IdoA residues, *J. Biol. Chem.* **249,** 3908-3915.

84. Lindahl, U., Jacobsson, I., Höök, M., Bäckström, G., and Feingold, D. S. (1976) Biosynthesis of heparin: loss of C-5 hydrogen during conversion of D-glucuronic acid to L-iduronic acid, *Biochem. Biophys. Res. Commun.* **70,** 492-499.

85. Jacobsson, I., Bäckström, G., Höök, M., Lindahl, U., Feingold, D. S., Malmström, A., and Rodén, L. (1979) Biosynthesis of heparin: assay and properties of the microsomal uronosyl C-5 epimerase, *J. Biol. Chem.* **254,** 2975-2982.

86. Malmström, A., Rodén, L., Feingold, D. S., Jacobsson, I., Bäckström, G., and Lindahl, U. (1980) Biosynthesis of heparin: partial purification of the uronosyl C-5 epimerase, *J. Biol. Chem.* **255,** 3878-3883.

87. Prihar, H. S., Campbell, P., Feingold, D. S., Jacobsson, I., Jensen, J. W., and Lindahl, U. A. R., L. (1980) Biosynthesis of heparin: hydrogen exchange at carbon 5 of the glycuronosyl residues, *Biochemistry* **19,** 495-500.

88. Jensen, J. W., Rodén, L., Jacobsson, I., Lindahl, U., Prihar, H., and Feingold, D. S. (1983) Biosynthesis of heparin: a new substrate for the heparosan N-sulfate-D-glucopyranosyluronate 5-epimerase, *Carbohyd. Res.* **117,** 241-253.

89. Jacobsson, I., Lindahl, U., Jensen, J. W., Rodén, L., Prihar, H., and Feingold, D. S. (1984) Biosynthesis of heparin: substrate specificity of heparosan N-sulfate D-glucuronosyl 5-epimerase, *J. Biol. Chem.* **259,** 1056-1063.

90. Lindahl, U., Thunberg, L., Bäckström, G., Riesenfeld, J., Nordling, K., and Björk, I. (1984) Extension and structural variability of the antithrombin-binding sequence in heparin, *J. Biol. Chem.* **259,** 12368-12376.

91. Kusche, M., Torri, G., Casu, B., and Lindahl, U. (1990) Biosynthesis of heparin: O-sulfation of D-glucuronic acid, *J. Biol. Chem.* **265,** 15403-15409.

92. Jacobsson, I., and Lindahl, U. (1980) Biosynthesis of heparin: concerted action of late polymer-modification reactions, *J. Biol. Chem.* **255,** 5094-5100.

93. Kusche, M., Bäckström, G., Riesenfeld, J., Petitou, M., Choay, J., and Lindahl, U. (1988) Biosynthesis of heparin: O-sulfation of the antithrombin binding region, *J. Biol. Chem.* **263,** 15474-15484.

94. Kusche, M., Torri, G., Casu, B., and Lindahl, U. (1990) Biosynthesis of heparin: availability of glucosaminyl 3-O-sulfation sites, *J. Biol. Chem.* **265,** 7292-7300.

95. Kusche, M., Oscarsson, L.-G., Reynertsson, R., Rodén, L., and Lindahl, U. (1991) Biosynthesis of heparin: enzymatic sulfation of pentasaccharides, *J. Biol. Chem.* **266,** 7400-7409.

96. Homer, A. A. (1971) Macromolecular heparin from rat skin: isolation, characterization and depolymerization with ascorbate, *J. Biol. Chem.* **246,** 231-239.

97. Oosta, G. M., Favreau, L. V., Beeler, D. L., and Rosenberg, R. D. (1982) Purification and properties of human platelet heparitinase, *J. Biol. Chem.* **257,** 11249-11255.

98. Thunberg, L., Bäckström, G., Wasteson, Å., Robinson, H. C., Ögren, S., and Lindahl, U. (1982) Enzymatic depolymerization of heparin-related polysaccharides. Substrate

specificities of mouse mastocytoma and human platelet endo-β-D-glucuronidases, *J.. Biol. Chem.* **257**, 10278-10282.

99. Höök, M., Lindahl, U., Hallén, A., and Bäckström, G. (1975) Biosynthesis of heparin: studies on the microsomal sulfation process, *J. Biol. Chem.* **250**, 6065-6071.

100. Cheigetz, S., Andres, J. L., and Massagué, J. (1988) The transforming growth factor receptor type III is a membrane proteoglycan, *J. Biol. Chem.* **263**, 16984-16991.

101. Riesenfeld, J., Höök, M., and Lindahl, U. (1982) Biosynthesis of heparan sulfate in rat liver: characterization of polysaccharides obtained with intact cells and with a cell-free extract, *J. Biol. Chem.* **257**, 7050-7055.

102. Vann, W. F., Schmidt, M. A., Jann, B., and Jann, K. (1981) The structure of the capsular polysaccharide (K5 antigen) of urinary tract-infective *Escherichia coli* O10:K5:H4, *Eur. J. Biochem.* **116**, 359-364.

103. van den Born, J., Gunnarsson, K., Bakker, M. A. H., Kjellén, L., Kusche-Gullberg, M., Maccarana, M., Berden, J. H. M., and Lindahl, U. (1995) Presence of N-unsubstituted glucosamine units in native heparan sulfate revealed by a monoclonal antibody, *J. Biol. Chem.* **270**, 31303-31309.

104. Lidholt, K., Weinke, J. L., Kiser, C. S., Lugemwa, F. N., Cheifetz, S., Bame, K. J., Massagué, J., Lindahl, U., and Esko, J. D. (1991) Chinese hamster ovary cell mutants defective in heparan sulfate biosynthesis, *Proc. Natl. Acad. Sci. U.S.A.* **89**, 2267-2271.

105. Wei, Z., Swiedler, S. J., Ishihara, M., Orellana, A., and Hirshberg, C. B. (1993) A single protein catalyzes both N-deacetylation and N-sulfation during the biosynthesis of heparan sulfate, *Proc. Natl. Acad. Sci. U.S.A.* **90**, 3885-3888.

106. Mandon, E., Kempner, E. S., Ishihara, M., and Hirschberg, C. B. (1994) A monomeric protein in the Golgi membrane catalyzes both the N-deacetylation and N-sulfation of heparan sulfate, *J. Biol. Chem.* **269**, 11729-11733.

107. Hashimoto, Y., Orellana, A., G., G., and Hirschberg, C. B. (1992) Molecular cloning and expression of rat liver N-heparan sulfate sulfotransferase, *J. Biol. Chem.* **267**, 15744-15750.

108. Orellana, A., Hirschberg, C. B., Z., W., Swiedler, S. J., and Ishihara, M. (1994) Molecular cloning and expression of a glycosaminoglycan N-acetylglucosaminyl N-deacetylase/N-sulfotransferase from a heparin-producing cell line, *J. Biol. Chem.* **269**, 2270-2276.

109. Eriksson, I., Sandbäck, D., Ek, B., Lindahl, U., and Kjellén, L. (1994) cDNA cloning and sequencing of mouse mastocytoma glucosaminyl N-deacetylase/N-sulfotransferase, an enzyme involved in the biosynthesis of heparin, *J. Biol. Chem.* **269**, 10438-10443.

110. Dixon, J., Loftus, S. K., Gladwin, A. J., Scambler, P. J., Wasmuth, J. J., and Dixon, M. J. (1995) Cloning of the human heparan sulfate-N-deacetylase/N-sulfotransferase gene from the Treacher Collins syndrome candidate region at 5q32-q33.1, *Genomics* **26**, 239-244.

111. Campbell, P., Hannesson, H. H., Sandbäck, D., Rodén, L., Lindahl, U., and Li, J.-p. (1994) Biosynthesis of heparin/heparan sulfate: purification of the D-glucuronosyl C-5 epimerase from bovine liver, *J. Biol. Chem.* **269**, 26953-26958.

112. Wlad, H., Maccarana, M., Eriksson, I., and Kjellén, L. (1994) Biosynthesis of heparin: different molecular forms of O-sulfotransferases, *J. Biol. Chem.* **269**, 24538-24541.

113. Kobayashi, M., Habuchi, H., Habuchi, O., Saito, M., and Kimata, K. (1996) Purification and characterization of heparan sulfate 2-sulfotransferase from cultured Chinese hamster ovary cells, *J. Biol. Chem.* **271**, 7645-7653.

114. Bai, X., and Esko, J. D. (1996) An animal cell mutant defective in heparan sulfate hexuronic acid 2-O-sulfation, *J. Biol. Chem.* **271**, 17711-17717.

115. Habuchi, H., Habuchi, O., and Kimata, K. (1995) Purification and characterization of heparan sulfate 6-sulfotransferase from the culture medium of Chinese hamster ovary cells, *J. Biol. Chem.* **270**, 4172-4179.

116. Razi, N., and Lindahl, U. (1995) Biosynthesis of heparin/heparan sulfate: the D-glucosaminyl 3-O-sulfotransferase reaction: target and inhibitor saccharides, *J. Biol. Chem.* **270**, 11267-11275.

117. Liu, J., Shworak, N. W., Fritze, L. M., Edelberg, J. M., and Rosenberg, R. D. (1996) Purification of heparan sulfate D-glucosaminyl 3-O-sulfotransferase, *J. Biol. Chem.* **271**, 27072-27082.

118. Oreste, P., and Torri, G. (1980) Fingerprinting of heparins by low-amperage electrophoresis in barium acetate, *J. Chromatog.* **195**, 398-401.

119. Bianchini, P., Nader, H. B., Takahashi, H. K., Osima, B., Straus, A. H., and Dietrich, C. P. (1980) Fractionation and identification of heparin and other acidic mucopolysaccharides by a new discontinuous electrophoretic method, *J. Chromatog.* **196**, 455-462.

120. Volpi, N. (1993) Characterization of heparins with different relative molecular masses (from 11,600 to 1,600) by various analytical techniques, *J. Chromatog.* **622**, 13-20.

121. Volpi, N. (1993) Fast moving and slow moving heparin, dermatan sulfate, and chondroitin sulfate: qualitative and quantitative analysis by agarose gel electrophoresis, *Carbohyd. Res.* **247**, 263-278.

122. Comper, W. D. (1981) Heparin (and Related Polysaccharides), in *Heparin (and related polysaccharides)* Gordon and Breach, New York, pp. 64-105.

123. Casu, B. (1984) Conformation of individual residues and chain segments of glycosaminoglycans in solution by spectroscopic methods, in *Molecular Biophysics of the Extracellular Matrix*, eds., Arnott, S., Rees, D. A. & Morris, E. R., Humana Press, Clifton, NJ, pp. 69-93.

124. Nieduszinski, I. (1989) General physical properties of heparin, in *Heparin: Chemical and Biological Properties, Clinical Applications*, eds., Lane, D. A. & Lindahl, U., CRC Press, Inc., Boca Raton, FL, pp. 52-63.

125. Casu, B. (1990) Heparin structure, *Haemostasis* **20**, 62-73.

126. Hounsel, E. F. (1994) Physicochemical analyses of oligosaccharide determinants of glycoproteins, *Adv. Carbohyd. Chem. Biochem.* **50**, 311-350.

127. Nieduszinski, I., and Atkins, E. D. T. (1975) Molecular conformations of heparan sulphate and heparin, in *Structure of Fibrous Biopolymers, Colston Papers No. 26*, eds., Atkins, E. D. T. & Keller, A., Butterworth, London, pp. 323-334.

128. Atkins, E. D. T., and Nieduszinski, I. (1976) Heparin: crystalline structures of sodium and calcium salts, in *Heparin Chemistry and Clinical Usage*, eds., Kakkar, V. V. & Thomas, D. P., Academic Press, London, pp. 21-35.

129. Elloway, H. F., and Atkins, E. D. T. (1977) Molecular conformation of sodium heparan sulphate in the condensed phase, *Biochem. J.* **161**, 495-498.

130. Rees, D. A., Morris, E. R., Thom, D., and Madden, J. K. (1982) Shapes and interactions of carbohydrate chains, in *The Polysaccharides*, ed. Aspinall, G. O., Academic Press, New York, pp. 196-290.

131. Gatti, P., Casu, B., Hamer, G. K., and Perlin, A. S. (1979) Studies on the conformation of heparin by ^{1}H- and ^{13}C-NMR spectroscopy, *Macromolecules* **12**, 824-828.

132. Mulloy, B., Forster, M. J., Jones, C., and Davies, D. B. (1993) N.m.r. and molecular-modeling studies of the solution conformation of heparin, *Biochem. J.* **293**, 849-858.

133. Stoddart, J. F. (1971) The Stereochemistry of Carbohydrates, in *The Stereochemistry of Carbohydrates* John Wiley & Sons, New York, pp. 1-19.

134. Forster, M. J., and Mulloy, B. (1993) Molecular dynamics study of iduronate ring conformation, *Biopolymers* **33**, 575-587.

135. Ragazzi, M., Ferro, D. R., and Provasoli, A. (1985) A force-field study of the conformational characteristics of the iduronate ring, *J. Comp. Chem.* **7**, 105-112.

136. Sanderson, P. N., Huckerby, T. N., and Nieduszynski, I. A. (1987) Conformational equilibria of α-L-iduronate residues in disaccharides derived from heparin, *Biochem. J.* **243**, 175-181.

137. Casu, B., Choay, J., Ferro, D. R., Gatti, G., Jacquinet, J.-C., Petitou, M., Provasoli, A., Ragazzi, J., Sinäy, P., and Torri, G. (1986) Controversial glycosaminoglycan conformations, *Nature* **332**, 215-216.

138. Ferro, D. R., Provasoli, A., Ragazzi, M., Torri, G., Casu, B., Gatti, G., Jacquinet, J.-C., Sinäy, P., Petitou, M., and Choay, J. (1986) Evidence of conformational equilibrium of

the sulfated L-iduronate residue in heparin and in synthetic heparin mono- and oligosaccharides: NMR and force-field studies, *J. Am. Chem. Soc.* **108**, 6773-6778.

139. Ferro, D. R., Provasoli, A., Ragazzi, M., Casu, B., Torri, G., Bossennec, V., Perly, B., Sinay, P., Petitou, M., and Choay, J. (1990) Conformer populations of L-iduronic acid residues in glycosaminoglycan sequences, *Carbohyd. Res.* **195**, 157-167.

140. French, A. D., and Brady, J. W. (1990) Computer modeling of carbohydrates: an introduction, in *Computer Modeling of Carbohydrate Molecules*, eds., French, A. D. & Brady, J. W., American Chemical Society, Washington, D.C., pp. 1-19.

141. Lindahl, U., and Höök, M. (1978) Glycosaminoglycans and their binding to biological macromolecules, *Annu. Rev. Biochem.* **47**, 385-417.

142. Poole, A. R. (1986) Proteoglycans in health and disease: structures and functions, *Biochem. J.* **236**, 1-14.

143. David, G. (1991) Biology and pathology of the pericellular heparan sulphate proteoglycans, *Biochem. Soc. Trans.* **19**, 816-820.

144. Bernfield, M., Kokenyesi, R., Kato, M., Hinkes, M. T., Spring, J., Gallo, R. L., and Lose, E. J. (1992) Biology of the syndecans: a family of transmembrane heparan sulfate proteoglycans, *Annu. Rev. Cell Biol.* **8**, 365-393.

145. David, G. (1992) Structural and functional diversity of the heparan sulfate proteoglycans, in *Heparin and Related Polysaccharides. Advances in Experimental Medicine and Biology* Plenum Press, New York, Vol. 313, pp. 69-78.

146. David, G. (1993) Integral membrane heparan sulfate proteoglycans, *FASEB J.* **7**, 1023-1030.

147. Margolis, R. K., and Margolis, R. U. (1993) Nervous tissue proteoglycans, *Experientia* **49**, 429-446.

148. Iozzo, R. V., and Murdock, A. D. (1996) Proteoglycans of the extracellular environment: clues from the gene and protein side offer novel perspectives in molecular diversity and function, *FASEB J.* **10**, 598-614.

149. Fransson, L.-Å., Silverberg, I., and Carlstedt, I. (1985) Structure of the heparan sulfate-protein linkage region: demonstration of the sequence galactosyl-galactosyl-xylose-2-phosphate, *J. Biol. Chem.* **260**, 14722-14726.

150. Zhang, L., and Esko, J. D. (1994) Amino acid sequences that drive heparan sulfate assembly in a proteoglycan, *J. Biol. Chem.* **269**, 19295-19299.

151. Zhang, L., David, G., and Esko, J. D. (1995) Repetitive Ser-Gly sequences enhance heparan sulfate assembly in proteoglycans, *J. Biol. Chem.* **270**, 27127-27135.

152. Krusius, T., Gehlsen, K. R., and Ruoslahti, E. (1987) A fibroblast chondroitin sulfate proteoglycan core protein contains lectin-like and growth factor-like sequences, *J. Biol. Chem.* **262**, 13120-13125.

153. Ahmed, I., and Piepkorn, M. (1991) The antiproliferative effects of enzymatic deglycosylation and metabolic undersulfation of proteoglycans from the cell surface, *J. Invest. Dermatol.* **97**, 43-49.

154. Kato, M., Wang, H., Bernfield, M., Gallagher, J. T., and Turnbull, J. E. (1994) Cell surface syndecan-1 on distinct cell types differs in fine structure and ligand binding of its heparan sulfate chains, *J. Biol. Chem.* **269**, 1881-1890.

155. Shworak, N. W., Shirakawa, M., Mulligan, R. C., and Rosenberg, R. D. (1994) Characterization of ryudocan glycosaminoglycan acceptor sites, *J. Biol. Chem.* **269**, 21204-21214.

156. Sanderson, R. D., Turnbull, J. E., Gallagher, J. T., and Lander, A. D. (1994) Fine structure of heparan sulfate regulates syndecan-1 function and cell behavior, *J. Biol. Chem.* **269**, 13100-13116.

157. Kato, M., Turnbull, J., Hooper, K., Gallagher, J. T., and Bernfield, M. (1991) Heparan sulfate chains on syndecan show cell type-specific fine structure, *J. Cell Biol.* **115**, 125a.

158. Filmus, J., Church, J. G., and Buick, R. N. (1988) Isolation of a cDNA corresponding to a developmentally regulated transcript in rat intestine, *Mol. Cell Biol.* **8**, 4243-4249.

159. Sanderson, R. D., Lalor, P., and Bernfield, M. (1989) B lymphocytes express and lose syndecan at specific stages of differentiation, *Cell Regul.* **1**, 27-35.

160. David, G., Bai, X. M., Van der Schueren, B., Marynen, P., Cassiman, J.-J., and Van den Berghe, H. (1993) Spatial and temporal changes in the expression of fibroglycan (syndecan-2) during mouse embryonic development, *Development* **119**, 841-854.

161. Inki, P., Larjava, H., Haapasalmi, K., Miettinen, H. M., Grenman, R., and Jalkanen, M. (1994) Expression of syndecan-1 is induced by differentiation and suppressed by malignant transformation of human keratinocytes, *Eur. J. Cell Biol.* **63**, 43-51.

162. Stipp, C. S., Litwack, E. D., and Lander, A. D. (1994) Cerebroglycan: an integral membrane heparan sulfate proteoglycan that is unique to the developing nervous system and expressed specifically during neuronal differentiation, *J. Cell Biol.* **124**, 149-160.

163. Watanabe, K., Yamada, H., and Yamaguchi, Y. (1995) K-Glypican: a novel GPI-anchored heparan sulfate proteoglycan that is highly expressed in developing brain and kidney, *J. Cell Biol.* **130**, 1207-1218.

164. Horiguchi, Y., Fine, J.-D., and Couchman, J. R. (1991) Human skin basement membrane-associated heparan sulfate proteoglycan: distinctive differences in ultrastructural localization as a function of developmental age, *Br. J. Dermatol.* **124**, 410-414.

165. Cizmeci-Smith, G., Asundi, V., Stahl, R. C., Teichman, L. J., Chernousov, M., Cowan, K., and Carey, D. J. (1992) Regulated expression of syndecan in vascular smooth muscle cells and cloning of rat syndecan core protein cDNA, *J. Biol. Chem.* **267**, 15729-15736.

166. Gould, S. E., Upholt, W. B., and Kosher, R. A. (1992) Syndecan 3: a member of the syndecan family of membrane-intercalated proteoglycans that is expressed in high amounts at the onset of chicken limb cartilage differentiation, *Proc. Natl. Acad. Sci. U.S.A.* **89**, 3271-3275.

167. Halfter, W. (1993) A heparan sulfate proteoglycan in developing avian axonal tracts, *J. Neurosci.* **13**, 2863-2873.

168. Gallagher, J. T., Lyon, M., and Steward, W. P. (1986) Structure and function of heparan sulphate proteoglycans, *Biochem. J.* **236**, 313-325.

169. Yanagishita, M., and Hascall, V. C. (1983) Characterization of heparan sulfate proteoglycans synthesized by rat ovarian granulosa cells in culture, *J. Biol. Chem.* **258**, 12858-12864.

170. Parthasarathy, N., and Spiro, R. G. (1984) Isolation and characterization of the heparan sulfate proteoglycan of the bovine glomerular basement membrane, *J. Biol. Chem.* **259**, 12749-12755.

171. Tanhravahi, R. V., Stevens, R. L., Austen, F., and Weis, J. H. (1986) A single gene in mast cells encodes the core peptides of heparin and chondroitin sulfate proteoglycans, *Proc. Natl. Acad. Sci. U.S.A.* **83**, 9207-9210.

172. Stevens, R. L., Otsu, K., Weis, J. H., Tantravahi, R. V., Austen, K. F., Henkart, P. A., Galli, M. C., and Reynolds, C. W. (1987) Co-sedimentation of chondroitin sulfate A glycosaminoglycans and proteoglycans with the cytolytic secretory granules of rat large granular lymphocyte (LGL) tumor cells, and identification of a mRNA in normal and transformed LGL that encodes proteoglycans, *J. Immunol.* **139**, 863-868.

173. Bourdon, M. A., Shiga, M., and Ruoslahti, E. (1987) Gene expression of the chondroitin sulfate proteoglycan core protein PG19, *Mol. Cell. Biol.* **7**, 33-40.

174. Pomerantz, J. L., Owen, W. F., Avraham, A., Soberman, R. J., Austen, K. F., and Stevens, R. L. (1988) Characterization of a human eosinophil proteoglycan, and augmentation of its biosynthesis and size by interleukin 3, interleukin 5, and granulocyte/macrophage colony stimulating factor, *J. Biol. Chem.* **263**, 13901-13908.

175. Stevens, R. L. (1987) Intracellular proteoglycans in cells of the immune system, in *Biology of Proteoglycans*, eds., Wight, T. N. & Mechan, R. P., Academic Press, New York, pp. 367-388.

176. Kusche, M., Lindahl, U., Enerbäch, L., and Rodén, L. (1988) Identification of oversulfated galactosaminoglycans in intestinal-mucosal mast cells of rats infected with the nematode worm, *Nippostrongylis brasiliensis*, *Biochem. J.* **253**, 885-893.

177. Metcalfe, D. D., Smith, J. A., Austen, K. F., and Silbert, J. E. (1980) Polydispersity of rat mast cell heparin: implications for proteoglycan assembly, *J. Biol. Chem.* **255**, 11753-11758.

178. Bland, C. E., Ginsburg, H., Silbert, J. E., and Metcalfe, D. D. (1982) Mouse heparin proteoglycan: synthesis by mast cell-fibroblast monolayers during lymphocyte-dependent mast cell proliferation, *J. Biol. Chem.* **257**, 8661-8666.

179. Avraham, S., Stevens, R. L., Gartner, M. C. S., Austen, K. F., Lalley, P. A., and Weis, J. H. (1988) Isolation of a cDNA that encodes the peptide core of the secretory granule proteoglycan of rat basophilic leukemia-1 cells and assessment of its homology to the human analogue, *J. Biol. Chem.* **263**, 7292-7296.

180. Stevens, R. L., Avraham, S., Gartner, M. C., Bruns, G. A. P., Austen, K. F., and Weis, J. H. (1988) Isolation and characterization of a cDNA that encodes the peptide core of the secretory granule proteoglycan of human promyelocytic leukemia HL-60 cells, *J. Biol. Chem.* **263**, 7287-7291.

181. Bourdon, M. A., Shiga, M., and Roushlati, E. (1986) Identification from cDNA of the precursor form of chondroitin sulfate proteoglycan core protein, *J. Biol. Chem.* **261**, 12534-12537.

182. Avraham, S., Austen, K. F., Nicodemus, C. F., Gartner, M. C., and Stevens, R. L. (1989) Cloning and characterization of the mouse gene that encodes the peptide core of secretory granule proteoglycans and expression of this gene in transfected rat-1 fibroblasts, *J. Biol. Chem.* **264**, 16719-16726.

183. Kjellén, L., Pettersson, I., Lillhager, P., Steen, M. L., Pettersson, U., Lehtonen, P., Karlsson, T., Ruoslahti, E., and Hellman, L. (1989) Primary structure of a mouse mastocytoma proteoglycan core protein, *Biochem. J.* **263**, 105-113.

184. Lohmander, L. S., Arnljots, K., and Yanagishita, M. (1990) Structure and synthesis of intracellular proteoglycan in HL-60 human leukemic promyelocytes, *J. Biol. Chem.* **265**, 5802-5808.

185. Avraham, S., Avraham, H., Austen, K. F., and Stevens, R. L. (1992) Negative and positive cis-acting elements in the promoter of the mouse gene that encodes the serine/glycine-rich peptide core of secretory granule proteoglycans, *J. Biol. Chem.* **267**, 610-617.

186. Humphries, D. E., Nicodemus, C. F., Schiller, V., and Stevens, R. L. (1992) The human serglycin gene: nucleotide sequence and methylation pattern in human promyelocytic leukemia HL-60 cells and T-lymphoblast Molt-4 cells, *J. Biol. Chem.* **267**, 13558-13563.

187. Elenius, K., and Jalkanen, M. (1994) Functions of syndecans—a family of cell surface proteoglycans, *J. Cell Sci.* **107**, 2975-2982.

188. Salmivirta, M., and Jalkanen, M. (1995) Syndecan family of cell surface proteoglycans: developmentally regulated receptors for extracellular effector molecules, *Experientia* **51**, 863-872.

189. Ktistakis, N. T., Thomas, D., and Roth, M. G. (1990) Characteristics of the tyrosine recognition signal for internalization of transmembrane surface glycoproteins, *J. Cell Biol.* **111**, 1393-1407.

190. Spring, J., Paine-Saunders, S. E., Hynes, R. O., and Bernfield, M. (1994) *Drosophila* syndecan: conservation of a cell-surface heparan sulfate proteoglycan, *Proc. Natl. Acad. Sci. U.S.A.* **91**, 3334-3338.

191. Saunders, S., Jalkanen, M., O'Farrell, S., and Bernfield, M. (1989) Molecular cloning of syndecan, an integral membrane proteoglycan, *J. Cell Biol.* **108**, 1547-1556.

192. Mali, M., Jaakkola, P., Arvilommi, A.-M., and Jalkanen, M. (1990) Sequence of human syndecan indicates a novel gene family of integral membrane proteoglycans, *J. Biol. Chem.* **265**, 6884-6889.

193. Kojima, T., Shworak, N. W., and Rosenberg, R. D. (1992) Molecular cloning and expression of two distinct cDNA-encoding heparan sulfate proteoglycan core proteins from a rat endothelial cell line, *J. Biol. Chem.* **267**, 4870-4877.

194. Kiefer, M. C., Stephans, J. C., Crawford, K., Okino, K., and Barr, P. J. (1992) Ligand-affinity cloning and structure of a cell surface heparan sulfate proteoglycan that binds basic fibroblast growth factor, *Proc. Natl. Acad. Sci. U.S.A.* **87**, 6985-6989.

195. Kovalszky, I., Gallai, M., Armbrust, T., and Ramadori, G. (1994) Syndecan-1 gene expression in isolated rat liver cells (hepatocytes, Kupffer cells, endothelial and Ito cells), *Biochem. Biophys. Res. Commun.* **204**, 944-949.

196. Rapraeger, A. C., Jalkanen, M., Endo, E., Koda, J., and Bernfield, M. (1985) The cell surface proteoglycan from mouse mammary epithelial cells bears chondroitin sulfate and heparan sulfate glycosaminoglycans, *J. Biol. Chem.* **20**, 11046-11052.

197. Kokenyesi, R., and Bernfield, M. (1994) Core protein structure and sequence determine the site and presence of heparan sulfate and chondroitin sulfate on syndecan-1, *J. Biol. Chem.* **269**, 12304-12309.

198. Vihinen, T., Auvinen, P., Alanen-Kurki, L., and Jalkanen, M. (1993) Structural organization and genomic sequence of mouse syndecan-1 gene, *J. Biol. Chem.* **268**, 17261-17269.

199. Hinkes, M. T., Goldberger, O. A., Neuman, P. E., Kokenyesi, R., and Bernfield, M. (1993) Organization and promoter activity of the mouse syndecan-1 gene, *J. Biol. Chem.* **268**, 11440-11448.

200. Marynen, P., Zhang, J., Cassiman, J.-J., van den Berghe, H., and David, G. (1989) Partial primary structure of the 48- and 90-kilodalton core proteins of cell surface-associated heparan sulfate proteoglycans of lung fibroblasts: prediction of an integral membrane domain and evidence for multiple distinct core proteins at the cell surface of human lung fibroblasts, *J. Biol. Chem.* **264**, 7017-7024.

201. Pierce, A., Lyon, M., Hampson, I. N., Cowling, G. J., and Gallagher, J. T. (1992) Molecular cloning of the major cell surface heparan sulfate proteoglycan from rat liver, *J. Biol. Chem.* **267**, 3894-3900.

202. Carey, D. J., Evans, D. M., Stahl, R. C., Asundi, V. K., Conner, K. J., Garbes, P., and Cizmeci-Smith, G. (1992) Molecular cloning and characterization of N-syndecan, a novel transmembrane heparan sulfate proteoglycan, *J. Cell Biol.* **117**, 191-201.

203. Baciu, P. C., Acastter, C., and Goettinck, P. F. (1994) Molecular cloning and genomic organization of chicken syndecan 4, *J. Biol. Chem.* **269**, 696-703.

204. Yeaman, C., and Rapraeger, A. C. (1993) Membrane-anchored proteoglycans of mouse macrophages: P388D1 cells express a syndecan-4-like heparan sulfate proteoglycan and a distinct chondroitin sulfate form, *J. Cell. Physiol.* **157**, 413-425.

205. David, G., van der Schueren, D., Marynen, P., Cassiman, J.-J., and van den Berghe, H. (1992) Molecular cloning of amphiglycan, a novel integral membrane heparan sulfate proteoglycan expressed by epithelial and fibroblastic cells, *J. Cell Biol.* **118**, 961-969.

206. Kojima, T., Leone, C. W., Marchildon, G. A., Marcum, J. A., and Rosenberg, R. D. (1992) Isolation and characterization of heparan sulfate proteoglycans produced by cloned rat microvascular endothelial cells, *J. Biol. Chem.* **267**, 4859-4869.

207. Kojima, T., Inazawa, J., Takamatsu, J. R., R. D., and Saito, H. (1993) Human ryudocan core protein: molecular cloning and characterization of the cDNA, and chromosomal localization of the gene, *Biochem. Biophys. Res. Commun.* **190**, 814-822.

208. Cross, G. A. M. (1990) Glycolipid anchoring of plasma membrane proteins, *Annu. Rev. Cell Biol.* **6**, 1-39.

209. Ferguson, M. A. J. (1992) Glycosyl-phosphatidylinositol membrane anchors: the tale of a tail, *Biochem. Soc. Trans.* **20**, 243-256.

210. Lewis, K. A., Garigapati, V. R., Zhou, C., and Roberts, M. F. (1993) Substrate requirements of bacterial phosphatidylinositol-specific phospholipase C, *Biochemistry* **32**, 8836-8841.

211. Toutant, J.-P., Roberts, W. L., Murray, N. R., and Rosenberry, T. L. (1989) Conversion of human erythrocyte acetylcholinesterase from an amphiphilic to a hydrophilic form by phosphatidylinositol-specific phospholipase C and serum phospholipase D, *Eur. J. Biochem.* **180**, 503-508.

212. Steiger, S., Diem, S., Jakob, A., and Brodbeck, U. (1991) Enzymatic properties of phosphatidylinositol-glycan-specific phospholipase C from rat liver and phosphatidylinositol-glycan-specific phospholipase D from rat serum, *Eur. J. Biochem.* **197**, 67-73.

213. Low, M. G., and Huang, K.-S. (1991) Factors affecting the ability of glycosylphosphatidylinositol-specific phospholipase D to degrade the membrane anchors of cell surface proteins, *Biochem. J.* **279**, 483-493.

214. Sevlever, D., and Rosenberry, T. L. (1993) Mannosamine inhibits the synthesis of putative glycoinositol phospholipid anchor precursors in mammalian cells without incorporating into an accumulated intermediate, *J. Biol. Chem.* **268**, 10938-10945.

215. David, G., Lories, V., Decock, B., Marynen, P., Cassiman, J.-J., and van den Berghe, H. (1990) Molecular cloning of a phosphatidylinositol-anchored membrane heparan sulfate proteoglycan expressed by epithelial and fibroblastic cells, *J. Cell Biol.* **111**, 3165-3176.

216. Carey, D. J., Stahl, R. C., Asundi, V. K., and Tucker, B. (1993) Processing and subcellular distribution of the Schwann cell lipid-anchored heparan sulfate proteoglycan and its identification as glypican, *Exp. Cell Res.* **208**, 10-18.

217. Stipp, C. S., Litwack, E. D., and Lander, A. D. (1992) Molecular cloning of a glypican-related heparan sulfate proteoglycan expressed in the developing rat brain, *Mol. Biol. Cell* **3**, 65a.

218. Vermeesch, J. R., Mertens, G., David, G., and Marynen, P. (1995) Assignment of the human glypican gene (GPC1) to 2q35-q37 by fluorescence *in situ* hybridization, *Genomics* **25**, 327-329.

219. Karthikeyan, L., Flad, M., Engel, M., Meyer-Puttlitz, B., Margolis, R. U., and Margolis, R. K. (1994) Immunocytocemical and *in situ* hybridization studies of the heparan sulfate proteoglycan, glypican, in nervous tissue, *J. Cell Sci.* **107**, 3213-3222.

220. Karthikeyan, L., Maurel, P., Rauch, U., Margolis, R. K., and Margolis, R. U. (1992) Cloning of a major heparan sulfate proteoglycan from brain and identification as the rat form of glypican, *Biochem. Biophys. Res. Commun.* **188**, 395-401.

221. Filmus, J., Shi, W., Wong, Z. M., and Wong, M. J. (1995) Identification of a new membrane-bound heparan sulphate proteoglycan, *Biochem. J.* **311**, 561-565.

222. Ishihara, M., Fedarko, N. S., and Conrad, H. E. (1987) Involvement of phosphatidylinositol and insulin in the coordinate regulation of proteoheparan sulfate metabolism and hepatocyte growth, *J. Biol. Chem.* **262**, 4708-4717.

223. Yanagishita, M. (1992) Glycosylphosphatidylinositol-anchored and core protein-intercalated heparan sulfate proteoglycans in rat ovarian granulosa cells have distinct secretary, endocytotic, and intracellular degradative pathways, *J. Biol. Chem.* **267**, 9505-9511.

224. Huang, L. R., and Hsu, H. C. (1995) Cloning and expression of CD24 gene in human hepatocellular carcinoma: a potential early tumor marker gene correlates with p53 mutation and tumor differentiation, *Cancer Res.* **55**, 4717-4721.

225. McQuillan, D. J., Midura, R. J., Hascall, V. C., and Yanagishita, M. (1992) Plasma-membrane-intercalated heparan sulphate proteoglycans in an osteogenic cell line (UMR 106-01 BSP), *Biochem. J.* **285**, 25-33.

226. Noonan, D. M., Fulle, A., Valente, P., Cai, S., Horigan, E., Sasaki, M., Yamada, Y., and Hassell, J. R. (1991) The complete sequence of perlecan, a basement membrane heparan sulfate proteoglycan, reveals extensive similarity with laminin A chain, low density lipoprotein receptor, and the neural cell adhesion molecule, *J. Biol. Chem.* **266**, 22939-22947.

227. van den Heuvel, L. P. W. J., van den Born, J., van de Velden, T. J. A. M., Veerkamp, J. H., Monnnens, L. A. H., Schroder, C. H., and Berden, J. H. M. (1989) Isolation and partial characterization of heparan sulphate proteoglycan from the human glomerular basement membrane, *Biochem. J.* **264**, 557-465.

228. SundarRaj, N., Fite, D., Ledbetter, S., Chakravarti, S., and Hassell, J. R. (1995) Perlecan is a component of cartilage matrix and promotes chondrocyte attachment, *J. Cell Sci.* **108**, 2663-2672.

229. Kallunki, P., and Tryggvason, K. (1992) Human basement membrane heparan sulfate proteoglycan core protein: a 467 kDa protein containing multiple domains resembling elements of the low density lipoprotein receptor, laminin, neural cell adhesion molecules, and epidermal growth factor, *J. Cell Biol.* **116**, 559-571.

230. Murdock, A. D., Dodge, G. R., Cohen, I., Tuan, R. S., and Iozzo, R. V. (1992) Primary structure of the human heparan sulfate proteoglycan from basement membrane (HSPG2/perlecan), a chimeric molecule with multiple domains homologous to the low density lipoprotein receptor, laminin, neural cell adhesion molecules, and epidermal growth factor, *J. Biol. Chem.* **267**, 8544-8557.

231. Schulze, B., Mann, K., Battistutta, R., Wiedemann, H., and Timpl, R. (1995) Structural properties of recombinant domain III-3 of prelecan containing a globular domain inserted into an epidermal-growth-factor-like motif, *Eur. J. Biochem.* **231**, 551-556.

232. Cohen, I. R., Grässel, S., Murdock, A. D., and Iozzo, R. V. (1993) Structural characterization of the complete human perlecan gene and its promoter, *Proc. Natl. Acad. Sci. U.S.A.* **90**, 10404-10408.

233. Kokenyesi, R., and Silbert, J. E. (1995) Formation of heparan sulfate or chondroitin/dermatan sulfate on recombinant domain I of mouse perlecan expressed in Chinese hamster ovary cells, *Biochem. Biophys. Res. Commun.* **211**, 262-267.

234. Murdock, A. D., and Iozzo, R. V. (1993) Perlecan: the multidomain heparan sulfate proteoglycan of basement membrane and extracellular matrix, *Virch. Arch. A. Pathol. Anat.* **423**, 237-242.

235. Iozzo, R., Cohen, I. R., Grässel, S., and Murdock, A. D. (1995) The biology of perlecan: the multifaceted heparan sulphate proteoglycan of basement membranes and pericellular matrices, *Biochem. J.* **302**, 625-639.

236. Noonan, D. M., Horigan, E. A., Ledbetter, S. R., Vogeli, G., Sasaki, M., Yamada, Y., and Hassell, J. R. (1988) Identification of cDNA clones encoding different domains of the basement membrane heparan sulfate proteoglycan, *J. Biol. Chem.* **263**, 16379-16387.

237. Kugelman, L. C., Ganguly, S., Haggerty, J. G., Weissman, S. M., and Milstone, L. M. (1992) The core protein of epican, a heparan sulfate proteoglycan on keratinocytes, is an alternative form of CD44, *J. Invest. Dermatol.* **99**, 887-891.

238. Segarini, P. R., and Seyedin, S. M. (1988) The high molecular weight receptor to transforming growth factor glycosaminoglycan chains, *J. Biol. Chem.* **263**, 8366-8370.

239. Lopez-Casillas, F., Cheifetz, S., Doody, S., Andres, J. T., Lane, W. S., and Massagué, J. (1991) Structure and function of the membrane proteoglycan betaglycan, a component of the TGF receptor system, *Cell* **67**, 785-795.

240. Wang, X.-F., Lin, J. Y., Ng-Eaton, E., Downward, J., Lodish, H. F., and Weinberg, R. A. (1991) Expression cloning and characterization of the TGF type III receptor, *Cell* **68**, 795-805.

241. Tsen, G., Halfter, W., Kröger, S., and Cole, G. J. (1995) Agrin is a heparan sulfate proteoglycan, *J. Biol. Chem.* **270**, 3392-3399.

242. Aliel, P. M., Perin, J., Jallès, P., and Bonnet, F. (1993) Testican, a multidomain testicular proteoglycan resembling modulators of cell social behaviour, *Eur. J. Biochem.* **214**, 347-350.

243. Hiscock, D. R. R., Canfield, A., and Gallagher, J. T. (1995) Molecular structure of heparan sulphate synthesized by bovine aortic endothelial cells, *Biochim. Biophys. Acta.* **1244**, 104-112.

244. Bienkowski, M. J. (1984) *Structure and metabolism of heparin and heparan sulfate, Ph.D. dissertation* (University of Illinois, Urbana, IL).

245. Elson, L. A., and Morgan, W. T. J. (1933) A colorimetric method for the determination of glucosamine and chondrosine, *Biochem. J.* **27**, 1824-1828.

246. Lagunoff, D., and Warren, G. (1962) Determination of 2-deoxy-2-sulfoaminohexose content of mucopolysaccharides, *Arch. Biochem. Biophys.* **99**, 396-400.
247. Bitter, T., and Muir, H. A. (1962) A modified uronic acid carbazole reaction, *Anal. Biochem.* **4**, 330-334.
248. Shively, J. E., and Conrad, H. E. (1976) Formation of anhydrosugars in the chemical depolymerization of heparin, *Biochemistry* **15**, 3932-3942.
249. Shively, J. E., and Conrad, H. E. (1970) Stoichiometry of the nitrous acid deaminative cleavage of model amino sugar glycosides and glycosaminoglycuronans, *Biochemistry* **9**, 33-43.
250. Conrad, H. E. (1980) The acid lability of the glycoside bonds of L-iduronic acid in glycosaminoglycans , *Biochem. J.* **19**, 1355-1363.
251. Gahan, L. C., Sandford, P. A., and Conrad, H. E. (1967) Structure of the serotype 2 capsular polysaccharide of *Aerobacter aerogenes*, *Biochemistry* **6**, 2755-2767.
252. Conrad, H. E., Bamburg, J. R., Epley, J. D., and Kindt, T. J. (1966) The structure of the *Aerobacter aerogenes* A3(Sl) polysaccharide. Sequence analysis and hydrolysis studies, *Biochemistry* **5**, 2802-2817.
253. Taylor, R. L., and Conrad, H. E. (1972) Stoichiometric depolymerization of polyuronides and glycosaminoglycuronans to monosaccharides following reduction of their carbodiimide-activated carboxyl groups, *Biochemistry* **11**, 1383-1388.
254. Perlin, A. S., and Sanderson, G. R. (1970) L-iduronic acid, a major constituent of heparin, *Carbohyd. Res.* **12**, 183-192.
255. Fischer, E., and Tiemann, F. (1894) Ueber das glucosamin, *Ber. dtsch. chem. Ges.* **27**, 138-147.
256. Foster, A. B. (1955) Deamination of D-glucosamine, D-glucosaminic acid, and D-glucosaminitol, *Chem. Ind.* , 627.
257. Shaklee, P. N., and Conrad, H. E. (1984) Hydrazinolysis of heparin and other glycosaminoglycans, *Biochem. J.* **217**, 187-197.
258. Shaklee, P. N., and Conrad, H. E. (1986) The disaccharides formed by deaminative cleavage of N-deacetylated glycosaminoglycans, *Biochem. J.* **235**, 225-236 .
259. Delaney, S. R., Leger, M., and Conrad, H. E. (1980) Quantitation of the sulfated disaccharides of heparin by high performance liquid chromatography, *Anal. Biochem.* **106**, 253-261.
260. Guo, Y., and Conrad, H. E. (1988) Analysis of oligosaccharides from heparin by reversed phase ion-pairing high pressure liquid chromatography, *Anal. Biochem.* **168**, 54-62.
261. Linhardt, R. J., Galliher, P. M., and Cooney, C. L. (1986) Polysaccharide lyases, *Appl. Biochem.Biotechnol.* **12**, 135-176.
262. Linhardt, R. J. (1994) Polysaccharide lyases for glycosaminoglycan analysis, in *Current Protocols Molec. Biol.*, ed. Ausubel, F. M., Wiley Interscience, New York, pp. 17.13.17-17.13.32.
263. Ernst, S., Langer, R., Cooney, C. L., and Sasisekharan, R. (1995) Enzymatic degradation of glycosaminoglycans, *Crit. Rev. Biochem. Mol. Biol.* **30**, 387-444.
264. Korn, E. D., and Payza, A. N. (1956) Bacterial degradation of heparin, *Nature* **177**, 88-89.
265. Galliher, P. M., Cooney, C. L., Langer, R., and Linhardt, R. J. (1981) Heparinase production of *Flavobacterium heparinum*, *Appl. Environ.l Microbiol.* **41**, 360-365.
266. Linker, A., and Hovingh, P. (1965) The enzymatic degradation of heparin and heparitin sulfate. I. The fractionation of a crude heparinase from *Flavobacteria*, *J. Biol. Chem.* **240**, 3724-3728.
267. Hovingh, P., and Linker, A. (1970) The enzymatic degradation of heparin and heparitin sulfate. III. Purification of a heparitinase and a heparinase from *Flavobacteria*, *J. Biol. Chem.* **245**, 6170-6175.
268. Dietrich, C. P., Silva, M. E., and Michelacci, Y. M. (1973) Sequential degradation of heparin in *Flavobactium heparinum*, *J. Biol. Chem.* **248**, 6408-6415.
269. Warnick, C. T., and Linker, A. (1972) Purification of an unusual α-glycuronidase from *Flavobacteria*, *Biochemistry* **11**, 568-572.

270. Silva, M. E., and Dietrich, C. P. (1974) Isolation and partial characterization of three induced enzymes from *Flavobacterium heparinum* involved in the degradation of heparin and heparitin sulfates, *Biochem. Biophys. Res. Commun.* **56**, 965-972.

271. Sasisekharan, R., Bulmer, M., Moremen, K. W., Cooney, C. L., and Langer, R. (1993) Cloning and expression of heparinase I gene from *Flavobacterium heparinum, Proc. Natl. Acad. Sci. U.S.A.* **90**, 3660-3664.

272. Sasisekharan, R., Leckband, D., Godavarti, R., Venkataraman, G., Cooney, C. L., and Langer, R. (1995) Heparinase I from *Flavobacterium heparinum*: the role of the cysteine residue in catalysis as probed by chemical modification and site-directed mutagenesis, *Biochemistry* **34**, 14441-14448.

273. Sasisekharan, R., Venkataraman, G., Godavarti, R., Ernst, S., Cooney, C. L., and Langer, R. (1996) Heparinase I from *Flavabacterium heparinum*: mapping and characterization of the heparin binding domain, *J. Biol. Chem.* **271**, 3124-3131.

274. Ernst, S., Venkataraman, G., Winkler, S., Godavarti, R., Langer, R., Cooney, C. L., and Sasisekharan, R. (1996) Expression in *Escherichia coli*, purification and characterization of heparinase I from *Flavobacterium heparinum, Biochem. J.* **315**, 589-597.

275. Su, H., Blain, F., Musil, R. A., Zimmermann, J. J. F., Gu, K., and Bennett, D. C. (1996) Isolation and expression in *Escherichia coli* of *hepB* and *hepC*, genes coding for the glycosaminoglycan-degrading enzymes heparinase II and heparinase III, respectively, from *Flavobacterium heparinum, Appl. Environ. Microbiol.* **62**, 2723-2734.

276. Lohse, D. L., and Linhardt, R. J. (1992) Purification and characterization of heparin lyases from *Flavobacterium heparinum, J. Biol. Chem.* **267**, 23347-23355.

277. Linker, A., and Hovingh, P. (1972) Isolation and characterization of oligosaccharides obtained from heparin by action of heparinase, *Biochemistry* **11**, 563-568.

278. Linhardt, R. J., Rice, K. G., Kim, Y. S., Lohse, D. L., Wang, H.-M., and Loganathhan, D. (1988) Mapping and quantification of the major oligosaccharide components of heparin, *Biochem. J.* **254**, 781-787.

279. Desai, U. R., Wang, H.-M., and Linhardt, R. J. (1993) Substrate specificity of the heparin lyases from *Flavobacterium heparinum, Arch. Biochem. Biophys.* **306**, 461-468.

280. Merchant, Z. M., Kim, Y. S., Rice, K. G., and Linhardt, R. J. (1985) Structure of heparin-derived tetrasaccharides, *Biochem. J.* **229**, 369-377.

281. Linhardt, R. J., Wang, H.-M., Loganathan, D., and Bae, J.-H. (1992) Search for the heparin antithrombin III-binding site precursor, *J. Biol. Chem.* **267**, 2380-2387.

282. Linhardt, R. J., Turnbull, J. E., Wang, H.-M., Loganathan, D., and Gallagher, J. T. (1990) Examination of the substrate specificity of heparin and heparan sulfate lyases, *Biochemistry* **29**, 2611-2617.

283. Rice, K. G., and Linhardt, R. J. (1989) Study of structurally defined oligosaccharide substrates of heparin and heparan monosulfate lyases, *Carbohyd. Res.* **190**, 219-233.

284. Gundlach, M. W., and Conrad, H. E. (1985) Glycosyl transferases in chondroitin sulfate biosynthesis, *Biochem. J.* **226**, 705-714.

285. Ampofo, S. A., Wang, H. M., and Linhardt, R. J. (1991) Disaccharide compositional analysis of heparin and heparan sulfate using capillary zone electrophoresis, *Anal. Biochem.* **199**, 249-255.

286. Desai, U. R., Wang, H.-M., Ampofo, S. A., and Linhardt, L. J. (1993) Oligosaccharide composition of heparin and low-molecular-weight heparins by capillary electrophoresis, *Anal. Biochem.* **213**, 120-127.

287. Pervin, A., Al-Hakim, A., and Linhardt, R. J. (1994) Separation of glycosaminoglycan-derived oligosaccharides by capillary electrophoresis using reverse polarity, *Anal. Biochem.* **221**, 182-1188.

288. Kitagawa, H., Kinoshita, A., and Sugahara, K. (1995) Microanalysis of glycosaminoglycan-derived disaccharides labeled with the fluorophore 2-aminoacridone by capillary electrophoresis and high-performance liquid chromatography, *Anal. Biochem.* **232**, 114-121.

289. Murata, K., Murata, A., and Yosida, K. (1995) High-performance liquid chromatographic identification of eight constitutional disaccharides from heparan sulfate isomers digested with heparitinases, *J. Chromatog. B* **670**, 3-10.

290. Linker, A., and Hovingh, P. (1984) Structural studies on heparin: tetrasaccharides obtained by heparinase degradation, *Carbohyd. Res.* **127**, 75-94.

291. Linhardt, R. J., Rice, K. G., Merchant, Z. M., Kim, Y. S., and Lohse, D. L. (1986) Structure and activity of a unique heparin-derived hexasaccharide, *J. Biol. Chem.* **261**, 14448-14454.

292. Rice, K. G., Rottink, M. J., and Linhardt, R. J. (1987) Fractionation of heparin-derived oligosaccharides by gradient polyacrylamide-gel electrophoresis, *Biochem. J.* **244**, 515-522.

293. Petitou, M., Lormeau, J. C., Perly, B., Berthault, P., Bossennec, V., Sié, P., and Choay, J. (1988) Is there a unique sequence in heparin for interaction with heparin cofactor II: structural and biological studies of heparin-derived oligosaccharides, *J. Biol. Chem.* **262**, 8685-8690.

294. Rice, K. G., Kim, Y. S., Grant, A. C., Merchant, Z. M., and Linhardt, R. J. (1985) High-pressure liquid chromatographic separation of heparin derived oligosaccharides, *Anal. Biochem.* **150**, 325-331.

295. Kosakai, M., and Yosizawa, Z. (1982) High pressure liquid chromatography of pyridylamino derivatives of sulfated oligosaccharides in the deamination products of heparin, *J. Biochem. Tokyo* **92**, 295-303.

296. Toomre, D. K., and Varki, A. (1994) Advances in the use of biotinylated diaminopyridine (BAP) as a versatile fluorescent tag for oligosaccharides, *Glycobiology* **4**, 653-663.

297. Hase, S., and Ikenaka, T. (1990) Estimation of elution times on reverse-phase high-pressure liquid chromatography of pyridylamino derivatives of sugar chains from glycoproteins, *Anal. Biochem.* **184**, 135-138.

298. Lee, Y. C., Lee, B. I., Tomiya, N., and Takahashi, N. (1990) Parameterization of contribution of sugar units to elution volumes in reverse-phase HPLC of 2-pyridylaminated oligosaccharides, *Anal. Biochem.* **188**, 259-266.

299. Cramer, J. A., and Bailey, L. C. (1991) A reversed-phase ion-pair high-performance liquid chromatography method for bovine testicular hyaluronidase digests using postcloumn derivatization with 2-cyanoacetamide and ultraviolet detection, *Anal. Biochem.* **196**, 183-191.

300. Scott, J. E. (1979) A reaction for the simple sensitive fluorimetric assay of heparin and 2-amino sugars, *Biochem. J.* **183**, 91-97.

301. Liu, J., Shirota, O., Wiesler, D., and Novotny, M. (1991) Ultrasensitive fluorometric detection of carbohydrates as derivatives in mixtures separated by capillary electrophoresis, *Proc. Natl. Acad. Sci. U.S.A.* **88**, 2302-2306.

302. Hopwood, J. J., and Muller, V. J. (1983) Selective depolymerisation of dermatan sulfate: production of radiolabelled substrates for α-L-iduronidase, sulfoiduronate sulfatase, and β-D-glucuronidase, *Carbohyd. Res.* **122**, 227-239.

303. Hopwood, J. J., and Elliott, H. (1983) Selective depolymerization of keratan sulfate: production of radiolabelled substrates for 6-O-sulfogalactose sulfatase and β-D-galactosidase, *Carbohyd. Res.* **117**, 263-274.

304. Edge, A. S. B., and Spiro, R. G. (1985) Structural elucidation of glycosaminoglycans through characterization of disaccharides obtained after fragmentation by hydrazine-nitrous acid treatment, *Arch. Biochem. Biophys.* **240**, 560-572.

305. Tyrrell, D. H., Ishihara, M., Rao, N., Horne, A., Kiefer, M. C., Stauber, G. B., Lam, L. H., and Stack, R. L. (1993) Structure and biological activities of a heparin-derived hexasaccharide with high affinity for basic fibroblast growth factor, *J. Biol. Chem.* **268**, 4684-4689.

306. Ludwigs, U., Elgavish, A., Esko, J. D., Meezan, E., and Rodén, L. (1986) Reaction of unsaturated uronic acid residues with mercuric salts: Cleavage of the hyaluronic acid

disaccharide 2-acetamido-2-deoxy-3-O-(-D-gluco-4-enepyranosyluronic acid)-D-glucose, *Biochem. J.* **245**, 795-804.

307. Glaser, J. H., and Conrad, H. E. (1979) Chick embryo liver β-glucuronidase: comparison of activity on natural and artificial substrates, *J. Biol. Chem.* **254**, 6588-6597.

308. Jansson, L., Ögren, S., and Lindahl, U. (1975) Macromolecular properties and end-group analysis of heparin isolated from bovine liver capsule, *Biochem. J.* **145**, 53-62.

309. Hopwood, J. J., and Robinson, H. C. (1973) The molecular-weight distribution of glycosaminoglycans, *Biochem. J.* **135**, 631-637.

310. Johnson, E. A., and Mulloy, B. (1976) The molecular-weight range of mucosal-heparin preparations, *Carbohyd. Res.* **51**, 119-127.

311. Desai, U. R., and Linhardt, R. J. (1995) Molecular weight of heparin using ^{13}C nuclear magnetic resonance spectroscopy, *J. Pharm. Sci.* **84**, 212-215.

312. Horenberg, J., and De Vries, J. X. (1983) Characterization of heparins by high-pressure size exclusion liquid chromatography, *J. Chromatog.* **261**, 287-292.

313. Kristensen, H. I., Trombrog, E. M., Nielsen, J. R., Nielson, J. I., Johansen, K. B., and Østegaard, P. B. (1991) Development and validation of a size exclusion chromatography method for determination of molecular masses and molecular mass distribution in low molecular weight heparin, *Thromb. Res.* **64**, 131-141.

314. Jeske, W., Ahsan, A., and Fareed, J. (1993) Molecular weight profiling of low molecular weight heparins utilizing a heparinase degraded oligosaccharide mixture as a calibrator, *Thromb. Res.* **E70**, 39-50.

315. Mulloy, B., Gee, C., Wheeler, S. F., Wait, R., Gray, E., and Barrowcliffe, T. W. (1997) Molecular weight measurements of low molecular weight heparins by gel permeation chromatography, *Thromb. Haemostas.* **73**, 668-674.

316. Knoblock, J. E., and Shaklee, P. N. (1997) Absolute molecular weight distribution of low molecular weight heparins by high performance size exclusion chromatography with multiangle lase light scattering detection, *Anal. Biochem.* **245**, 231-241.

317. Laurent, T. C., Tengblad, A., Thunberg, L., Höök, M., and Lindahl, U. (1978) The molecular-weight-dependence of the anti-coagulant activity of heparin, *Biochem. J.* **175**, 691-701.

318. Johnson, E. A. (1982) Characterization and separation of sulphated glycosaminoglycuronans, *Pharmacol. Res. Commun.* **14**, 289-292.

319. Ahsan, A., Hoppensteadt, D., Lormeau, J. C., Wolf, H., and Fareed, J. (1995) Molecular profiling and weight determination of heparins and depolymerized heparins, *J. Pharm. Sci.* **86**, 724-727.

320. Nielsen, J.-I. (1992) A convenient method for molecular mass determination of heparin, *Thromb. Haemostas.* **68**, 478-480.

321. Atkins, E. D. T., Isaac, D. H., Nieduszinski, I., Phelps, C. F., and Sheehan, J. K. (1974) The polyuronides: their molecular architecture, *Polymer* **15**, 263-271.

322. Atkins, E. D. T., and Nieduszinski, I. (1975) Crystalline structure of heparin, in *Heparin. Structure, Function and Clinical Implications*, eds., Bradshaw, R. A. & Wessler, S., Plenum Press, New York, pp. 19-36.

323. Perlin, A. S., Mazurek, M., Jaques, L. B., and Kavanagh, L. W. (1968) A proton magnetic resonance spectral study of heparin. L-Iduronic acid residues in commercial heparins, *Carbohyd. Res.* **7**, 369-379.

324. Perlin, A. S., Casu, B., and Sanderson, G. R. (1970) 220 MHz spectra of heparin, chondroitins, and other mucopolysaccharides, *Can. J. Chem.* **48**, 2260-2268.

325. van Boeckel, C. A. A., Lucas, H., van Aelst, S. F., van den Nieuwenhof, M. W. P., Wagenaars, G. N., and Mellema, J.-R. (1987) Synthesis and conformational analysis of an analogue of the antithrombin-binding region of heparin: the role of the carboxylate function of α-L-idopyranuronate, *Recl. Trav. Chim. Pays-Bas* **106**, 581-591.

326. Ragazzi, M., Ferro, D. R., Perly, B., Torri, G., Casu, B., Sinäy, P., Petitou, M., and Choay, J. (1987) Conformation of the pentasaccharide corresponding to the binding site of heparin to antithrombin III, *J. Carbohyd. Res.* **165**, C1-C5.

327. van Boeckel, C. A. A., Beetz, T., and van Aelst, S. F. (1988) Synthesis of a potent antithrombin activating pentasaccharide: a new heparin-like fragment containing two 3-O-sulphated glucosamines, *Tetrahedron Lett.* **29**, 803-806.

328. van Boeckel, C. A. A., van Aelst, S. F., Beetz, T., Meeuleman, D. G., van Dinther, T. G., and Moelker, H. C. T. (1989) Structure-activity relationships of synthetic heparin fragments, *Ann. N. Y. Acad. Sci.* **556**, 489-491.

329. Ragazzi, M., Ferro, D. R., Perly, B., Sinäy, P., Petitou, M., and Choay, J. (1990) Conformation of the pentasaccharide corresponding to the binding site of heparin for antithrombin III, *Carbohyd. Res.* **195**, 169-185.

330. Mulloy, B., and Johnson, E. A. (1987) Assignment of the ^{1}H-N.M.R. spectra of heparin and heparan sulphate, *Carbohydr. Res.* **170**, 151-165.

331. Desai, U. R., Wang, H.-M., Kelly, T. R., and Linhardt, R. J. (1993) Structure elucidation of a novel acidic tetrasaccharide and hexasaccharide derived from a chemically modified heparin, *Carbohydr. Res.* **241**, 249-259.

332. Liu, J., Desai, U. R., Han, X.-J., Toida, T., Toshihiko, T., and Linhardt, R. J. (1995) Strategy for the sequence analysis of heparin, *Glycobiology* **5**, 765-774.

333. Gatti, G., Casu, B., Hamer, G. K., and Perlin, A. S. (1979) Studies on the conformation of heparin by ^{1}H and ^{13}C NMR spectroscopy, *Macromolecules* **12**, 1001-1007.

334. Perlin, A. S., Mackie, D. M., and Dietrich, C. P. (1971) Evidence for a (1–4)-linked 4-O-(-L-idopyranosyl-uronic acid 2-sulfate)-(2-O-sulfoamino-D-glucopyranosyl 6-sulfate) sequence in heparin: long range H-H coupling in 4-deoxy-hex-4-enopyranosides, *Carbohyd. Res.* **18**, 185-194.

335. Perlin, A. S., Ng Ying King, N. M. K., Bhattacharjee, S. S., and Johnson, L. F. (1972) The ^{13}C Fourier transform spectrum of heparin: evidence for a bios repeating sequence of residues, *Can. J. Chem* **50**, 2437-2442.

336. Patt, S. L., Sauriol, F., and Perlin, A. S. (1982) Determination of the positions of glycosidic linkages from ^{13}C-^{13}C connectivity plots, *Carbohyd. Res.* **107**, C1-C4.

337. Wang, H.-M., Loganathan, D., and Linhardt, R. J. (1991) Determination of the pKa of glucuronic acid and the carboxy groups of heparin by ^{13}C-nuclear-magnetic-resonance spectroscopy, *Biochem. J.* **278**, 689-695.

338. Dais, P., and Perlin, A. S. (1982) An examination of ^{13}C- and ^{1}H-chemical shifts in relation to the conformational stabilities of D-glucopyranose disaccharides and polysaccharides, *Carbohyd. Res.* **107**, 263-269.

339. McNeil, C. J., Macfarlane, R. D., and Jardine, I. (1986) A novel mass spectrometric procedure to rapidly determine the partial structure of heparin fragments, *Biochem. Biophys. Res. Commun.* **139**, 18-24.

340. McNeil, C. J., and Macfarlane, R. D. (1986) The use of a stationary cationic surfactant as a selective matrix in ^{252}Cf-plasma desorption mass spectrometry, *J. Am. Chem. Soc.* **108**, 2132-2139.

341. Reinhold, V. N., Carr, S. A., Green, B. N., Petitou, M., Choay, J., and Sinäy, P. (1987) Structural characterization of sulfated glycosaminoglycans by fast atom bombardment mass spectrometry: application to heparin fragments prepared by chemical synthesis, *Carbohyd. Res.* **161**, 305-313.

342. Mallis, L. M., Wang, H.-M., Loganathan, D., and Linhardt, R. J. (1989) Sequence analysis of highly sulfated, heparin-derived oligosaccharides using fast atom bombardment mass spectroscopy, *Anal. Chem.* **61**, 1453-1458.

343. Yamada, S., Sakamoto, K., Tsuda, H., Yoshida, K., Sugahara, K., Khoo, K.-H., Morris, H. R., and Dell, A. (1993) Structural studies on the tri- and tetrasaccharides isolated from porcine intestinal heparin and characterization of heparinase/heparitinases using them as substrates, *Glycobiology* **4**, 69-78.

344. Dell, A., Rogers, M. E., Thomas-Oates, J. E., Huckerby, T. N., Sanderson, P. N., and Nieduszynski, I. A. (1988) Fast-atom bombardment mass-spectrometric strategies for sequencing sulphated oligosaccharides, *Carbohyd. Res.* **179**, 7-19.

345. Jalkanen, M., Nguyen, H., Rapraeger, A., Kurn, N., and Bernfield, M. (1985) Heparan sulfate proteoglycans from mouse mammary epithelial cells: localization on the cell surface with a monoclonal antibody, *J. Cell Biol.* **101**, 976-984.

346. de Boeck, H., Lories, V., David, G., Cassiman, J.-J., and vanden Berghe, H. (1987) Identification of a 64 kDa heparan sulphate proteoglycan core protein from human lung fibroblast plasma membranes with a monoclonal antibody, *Biochem. J.* **247**, 765-771.

347. van den Born, J., van den Heuvel, L. P. W. J., Bakker, M. A. H., Veerkamp, J. H., Assmann, J. K. M., and Berden, J. H. M. (1994) Monoclonal antibodies against the protein core and glycosaminoglycan side chain of glomerular basement membrane heparan sulfate proteoglycan: characterization and immunohistological application in human tissues, *J. Histochem. Cytochem.* **42**, 89-102.

348. Straus, A. H., Tavassos, L. R., and Takahashi, H. K. (1992) A monoclonal antibody (ST-1) directed to the native heparin chain, *Anal. Biochem.* **201**, 1-8.

349. Shibata, S., Harpel, P., Bona, C., and Fillit, H. (1993) Monoclonal antibodies to heparan sulfate inhibit the formation of thrombin-antithrombin III complexes, *Clin. Immunol. Immonopathol.* **68**, 264-272.

350. Shibata, S., Harpel, P. C., Gharavi, A., Rand, J., and Fillit, H. (1994) Autoantibodies to heparin from patients with antiphospholipid antibody syndrome inhibit formation of antithrombin III-thrombin complexes, *Blood* **83**, 2532-2540.

351. van den Born, J., van den Heuvel, L. P. W. J., Bakker, M. A. H., Veerkamp, J. H., Assmann, K. J. M., and Berden, J. H. M. (1992) A monoclonal antibody against GBM heparan sulfate induces an acute selective proteinuria in rats, *Kidney Int.* **41**, 115-123.

352. van den Born, J., Jann, K., Assmann, K. J. M., Lindahl, U., and Berden, J. H. M. (1996) N-Acetylated domains in heparan sulfates revealed by a monoclonal antibody against the *Escherichia coli* K5 capsular polysaccharide: distribution of the cognate epitope in normal human kidney and transplant kidney with chronic vascular rejection, *J. Biol. Chem.* **271**, 22802-22809.

353. Pejler, G., Lindahl, U., Larm, O., Scholander, E., Sandgren, E., and Lundblad, A. (1988) Monoclonal antibodies specific for oligosaccharides prepared by partial nitrous acid deamination of heparin, *J. Biol. Chem.* **263**, 5197-5201.

354. Conrad, H. E., and Guo, Y. (1992) Structural analysis of periodate-oxidized heparin, in *Heparin and Related Polysaccharides*, eds., Lane, D. A. & Lindahl, U., Plenum Press, New York, pp. 31-36.

355. Casu, B., Diamantini, G., Fedeli, G., Mantovani, M., Oreste, P., Pescador, R., Porta, R., Prino, G., Torri, G., and Zoppetti, G. (1986) Retention of antilipemic activity by periodate-oxidized non-anticoagulant heparin, *Arzneim-Forsch./Drug Res.* **36**, 637-642.

356. Nagasawa, K., and Inouye, Y. (1974) Solvolytic desulfation of 2-deoxy-2-sulfamimo-D-glucose and D-glucose 6-sulfate, *Carbohyd. Res.* **36**, 265-271.

357. Höök, M., Riesenfeld, J., and Lindahl, U. (1982) N-[^{3}H]Acetyl-labeling, a convenient method for radiolabeling of glycosaminoglycans, *Anal. Biochem.* **119**, 236-245.

358. Levy, L., and Petracek, F. J. (1962) Chemical and pharmacological studies on N-resulfated heparin, *Proc. Soc. Expt. Biol. Med.* **109**, 901-905.

359. Rej, R. N., Ludwig-Baxter, K. G., and Perlin, A. S. (1991) Sulfation of some chemically-modified heparins. Formation of a 3-sulfate analog of heparin, *Carbohyd. Res.* **210**, 299-310.

360. Rej, R., Jaseja, M., and Perlin, A. S. (1989) Importance for blood anticoagulant activity of a 2-sulfate group on L-iduronic acid residues in heparin, *Thromb. Haemostas.* **61**, 540-541.

361. Jaseja, M., Rej, R. N., Sauriol, F., and Perlin, A. S. (1989) Novel regio- and stereoselective modifications of heparin in alkaline solution: nuclear magnetic resonance spectroscopic evidence, *Can. J. Chem.* **67**, 1449-1456.

362.   Fraidenraich y Waisman, D., and Fernandez Cirelli, A. (1992) Selective O-desulphation of heparin in triethylamine, *Carbohyd. Polymers* **17**, 111-114.

363.   Piani, S., Casu, B., Marchi, E. G., Torri, C. G., and Ungarelli, F. (1993) Alkali-induced optical rotation changes in heparins and heparan sulfates, and their relation to iduronic acid-containing sequences, *J. Carbohyd. Chem.* **12**, 507-521.

364.   Tiozzo, R., Cingi, M. R., Reggiana, D., Andreoli, T., Calandra, S., Milani, M. R., Piani, S., Marchi, E., and Barbanti, M. (1993) Effect of desulfation of heparin on its anticoagulant and anti-proliferative activity, *Thromb. Res.* **70**, 99-106.

365.   Liu, Z., and Perlin, A. S. (1992) Adverse effects of alkali and acid on the anticoagulant potency of heparin, evaluated with methyl 2-deoxy-2-sulfamino-(-D-glucopyranoside-3-sulfate as a model compound, *Carbohyd. Res.* **228**, 29-36.

366.   Ishihara, M. (1994) Structural requirements in heparin for binding and activation of FGF-1 and FGF-4 are different from that for FGF-2, *Glycobiology* **4**, 817-824.

367.   Ishihara, M., Shaklee, P. N., Yang, Z., Liang, W., Wei, Z., Stack, R. J., and Holme, K. (1994) Structural features in heparin which modulate specific biological activities mediated by basic fibroblast growth factor, *Glycobiology* **4**, 451-458.

368.   Inoue, Y., and Nagasawa, K. (1976) Selective N-desulfation of heparin with dimethylsulfoxide containing water or methanol, *Carbohyd. Res.* **46**, 87-95.

369.   Nagasawa, K., Inoue, Y., and Kamata, T. (1977) Solvolytic desulfation of glycosaminoglycuronan sulfates with dimethyl sulfoxide containing water or methanol, *Carbohyd. Res.* **58**, 47-55.

370.   Matuo, M., Takano, R., Kamei-Hayashi, K., and Hara, S. (1993) A novel regioselective desulfation of polysaccharide sulfate: specific 6-O-desulfation with N,O-bis(trimethylsilyl) acetamide, *Carbohyd. Res.* **241**, 209-215.

371.   Takano, R., Kanda, T., Hayashi, K., Yoshida, K., and Hara, S. (1995) Desulfation of sulfated carbohydrates mediated by silylating reagents, *J. Carbohyd. Res.* **14**, 885-888.

372.   Taylor, R. L., Shively, J. E., and Conrad, H. E. (1976) Stoichiometric reduction of uronic acid carboxyl groups in polysaccharides, *Meth. Carbohyd. Chem.* **7**, 149-151.

373.   Inoue, Y., and Nagasawa, K. (1982) On the reaction of N-acetylchondrosine, N-acetylchondrosine-6-sulfate, chondroitin-6-sulfate, and heparin with 1-(3-dimethylaminopropyl)-3-ethylcarbodiimide, *Carbohyd. Res.* **111**, 113-125.

374.   Shively, J. E., and Conrad, H. E. (1976) Nearest neighbor analysis of heparin: identification and quantitation of the products formed by selective depolymerization procedures, *Biochemistry* **15**, 3943-3950.

375.   Hoare, D. G., and Koshland, D. E. (1967) A method for the quantitative modification and estimation of carboxylic acid groups in proteins, *J. Biol. Chem.* **242**, 2447-2453.

376.   Danishefsky, I., and Siskovic, E. (1971) Conversion of carboxyl groups of mucopolysaccharides into amides of amino acid esters, *Carbohyd. Res.* **16**, 199-205.

377.   Shaklee, P. N., Yang, Z., and Herrmann, J. (1993) Synthesis and analysis of carboxy-reduced and amidomethylsulfonated heparins, *Glycobiology* **3**, 541.

378.   Freeman, C., and Hopwood, J. J. (1989) Human liver glucuronic acid 2-sulphatase: purification, characterization and catalytic properties, *Biochem. J.* **279**, 399-405.

379.   Skåk-Bræk, G., Eklund, T., von Husby, K., and Smidsrød, O. (1988) Modification of alginates or other uronic acid compounds by treatment with CO_2, in *International Patent No. WO 88/02758.*

380.   Skåk-Bræk, G., Eklund, T., von Husby, K. O., Kvam, B. J., and Smidsrød, O. (1991) Modification of alginates or other uronic acid compounds by treatment with CO_2, in *United States Patent No. 4990601.*

381.   Uchiyama, H., and Nagasawa, K. (1991) Changes in the structure and biological properties of N→S sulfate-transferred, N-resulfated heparin, *J. Biol. Chem.* **266**, 6756-6760.

382.   Soeda, S., Sakaguchi, S., Shimeno, H., and Nagamatsu, A. (1992) Fibrinolytic and anticoagulant activities of highly sulfated fucoidin, *Biochem. Pharmacol.* **43**, 1853-1858.

383. Nagasawa, K., Uchiyama, H., and Wajima, N. (1986) Chemical sulfation of preparations of chondroitin 4- and 6-sulfate and dermatan sulfate: preparation of chondroitin sulfate E-like materials from chondroitin 4-sulfate, *Carbohyd. Res.* **158**, 183-190.

384. Casu, B., Grazioli, G., Razi, N., Guerrini, Naggi, A., Torri, G., Oreste, P., Tursi, Zoppetti, G., and Lindahl, U. (1994) Heparin-like compound prepared by chemical modification of capsular polysaccharide from *E. coli* K5, *Carbohyd. Res.* **263**, 271-284.

385. Colliec-Jouault, S., Shworak, N. W., Liu, J., de Agostini, A. I., and Rosenberg, R. D. (1994) Characterization of a cell mutant specifically defective in the synthesis of anticoagulantly active heparan sulfate, *J. Biol. Chem.* **269**, 35-43.

386. Lloyd, P. F., and Forrester, P. F. (1971) Facile acid hydrolysis of glycoside 2-sulphates, *Carbohyd. Res.* **19**, 430-431.

387. Razi, N., Feyzi, E., Björk, I., Naggi, A., Casu, B., and Lindahl, U. (1995) Structural and functional properties of heparin analogues obtained by chemical sulphation of *Escherichia coli* K5 polysaccharide, *Biochem. J.* **309**, 465-472.

388. Volpi, N., Mascellani, G., and Bianchini, P. (1992) Low molecular weight heparins (5 kDa) and oligoheparins ((2 kDa) produced by gel permeation enrichment or radical process: comparison of structures and physicochemical and biological properties, *Anal. Biochem.* **200**, 100-107.

389. Choay, J. (1989) Chemically synthesized heparin-derived oligosaccharides, in *Heparin and Related Polysaccharides. Structure and Activities, Annals of the New York Academy of Sciences*, eds., Ofosu, F. A., Danishefsky, I. & Hirsh, J., New York Acad. Sci., New York, Vol. 556, pp. 61-74.

390. Petitou, M. (1989) Chemical synthesis of heparin, in *Heparin. Chemical and Biological Properties, Clinical Applications*, eds., Lane, D. A. & Lindahl, U., CRC Press, Inc., Boca Raton, FL, pp. 65-80.

391. Petitou, M., Lormeau, J. C., and Choay, J. (1991) Chemical synthesis of glycosaminoglycans: new approaches to antithrombotic drugs, *Nature* **350**, 30-33.

392. Petitou, M. (1992) Partial synthesis and hemisynthesis in the field of glycosaminoglycans, in *Heparin and Related Polysaccharides, Adv. Experimental Med. Biol.*, eds., Lane, D. A., Björk, I. & Lindahl, U., Plenum Press, New York, Vol. 313, pp. 21-30.

393. Kim, J. J., and Conrad, H. E. (1976) Kinetics of mucopolysaccharide and glycoprotein synthesis by chick embryo chondrocytes: effect of D-glucose concentration in the culture medium, *J. Biol. Chem.* **251**, 6210-6217.

394. Sobue, M., Takeuchi, J., Ito, K., Kimata, K., and Suzuki, S. (1978) Effect of environmental sulfate concentration on the synthesis of low and high sulfate chondroitin sulfates by chick embryo cartilage, *J. Biol. Chem.* **253**, 6190-6196.

395. Drake, M. R., De la Rosa, J., and Stipanuk, M. H. (1987) Metabolism of cysteine in rat hepatocytes: evidence for cysteinesulphinate-independent pathways, *Biochem. J.* **244**, 279-286.

396. Humphries, D. E., Silbert, C. K., and Silbert, J. E. (1988) Sulfation by cultured cells: cysteine, cysteinsulfinic acid, and sulfite as sources for proteoglycan sulfate, *Biochem. J.* **252**, 305-308.

397. Esko, J. D., Elgavish, A., Prasthofer, T., Taylor, W. H., and Weinke, J. L. (1986) Sulfate transport-deficient mutants of Chinese hamster ovary cells: sulfation of glycosaminoglycans dependent on cysteine, *J. Biol. Chem.* **261**, 15725-15733.

398. Keller, J. M., and Keller, K. M. (1987) Amino acid sulfur as a source of sulfate for sulfated proteoglycans produced by Swiss mouse 3T3 cells, *Biochim. Biophys. Acta.* **926**, 139-144.

399. Templeton, D. M., and Wang, A. (1992) Conserved charge of glomerular and mesangial cell proteoglycans: possible role of amino acid-derived sulphate, *Can. J. Physiol. Pharmacol.* **70**, 843-852.

400. Imai, Y., Yanagishita, M., and Hascall, V. C. (1994) Measurement of contribution from intracellular cysteine to sulfate in phosphoadenosine phosphosulfate in rat ovarian granulosa cells, *Arch. Biochem. Biophys.* **312**, 392-400.

401. Humphries, D. E., Silbert, C. K., and Silbert, J. E. (1986) Glycosaminoglycan production by bovine aortic endothelial cells cultured in sulfate-depleted medium, *J. Biol. Chem.* **261**, 9122-9127.
402. Silbert, C. K., Humphries, D. E., Palmer, M. E., and Silbert, J. E. (1991) Effects of sulfate deprivation on the production of chondroitin/dermatan sulfate by cultures of skin fibroblasts from normal and diabetic individuals, *Arch. Biochem. Biophys.* **285**, 137-141.
403. Sekler, I., Lo, R. S., Mastrocola, T., and Kopito, R. R. (1995) Sulfate transport mediated by the mammalian anion exchangers in reconstituted proteoliposomes, *J. Biol. Chem.* **270**, 11251-11256.
404. Bienkowski, M. J., and Conrad, H. E. (1984) Kinetics of proteoheparan sulfate synthesis, secretion, endocytosis, and catabolism by a hepatocyte cell line, *J. Biol. Chem.* **259**, 12989-12996.
405. Yanagishita, M., and Hascall, V. C. (1984) Proteoglycans synthesized by rat ovarian granulosa cells in culture. Isolation, fractionation, and characterization of proteoglycans associated with the cell layer, *J. Biol. Chem.* **259**, 10260-10269.
406. Lories, V., Cassiman, J. J., Van den Berghe, H., and David, G. (1992) Differential expression of cell surface heparan sulfate proteoglycans in human mammary epithelial cells and lung fibroblasts, *J. Biol. Chem.* **267**, 1116-1122.
407. Müller, G., Korndörfer, A., Saar, K., Karbe-Thönges, B., Fasold, H., and Müllner, S. (1994) 4'-Amino-benzamido-taurocholic acid selectively solubilizes glycosyl-phosphatidylinositol-anchored membrane proteins and improves lipolytic cleavage of their membrane anchors by specific phospholipases, *Arch. Biochem. Biophys.* **309**, 329-340.
408. Baenziger, J. U., Kumar, S., Bradbeck, R. M., Smith, P. L., and Beranek, M. C. (1992) Circulatory half-life but not interaction with the lutropin/chorionic gonadotropin receptor is modulated by sulfation of bovine lutropin oligosaccharides, *Proc. Natl. Acad. Sci. U.S.A.* **89**, 334-338.
409. Mayo, J. W., and Carlson, D. M. (1970) Effect of alkali and sodium borohydride at alkaline pH on N-acetylchondrosine: reduction versus cleavage, *Carbohyd. Res.* **15**, 300-303.
410. Seno, N., and Sekizuka, E. (1978) Quantitative β-elimination of O-glycosyl linkages in chondroitin sulfates, *Carbohyd. Res.* **62**, 271-299.
411. Ogata, S.-I., and Lloyd, K. O. (1982) Mild alkaline borohydride treatment of glycoproteins: a method for liberating both N- and O-linked carbohydrate chains, *Anal. Biochem.* **119**, 351-359.
412. Fedarko, N. S., and Conrad, H. E. (1986) A unique heparan sulfate in the nuclei of hepatocytes: structural changes with the growth state of the cells, *J. Cell Biol.* **102**, 587-599.
413. Conrad, H. E., James, M. E., and Varboncoeur, E. (1973) Qualitative and quantitative analysis of reducing carbohydrates by radiochromatography on ion exchange papers, *Anal. Biochem.* **51**, 486-500.
414. Mertens, G., Cassiman, J.-J., Van den Berghe, H., Vermylen, J., and David, G. (1992) Cell surface heparan sulfate proteoglycans from human vascular endothelial cells: core protein characterization and antithrombin III binding properties, *J. Biol. Chem.* **267**, 20435-20443.
415. Bame, K. J., Zhang, L., David, G., and Esko, J. D. (1994) Sulphated and undersulphated heparan sulphate proteoglycans in a Chinese hamster ovary cell mutant defective in N-sulphotransferase, *Biochem. J.* **303**, 81-87.
416. Chernousov, M. A., and Carey, D. J. (1993) N-Syndecan (syndecan 3) from neonatal rat brain binds basic fibroblast growth factor, *J. Biol. Chem.* **268**, 16810-16814 .
417. Bourdon, M. A., Oldberg, Å., Pierschbacher, M., and Ruoslahti, E. (1985) Molecular cloning and sequence analysis of a chondroitin sulfate proteoglycan cDNA, *Proc. Natl. Acad. Sci. U.S.A.* **261**, 12534-12537.
418. Alliel, P. M., Périn, J.-P., Bonnet, F., Rosa, J.-C., and Jollés, P. (1988) Complete amino acid sequence of a human platelet proteoglycan, *FEBS Lett.* **236**, 123-126.

419. Nicodemus, C. F., Avraham, S., Austen, K. F., Purdy, S., Jablonski, J., and Stevens, R. L. (1990) Characterization of the human gene that encodes the peptide core of secretory granule proteoglycans in promyelocytic leukemia HL-60 cells and analysis of the translated product, *J. Biol. Chem.* **265**, 5889-5896.

420. Stellrecht, C. M., and Saunders, G. F. (1989) Nucleotide sequence of a cDNA encoding a hemopoietic proteoglycan core protein, *Nucleic Acids Res.* **17**, 7523.

421. Giorda, R., Chambers, W. H., Dahl, C. A., and Trucco, M. (1990) Isolation and characterization of a cDNA that encodes the core protein of the cytolytic granule proteoglycan of rat natural killer cells, *Nat. Immun. Cell Growth Regul.* **9**, 91-102.

422. Stellrecht, C. M. M., Fraizer, G., Selvanayagam, C., Chao, L.-Y., Lee, A., and Saunders, G. F. (1993) Transcriptional regulation of a hematopoietic proteoglycan core protein gene during hematopoiesis, *J. Biol. Chem.* **268**, 4078-4084 .

423. Zheng, N., and Gierasch, L. M. (1996) Signal sequences: the same yet different, *Cell* **86**, 849-852.

424. Masterson, W. J., Doering, T. L., Hart, G. W., and Englund, P. T. (1989) A novel pathway for glycan assembly: biosynthesis of the glycosyl-phosphatidylinositol anchor of the trypanosome variant surface glycoprotein, *Cell* **56**, 793-800.

425. Menon, A. K., and Stevens, V. L. (1992) Phosphatidylethanolamine is the donor of the ethanolamine residue linking a glycosylphosphatidylinositol anchor to protein, *J. Biol. Chem.* **267**, 15277-15280.

426. Tartakoff, A. M., and Singh, N. (1992) How to make a glycosylinositol phospholipid anchor, *Trends Biochem. Sci.* **17**, 470-473.

427. Takeda, J. K., and Kinoshita, T. (1995) GPI-anchor-anchor synthesis, *TIBS* **20**, 367-371.

428. Udenfriend, S., and Kodukula, K. (1995) How glycosyl-phosphatidylinositol-anchored membrane proteins are made, *Annu. Rev. Biochem.* **1964**, 563-591.

429. Medof, M. E., Nagarajan, S., and Tykocinski, M. L. (1996) Cell-surface engineering with GPI-anchored proteins, *FASEB J.* **10**, 574-586.

430. Takahashi, M., Inoue, N., Ohishi, K., Maeda, Y., Nakamura, N., Endo, Y., Fujita, T., Takeda, J., and Kinoshita, T. (1996) PIG-B, a membrane protein of the endoplasmic recitulum with a large lumenal domain, is involved in transferring the third mannose of the GPI anchor, *EMBO J.* **15**, 4254-4261.

431. Moran, P., and Caras, I. W. (1991) A nonfunctional sequence converted to a signal for glycophosphatidylinositol membrane anchor attachment, *J. Cell Biol.* **115**, 329-336 .

432. Møller, L. B., Ploug, M., and Blasi, F. (1992) Structural requirements for glycosyl-phosphatidylinositol-anchor attachment in the cellular receptor for urokinase plasminogen activator, *Eur. J. Biochem.* **208**, 493-500.

433. Beghdadi-Rais, C., Schreyer, M., Rousseaux, M., Borel, P., Eisenberg, R. J., Cohen, G. H., Bron, C., and Fasel, N. (1993) Carboxyl terminus structural requirements for glycosyl-phosphatidylinositol anchor addition to cell surface proteins, *J. Cell Sci.* **105**, 831-840.

434. Coyne, K. E., Crisci, A., and Lubin, D. M. (1993) Construction of synthetic signals for glycosyl-phosphatidylinositol anchor attachment: analysis of amino acid sequence requirements for anchoring, *J. Biol. Chem.* **268**, 6689-6693.

435. Kodukula, K., Gerber, L. D., Amthauer, R., Brink, L., and Udenfriend, S. (1993) Biosynthesis of glycosylphosphatidylinositol (GPI)-anchored membrane proteins in intact cells: specific amino acid requirements adjacent to the site of cleavage and GPI attachment, *J. Cell Biol.* **120**, 657-664.

436. Esko, J. D., Weinke, J. L., Taylor, W. H., Ekborg, G., Rodén, L., Anantharamaiah, G., and Gawish, A. (1987) Inhibition of chondroitin and heparan sulfate biosynthesis in Chinese hamster ovary cell mutants defective in galactosyltransferase I, *J. Biol. Chem.* **262**, 12189-12195.

437. Kearns, A. E., Vertel, B. M., and Schwartz, N. B. (1993) Topography of glycosylation and UDP-xylose production, *J. Biol. Chem.* **268**, 11097-11104.

438. Vertel, B. M., Walters, L. M., Flay, N., Kearns, A. E., and Schwartz, N. B. (1993) Xylosylation is an endoplasmic reticulum to Golgi event, *J. Biol. Chem.* **268**, 11105-11112.

439. Nuwaydid, N., Glaser, J. H., Johnson, J. C., Conrad, H. E., Hauser, S. C., and Hirschberg, C. B. (1986) Xylosylation and glucuronosylation reactions in rat liver Golgi apparatus and endoplasmic reticulum, *J. Biol. Chem.* **261**, 12936-12941.

440. Lidholt, K., Weinke, J. L., Kiser, C. S., Lugemwa, F. N., Bame, K. J., Cheifetz, S., Massagué, J., Lindahl, U., and Esko, J. D. (1992) A single mutation affects both N-acetylglucosaminyltransferase and glucuronosyltransferase activities in a Chinese hamster ovary cell mutant defective in heparan sulfate biosynthesis, *Proc. Natl. Acad. Sci. U.S.A.* **89**, 2267-2271.

441. Kjellén, L., Pettersson, I., Unger, E., and Lindahl, U. (1992) Two enzymes in one: N-deacetylation and N-sulfation in heparin biosynthesis are catalyzed by the same protein, in *Heparin and Related Polysaccharides, Advances in Experimental Medicine and Biology, Vol. 313,*, eds., Lane, D. A., Björk, I. & Lindahl, U., Plenum Press, New York, pp. 107-111.

442. Cheung, W.-F., Eriksson, I., Kusche-Gullberg, M., Lindahl, U., and Kjellén, L. (1996) Expression of the mouse mastocytoma glucosaminyl N-deacetylase/N-sulfotransferase in human kidney 293 cells results in increased N-sulfation of heparan sulfate, *Biochemistry* **35**, 5250-5256.

443. Capasso, J. M., and Hirschberg, C. B. (1985) Mechanisms of glycosylation and sulfation in the Golgi apparatus: evidence for nucleotide sugar/nucleoside monophosphate antiports in the Golgi apparatus membrane, *Proc. Natl. Acad. Sci. U.S.A.* **81**, 7051-7055.

444. Yanagishita, M., and Hascall, V. C. (1985) Effects of monensin on the synthesis, transport, and intracellular degradation of proteoglycans in rat ovarian granulosa cells in culture, *J. Biol. Chem.* **260**, 5445-5455.

445. Milla, M. E., and Hirschberg, C. B. (1989) Reconstitution of Golgi vesicle CMP-sialic acid and adenosine 3'-phosphate 5'-phosphosulfate transport into proteoliposomes, *Proc. Natl. Acad. Sci. U.S.A.* **86**, 1786-1790 .

446. Ozeran, J. D., Westley, J., and Schwartz, N. B. (1996) Kinetics of PAPS translocase: evidence for an antiport mechanism, *Biochemistry* **35**, 3685-3694.

447. Ozeran, J. D., Westley, J., and Schwartz, N. B. (1996) Identification and partial purification of the PAPS translocase, *Biochemistry* **35**, 3695-3703.

448. Toma, L., Pinhal, M. A. S., Dietrich, C. P., Nader, H. B., and Hirschberg, C. D. (1996) Transport of UDP-galactose into the Golgi lumen regulates the biosyntheis of proteoglycans, *J. Biol. Chem.* **271**, 3897-3901.

449. Abeijon, C., and Hirschberg, C. B. (1992) Topography of glycosylation reactions in the endoplasmic reticulum, *Trends Biochem. Sci.* **17**, 32-36.

450. Prinz, R., Klein, U., Sudhakaran, P. R., Sinn, W., Ullrich, K., and Von Figura, K. (1980) Metabolism of sulfated glycosaminoglycans in rat hepatocytes. Synthesis of heparan sulfate and distribution into cellular and extracellular pools, *Biochim. Biophys. Acta.* **630**, 402-413 .

451. Ishihara, M., and Conrad, H. E. (1989) Correlations between heparan sulfate metabolism and hepatoma growth, *J. Cell. Physiol.* **138**, 467-476.

452. Bame, K. J., and Esko, J. D. (1989) Undersulfated heparan sulfate in a Chinese hamster ovary cell mutant defective in heparan sulfate N-sulfotransferase, *J. Biol. Chem.* **264**, 8059-8065.

453. Bame, K. J., Lidhholt, K., Lindahl, U., and Esko, J. D. (1991) Biosynthesis of heparan sulfate: coordination of polymer-modification reactions in a Chinese hamster ovary cell mutant defective in N-sulfotransferase, *J. Biol. Chem.* **266**, 10287-10293.

454. Lidholt, K., Riesenfeld, J., Jacobsson, J.-G., Feingold, D. S., and Lindahl, U. (1988) Biosynthesis of heparin: modulation of polysaccharide chain length in a cell-free system, *Biochem. J.* **254**, 571-578.

455. Lidholt, K., Kjellén, L., and Lindahl, U. (1989) Biosynthesis of heparin: relationship between the polymerization and sulfation processes, *Biochem. J.* **261**, 999-1007.

456. Lisante, M. P., Sargiacomo, M., Graeve, L., Saltiel, A. R., and Rodriguez-Boulan, E. (1988) Preferred apical distribution of glycosyl phosphatidylinositol (GPI)-anchored proteins: a highly conserved feature of the polarized epithelial cell phenotype, *J. Memb. Biol.* **113**, 155-167.

457. Brown, D. A., and Rose, J. K. (1992) Sorting of GPI-anchored proteins to glycolipid-enriched membrane subdomains during transport to the apical cell surface, *Cell* **68**, 533-544.

458. Lisanti, M. P., Caras, I. W., Gilbert, T., Hanzel, D., and Rodriguez-Boulan, E. (1990) Vectorial apical delivery and slow endocytosis of a glycolipid-anchored fusion protein in transfected MDCK cells, *Proc. Natl. Acad. Sci. U.S.A.* **87**, 7419-7423.

459. Zurzolo, C., Lisante, M. P., Caras, I. W., Nitsch, L., and Rodriguez-Boulan, E. (1993) Glycosylphosphatidylinositol-anchored proteins are preferentially targeted to the basolateral surface in Fischer rat thyroid epithelial cells, *J. Cell Biol.* **121**, 1031-1039.

460. Mertens, G., Van der Schueren, B., van den Berghe, H., and David, G. (1996) Heparan sulfate expression in polaryzed epithelial cells: the apical sorting of glypican (GPI-anchored proteoglycan) is inversely related to its heparan sulfate content, *J. Cell Biol.* **132**, 487-497.

461. Robinson, J. M., Viti, M., and Höök, M. (1984) Structure and properties of an undersulfated heparan sulfate proteoglycan synthesized by a rat hepatoma cell line, *J. Cell Biol.* **98**, 946-953.

462. Carey, D. J., and Evans, D. (1989) Membrane anchoring of heparan sulfate proteoglycans by phosphatidylinositol and kinetics of synthesis of peripheral and detergent-solubilized proteoglycans in Schwann cells, *J. Cell Biol.* **108**, 1891-1897.

463. Kraemer, P. M. (1968) Production of heparin related glycosaminoglycans by an established cell line, *J. Cell. Physiol.* **69**, 109-119.

464. Kraemer, P. M. (1971) Heparan sulfates of cultured cells. I. Membrane-associated and cell-sap species in Chinese hamster cells, *Biochemistry* **10**, 1437-1445.

465. Kraemer, P. M. (1971) Heparan sulfates of cultured cells. II. Acid-soluble and precipitable species of different cell lines, *Biochemistry* **10**, 1445-1451.

466. Kraemer, P. M., and Tobey, R. A. (1972) Cell-cycle dependent desquamation of heparan sulfate from the cell surface, *J. Cell Biol.* **55**, 713-717.

467. Kraemer, P. M. (1974) High molecular weight heparan sulfate from the cell surface, *Biochem. Biophys. Res. Commun.* **56**, 423-430.

468. Kraemer, P. M. (1977) Heparin releases heparan sulfate from the cell surface, *Biochem. Biophys. Res. Commun.* **78**, 1334-1340.

469. Ishihara, M., Fedarko, N. S., and Conrad, H. E. (1986) Transport of heparan sulfate into the nuclei of hepatocytes, *J. Biol. Chem.* **261**, 13575-13580.

470. Yanagishita, M., and Hascall, V. C. (1984) Metabolism of proteoglycans in rat ovarian granulosa cell culture. Multiple intracellular degradative pathways and the effect of chloroquine, *J. Biol. Chem.* **259**, 10270-10283.

471. Kjellén, L., Pertoft, H., Oldberg, Å., and Höök, M. (1985) Oligosaccharides generated by an enodglucuronidase are intermediates in the intracellular degradation of heparan sulfate proteoglycans, *J. Biol. Chem.* **260**, 8416.

472. Iozzo, R. V., and Hassell, J. R. (1989) Identification of the precursor protein for the heparan sulfate proteoglycan of human colon carcinoma cells and its post-translation modifications, *Arch. Biochem. Biophys.* **269**, 239-249.

473. Owens, R. T., and Wagner, W. D. (1991) Metabolism and turnover of cell surface-associated heparan sulfate proteoglycan and chondroitin sulfate proteoglycan in normal and cholesterol-enriched macrophages, *Arterioscler. Thromb.* **11**, 1752-1758.

474. Takeuchi, Y., Yanagishita, M., and Hascall, V. C. (1992) Metabolic pathways of heparan sulfate proteoglycans in a rat parathyroid cell line, *J. Biol. Chem.* **267**, 14677-14684.

475. Takeuchi, Y., Yanagishita, M., and Hascall, V. C. (1992) Effects of MnCl$_2$ on the release and recycling of heparan sulfate proteoglycans in a rat parathyroid cell line, *Arch. Biochem. Biophys.* **298**, 371-379.

476. Yanagishita, M., and Hascall, V. C. (1992) Cell surface heparan sulfate proteoglycans, *J. Biol. Chem.* **267**, 9451-9454.

477. Sivaram, P., Obunike, J. C., and Goldberg, I. J. (1995) Lysolecithin-induced alteration of subendothelial heparan sulfate proteoglycans increases monocyte binding to matrix, *J. Biol. Chem.* **270**, 29760-29765.

478. Gallagher, J. T., Walker, A., Lyon, M., and Evans, W. M. (1988) Heparan sulphate-degrading endoglycosidase in liver plasma membranes, *Biochem. J.* **250**, 719-726.

479. Caras, I. G. (1991) Probing the signal for glycophosphatidylinositol anchor attachment using decay accelerating factor as a model system, *Cell Biol. International Reports* **15**, 815-826.

480. Brabeer, M., Robinson, J. A., and Robinson, H. C. (1990) The mechanism of initiation of chondroitin sulphate synthesis by β-D-galactosides, *Biochem. Int.* **21**, 1161-1168.

481. Höök, M., Wasteson, A., and Oldberg, Å. (1975) A heparan sulfate degrading endoglycosidase from rat liver tissue, *Biochem. Biophys. Res. Commun.* **67**, 1422-1428.

482. Klein, U., and Von Figura, K. (1976) Partial purification and characterization of a heparan sulfate specific endoglucuronidase, *Biochem. Biophys. Res. Commun.* **73**, 569-576.

483. Young, E., and Horner, A. A. (1979) The assay and partial characterization of macromolecular heparin depolymerase activity in rat small intestine, *Biochem. J.* **180**, 587-596 .

484. Oldberg, Å., Heldin, C.-H., Wasteson, Å., Busch, C., and Höök, M. (1980) Characterization of a platelet endoglycosidase degrading heparin-like polysaccharides, *Biochemistry* **19**, 5755 .

485. Najajima, N., Irimura, T., Di Ferrante, N., and Nicolson, G. L. (1984) Metastatic melanoma cell heparanase: characterization of heparan sulfate degradation fragments produced by B16 melanoma endoglucuronidase, *J. Biol. Chem.* **259**, 2283-2290.

486. Krüger, U., and Kresse, H. (1986) Endocytosis of proteoheparan sulfate by cultured skin fibroblasts, *Hoppe-Seyler's Zeit. Physiol. Chem.* **367**, 465-471.

487. Bârzu, T., Van Rijn, J. L. M. L., Petitou, M., Tobelem, G., and Caen, J. (1987) Heparin degradation in endothelial cells, *Thromb. Res.* **47**, 601-609.

488. Iozzo, R. (1987) Turnover of heparan sulfate proteoglycan in human colon carcinoma cells: a quantitative biochemical and autoradiographic study, *J. Biol. Chem.* **262**, 1888-1900.

489. Nakajima, B., Irimura, T., and Nicoloson, G. L. (1988) Heparanase and tumor metastasis, *J. Cell Biochem.* **36**, 157-167.

490. Sewell, R. F., Brenchley, P. E. C., and Mallick, N. P. (1989) Human mononuclear cells contain an endoglycosidase specific for heparan sulphate glycosaminoglycan demonstrable with the use of a specific solid-phase metabolically radiolabelled substrate, *Biochem. J.* **264**, 777-783.

491. Jin, L., Nakajima, M., and Nicolson, G. L. (1990) Immunochemical localization of heparanase in mouse and human melanomas, *Int. J. Cancer* **45**, 1088-1095.

492. Bame, K. J. (1993) Release of heparan sulfate glycosaminoglycans from proteoglycans in Chinese hamster ovary cells does not require proteolysis of the core protein, *J. Biol. Chem.* **28**, 19956-19964.

493. Hopwood, J. J. (1989) Enzymes that degrade heparin and heparan sulfate, in *Heparin. Chemical and Biological Properties, Clinical Applications*, eds., Lane, D. A. & Lindahl, U., CRC Press, Inc., Boca Raton, FEBS Lett., pp. 191-287.

494. Freeman, C., and Hopwood, J. J. (1992) Lysosomal degradation of heparin and heparan sulfate, in *Heparin and Related Polysaccharides*, eds., Lane, D. A., Björk, I. & Lindahl, U., Advances in Experimental Medicine and Biology, Vol. 313 Plenum Press, New York, pp. 121-134.

495. Bame, K. J., and Rome, L. H. (1985) Acetyl coenzyme A:α-glucosaminide N-acetyltransferase: evidence for a transmembrane acetylation mechanism, *J. Biol. Chem.* **260**, 11293-11299.

496. Rome, L. H., and Hill, D. F. (1986) Lysosomal degradation of glycoproteins and glycosaminoglycans: efflux and recycling of sulphate and N-acetylhexosamines, *Biochem. J.* **235**, 707-713.

497. Jonas, A. J., Speller, R. J., Conrad, P. B., and Dubinsky, W. P. (1989) Transport of N-acetyl-D-glucosamine and N-acetyl-D-galactosamine by rat liver lysosomes, *J. Biol. Chem.* **264**, 4953-4956.

498. Jonas, A. J., and Jobe, H. (1990) N-Acetyl-D-glucosamine countertransport in lysosomal membrane vesicles, *Biochem. J.* **268**, 41-45.

499. Jonas, A. J., and Jobe, H. (1990) Sulfate transport by rat liver lysosomes, *J. Biol. Chem.* **265**, 17545-17549.

500. Mancini, G. M. S., De Jong, H. R., Galjaard, H., and Verheijen, F. W. (1989) Characterization of a proton-driven carrier for sialic acid in the lysosomal membrane. Evidence for a group-specific transport system for acidic monosccccharides, *J. Biol. Chem.* **264**, 15247-15254.

501. Chou, H.-F., Passage, M., and Jonas, A. J. (1994) Regulation of lysosomal sulfate transport by thyroid hormone, *J. Biol. Chem.* **269**, 23524-23529.

502. Kinoshita, S. (1969) Periodical release of heparin-like polysaccharide within cytoplasm during cleavage of sea urchin egg, *Exp. Cell Res.* **56**, 39-43 .

503. Fedarko, N. S., Ishihara, M., and Conrad, H. E. (1989) Control of cell division in hepatoma cells by exogenous heparan sulfate proteoglycan, *J. Cell. Physiol.* **139**, 287-294.

504. Preston, S. F., Regula, C. S., Sager, P. R., Pearson, C. B., Brown, L. S., and Berlin, R. D. (1985) Glycosaminoglycan synthesis is depressed during mitosis and elevated during early G1, *J. Cell Biol.* **101**, 1086-1093.

505. Berlin, R. B., Oliver, J. M., and Walter, R. J. (1978) Surface functions during mitosis. 1. Phagocytosis, pinocytosis, and mobility of surface bound concanavalin A, *Cell* **15**, 327-341.

506. Berlin, R. B., and Oliver, J. M. (1980) Surface functions during mitosis. 2. Quantitation of pinocytosis and kinetic characterization of the mitotic cycle with a new fluorescence technique, *J. Cell Biol.* **85**, 660-671.

507. Warren, G., C. Featherstone, G. G., and Burke, B. (1983) Newly synthesized G protein of vesicular stomatitis virus is not transported to the cell surface during mitosis, *J. Cell Biol.* **97**, 1623-1628.

508. Breton, M., Berrou, E., Brahimi-Horn, M.-C., Devdon, E., and Picard, J. (1986) Synthesis of sulfated proteoglycans throughout the cell cycle in smooth muscle cells from pig aorta, *Exp. Cell Res.* **166**, 416-426 .

509. Reilly, C. F., Kindy, M. S., Brown, K. E., Rosenberg, R. D., and Sonenshein, G. E. (1989) Heparin prevents vascular smooth muscle cells progression through the G1 phase of the cell cycle, *J. Biol. Chem.* **264**, 6990-6995.

510. Stein, G. S., Roberts, R. M., Davis, J. L., Head, W. J., Stein, J. L., Thrall, C. L., Van Veen, J., and Welch, D. W. (1975) Are glycoproteins and glycosaminoglycans components of the eukaryotic genome?, *Nature* **258**, 639-641.

511. Fromme, H. G., Buddecke, E., Von Figura, K., and Kresse, H. (1976) Localization of sulfated glycosaminoglycans within cell nuclei by high resolution autoradiography, *Exp. Cell Res.* **102**, 445-449.

512. Margolis, R. K., Crockett, C. P., Kiang, W. L., and Margolis, R. U. (1976) Glycosaminoglycans and glycoproteins associated with rat brain nuclei, *Biochim. Biophys. Acta* **451**, 465-469.

513. Furakawa, K., and Terayama, H. (1977) Isolation and identification of glycosaminoglycans associated with purified nuclei from rat liver, *Biochim. Biophys. Acta.* **499**, 278-289.

514. Furakawa, K., and Terayama, H. (1979) Pattern of glycosaminoglycans and glycoproteins associated with nuclei of regenerating liver of rat, *Biochim. Biophys. Acta* **585**, 575-588.

515. Milla, M. E., Clairmont, C. A., and Hirschberg, C. B. (1992) Reconstitution into proteoliposomes and partial purification of the Golgi apparatus membrane UDP-galactose, UDP-xylose, and UDP-glucuronic acid transport activities, *J. Biol. Chem.* **267**, 103-107.

516. Hiscock, D. R. R., Yanagishita, M., and Hascall, V. C. (1994) Nuclear localization of glycosaminoglycans in rat ovarian granulosa cells, *J. Biol. Chem.* **269**, 4539-4546.

517. Castellot, J. J., Addonizio, M. L., Rosenberg, R. D., and Karnovsky, M. J. (1981) Cultured endothelial cells produce a heparin-like inhibitor of smooth muscle cell growth, *J. Cell Biol.* **90**, 372-379.

518. Kawakami, H., and Terayama, H. (1981) Liver plasma membranes and proteoglycan prepared therefrom inhibit the growth of hepatoma cells *in vitro*, *Biochim. Biophys. Acta* **646**, 161-166.

519. Thornton, Mueller, S. N., and Levine, E. M. (1983) Human endothelial cells: use of heparin in cloning and long-term serial cultivation, *Science* **222**, 623-625.

520. Castellot, J. J., Beeler, B. L., Rosenberg, R. D., and Karnovsky, M. J. (1984) Structural determinants of the capacity of heparin to inhibit the proliferation of vascular smooth muscle cells, *J. Cell. Physiol.* **120**, 315-320.

521. Maciag, T., Mehlman, T., Friesel, R., and Schreiber, A. B. (1984) Heparin binds endothelial cell growth factor, the principal endothelial cell mitogen in bovine brain, *Science* **225**, 932-935.

522. Matuoka, K., Mitsui, Y., and Murota, S. (1984) Heparan sulfate enhances growth of transformed cells, *Cell Struct. Funct.* **9**, 357-367.

523. Castellot, J. J., Cochran, D. L., and Karnovsky, M. J. (1985) Effect of heparin on vascular smooth muscle cells. 1. Cell metabolism, *J. Cell. Physiol.* **124**, 21-18.

524. Castellot, J. J., Hoover, R. L., Harper, P. A., and Karnovsky, M. J. (1985) Heparin and glomerular epithelial cell-secreted heparin-like species inhibit mesangial cell proliferation, *Am. J. Pathol.* **120**, 427-435.

525. Majack, R. A., Cook, S. C., and Bornstein, P. (1985) Platelet-derived growth factor and heparin-like heparan sulfate proteoglycans regulate thrombospondin synthesis and deposition in the matrix by smooth muscle cells, *J. Cell Biol.* **101**, 1059-1070.

526. Folkman, J. (1985) Regulation of angiogenesis: a new function of heparin, *Biochem. Phamacol.* **34**, 905-909.

527. Matuoka, K., Mitsui, Y., Murota, S., and Namba, M. (1985) Actions of exogenous heparan sulfate and hyaluronic acid on growth and thymidine incorporation of normal and transformed human fibroblasts, *Cell Biol. Int. Rept.* **9**, 815-824.

528. Schreiber, A. B., Kenney, J., Kowalski, W. J., Friesel, R., Mehlman, T., and Maciag, T. (1985) Interaction of endothelial cell growth factor with heparin: characterization by receptor and antibody recognition, *Proc. Natl. Acad. Sci. U.S.A.* **82**, 6138-6142.

529. Benitz, W. E., Lessler, D. S., Coulson, J. B., and Bernfield, M. (1986) Heparin inhibits proliferation of fetal vascular smooth muscle cells in the absence of platelet-derived growth factor, *J. Cell. Physiol.* **127**, 1-7.

530. Majack, R. A., Cook, S. C., and Bornstein, P. (1986) Control of smooth muscle cell growth by components of the extracellular matrix: autocrine role for thrombospondin, *Proc. Natl. Acad. Sci. U.S.A.* **83**, 9050-9054.

531. Ullrich, S., Lagente, O., Choay, J., Courtos, Y., and Lenfant, M. (1986) Structure activity relationship in heparin: stimulation of non-vascular cells by a synthetic heparin pentasaccharide in cooperation with human acidic fibroblast growth factors, *Biochem. Biophys. Res. Commun.* **139**, 728-732.

532. Imamura, T., and Mitsui, Y. (1987) Heparan sulfate and heparin as a protentiator or a suppressor of growth of normal and transformed vascular endothelial cells, *Exp. Cell Res.* **172**, 92-100.

533. Reilly, C. F., Fritze, L. M., and Rosenberg, R. D. (1987) Antiproliferative effects of heparin on vascular smooth muscle cells are reversed by epidermal growth factor, *J. Cell. Physiol.* **131**, 149-157.

534. Neufeld, G., Gospodarowicz, D., Dodge, L., and Fujii, D. J. (1987) Heparin modulation of the neurotropic effects of acidic and basic fibroblast growth factors and nerve growth factor on PC12 cells, *J. Cell. Physiol.* **131**, 131-140.

535. Baird, A., Schubert, D., Ling, N., and Guillemin, R. (1988) Receptor- and heparin-binding domains of basic fibroblast growth factor, *Proc. Natl. Acad. Sci. U.S.A.* **85**, 2324-2328.

536. Lankes, W., Griesmacher, A., Grunwald, J., Schartz-Albiez, R., and Keller, R. (1988) A heparin-binding protein involved in inhibition of smooth-muscle cell proliferation, *Biochem. J.* **251**, 831-842.

537. Dupuy, E., Rohrlich, P.-S., and Tobelem, G. (1988) Heparin stimulates fibroblasts growth induced by platelet derived growth factor, *Cell Biol. Int. Reports* **12**, 17-28.

538. Murphy-Ullrich, J. E., Westrick, L. G., Esko, J. D., and Mosher, D. F. (1988) Altered metabolism of thrombospondin by Chinese hamster ovary cells defective in glycosaminoglycan synthesis, *J. Biol. Chem.* **263**, 6400-6406.

539. Delahunty, M. D., Stafford, F. J., Yuan, L. C., Shaz, D., and Boniface, J. S. (1993) Uncleaved signals for glycosylphosphatidylinositol anchoring cause retention of precursor proteins in the endoplasmic reticulum, *J. Biol. Chem.* **268**, 12017-12027.

540. Buscà, R., Peinado, J., Vilella, E., Auwerx, J., Deeb, S. S., Vilaró, S., and Reina, M. (1995) The mutant $Asn^{291} \rightarrow$ Ser human lipoprotein lipase is associated with reduced catalytic activity and does not influence binding to heparin, *FEBS Lett.* **367**, 257-262.

541. Sing, J. P., Wiernicki, T. R., and Gupta, S. K. (1993) Role of serine/threonine kinase casein kinase-II in vascular smooth muscle cell proliferation and inhibition by heparin, *Drug Dev. Res.* **29**, 129-136.

542. Bârzu, T., Pascal, M., Maman, M., Roque, C., Lafont, F., and Rousselet, A. (1996) Entry and distribution of fluorescent antiproliferative heparin derivatives into rat vascular smooth cells: comparison between heparin-sensitive and heparin-resistant cultures, *J. Cell. Physiol.* **167**, 8-21.

543. Mintz, K. P., Fisher, L. W., Grzesik, W. J., Hascall, V. C., and Midura, R. J. (1994) Chlorate-induced inhibition of tyrosine sulfation on bone sialoprotein synthesized by a rat osteoblast-like cell line (UMR 106-01 BSP), *J. Biol. Chem.* **269**, 4845-4852.

544. Schwartz, N. B. (1977) Regulation of chondroitin sulfate synthesis: effect of β-xylosides on synthesis of chondroitin sulfate proteoglycan, chondroitin sulfate chains, and core protein, *J. Biol. Chem.* **252**, 6316-6321.

545. Robinson, J., and Gospodarowicz, D. (1984) Effect of p-nitrophenyl-β-D-xyloside on proteoglycan synthesis and extracellular matrix formation by bovine corneal endothelial cell, *J. Biol. Chem.* **259**, 3818-3824.

546. Fritz, T. A., Lugemwa, F. N., Sarkar, A. K., and Esko, J. D. (1994) Biosynthesis of heparan sulfate on β-D-xylosides depends on aglycone structure, *J. Biol. Chem.* **269**, 300-307.

547. Masson, P. J., Coup, D., Millet, J., and Brown, N. L. (1995) The effect of the β-D-xyloside naroparcil on circulating plasma glycosaminoglycans: an explanation for its anthrombotic activity in the rabbit, *J. Biol. Chem.* **270**, 2662-2668.

548. Lugemwa, F. N., and Esko, J. D. (1991) Estradiol β-D-xyloside, an efficient primer for heparan sulfate biosynthesis, *J. Biol. Chem.* **266**, 6674-6677.

549. Lugemwa, F. N., Sarkar, A. K., and Esko, J. D. (1996) Unusual β-D-xylosides that prime glycosaminoglycans in animal cells, *J. Biol. Chem.* **271**, 19159-19165.

550. Zhuang, D., Grey, A., Harris Brandts, M., Higgins, E., Kashem, M. A., and Dennis, J. W. (1991) Characterization of O-linked oligosaccharide biosynthesis in cultured cells using paranitrophenyl α-D-GalNAc as an acceptor, *Glycobiology* **1**, 425-433.

551. Geetha-Habib, M., Park, H. R., and Lennarz, W. J. (1990) *In vivo* N-glycosylation and fate of Asn-X-Ser/Thr tripeptides, *J. Biol. Chem.* **265**, 13655-13660.

552. Davis, S. J., and Wheldrake, J. F. (1985) The sulphation inhibitor sodium selenate arrests the growth of *Dictyostelium discoideum*, *FEMS Micro Lett.* **30**, 353-358.
553. Baeuerle, P. A., and Huttner, W. B. (1986) Chlorate-a potent inhibitor of protein sulfation in intact cells, *Biochem. Biophys. Res. Commun.* **141**, 870-877.
554. Greve, H., Cully, Z., Brumborg, P., and Kresse, H. (1988) Influence of chlorate on proteoglycan biosynthesis by cultured human fibroblasts, *J. Biol. Chem.* **263**, 128865-12892.
555. Humphries, D. E., and Silbert, J. E. (1988) Chlorate: a reversible inhibitor of proteoglycan sulfation, *Biochem. Biophys. Res. Commun.* **14**, 365-371.
556. Keller, K. M., Brauer, P. B., and Keller, J. M. (1989) Modulation of cell surface heparan sulfate structure by growth of cells in the presence of chlorate, *Biochemistry* **28**, 8100-8107.
557. Schmidt, S., and Wheldrake, J. F. (1993) Accumulation of unsulfated precursors in *Dictyostelium discoideum* during selenate inhibition of growth, *Mol. Cell. Biochem.* **126**, 109-114.
558. Hoogewerf, A. J., Cisar, L. A., Evans, D. C., and Bensadoun, A. (1991) Effect of chlorate on the sulfation of lipoprotein lipase and heparan sulfate proteoglycans: sulfation of heparan sulfate proteoglycans affect lipoprotein lipase degradation, *J. Biol. Chem.* **266**, 16564-16571.
559. Yayon, A., Klagsbrun, M., Esko, J. D., Leder, P., and Ornitz, D. M. (1991) Cell surface, heparin-like molecules are required for binding of basic fibroblast growth factor to its high affinity receptor, *Cell* **64**, 841-848.
560. Ornitz, D. M., Yayon, A., Flanagan, J. G., Svahn, C. M., Levi, E., and Leder, P. (1992) Heparin is required for cell-free binding of basic fibroblast growth factor to a soluble receptor and for mitogenesis in whole cells, *Mol. Cell. Biol.* **12**, 240-247.
561. Ishihara, M., Tyrrell, D. J., Stauber, G. B., Brown, S., Cousens, L. S., and Stack, R. J. (1993) Preparation of affinity-fractionated, heparin-derived oligosaccharides and their effects on selected biological activities mediated by basic fibroblast growth factor, *J. Biol. Chem.* **268**, 4675-4683.
562. Hidaka, M., Nagakura, S., Horikawa, K., Kawaguchi, T., Iwamoto, N., Kagimoto, T., Takatsuki, K., and Nakakuma, H. (1993) Impaired glycosylation of glycosylphosphatidylinositol-anchor synthesis in paroxysmalnocturnal hemoglobinuria leucocytes, *Biochem. Biophys. Res. Commun.* **191**, 571-579 .
563. Thomas, L. J., DeGasperi, R., Sugiyama, E., Chang, H.-M., Beck, P. J., Orlean, P., Urakaze, M., Kamitani, T., Sambrook, J. F., Warren, C. D., and Yeh, E. T. H. (1991) Functional analysis of T-cell mutants defective in the biosynthesis of glycosylphosphatidylinositol anchor. Relative importance of glycosylphosphatidylinositol anchor vs N-linked glycosylation in T-cell activation, *J. Biol. Chem.* **266**, 23175-23184.
564. Camp, L. A., Chauhan, P., Farrar, J. D., and Lehrman, M. A. (1993) Defective mannosylation of glycosylphosphatidylinositol in Lec35 Chinese hamster ovary cells, *J. Biol. Chem.* **268**, 6721-6728.
565. Lisanti, M. P., Field, M. C., Caras, I. W., Menon, A. K., and Rodriguez-Boulan, E. (1991) Mannosamine, a novel inhibitor of glycosylphosphatidylinositol incorporation into proteins, *EMBO J.* **10**, 1969-1977.
566. Pan, Y.-T., Kamtani, T., Bhuvaneswaran, C., Hallaq, Y., Warren, C. D., Yeh, E. T. H., and Elbein, A. D. (1992) Inhibition of glycosylphosphatidylinositol anchor formation by mannosamine, *J. Biol. Chem.* **267**, 21250-21255.
567. Ralton, J. E., Milne, K. G., Güther, M. L. S., Field, R. A., and Ferguson, M. A. J. (1993) The mechanism of inhibition of glycosylphosphatidylinositol anchor biosynthesis in *Trypanosoma brucei* by mannosamine, *J. Biol. Chem.* **268**, 24183-24189.
568. Schwarz, R. T., Mayor, S., Menon, A. K., and Cross, G. A. M. (1988) Biosynthesis of the glycolipid anchor of *Trypanosoma brucei* variant surface glycoproteins: involvement of Dol-P-Man, *Biochem. Soc. Trans.* **17**, 746-748.

569. Takami, N., Oda, K., and Ikehara, Y. (1992) Abherrant processing of alkaline phosphatase precusor caused by blocking the synthesis of glycosylphosphatidylinositol, *J. Biol. Chem.* **267**, 1042-104 .

570. Kang, M. S., Spencer, J. P., and Elbein, A. D. (1978) Amphomycin inhibition of mannose and GlcNAc incorporation into lipid-linked saccharides, *J. Biol. Chem.* **253**, 8860-8866.

571. Menon, A. K., Major, S., and Schwarz, R. T. (1990) Biosynthesis of glycosyl-phosphatidylinositol lipids in *Trypanosoma brucei*: involvement of mannosyl-phosphoryldolichol as the mannose donor, *EMBO J.* **9**, 4249-4258.

572. Fedders, G., Kock, R., Van de Leur, E., and Greiling, H. (1994) The metabolism of 2-fluoro-2-deoxy-D-glucose in human chondrocytes and its incorporation into keratan sulfate proteoglycans, *Eur. J. Biochem.* **219**, 1063-1071.

573. Noort, D., Coughtrie, M. W. H., Burchell, B., Van der Marel, G. A., Van Boom, J. H., Van der Gen, A., and Mulder, G. J. (1990) Inhibition of UDP-glucuronosyltransferase activity by possible transition-state analogues in rat-liver microsomes, *Eur. J. Biochem.* **188**, 390-312.

574. Gupta, D. K., Gieselmann, V., Hasilik, A., and Von Figura, K. (1984) Tilorone acts as a lysosomotropic agent in fibroblasts, *Hoppe-Seyler's Z. Physiol. Chem.* **365**, 859-856.

575. Dunn, W. A., Hubbard, A. L., and Aronson, N. N. (1980) Low temperature selectively inhibits fusion between pinocytotic vesicles and lysosomes during heterophagy of ^{125}I-asialofetuin by perfused rat liver, *J. Biol. Chem.* **255**, 5971-5978.

576. Yanagishita, M. (1985) Inhibition of intracellular degradation of proteoglycans by leupeptin in rat ovarian granulosa cells, *J. Biol. Chem.* **260**, 11075-10082.

577. Esko, J. D. (1992) Animal cell mutants defective in heparan sulfate polymerization, in *Heparin and Related Polysaccharides*, eds., Lane, D. A., Björk, I. & Lindahl, U., Plenum Press, New York, pp. 97-106.

578. Esko, J. D. (1986) Detection of animal cell LDL mutants by replica plating, *Methods, Enzymol.* **129**, 237-253.

579. Esko, J. D. (1989) Replica plating of animal cells, *Methods Cell Biol.* **32**, 387-422.

580. Esko, J. D., Stewart, T. E., and Taylor, W. H. (1985) Animal cell mutants defective in glycosaminoglycan biosynthesis, *Proc. Natl. Acad. Sci. U.S.A.* **82**, 3197-3201 .

581. Elgavish, A., Esko, J. D., and Knurr, A. (1988) Chinese hamster ovary cell mutants deficient in an anion exchanger functionally similar to the erythroid band 3, *J. Biol. Chem.* **263**, 18607-18613.

582. Ishihara, M., Guo, Y., and Swiedler, S. J. (1993) Selective impairment of the synthesis of basic fibroblast growth factor binding domains of heparan sulfate in a COS cell mutant defective in N-sulphotransferase, *Glycobiology* **3**, 83-88 .

583. De Agostini, A. L., Lau, H. K., Leone, C., Youssoufian, H., and Rosenberg, R. D. (1990) Cell mutants defective in synthesizing a heparan sulfate proteoglycan with regions of defined monosaccharide sequence, *Proc. Natl. Acad. Sci. U.S.A.* **87**, 9784-9788.

584. LeBaron, R. G., Esko, J. D., Woods, A., Johansson, S., and Höök, M. (1989) Binding of heparan sulfate to type V collagen. A mechanism of cell substrate adhesion, *J. Biol. Chem.* **264**, 7950-7956.

585. LeBaron, R. G., Esko, J. D., Woods, A., Johansson, S., and Höök, M. (1988) Adhesion of glycosaminoglycan-deficient Chinese hamster ovary cell mutants to fibronectin substrata, *J. Cell Biol.* **106**, 945-952.

586. Rapraeger, A. C., Krufka, A., and Olwin, B. B. (1991) Requirement of heparan sulfate for bFGF-mediated fibroblast growth and myoblast differentiation, *Science* **252**, 1705-1708.

587. Dunn, W. U., and Spear, P. G. (1989) Initial interaction of herpes simplex virus with cells is binding to heparan sulfate, *J. Virol.* **63**, 52-58.

588. Herold, B. C., Visalli, R. J., Susmarski, N., Brandt, C. R., and Spear, P. G. (1994) Glycoprotein C-independent binding of herpes simplex virus to cells requires cell surface heparan sulphate and glycoprotein B, *J. Gen. Virol.* **75**, 1211-1222.

589. Rider, C. C., Coombe, D. R., Harrop, H. A., Hounsell, E. F., Bauer, C., Feeney, J., Mulloy, B., Mahmood, N., Hay, A., and Parish, C. R. (1994) Anti-HIV-1 activity of

chemically modified heparins: correlation between binding to the V3 loop of gp120 and inhibition of cellular HIV-1 infection in vitro, *Biochemistry* **33**, 6974-6980.

590. Esko, J. D., Rostand, K. S., and Weinke, J. L. (1988) Tumor formation dependent on proteoglycan biosynthesis, *Science* **41**, 1092-1096.

591. Castellot, J. J., Wong, J., Herman, B., Hoover, R. L., Albertini, D. F., Wright, T. C., Caleb, B. L., and Karnovsky, M. J. (1985) Binding and internalization of heparin by vascular smooth muscle cells, *J. Cell. Physiol.* **124**, 13-20.

592. Letourneur, D., Caleb, B. L., and Castellot, J. J. (1995) Heparin binding, internalization, and metabolism in vascular smooth muscle cells. 2. Degradation and secretion in sensitive and resistant cells, *J. Cell. Physiol.* **165**, 687-695.

593. Letourneur, D., Caleb, B. L., and Castellot Jr., J. J. (1995) Heparin binding, internalization, and metabolism in vascular smooth muscle cells. 1. Upregulation of heparin binding correlates with antiproliferative activity, *J. Cell. Physiol.* **165**, 676-686.

594. Hirsh, J. (1991) Heparin, *N. Engl. J. Med.* **324**, 1565-1574.

595. Hirsh, J., and Levine, M. N. (1992) Low molecular weight heparin, *Blood* **79**, 1-17.

596. Hirsh, J., Dalen, J. E., Deykin, D., and Poller, L. (1992) Heparin: mechanism of action, pharmacokinetics, dosing considerations, monitoring, efficacy, and safety, *Chest* **102**, 337S-351S.

597. Hirsh, J., and Fuster, V. (1994) Guide to anticoagulant therapy. 1. Heparin, *Circulation* **89**, 1449-1468.

598. Mahadoo, J., Hiebert, L., and Jaques, L. B. (1977) Vascular sequestration of heparin, *Thromb. Res.* **12**, 79-90.

599. McAllister, B. M., and Demis, D. J. (1966) Heparin metabolism: isolation and characterization of uroheparin, *Nature* **212**, 293-294.

600. Dawes, J., and Pepper, D. S. (1979) Catabolism of low-dose heparin in man, *Thrombosis Res.* **14**, 845-860.

601. Bendetowicz, A. V., Béguin, S., Caplain, H., and Hemker, H. C. (1994) Pharmacokinetics and pharmacodynamics of a low molecular weight heparin (Enoxaparin) after subcutaneous injection, comparison with unfractionated heparin-a three way crossover study in human volunteers, *Thromb. Haemost.* **71**, 305-313.

602. Sue, T. K., Jaques, L. B., and Yuen, E. (1976) Effects of acidity, cations, and alcoholic fractionation on absorption of heparin from gastrointestinal tract, *Can. J. Physiol. Pharmacol.* **54**, 613-617.

603. Doutremépuich, C., Toulemonde, F., and Lormeau, J.-C. (1985) Oral administration of low molecular weight heparin fractions in rabbits, *Sem. Thromb. Hemost.* **11**, 323-325.

604. Pozzo, A. D., Acquasaliente, M., and Geron, M. R. (1989) New heparin complexes active by intestinal absorption. I.Multiple ion pairs with basic organic compounds, *Thromb. Res.* **56**, 119-124.

605. Vasdev, S., Sampson, C. A., Longerich, L., and Parai, S. (1992) Oral heparin prevents hypertension and elevated cytosolic calcium in salt-sensitive rate, *Artery* **19**, 225-245.

606. Zoppetti, G., Caramazza, I., Murakami, Y., and Ohno, T. (1992) Structural requirements for duodenal permeability of heparin-diamine complexes, *Biochim. Biophys. Acta* **1156**, 92-98.

607. Scagnol, I., Fumagalli, G., and Andriuoli, G. (1992) Anticoagulant and antithrombotic activity of heparin salts by intraduodenal route in rabbits, *Thromb. Res.* **6**, 195-200.

608. Jaques, L. B., Hiebert, L. M., and Wice, S. M. (1991) Evidence from endothelium of gastric absorption of heparin and of dextran sulphaset 8000, *J. Lab. Clin. Med.* **117**, 122-130.

609. Hiebert, L. M., Wice, S. M., and Jaques, L. B. (1996) Antithrombic activity of oral unfractionated heparin, *J. Cardiovasc. Pharmacol.* **28**, 26-29.

610. Larsen, A. K., P., L. D., Langer, R., and Folkman, J. (1986) Oral heparin results in the appearance of heparin fragments in the plasma of rats, *Proc. Natl. Acad. Sci. U.S.A.* **83**, 2964-2968 .

611. Wells, X. E., and Dawes, J. (1995) Role of liver and kidney in the desulphation of heparin *in vivo, Thromb. Haemostas.* **74,** 667-672.

612. Larsen, A. K., Rice, K. G., Linhardt, R. J., Wogan, G., and Langer, R. (1989) Resistance of heparinase-derived heparin fragments to biotransformation, *J. Biol. Chem.* **264,** 1570-1577.

613. Yanagishita, M., and McQuillan, D. J. (1989) Two forms of plasma membrane-intercalated heparan sulfate proteoglycan in rat ovarian granulosa cells: labeling of proteoglycans with a photoactivatable hydrophobic probe and effect of the membrane anchor-specific phospholipase C, *J. Biol. Chem.* **264,** 17551-17558.

614. Cizmeci-Smith, G., Stahl, R. C., Showalter, L. J., and Carey, D. J. (1993) Differential expression of transmembrane proteoglycans in vascular smooth muscle cells, *J. Biol. Chem.* **268,** 18740-18747.

615. Hampson, I. N., Kumar, S., and Gallagher, J. T. (1983) Differences in the distribution of O-sulphate of cell-surface and secreted heparan sulphate produced by human neuroblastoma cells in culture, *Biochim. Biophys. Acta* **763,** 183-190.

616. Morris, J. E., Potter, S. W., and Gaza-Bulseco, G. (1988) Estradiol-stimulated turnover of heparan sulfate proteoglycan in mouse uterine epithelium, *J. Biol. Chem.* **263,** 4712-4718.

617. Elenius, K., Määttä, A., Salmivirta, M., and Jalkanen, M. (1992) Growth factors induce 3T3 cells to express bFGF-binding syndecan, *J. Biol. Chem.* **267,** 6435-6441.

618. Grässel, S., Cohen, I. R., Murdock, A. D., Eichsteter, I., and Iozzo, R. V. (1995) The proteoglycan perlecan is expressed in the erythroleukemia cell line K562 and is upregulated by sodium butyrate and phorbol ester, *Mol. Cell Biol.* **145,** 61-68.

619. Schmidt, A., Skaletz-Rorowski, A., and Buddecke, E. (1995) Basic fibroblast growth factor controls the expression and molecular structure of heparan sulfate in corneal endothelial cells, *Eur. J. Biochem.* **234,** 479-484.

620. Gershman, H., and Robbins, P. W. (1981) Transitory effects of glucose starvation on the synthesis of dolichol-linked oligosaccharides in mammalian cells, *J. Biol. Chem.* **256,** 7774-7780.

621. Rearick, J. I., Chapman, A., and Kornfeld, S. (1981) Glucose starvation alters lipid linked oligosaccharide biosynthesis in Chinese hamster ovary cells, *J. Biol. Chem.* **256,** 6255-6261.

622. Rosengart, T. K., Johnson, W. V., Friesel, R., Clark, R., and Maciag, T. (1988) Heparin protects heparin-binding growth factor-1 from proteolytic inactivation *in vitro, Biochem. Biophys. Res. Commun.* **152,** 432-440.

623. Lindahl, U., Bäckström, G., Höök, M., Thunberg, L., Fransson, L.-Å., and Linker, A. (1979) Structure of the antithrombin-binding site in heparin, *Proc. Natl. Acad. Sci. U.S.A.* **76,** 3198-3202.

624. Guimond, S., Maccarana, M. Olwin, B. B., Lindahl, U., and Rapraeger, A. C. (1993) Activating and inhibitor heparin sequences for FGF-2 (basic FGF): distinct requirements for FGF-1, FGF-2, and FGF-4, *J. Biol. Chem.* **268,** 23906-23914.

625. Habuchi, H., Suzuki, S., Saito, T., Tamura, T., Harada, T., Yoshida, K., and Kimata, K. (1992) Structure of a heparan sulphate oligosaccharide that binds to basic fibroblast growth factor, *Biochem. J.* **285,** 805-813.

626. Turnbull, J. E., Fernig, D. G., Ke, Y., Wilkinson, M. C., and Gallagher, J. T. (1992) Identification of the basic fibroblast growth factor binding sequence in fibroblast heparan sulfate, *J. Biol. Chem.* **267,** 10337-10341.

627. Maccarana, M., Casu, B., and Lindahl, U. (1993) Minimal sequence in heparin/heparan sulfate required for binding of basic fibroblast growth factor, *J. Biol. Chem.* **268,** 23898-23905.

628. Walker, A., Turnbull, J. E., and Gallagher, J. T. (1994) Specific heparan sulfate saccharides mediate the activity of basic fibroblast growth factor, *J. Biol. Chem.* **269,** 931-935.

629. Marcum, J. A., and Rosenberg, R. D. (1984) Anticoagulantly active heparin-like molecules from vascular tissue, *Biochemistry* **23,** 1730-1737.

630. Marcum, J. A., and Rosenberg, R. D. (1985) Heparinlike molecules with anticoagulant activity are synthesized by cultured endothelial cells, *Biochem. Biophys. Res. Commun.* **126**, 365-372.

631. Marcum, J. A., Atha, D. H., Fritze, L. M. S., Nawroth, P., Stern, D., and Rosenberg , R. D. (1986) Cloned bovine aortic endothelial cells synthesize anticoagulantly active heparan sulfate proteoglycan, *J. Biol. Chem.* **261**, 7507-7517.

632. Pejler, G., and David, G. (1987) Basement-membrane heparan sulphate with high affinity for antithrombin synthesized by normal and transformed mouse mammary epithelial cells, *Biochem. J.* **248**, 69-77 .

633. Sun, X., Mosher, D. F., and Rapraeger, A. (1989) Heparan sulfate-mediated binding of epithelial cell surface proteoglycan to thrombospondin, *J. Biol. Chem.* **264**, 2885-2889 .

634. Cisar, L. A., Hoogewerf, A. J., Cupp, M., Rapport, C. A., and Bensadoun, A. (1989) Secretion and degradation of lipoprotein lipase in cultured adipocytes: binding of lipoprotein lipase to membrane heparan sulfate proteoglycans is necessary for degradation, *J. Biol. Chem.* **264**, 1767-1774.

635. Saxena, U., Klein, M. G., and Goldberg, I. J. (1990) Metabolism of endothelial cell-bound lipoprotein lipase: evidence for heparan sulfate proteoglycan-mediated internalization and recycling, *J. Biol. Chem.* **265**, 12880-12886.

636. Murono, E. P., Washburn, A. L., Goforth, D. P., and Wu, N. (1993) Evidence that both receptor- and heparan sulfate proteoglycan-bound basic fibroblast growth factor are internalized by cultured immature Leydig cells, *Mol. Cell. Endrocrinol.* **98**, 81-90.

637. Friesel, R., and Maciag, T. (1988) Internalization and degradation of heparin binding growth factor-I by endothelial cells, *Biochem. Biophys. Res. Commun.* **151**, 957-964.

638. Renko, M., Quarto, N., Morimoto, T., and Rifkin, D. B. (1990) Nuclear and cytoplasmic localization of different basic fibroblast growth factor species, *J. Cell. Physiol.* **144**, 108-114 .

639. Acland, P., Dixon, M., Peters, G., and Dickson, C. (1990) Subcellular fate of Int-2 protein is determined by choice of initiation codon, *Nature* **343**, 662-665.

640. Powell, P. P., and Klagsbum, M. (1991) Three forms of basic fibroblast growth factor are made from a single mRNA and localize to the nucleus, *J. Cell. Physiol.* **148**, 202-210.

641. Prudovsky, I., Savion, N., Zhan, X., Friesel, R., Xu, J., Hou, J., McKeehan, W. L., and Maciag, T. (1994) Intact and functional fibroblast growth factor (FGF) receptor-1 trafficks near the nucleus in response to FGF-1, *J. Biol. Chem.* **269**, 31720-31724.

642. Maher, P. A. (1996) Nuclear translocation of fibroblast growth factor (FGF) receptors in response to FGF-2, *J. Cell Biol.* **134**, 529-536.

643. Jackson, R. I., Busch, S. J., and Cardin, A. D. (1991) Glycosaminoglycans: molecular properties, protein interactions, and role in physiological processes, *Physiol. Rev.* **71**, 481-539.

644. Wang, H. M., Loganathan, D., and Linhardt, R. J. (1991) Determination of the pKa of glucuronic acid and the carboxy groups of heparin by ^{13}C-nuclear-magnetic-resonance spectroscopy, *Biochem. J.* **278**, 689-695.

645. Arnott, S., and Mitra, A. K. (1984) X-ray diffraction analyses of glycosaminoglycans, in *Molecular Biophysics of the Extracellular Matrix*, eds., Arnott, S., Rees, D. A. & Morris, E. R., Humana Press, Clifton, NJ, pp. 41-67.

646. Cardin, A. D., and Weintraub, H. J. R. (1989) Molecular modeling of protein-glycosaminoglycan interactions, *Atherosclerosis* **9**, 21-32.

647. Wong, P., Hampton, B., Szylobryt, E., Gallagher, A. M., Jaye, M., and Burgess, W. H. (1995) Analysis of putative heparin-binding domains of fibroblast growth factor-1, *J. Biol. Chem.* **270**, 25805-25811.

648. Lellouch, A. C., and Lansbury, P. T. (1992) A peptide model for the heparin binding site of antithrombin III, *Biochemistry* **31**, 2279-2285.

649. Tyler-Cross, R., Sobel, M., Marques, D., and Harris, R. B. (1994) Heparin binding domain peptides of antithrombin III: analysis by isothermal titration calorimetry and circular dichroism spectroscopy, *Protein Sci.* **3**, 620-627.

650. Margalit, H., Fischer, N., and Ben-Sasson, S. A. (1993) Comparative analysis of structurally defined heparin binding sequences reveals a distinct spatial distribution of basic residues, *J. Biol. Chem.* **268**, 19228-19231.

651. Suzuki, M. (1994) A framework for the DNA-protein recognition code of the probe helix in transcription factors: the chemical and stereo chemical rules, *Structure* **4**, 317-326.

652. Steitz, T. A. (1990) Structural studies of protein-nucleic acid interaction: the sources of sequence-specific binding, *Quart. Rev. Biophys.* **23**, 205-280.

653. Grant, D., Long, W. F., and Williamson, F. B. (1987) Infrared spectroscopy of heparin-cation complexes, *Biochem. J.* **244**, 143-149.

654. Liang, J. N., Chakrabarti, B., Ayotte, L., and Perlin, A. S. (1982) An essential role for the 2-sulfamino group in the interaction of calcium ion with heparin, *Carbohyd. Res.* **106**, 101-109.

655. Ayotte, L., and Perlin, A. S. (1986) N.M.R. spectroscopic observations related to the function of sulfate groups in heparin: calcium binding vs. biological activity, *Carbohyd. Res.* **145**, 267-277.

656. Grant, D., Long, W. F., and Williamson, F. B. (1992) Zn^{2+}-heparin interaction studied by potentiometric titration, *Biochem. J.* **287**, 849-853.

657. Grant, D., Long, W. F., and Williamson, F. B. (1992) A potentiometric titration study of the interaction of heparin with metal cations, *Biochem. J.* **285**, 477-480.

658. Grant, D., Long, W. F., Moffat, C. F., and Williamson, F. B. (1992) A study of Ca^{2+}-heparin complex-formation by polarimetry, *Biochem. J.* **282**, 601-604.

659. Kan, M., Wang, F., Kan, M., To, B., Gabriel, J. L., and McKeehan, W. L. (1996) Divalent cations and heparin/heparan sulfate cooperate to control assembly and activity of the fibroblast growth factor receptor complex, *J. Biol. Chem.* **271**, 26143-26148.

660. Tyler-Cross, R., Sobel, M., Marques, D., Soler, D. F., and Harris, R. B. (1993) Heparin-von Willebrand factor binding as assessed by isothermal titration calorimetry and by affinity fractionation of heparins using synthetic peptides, *Arch. Biochem. Biophys.* **306**, 528-533.

661. Pantoliano, M. W., Horlick, R. A., Springer, B. A., Van Dyk, D. E., Tobery, T., Witmore, D. R., Lear, J. D., Nahapetian, A. T., Bradley, J. D., and Sisk, W. P. (1994) Multivalent ligand-receptor binding interactions in the fibroblast growth factor system produce a cooperative growth factor and heparin mechanism for receptor dimerization, *Biochemistry* **33**, 10229-10248.

662. Thompson, L. D., Pantoliano, M. W., and Springer, B. A. (1994) Energetic characterization of the basic fibroblast growth factor-heparin interaction: identification of the heparin binding domain, *Biochemistry* **33**, 3831-3840.

663. Tyler-Cross, R., Sobel, M., McAdory, L. E., and Harris, R. B. (1996) Structure-function relationships of antithrombin III-heparin interactions as assessed by biophysical and biological assays and molecular modeling of peptide-pentasaccharide-docked complexes, *Arch. Biochem. Biophys.* **334**, 206-213.

664. Lee, M. K., and Lander, A. D. (1991) Analysis of affinity and structural selectivity in the binding of proteins to glycosaminoglycans: development of a sensitive electrophoretic approach, *Proc. Natl. Acad. Sci. U.S.A.* **88**, 2768-2772.

665. Shing, Y. (1989) Heparin-copper biaffinity chromatography of fibroblast growth factors, *J. Biol. Chem.* **263**, 9059-9062 .

666. Goldsmith, E. J., and Mottonen, J. (1994) Serpins: the uncut version, *Structure* **2**, 241-244.

667. Madison, E. L. (1994) Studies of serpins unfold at a feverish pace, *J. Clin. Invest.* **94**, 2174-2175.

668. Lawrence, D. A., Ginsburg, D., Day, D. E., Berkenpas, M. B., Verhamme, I. M., Kvassman, J.-O., and Shore, J. D. (1995) Serpin-protease complexes are trapped as stable acyl-enzyme intermediates, *J. Biol. Chem.* **270**, 25309-25312.

669. Wilczynska, M., Fa, M., Ohlsson, P.-I., and Ny, T. (1995) The inhibition mechanism of serpins: evidence that the mobile reactive center loop is cleaved in the native protease-inhibitor complex, *J. Biol. Chem.* **270**, 29652-29655.

670. Rubin, H. (1996) Serine protease inhibitors (SERPINS): where mechanism meets medicine, *Nature Med.* **2**, 632-633.

671. Pemberton, P. A., Wong, D. T., Gibson, H. L., Kiefer, M. C., Fitzpatrick, P. A., Sager, R., and Barr, P. J. (1995) The tumor suppressor maspin does not undergo the stressed to relaxed transition or inhibit trypsin-like serine proteases: evidence that maspin is not a protease inhibitory serpin, *J. Biol. Chem.* **279**, 15832-15837.

672. Mathialagan, N., and Hansen, T. R. (1996) Pepsin-inhibitory activity of the uterine serpins, *Proc. Natl. Acad. Sci. U.S.A.* **93**, 13653-13658.

673. Schechter, I., and Berger, A. (1967) On the size of the active site in proteases, *Nature* **27**, 157-162.

674. Damus, P. S., Hicks, M., and Rosenberg, R. D. (1973) Anticoagulant action of heparin, *Nature* **246**, 355-357.

675. Rosenberg, R. D., and Damus, P. S. (1973) The purification and mechanism of action of human antithrombin-heparin cofactor, *J. Biol. Chem.* **248**, 6490-6505.

676. Kurachi, K., Fujikawa, K., Schmer, G., and Davie, E. W. (1976) Inhibition of bovine factor IXa and factor Xa_β by antithrombin III, *Biochemistry* **15**, 373-377.

677. Rosenberg, R. D., Armand, G., and Lam, L. (1978) Structure-function relationship of heparin species, *Proc. Natl. Acad. Sci. U.S.A.* **75**, 3065-3069.

678. Stead, N., Kaplan, A., and Rosenberg, R. D. (1976) Inhibition of activated factor XII by antithrombin-heparin cofactor, *J. Biol. Chem.* **251**, 6481-6488.

679. Highsmith, R. F., and Rosenberg, R. D. (1974) The inhibition of human plasmin by human antithrombin-heparin cofactor, *J. Biol. Chem.* **249**, 4339-4338.

680. Travis, J., and Salvesen, G. S. (1983) Human plasma proteinase inhibitors, *Annu. Rev. Biochem.* **52**, 655-709.

681. Rosenberg, R. D. (1987) Regulation of the hemostatic mechanism, in *The Molecular Basis of Blood Diseases*, eds., Stamatoyannopoulos, G., Nienhuis, A. W., Leder, P. & Majerus, P., W. B. Saunders Co., Philadelphia, PA, pp. 534-574.

682. Björk, I., and Lindahl, U. (1982) Mechanism of the anticoagulant activity of heparin, *Mol. Cell. Biochem.* **48**, 161-182.

683. Brinkhaus, K. M., Smith, H. P., Warner, E. D., and Seegers, W. H. (1939) The inhibition of blood clotting: an unidentified substance which acts in conjunction with heparin to prevent the conversion of prothrombin to thrombin, *Am. J. Phys.* **125**, 683-687.

684. Marcum, J. A., and Rosenberg, R. D. (1989) The biochemistry, cell biology, and pathophysiology of anticoagulantly active heparin-like molecules of the vessel wall, in *Heparin: Chemical and Biological Properties, Clinical Applications*, eds., Lane, D. A. & Lindahl, U., CRC Press, Inc., Boca Raton, FL, pp. 275-294.

685. Lane, D. A., Ireland, H., Olds, R. J., Thein, S. L., Perry, D. J., and Aiach, M. (1991) Antithrombin III: a database of mutations, *Thromb. Haemost.* **66**, 657-661.

686. Lane, D. A., Olds, R. J., Boisclair, M., Chowdhury, V., Thein, S. L., Cooper, D. N., Blajchman, M., Perry, D., Emmerich, J., and Aiach, M. (1993) Antithrombin III mutation database: first update, *Thromob. Haemost.* **70**, 361-369.

687. Barrowcliffe, T. W., Johnson, E. A., and Thomas, D. (1978) Antithrombin III and heparin, *Br. Med. Bull.* **34**, 143-150.

688. Miller-Andersson, M., Borg, H., and Andersson, L.-O. (1974) Purification of antithrombin III by chromatography, *Thromb. Res.* **5**, 439-452.

689. Nordenman, B., Nyström, C., and Björk, I. (1977) The size and shape of human bovine antithrombin III, *Eur. J. Biochem.* **78**, 195-204.

690. Petersen, E. T., Dudek-Wojchiechowska, G., Sottrup-Jensen, L., and Magnusson, S. (1979) Primary structure of antithrombin III (heparin cofactor): partial homology between α_1-antitrypsin and antithrombin III, in *The Physiological Inhibitors of Blood Coagulation*

and Fibrinolysis, eds., Collen, D., Wiman, B. & Verstraete, M., Elsevier, Amsterdam, pp. 43-54.

691. Murano, G., Williams, L., Miller-Andersson, M., Aronson, D., and King, C. (1980) Some properties of antithrombin III and its concentration in human plasma, Thromb. Res. 18, 259-262.

692. Prochownik, E. V., Markman, A. F., and Orkin, S. H. (1983) Isolation of a cDNA clone for antithrombin III, J. Biol. Chem. 258, 8389-8394.

693. Prochownik, E. V., and Orkin, S. H. (1984) In vivo transcription of a human antithrombin III "minigene", J. Biol. Chem. 259, 15386-15392.

694. Andersson, L.-O., Engman, L., and Henningsson, E. (1977) Crossed immuno-electrophoresis as applied to studies on complex formation: the binding of heparin to antithrombin III and the antithrombin III-thrombin complex, J. Immunol. Meth. 14, 271-281.

695. Carlström, A.-S., Liedén, K., and Björk, I. (1977) Decreased binding of heparin to antithrombin following the interaction between antithrombin and thrombin, Thromb. Res. 11, 785-797.

696. Fish, W. W., Danielson, Å., Nordling, K., Miller, S. H., Lam, C. F., and Björk, I. (1985) Denaturation behavior of antithrombin in guanidinium chloride: irreversibility of unfolding caused by aggregation, Biochemistry 24, 1510-1517.

697. Björk, I., and Danielsson, Å. (1986) Antithrombin and related inhibitors of coagulation proteases, in Proteinase Inhibitors, eds., Barrett, A. J. & Salvesen, G., Elsevier, Amsterdam, pp. 489-513.

698. Fish, W. W., and Björk, I. (1979) Release of a two-chain form of antithrombin from the antithrombin-thrombin complex, Eur. J. Biochem. 101, 31-38.

699. Fish, W. W., Orre, K., and Björk, I. (1979) The production of an inactive form of antithrombin through limited proteolysis by thrombin, FEBS Lett. 98, 103-106.

700. Jesty, J. (1979) The kinetics of formation and dissociation of the bovine thrombin-antithrombin III complex, J. Biol. Chem. 254, 10044-10050.

701. Longas, M. O., and Finlay, T. H. (1980) The covalent nature of he human antithrombin III-thrombin bond, Biochem. J. 189, 481-489.

702. Griffith, M. J., and Lundblad, R. L. (1981) Dissociation of antithrombin III-thrombin complex: formation of active and inactive antithrombin III, Biochemistry 20, 105-110.

703. Danielsson, Å., and Björk, I. (1982) Mechanism of inactivation of trypsin by antithrombin, Biochem. J. 207, 21-28.

704. Owen, W. G. (1975) Evidence for the formation of an ester between thrombin and heparin cofactor, Biochim. Biophys. Acta. 405, 380-387.

705. Björk, I., and Fish, W. W. (1982) Production in vitro and properties of a modified form of bovine antithrombin, cleaved at the active site by thrombin, J. Biol. Chem. 257, 9487-9493.

706. Andersson, L.-O., Barrowcliffe, T. W., Holmer, E., Johnson, E. A., and Sims, G. E. C. (1976) Anticoagulant properties of heparin fractionated by affinity chromatography on matrix-bound antithrombin III and by gel filtration, Thromb. Res. 9, 575-583.

707. Höök, M., Björk, I., Hopwood, J. J., and Lindahl, U. (1976) Anticoagulant activity of heparin: separation of high-activity and low-activity heparin species by affinity chromatography on immobilized antithrombin, FEBS Lett. 66, 90-93.

708. Lam, L. H., Silbert, J. E., and Rosenberg, R. D. (1976) The separation of active and inactive forms of heparin, Biochem. Biophys. Res. Commun. 69, 570-577.

709. Villanueva, G. B., and Danishefsky, I. (1977) Evidence for a heparin-induced conformational change on antithrombin III, Biochem. Biophys. Res. Commun. 74, 803-809.

710. Nordenman, B., Danielsson, Å., and Björk, I. (1978) The binding of low affinity and high affinity heparin to antithrombin: fluorescent studies, Eur. J. Biochem. 90, 1-6.

711. Olson, S. T., and Shore, J. D. (1981) Binding of high affinity heparin to antithrombin III: characterization of the protein fluorescence enhancement, *J. Biol. Chem.* **256**, 11065-11072.

712. Olson, S. T., and Shore, J. D. (1986) Transient kinetics of heparin-catalyzed protease inactivation by antithrombin III: the reaction step limiting heparin turnover in thrombin neutralization, *J. Biol. Chem.* **261**, 13151-13159.

713. Gettins, P., Choay, J., Crews, B. C., and Zettlmeissl, G. (1992) Role of tryptophan 49 in the heparin cofactor activity of human antithrombin III, *J. Biol. Chem.* **267**, 21946-21953.

714. Streusand, V. J., Björk, I., Gettins, P. G. W., Petitou, M., and Olson, S. T. (1995) Mechanism of acceleration of antithrombin-proteinase reactions by low affinity heparin: role of the antithrombin binding pentasaccharide in heparin rate enhancement, *J. Biol. Chem.* **270**, 9043-9051.

715. Shore, J. D., Olson, S. T., Craig, P. A., Choay, J., and Björk, I. (1989) Kinetics of heparin action, in *Heparin and Related Polysaccharides. Structure and Activities, Annals of the New York Academy of Sciences*, eds., Ofosu, F. A., Danishefsky, I. & Hirsh, J., New York Acad. Sci., New York, Vol. 556, pp. 75-80.

716. Hunt, T., and Dayhoff, M. O. (1980) A surprising new protein superfamily containing ovalbumin, antithrombin III, and α_1-protease inhibitor, *Biochem. Biophys. Res. Commun.* **95**, 864-871.

717. Beresford, C. H., and Owen, M. C. (1980) Antithrombin and related inhibitors of coagulation proteases, *Int. J. Biochem.* **22**, 121-128.

718. Björk, I., Olson, S. T., and Shore, J. D. (1989) Molecular mechanisms of the accelerating effect of heparin on the reactions between antithrombin and clotting proteinases, in *Heparin: Chemical and Biological Properties, Clinical Applications*, eds., Lane, D. A. & Lindahl, U., CRC Press, Inc., Boca Raton, FL, pp. 229-255.

719. Einarsson, R., and Anderssson, L.-O. (1977) Binding of heparin to human antithrombin III as studied by measurement of tryptophane fluorescence, *Biochim. Biophys. Acta* **490**, 104-111.

720. Nordenman, B., and Björk, I. (1978) Binding of low-affinity and high-affinity heparin to antithrombin: ultraviolet difference spectroscopy and circular dichroism studies, *Biochemistry* **17**, 3339-3344.

721. Jordan, R., Beeler, D., and Rosenberg, R. D. (1979) Fractionation of low molecular weight heparin species and their interaction with antithrombin, *J. Biol. Chem.* **254**, 2902-2913.

722. Olson, S. T., Sheffer, R., Stephens, A. W., and Hirs, C. H. W. (1991) Molecular basis of the reduced activity of antithrombin-Denver with thrombin and factor Xa: role of the P'1 residue (Abstract), *Thromb. Haemostas.* **65**, 670.

723. Jordan, R. E., Oosta, G. M., Gardner, W. T., and Rosenberg, R. D. (1980) The binding of low molecular weight heparin to hemostatic enzymes, *J. Biol. Chem.* **255**, 10073-10080.

724. Jordan, R. E., Oosta, G. M., Gardner, W. T., and Rosenberg, R. D. (1980) The kinetics of haemostatic enzyme-antithrombin interactions in the presence of low molecular weight heparin, *J. Biol. Chem.* **255**, 10081-10090.

725. Nordenman, B., and Björk, I. (1981) Influence of ionic strength and pH on the interaction between high-affinity heparin and antithrombin, *Biochim. Biophys. Acta.* **672**, 227-238.

726. Olson, S. T., Srinivasan, K. R., Björk, I., and Shore, J. D. (1981) Binding of high affinity heparin to antithrombin III: stopped flow kinetic studies of the binding interaction, *J. Biol. Chem.* **256**, 11073-11079.

727. Olson, S. T., and Shore, J. D. (1982) Demonstration of a two-step reaction mechanism for inhibition of α-thrombin by antithrombin III and identification of the step affected by heparin, *J. Biol. Chem.* **257**, 14891-14895.

728. Peterson, C. D., and Blackburn, M. N. (1987) Antithrombin conformation and the catalytic role of heparin. II: is the heparin-induced conformational change in antithrombin required for rapid inactivation of thrombin?, *J. Biol. Chem.* **262**, 7559-7566.

729. Rosenberg, R. D. (1985) Role of heparin and heparinlike molecules in thrombosis and atherosclerosis, *Fed. Proc.* **44**, 404-409 .

730. Holmer, E., Lindahl, U., Bäckström, G., Thunberg, L., Sandberg, H., Söderström, G., and Andersson, L.-O. (1980) Anticoagulant activities and effects on platelets of a heparin fragment with high affinity for antithrombin, *Thromb. Res.* **18**, 861-869.

731. Oosta, G. M., Gardner, W. T., Beeler, D. L., and Rosenberg, R. D. (1981) Multiple functional domains of the heparin molecule, *Proc. Natl. Acad. Sci. U.S.A.* **78**, 829-833.

732. Holmer, E., Kurachi, K., and Söderström, G. (1981) The molecular-weight dependence of the rate-enhancing effect of heparin on the inhibition of thrombin, factor Xa, factor IXa, factor XIa, factor XIIa, and kallikrein by antithrombin, *Biochem. J.* **193**, 395-400.

733. Choay, J., Petitou, M., Lormeau, J. C., Sinäy, P., Casu, B., and Gatti, G. (1983) Structure-activity relationship in heparin: a synthetic pentasaccharide with high affinity for antithrombin III and eliciting high antifactor Xa activity, *Biochem. Biophys. Res. Commun.* **116**, 492-499.

734. Lane, D. A., Denton, J., Flynn, A. M., Thunberg, L., and Lindahl, U. (1984) Anticoagulant activities of heparin oligosaccharides and their neutralization by platelet factor 4, *Biochem. J.* **218**, 725-732.

735. Bourin, M.-C., and Lindahl, U. (1993) Glycosaminoglycans and the regulation of blood coagulation, *Biochem. J.* **289**, 313-330.

736. Machovich, R., Blásko, G., and Pálos, L. A. (1975) Action of heparin on thrombin-antithrombin reaction, *Biochim. Biophys. Acta* **379**, 193-197.

737. Pomerantz, M. W., and Owen, W. G. (1978) A catalytic role for heparin: evidence for a ternary complex of heparin cofactor, thrombin and heparin, *Biochim. Biophys. Acta.* **535**, 66-77.

738. Griffith, M. J., Kingdon, H. S., and Lundblad, R. L. (1978) Fractionation of heparin by affinity chromatography on covalently-bound human α-thrombin, *Biochem. Biophys. Res. Commun.* **83**, 1198-1205.

739. Gentry, P. W., and Alexander, B. (1973) Specific coagulation factor adsorption to insoluble heparin, *Biochem. Biophys. Res. Commun.* **50**, 500-509.

740. Danishefsky, I., Tzeng, F., Ahrens, M., and Klein, S. (1976) Synthesis of heparin-Sepharoses and their binding with thrombin and antithrombin-heparin-cofactor, *Thromb. Res.* **8**, 131-140.

741. Nordenman, B., and Björk, I. (1977) Purification of thrombin by affinity chromatography on immobilized heparin, *Thromb. Res.* **11**, 799-808.

742. Nordenman, B., and Björk, I. (1978) Studies of the binding of heparin to prothrombin and thrombin and the effect of heparin binding on thrombin, *Thromb. Res.* **12**, 757-765.

743. Holmer, E., Söderström, G., and Andersson, L.-O. (1979) Studies of the mechanism of the rate-enhancing effect of heparin on the thrombin-antithrombin III reaction, *Eur. J. Biochem.* **93**, 1-5.

744. Longas, M. O., Ferguson, W. S., and Finlay, T. H. (1980) Studies on the interaction of heparin with thrombin, antithrombin, and other plasma proteins, *Arch. Biochem. Biophys.* **200**, 595-602.

745. Olson, S. T., and Björk, I. (1991) Predominant contribution of surface approximation to the mechanism of heparin acceleration of the antithrombin-thrombin reaction: elucidation of the salt concentration effects, *J. Biol. Chem.* **266**, 633-636.

746. Gan, Z.-R., Li, Y., Chen, Z., Lewis, S. D., and Shafer, J. A. (1994) Identification of basic amino acid residues in thrombin essential for heparin-catalyzed inactivation by antithrombin III, *J. Biol. Chem.* **269**, 1301-1305.

747. Villanueva, G. B., and Danishefsky, I. (1979) Conformational changes accompanying the binding of antithrombin III to thrombin, *Biochemistry* **18**, 810-817.

748. Griffith, M. J., Kingdon, H. S., and Lundblad, R. L. (1979) The interaction of heparin with human α-thrombin: effect on the hydrolysis of anilide tripeptide substrates, *Arch. Biochem. Biophys.* **195**, 378-384.

749. Smith, G. F., and Sundboom, J. L. (1981) Heparin and protease inhibition. Heparin complexes with thrombin, plasmin, and trypsin, *Thromb. Res.* **22**, 103-114.

750. Griffith, M. J. (1982) Kinetics of the heparin-enhanced antithrombin III/thrombin reaction: evidence for a template model for the mechanism of action of heparin, *J. Biol. Chem.* **257**, 7360-7365.

751. Nesheim, M. E. (1983) A simple rate law that the describes the kinetics of the heparin-catalyzed reaction, *J. Biol. Chem.* **258**, 14708-14717.

752. Hoylaertz, M., Owen, W. G., and Collen, D. (1984) Involvement of heparin chain length in the heparin-catalyzed inhibition of thrombin by antithrombin III, *J. Biol. Chem.* **259**, 5670-5677.

753. Horner, A. A., Kusche, M., Lindahl, U., and Peterson, C. B. (1988) Determination of the range in binding-site densities of rat skin heparin chains with high binding affinities for antithrombin, *Biochem. J.* **251**, 141-145.

754. Danielsson, Å., E., R., Lindahl, U., and Björk, I. (1986) Role of ternary complexes, in which heparin binds both antithrombin and proteinase, in the acceleration of the reactions between antithrombin and thrombin or factor Xa, *J. Biol. Chem.* **261**, 15467-15473.

755. Collen, D., Schetz, J., deCock, F., Holmer, E. H., and Verstraete, M. (1977) Metabolism of antithrombin III (heparin cofactor) in man: effects of venous thrombosis and of heparin administration, *Eur. J Clin. Invest.* **7**, 27-35.

756. Banfield, D. K., and MacGillivray, R. T. A. (1992) Partial characterization of vertebrate prothrombin cDNAs: amplification and sequence analysis of the B chain of thrombin from nine different species, *Proc. Natl. Acad. Sci. U.S.A.* **89**, 2779-2783.

757. Bock, S. C., Wion, K. L., Vehar, G. A., and Lawn, R. M. (1982) Cloning and expression of the cDNA for human antithrombin III, *Nucleic Acids Res.* **10**, 8113-8125.

758. Chandra, T., Stackhouse, R., Kidd, V. J., and Woo, S. L. W. (1983) Isolation and sequence characterization of human antithrombin III, *Proc. Natl. Acad. Sci. U.S.A.* **80**, 259-262.

759. Peterson, C. D., and Blackburn, M. N. (1985) Isolation and characterization of an antithrombin III variant with reduced carbohydrate content and enhanced heparin binding, *J. Biol. Chem.* **260**, 610-615.

760. Brennan, S. O., George, P. M., and Jordan, R. E. (1987) Physiological variant of antithrombin III lacks carbohydrate side chain at Asn 135, *FEBS Lett.* **219**, 431-436.

761. Ersdal-Badju, E., Lu, A., Peng, X., Picard, V., Zendehrouh, J., Turk, B., Björk, I., Olson, S. T., and Bock, S. C. (1995) Elimination of glycosylation heterogeneity affecting heparin affinity of recombinant human antithrombin III by expression of a β-like variant in baculovirus-infected insect cells, *Biochem. J.* **310**, 323-330.

762. Picard, V., Ersdals-Badju, E., and Bock, S. C. (1995) Partial glycosylation of antithrombin III asparagine-135 is caused by the serine in the third position of its N-glycosylation consensus sequence and is responsible for production of the β-anththrombin III isoform with enhanced heparin affinity, *Biochemistry* **34**, 8433-8440.

763. Garone, L., Edmunds, T., Hanson, E., Bernasconi, R., Huntington, J. A., Meagher, J. L., Fan, B., and Gettins, P. G. W. (1996) Antithrombin-heparin affinity reduced by fucosylation of carbohydrate at asparagine 155, *Biochemistry* **35**, 8881-8889.

764. Jörnvall, H., Fish, W. W., and Björk, I. (1979) The thrombin cleavage site in bovine antithrombin, *FEBS Lett.* **106**, 358-362.

765. Björk, I., Danielsson, Å., Fenton II, J. W., and Jörnvall, H. (1981) The site in human antithrombin for functional proteolytic cleavage by human thrombin, *FEBS Lett.* **126**, 257-260.

766. Björk, I., Jackson, C. M., and Jönvall, H. (1981) The site in human antithrombin for functional proteolytic cleavage by human thrombin, *FEBS Lett.* **126**, 257-260.

767. Carrell, R. W., Stein, P. E., Fermi, G., and Wardell, M. R. (1994) Biological implications of a 3 Å structure of dimeric antithrombin, *Structure* **2**, 257-270.

768. Erdjument, H., Lane, D. A., Panico, M., Di Marzo, V., and Morris, H. R. (1988) Single amino acid substitutions in the reactive site of antithrombin leading to thrombosis:

congenital substitution of arginine 393 to cysteine in antithrombin Northwick Park and to histidine in antithrombin Glasgow, *J. Biol. Chem.* **263**, 5589-5593.

769. Schreuder, H. A., de Boer, B., Dijkema, R., Mulders, J., Theunissen, H. J. M., Grootenhuis, P. D. J., and Hol, W. G. J. (1994) The intact and cleaved human antithrombin III complex as a model for serpin-proteinase interactions, *Nat. Struct. Biol.* **1**, 48-54.

770. Mourey, L., Samama, J.-P., Delarue, M., Petitou, M., Choay, J., and Moras, D. (1993) Crystal structure of cleaved bovine antithrombin III at 3.2 Å resolution, *J. Mol. Biol.* **232**, 223-241.

771. Wardell, M. R., Abrahams, J.-P., Bruce, D., Skinner, R., and Leslie, A. G. W. (1993) Crystallization and preliminary X-ray diffraction analysis of two conformations of intact human antithrombin, *J. Mol. Biol.* **234**, 1253-1258.

772. Carrell, R. W., Boswell, D. R., Brennan, S. O., and Owen, M. C. (1980) Active site of α_1-antitrypsin: homologous site in antithrombin III, *Biochem. Biophys. Res. Commun.* **93**, 399-402.

773. Löbermann, H., Tokuoka, R., Deisenhofer, J., and Huber, R. (1984) Human α_1-protease inhibitor: crystal structure analysis of two crystal modifications, molecular model and preliminary analysis of the implications for function, *J. Mol. Biol.* **177**, 531-556.

774. Huber, R., and Carrell, R. W. (1989) Implications of the three-dimensional structure of α_1-antitrypsin for structure and function of serpins, *Biochemistry* **28**, 8951-8966.

775. Potempa, J., Korzus, E., and Travis, J. (1994) The serpin superfamily of proteinase inhibitors: structure, function, and regulation, *J. Biol. Chem.* **269**, 15957-15960.

776. Mourey, L., Samana, J. P., Delarue, M., Choay, J., Lormeau, J. C., Petitou, M., and Moras, D. (1990) Antithrombin III: structural and functional aspects, *Biochimie* **72**, 599-608.

777. Stein, P., and Chothia, C. (1991) Serpin tertiary structure transformation, *J. Mol. Biol.* **221**, 615-621.

778. Pratt, C. W., Whinna, H. C., and Church, F. C. (1992) A comparison of three heparin-binding serpin protease inhibitors, *J. Biol. Chem.* **267**, 8795-8801.

779. Baumann, U., Huber, R., Bode, W., Grosse, D., Lesjak, M., and Laurell, C. B. (1991) Crystal structure of cleaved human α_1-antichymotrypsin at 2.71 Å resolution, *J. Mol. Biol.* **218**, 595-606.

780. Baumann, U., Bode, W., Huber, R., Travis, J., and Potempa, J. (1992) Crystal structure of cleaved equine leukocyte elastase inhibitor determined at 1.95 Å resolution, *J. Mol. Biol.* **226**, 1207-1218.

781. Mottonen, J., Strand, A., Symersky, J., Sweet, R. M., Danley, D. E., Geoghegan, K. F., Gerard, R. D., and Goldsmith, E. J. (1992) Structural basis of latency in plasminogen-activator inhibitor-1, *Nature* **355**, 270-273.

782. Wright, H. T., Qian, H. K., and Huber, R. (1990) Crystal structure of plakalbumin, a proteolytically nicked form of ovalbumin. Its relationship to the structure of cleaved α_1-protease inhibitor, *J. Mol. Biol.* **213**, 513-528.

783. Stein, P. E., Leslie, A. G. W., Finch, J. T., Turnell, W. G., McLaughlin, P. J., and Carrell, R. W. (1990) Crystal structure of ovalbumin as a model for the reactive center for serpins, *Nature* **347**, 99-102.

784. Ehrlich, H. J., Keijel, J., Preissner, K. T., Gebbink, R. K., and Pannekoek, H. (1991) Functional interaction of plasminogen activator inhibitor type 1 (PAI-1) and heparin, *Biochemistry* **30**, 1021-1028.

785. Kuhn, L. A., Griffin, J. H., Fisher, C. L., Greengard, J. S., Bouma, B. N., España, F., and Tainer, J. A. (1990) Elucidating the structural chemistry of glycosaminoglycan recognition by protein C inhibitor, *Proc. Natl. Acad. Sci. U.S.A.* **87**, 8506-8510.

786. Brennan, S. O., Borg, J.-Y., George, P. M., Soria, C., Soria, J., Caen, J., and Carrell, R. W. (1988) New carbohydrate site in mutant antithrombin (7Ile → Asn) with decreased heparin affinity, *FEBS Lett.* **237**, 118-122.

787. Lane, D. A., Olds, R. R., and Thein, S.-L. (1992) Antithrombin and its deficiency states, *Blood Coag. Fibrinol.* **3**, 315-341.

788. Lawrence, D. A., Olson, S. T., Palaniappan, S., and Ginsburg, D. (1994) Serpin reactive center loop mobility is required for inhibitor function but not for enzyme recognition, *J. Biol. Chem.* **269**, 27657-27662.

789. Crowther, D. C., Evans, D. L. I., and Carrell, R. W. (1992) Serpins: implications of a mobile reactive centre, *Curr. Opin. Biotech.* **3**, 7560-7565.

790. Mast, A. E., Enghild, J. J., and Salvesen, G. (1992) Conformation of the reactive site loop of α_1-proteinase inhibitor probed by limited proteolysis, *Biochemistry* **31**, 2720-2728.

791. Björk, I., Nordling, K., and Olson, S. T. (1993) Immunologic evidence for insertion of the reactive-bond loop of antithrombin into the A β-sheet of the inhibitor during trapping of target proteinases, *Biochemistry* **32**, 6501-6505.

792. Hopkins, P. C. R., Carrell, R. W., and Stone, S. R. (1993) Effects of mutations in the hinge region of serpins, *Biochemistry* **32**, 7650-7657.

793. Olson, S. T., Stephens, A. W., Hirs, C. H. W., Bock, P. E., and Björk, I. (1995) Kinetic characterization of the proteinase binding defect in a reactive site variant of the serpin, antithrombin: role of the P1' residues in transition-state stabilization of the antithrombin-proteinase complex formation, *J. Biol. Chem.* **270**, 9717-9724.

794. Huntington, J. A., Olson, S. T., Fan, B., and Gettins, P. G. W. (1996) Mechanism of heparin activation of antithrombin: evidence for reactive center loop preinsertion with expulsion upon heparin binding, *Biochemistry* **35**, 8495-8503.

795. Carrell, R. W., Evans, D. L. I., and Stein, P. E. (1990) Mobile reactive centre of serpins and the control of thrombosis, *Nature* **353**, 576-578.

796. Lomas, D. A., Elliott, P. R., Chang, W.-S. W., Wardell, M. R., and Carrell, R. W. (1995) Preparation and characterization of latent α_1-antitrypsin, *J. Biol. Chem.* **270**, 5282-5288.

797. Grootenhuis, P. D. J., and van Boeckel, C. A. A. (1991) Constructing a molecular model of the interaction between antithrombin III and a potent heparin analogue, *J. Am. Chem. Soc.* **113**, 2743-2747.

798. Borg, J. Y., Brennan, S. O., Carrell, R. W., George, P., Perry, D. J., and Shaw, J. (1990) Antithrombin Rouen IV 24 Arg to Cys. The amino terminal contribution to heparin binding, *FEBS Lett.* **266**, 163-166.

799. Gandrill, S., M. Aiach, M., Lane, D. A., Vadaud, D., Molho-Sabatier, P., Caso, R., De Moerloose, P., Fiessinger, J.-N., and Clauser, E. (1990) Important role of arginine 129 in heparin-binding site of antithrombin III: identification of a novel mutation of arginine 129 to glutamine, *J. Biol. Chem.* **265**, 18997-19001.

800. Gettins, P., and Wooten, E. W. (1987) On the domain structure of antithrombin III: tentative localization of the heparin binding region using ^{1}H NMR spectroscopy, *Biochemistry* **26**, 4403-4408.

801. Chang, J.-Y. (1989) Binding of heparin to human antithrombin III activates selective chemical modification at lysine 236: Lys 107, Lys 125, and Lys 136 are situated within the heparin- binding site of antithrombin III, *J. Biol. Chem.* **264**, 3111-3115.

802. Rosenfeld, L., and Danishefsky, I. (1986) A fragment of antithrombin that binds both heparin and thrombin, *Biochem. J.* **237**, 639-646.

803. Vaughan, S. A., Smith, J. W., and Knauer, D. J. (1988) Structural features of heparin binding domains in heparin activated serine protease inhibitors, *J. Cell Biol.* **107**, 833a.

804. Sun, X.-J., and Chang, J.-Y. (1990) Evidence that arginine-129 and arginine-145 are located within the heparin binding site of human antithrombin III, *Biochemistry* **29**, 8957-8962.

805. Chang, J.-Y., and Tran, T. H. (1986) Antithrombin III$_{Basel}$, *J. Biol. Chem.* **261**, 1174-1176.

806. Koide, T., Odani, S., Takahashi, K., Ono, T., and Sukuragawa, N. (1984) Antithrombin III Toyama: replacement of arginine-47 by cysteine in hereditary abnormal antithrombin III that lacks heparin-binding ability, *Proc. Natl. Acad. Sci. U.S.A.* **81**, 289-293.

807. Blackburn, M. N., and Sibley, C. C. (1980) The heparin binding site of antithrombin III, *J. Biol. Chem.* **255**, 824-826.

808. Blackburn, M. N., Smith, R. L., Carson, J., and Sibley, C. C. (1984) The heparin- binding site of antithrombin III: identification of a critical tryptophane in the amino acid sequence, *J. Biol. Chem.* **259**, 939-941.

809. Duchange, N., Chasse, J.-F., Cohen, G. N., and Zakin, M. M. (1986) Antithrombin III Tours gene: identification of a point mutation leading to an arginine-cysteine replacement results in a silent deficiency, *Nucl. Acids Res.* **14**, 2408.

810. Brunel, F., Duchange, N., Fischer, A. M., Cohen, G. U., and Zakin, M. M. (1987) Antithrombin III Alger: a new case of Arg 47 to Cys mutation, *Am. J. Hematol.* **25**, 223-224.

811. Owen, M. C., Borg, J. Y., Soria, C., Soria, J., Caen, J., and Carrell, R. W. (1987) Heparin binding defect in a new antithrombin III variant: Rouen, 47 Arg to His, *Blood* **69**, 1275-1279.

812. Sun, I.-J., and Chang, J.-Y. (1989) Heparin binding domain of human antithrombin III inferred from the sequential reduction of its three disulfide linkages: an efficient method for structural analysis of partially reduced proteins, *J. Biol. Chem.* **264**, 11288-11293.

813. Villanueva, G. B., and Allen, N. (1983) Demonstration of a two-domain structure of antithrombin III during its denaturation in guanidinium chloride, *J. Biol. Chem.* **258**, 11010-11013.

814. Liu, C.-S., and Chang, J.-Y. (1987) The heparin binding site of human antithrombin III: selective chemical modification at Lys 114, Lys 125, and Lys 287 impairs its heparin cofactor activity, *J. Biol. Chem.* **262**, 17356-17361.

815. Peterson, C. B., Noyes, C. M., Pecon, J. M., Church, F. C., and Blackburn, M. N. (1987) Identification of a lysyl residue in antithrombin which is essential for heparin binding, *J. Biol. Chem.* **262**, 8061-8065.

816. Smith, J. W., and Knauer, D. J. (1987) A heparin binding site in antithrombin III: identification, purification and amino acid sequence, *J. Biol. Chem.* **262**, 11964-11973.

817. Albert, J. S., and Hamilton, A. D. (1995) Stabilization of helical domains in short peptides using hydrophobic interactions, *Biochemistry* **34**, 984-990.

818. Owen, M. C., George, P. M., Lane, D. A., and Boswell, D. R. (1991) P1 variant antithrombins Glasgow (393 Arg to His) and Pescara (393 Arg to Pro) have increased heparin affinity and are resistant to catalytic cleavage by elastase: implications for the heparin activation mechanism, *FEBS Lett.* **280**, 216-220.

819. Okajima, K., Abe, H., Wagatsuma, M., Okabe, H., and Takatsuki, K. (1995) Antithrombin III Kumamoto; a single mutation at Arg393-His increased the affinity of antithrombin III for heparin, *Am. J. Hematol.* **48**, 12-18.

820. Villanueva, G. B. (1984) Predictions of the secondary structure of antithrombin III and the location of the heparin- binding site, *J. Biol. Chem.* **259**, 2531-2536.

821. Fan, B., Turko, I. V., and Gettins, P. G. W. (1994) Lysine-heparin interactions in antithrombin: properties of K125M and K290M, K294M, K297M variants, *Biochemistry* **33**, 14156-14161.

822. Fenton, J. W. I., Ofosu, F. A., Moon, D. G., and Maraganore, J. M. (1991) Thrombin structure and function: why thrombin is the primary target for antithrombotics, *Blood Coag. Fibrinol.* **2**, 69-75.

823. Stubbs, M. T., and Bode, W. (1993) A player of many parts: the spotlight falls on thrombin's structure, *Thromb. Res.* **69**, 1-58.

824. Tulinsky, A., and Qiu, X. (1993) Active site and exosite binding of α-thrombin, *Blood Coag. Fibrinol.* **4**, 305-312.

825. DiBella, E. E., Maurer, M. C., and Scheraga, H. A. (1995) Expression and folding of recombinant bovine prethrombin-2 and its activation to thrombin, *J. Biol. Chem.* **270**, 163-169.

826. Degen, S. J. F., MacGillivray, R. T. A., and Davie, E. W. (1983) Characterization of the complementary deoxyribonucleic acid and gene coding for human pro-thrombin, *Biochemistry* **22**, 2087-2097.

827. Degen, S. J. E., and Davie, E. W. (1987) Nucleotide sequence of the gene for human prothrombin, *Biochemistry* **26**, 6165-61778.

828. Machovich, R., and Horváth, I. (1981) Thrombin and haemostasis: regulation of the biological functions of thrombin, *Haematologia* **14**, 339-359.

829. Bode, W., Mayr, I., Baumann, U., Huber, R., Stone, S. R., and Hofsteenge, J. (1989) The refined 1.9 Å crystal structure of human α-thrombin: interaction with d Phe-Pro-Arg chloromethylketone and significance of the Tyr-Pro-Pro-Trp insertion segment, *EMBO J.* **8**, 3467-3475.

830. Tsiang, M., Jain, A. K., Dunn, K. E., Rojas, M. E., Leung, L. L., and Gibbs, C. S. (1995) Functional mapping of the surface residues of human thrombin, *J. Biol. Chem.* **270**, 16854-16863.

831. Matthews, J. H., Krishnan, R., Costanzo, M. J., Maryanoff, B. E., and Tulinsky, A. (1996) Crystal structures of thrombin with thiazole-containing inhibitors: probes of the S1' binding site, *Biophys. J.* **71**, 2830-2839.

832. Le Bonniec, B. F., Guinto, E. R., and Stone, S. R. (1995) Identification of thrombin residues that modulate its interactions with antithrombin III and α1-antitrypsin, *Biochemistry* **34**, 12241 12248.

833. Rezaie, A. R. (1996) Tryptophan 60-D in the B-insertion loop of thrombin modulates the thrombin-antithrombin reaction, *Biochemistry* **35**, 1918-1924.

834. Sheenan, J. P., and Sadler, J. E. (1994) Molecular mapping of the heparin-binding exosite of thrombin, *Proc. Natl. Acad. Sci. U.S.A.* **91**, 5518-5522.

835. Church, F. C., Pratt, C. W., Noyes, C. M., Kalayanamit, T., Sherril, G. B., Tobin, R. B., and Meade, B. (1989) Structural and functional properties of human α-thrombin, phosphopyridoxylated α-thrombin, and γT-thrombin: identification of lysyl residues in α-thrombin that are critical for heparin and fibrin(ogen) interactions, *J. Biol. Chem.* **264**, 18419-18425.

836. White, G. C., Lundblad, R. L., and Griffith, M. J. (1981) Structure-function relations in platelet-thrombin reactions, *J. Biol. Chem.* **256**, 1763-1766 .

837. Wu, Q., Picard, V., Aiach, M., and Sandler, J. E. (1994) Activation-induced exposure of the thrombin anion-binding exosite: interactions of recombinant mutant prothrombins with thrombomodulin and a thrombin exosite-specific antibody, *J. Biol. Chem.* **269**, 3724-3730.

838. Rosenberg, R. D., and Lam, L. (1979) Correlation between structure and function of heparin, *Proc. Natl. Acad. Sci. U.S.A.* **76**, 1218-1222.

839. Rosenberg, R. D., Jordan, R. E., Favreau, L. V., and Lam, L. (1979) Highly active heparin species with multiple binding sites for antithrombin, *Biochem. Biophys. Res. Commun.* **86**, 1319-1324.

840. Choay, J., Lormeau, J. C., Petitou, M., Sinäy, P., Casu, B., Oreste, P., Torri, G., and Gatti, G. (1980) Anti factor-Xa active heparin oligosaccharides, *Thromb. Res.* **18**, 573-578.

841. Thunberg, L., Bäckström, G., Grundberg, H., Riesenfeld, J., and Lindahl, U. (1980) The molecular size of the antithrombin-binding sequence in heparin, *FEBS Lett.* **117**, 203-206.

842. Rosenberg, R. D., Oosta, G. M., Jordan, R. E., and Gardner, W. T. (1980) The interaction of heparin with thrombin and antithrombin, *Biochem. Biophys. Res. Commun.* **96**, 1200-1208.

843. Lindahl, U., Bäckström, G., Thunberg, L., and Leder, I. G. (1980) Evidence of a 3-O-sulfated D-glucosamine residue in the antithrombin-binding sequence of heparin, *Proc. Natl. Acad. Sci. U.S.A.* **77**, 6551-6555.

844. Meyer, B., Thunberg, L., Lindahl, U., Larm, O., and Leder, I. R. (1981) The antithrombin-binding sequence of heparin studied by n.m.r. spectroscopy, *Carbohyd. Res.* **88**, C1-C4.
845. Stone, A. L., Beeler, D., Oosta, G., and Rosenberg, R. D. (1982) Circular dichroism spectroscopy of heparin- antithrombin interactions, *Proc. Natl. Acad. Sci. U.S.A.* **79**, 7190-7194.
846. Lindahl, U., Bäckström, G., and Thunberg, L. (1983) The antithrombin- binding sequence in heparin, *J. Biol. Chem.* **258**, 9826-9830.
847. Torri, G., Casu, B., Gatti, G., Petitou, M., Choay, J., Jacquinet, J.-C., and Sinäy, P. (1985) Monodimensional and bidimensional 500 megahertz proton NMR spectra of a synthetic pentasaccharide corresponding to the binding sequence of heparin to antithrombin III: evidence for conformational peculiarity of the sulfated iduronate residue, *Biochem. Biophys. Res. Commun.* **128**, 134-140.
848. Petitou, M. (1984) Synthetic heparin fragments: new and efficient tools for the study of heparin and its interactions, *Nouv. Rev. Fr. Hematol.* **26**, 221-226.
849. Sinäy, P., Jacquinet, J., Petitou, M., Duchaussoy, P., Lederman, I., Choay, J., and Torri, G. (1984) The synthesis of a heparin pentasaccharide fragment having high affinity for antithrombin III, *Carbohyd. Res.* **132**, C5-C9.
850. van Boeckel, C. A. A., Beetz, T., Vos, J. N., de Jong, A., van Aelst, S. F., van der Bosch, R. H., Mertens, J. M. R., and Van der Vlugt, F. A. (1985) Synthesis of a pentasaccharide corresponding to the antithrombin III binding fragment of heparin, *Carbohyd. Chem.* **4**, 293-321.
851. Beetz, T., and van Boeckel, C. A. A. (1986) Synthesis of an antithrombin binding heparin-like pentasaccharide lacking 6-O sulphate at its reducing end, *Tetrahedron Lett.* **27**, 5889-5892.
852. van Aelst, S. F., and van Boeckel, C. A. A. (1987) Synthesis of an analogue of the antithrombin binding region of heparin containing α-L-idopyranose, *Recl. Trav. Chim. Pays-Bas* **106**, 593-595.
853. Choay, J., Lormeau, J. C., Petitou, M., Sinäy, P., and Fareed, J. (1981) Structural studies on a biologically active hexasaccharide obtained from heparin, *Ann. N. Y. Acad. Sci.* **370**, 644-649.
854. Atha, D. H., Lormeau, J.-C., Petitou, M., Rosenberg, R. D., and Choay, J. (1985) Contribution of monosaccharide residues in heparin binding to antithrombin III, *Biochemistry* **24**, 6723-6729.
855. Leder, I. G. (1980) A novel 3-O-sulfatase from human urine acting on methyl-2-deoxy-2-sulfamino-alpha-d-glucopyranoside 3-sulfate, *Biochem. Biophys. Res. Commun.* **94**, 1183-1189.
856. Lucas, H., Basten, J. E. M., van Dinther, T., Meuleman, D. G., van Aelst, S. F., and van Boeckel, C. A. A. (1987) Synthesis of heparin-like pentamers containing "opened" uronic acid moieties, *Tetrahedron* **46**, 8207-8228.
857. Hovingh, P., Piepkorn, M., and Linker, A. (1986) Biological implications of the structural antithrombin affinity and anticoagulant activity relationships among vertebrate heparins and heparan sulphate, *Biochem. J.* **237**, 573-581.
858. Horner, A. A. (1986) Rat heparins: a study of the relative sizes and antithrombin-binding characteristics of heparin proteoglycans, chains and depolymerization products from rat adipose tissue, heart, lungs, peritoneal cavity and skin, *Biochem. J.* **240**, 171-179.
859. Marcum, J. A., McKenney, J. B., Galli, S. J., Jackman, R. W., and Rosenberg, R. D. (1986) Anticoagulantly active heparin-like molecules from mast cell-deficient mice, *Am. J. Physiol.* **250**, H879-H888.
860. Horner, A. A. (1990) Rat heparan sulphates: a study of the antithrombin-binding properties of heparan sulphate chains from rat adipose tissue, brain, carcass, heart, intestine, kidneys, liver, lungs, skin and spleen, *Biochem. J.* **266**, 553-559.
861. Petitou, M., Lormeau, J. C., and Choay, J. (1988) Interaction of heparin and antithrombin III: the role of O-sulfate groups, *Eur. J. Biochem.* **176**, 637-640.

862. Shworak, N. W., Fritze, L. M. S., Liu, J., Butler, L. D., and Rosenberg, R. D. (1996) Cell-free synthesis of anticoagulant heparan sulfate reveals a limiting converting activity that modifies an excess precursor pool, *J. Biol. Chem.* **271**, 27063-27071.

863. Toida, T., Hileman, R. E., Smith, A. E., Vlahova, P. I., and Linhardt, R. J. (1996) Enzymatic preparation of heparin oligosaccharides containing antithrombin III binding sites, *J. Biol. Chem.* **271**, 32040-32047.

864. Carlson, T. H., Babcock, T., Atencio, A. C., Levinson, C., and Mora, H. R. (1988) Behavior of antithrombin III isoforms on immobilized heparins: evidence that the isoforms bind to different numbers of low-affinity heparin sites, *J. Biol. Chem.* **263**, 2187-2194.

865. Hol, W. G. J., van Duijnen, P. T., and Berendsen, H. J. C. (1978) The α-helix dipole and the properties of proteins, *Nature* **273**, 443-446.

866. Volpi, N., Cusmano, M., and Venturelli, T. (1995) Qualitative and quantitative studies of heparin and chondroitin sulfates in normal human plasma, *Biochim. Biophys. Acta* **1243**, 49-58.

867. Yurt, R. W., Leid, R. W., Austen, K. F., and Silbert, J. E. (1977) Native heparin from rat peritoneal mast cells, *J. Biol. Chem.* **252**, 528-521 .

868. Berlin, G., and Enerbäck, L. (1978) Changes in numbers and heparin content of peritoneal fluid mast cells of growing rats measured by flow cytometry, *J. Histochem. Cytochem.* **26**, 14-21.

869. Marcum, J. A., Fritze, L., Galli, S. J., Karp, G., and Rosenberg, R. D. (1983) Microvascular heparinlike species with anticoagulant activity, *Am. J. Physiol.* **245**, H725-H733.

870. Marcum, J. A., McKenney, J., B., and Rosenberg, R. D. (1984) Acceleration of thrombin-antithrombin complex formation in rat hindquarters via heparinlike molecules bound to the endothelium, *J. Clin. Invest.* **74**, 341-350.

871. Stern, D., Nawroth, P., Marcum, J. A., Handley, D., Rosenberg, R. D., and Stern, K. (1985) Interaction of antithrombin III with bovine aortic segments: role of heparin in binding and enhanced anticoagulant activity, *J. Clin. Invest.* **75**, 272-279.

872. Lane, D. A., Pejler, G. G., Flynn, A. M., Thompson, E. A., and Lindahl, U. (1986) Neutralization of heparin-related saccharides by histidine-rich glycoprotein and platelet factor 4, *J. Biol. Chem.* **261**, 3980-3986.

873. Halkier, T. (1991) *Mechanisms in Blood Coagulation, Fibrinolysis, and the Complement System* (Cambridge University Press, Cambridge).

874. Sherry, S. (1992) *Fibrinolysis, Thrombosis, and Hemostasis. Concepts, Perspectives, and Clinical Applications* (Lea & Febiger, Philadelphia).

875. Broze Jr., G. J., and Tollefsen, D. M. (1994) Regulation of blood coagulation by protease inhibitors, in *The Molecular Basis of Blood Diseases, Second Edition*, eds., Stamatoyannopoulos, G., Nienhuis, A. W., Majerus, P. & Varmus, H., W. B. Saunders Co., Philadelphia, pp. 629-656.

876. Hawiger, J. (1987) Formation and regulation of platelet and fibrin hemostatic plug, *Hum. Pathol.* **18**, 111-122.

877. Cadroy, Y., and Harker, L. A. (1990) Platelets, thrombosis, and antithrombotic therapies, in *Cardiovascular Pharmacology, 3rd Edition*, ed. Antonaccio, M., Raven Press, Ltd., New York, pp. 515-539.

878. Roberts, H. R., and Lozier, J. N. (1992) New perspectives on the coagulation cascade, *Hosp. Pract.* , 97-112.

879. Blombäck, B. (1996) Fibrinogen and fibrin--proteins with complex roles in hemostasis and thrombosis, *Thomb. Res.* **83**, 1-75.

880. Davie, E. W., Fujikawa, K., and Kisiel, W. (1991) The coagulation cascade: initiation, maintenance, and regulation, *Biochemistry* **30**, 10363-10370.

881. Broze Jr., G. J. (1992) Tissue factor pathway inhibitor and the revised hypothesis of blood coagulation, *Trends Cardiovasc. Med.* **2**, 72-77.

882. Voorberg, J., van Stempvoort, G., Bos, J. M. K., Mertens, K., van Mourik, J. A., and Donath, M.-J. S. H. (1996) Enhanced thrombin sensitivity of a factor VIII-heparin cofactor II hybrid, *J. Biol. Chem.* **271**, 20985-20988.

883. Edelberg, J. M., and Pizzo, S. V. (1991) Lipoprotein (a) inhibits plasminogen activation in a template-dependent manner, *Blood Coag. Fibrinol.* **2**, 759-764.

884. Edelberg, J. M., and Pizzo, S. V. (1991) Lipoprotein (a): the link between impaired fibrinolysis and atheroschlerosis, *Fibrinolysis* **5**, 135-143.

885. Hynes, R. O. (1991) The complexity of platelet adhesion to extracellular matrices, *Thromb. Haemostas.* **66**, 40-43.

886. von dem Borne, P. A. K., Meijers, J. C. M., and Bouma, B. N. (1996) Effect of heparin on the activation of factor XI by fibrin-bound thrombin, *Thromb. Haemostas.* **76**, 347-353.

887. Marciniak, E. (1973) Factor-Xa inactivation of antithrombin III: evidence of biological stabilization of factor Xa by factor V-phospholipid complexes, *Br. J. Haematol.* **24**, 391-400.

888. Miletich, J. P., Jackson, C. M., and Majerus, P. W. (1978) Properties of the factor Xa binding site of platelets, *J. Biol. Chem.* **253**, 6908-6913.

889. Hembrough, T. A., Li, L., and Gonias, S. L. (1996) Cell-surface cytokeratin 8 is the major plasminogen receptor on breast cancer cells and is required for the accelerated activation of cell-associated plasminogen by tissue-type plasminogen activator, *J. Biol. Chem.* **271**, 25684-25691.

890. Kim, S.-O., Plow, E. F., and Miles, L. A. (1996) Regulation of plasminogen receptor expression on monocytoid cells by β_1-integrin-dependent cellular adherence to extracellular matrix proteins, *J. Biol. Chem.* **271**, 23761-23767.

891. Ofosu, F. A., Sie, P., Modi , G. J., Fernandez, F., Buchanan, M. R., Blajchman, M. A., Boneu, B., and Hirsh, J. (1987) The inhibition of thrombin-dependent positive-feedback reactions is critical to the expression of the anticoagulant effect of heparin, *Biochem. J.* **243**, 579-588.

892. Ofosu, F. A., Hirsh, J., Esmon, C. T., Modi, G. J., Smith, L. M., Anvari, N., Buchanan, M. R., Fenton 2nd, J. W., and Blajchman, M. A. (1989) Unfractionated heparin inhibits thrombin-catalyzed amplification reactions of coagulation more efficiently than those catalyzed by factor Xa, *Biochem. J.* **257**, 147-150.

893. Fuchs, H. E., Shifman, M. A., and Pizzo, S. V. (1982) *In vivo* catabolism of α_1-proteinase inhibitor-trypsin, antithrombin III-thrombin, and α_2-macroglobulin-methylamine, *Biochim. Biophys. Acta* **716**, 151-157.

894. Gonias, S. L., Fuchs, H. E., and Pizzo, S. V. (1982) A unique pathway for the plasma elimination of α_2-antiplasmin-protease complexes in mice, *Thromb. Haemostas.* **48**, 208-210.

895. Pizzo, S. V., Mast, A. E., Feldman, S. R., and Salvesen, G. (1988) *In vivo* catabolism of α_1-antichymotrypsin is mediated by the serpin receptor which binds α_1-proteinase inhibitor, antithrombin III and heparin cofactor II, *Biochim. Biophys. Acta* **967**, 158-162.

896. Pratt, C. W., Church, F. C., and Pizzo, S. V. (1988) *In vivo* catabolism of heparin cofactor II and its complex with thrombin: evidence for a common receptor-mediated clearance pathway for three serine proteinase inhibitors, *Arch. Biochem. Biophys.* **262**, 111-117.

897. Takeya, H., Hamada, T., Kume, M., and Suzuki, K. (1994) Receptor-mediated endocytosis of thrombin-antithrombin III complex by the human monocytoid cell line U937, *Biochem. Biophys. Res. Commun.* **200**, 1334-1340.

898. Borth, W. (1992) α_2-Macroglobulin, a multifunctional binding protein with targeting characteristics, *FASEB J.* **6**, 3345-3353.

899. Béguin, S., Kessels, H., Dol, F., and Hemker, H. C. (1992) The consumption of antithrombin III during coagulation, its consequences for the calculation of prothrombinase activity and standardization of heparin activity, *Thromb. Haemostas.* **68**, 136-142.

900. Rosing, J., Hoekema, L., Nicolaes, G. A. F., Cristella, M., Thomassen, L. G. D., Hemker, H. C., Varadi, K., Schwarz, H. P., and Tans, G. (1995) Effects of protein S and factor Xa on peptide bond cleavages during inactivation of factor Va and factor Va^{R506Q} by activated protein C, *J. Biol. Chem.* **270**, 27852-27858.

901. Suzuki, K., Nishioka , J., and Hashimoto, S. (1983) Protein C inhibitor: purification from human plasma and characterization, *J. Biol. Chem.* **258**, 163-168.

902. Suzuki, K., Nishioka, J., Kusumoto, H., and Hashimoto, S. (1984) Mechanism of inhibition of activated protein C inhibitor, *J. Biochem. Tokyo* **95**, 187-195.

903. Broze Jr., G. J., and D. Gailani, D. (1993) The role of factor XI in coagulation, *Thromb. Haemost.* **70**, 72-74.

904. Odrljin, T. M., Shainoff, J. R., Lawrence, S. O., and Simpson-Haidaris, P. J. (1996) Thrombin cleavage enhances exposure of a heparin binding domain in the N-terminus of the fibrin β chain, *Blood* **88**, 2050-2061.

905. Odrljin, T. M., Francis, C. W., Sporn, L. A., Bunce, L. A., Marder, V. J., and Simpson-Haidaris, P. J. (1996) Heparin-binding domain of fibrin mediates its binding to endothelial cells, *Arterioscler. Thromb. Vasc. Biol.* **16**, 1544-1551.

906. Lane, D. A. (1989) Heparin binding and neutralizing proteins, in *Heparin: Chemical and Biological Properties, Clinical Applications*, eds., Lane, D. A. & Lindahl, U., CRC Press, Inc., Boca Raton, FL, pp. 363-391.

907. Young, E., Prins, M., Levine, M. N., and Hirsh, J. (1992) Heparin binding to plasma proteins, an important mechanism for heparin resistance, *Thromb. Haemostas.* **67**, 639-643.

908. Cosmi, B., Fredenburgh, J. C., Rischke, J., Hirsh, J., Young, E., and Weitz, J. I. (1997) Effect of nonspecific binding to plasma proteins on the antithrombin activities of unfractionated heparin, low-molecular-weight heparin and dermatan sulfate, *Circulation* **95**, 118-124.

909. Luscinskas, F. W., and Lawler, J. (1994) Integrins as dynamic regulators of vascular function, *FASEB J.* **8**, 929-938.

910. Rogers, S. J., Pratt, C. W., Whinna, H. C., and Church, F. C. (1992) Role of thrombin exosites in inhibition by heparin cofactor II, *J. Biol. Chem.* **267**, 3613-3617.

911. Tollefsen, D. M. (1989) Heparin cofactor II, in *Heparin. Chemical and Biological Properties, Clinical Applications*, eds., Lane, D. A. & Lindahl, U., CRC Press, Inc., Boca Raton, FL, pp. 257-273.

912. Blinder, M. A., Marasa, J. C., Reynolds, C. H., Deavon, L. L., and Tollefsen, D. M. (1988) Heparin cofactor II: cDNA sequence, chromosome localization, restriction fragment length polymorphism, and expression in *Escherichia coli, Biochemistry* **27**, 752-759.

913. Griffith, M. J., Noyes, C. M., and Church, F. C. (1985) Reactive site peptide structural similarity between heparin cofactor II and antithrombin III, *J. Biol. Chem.* **260**, 2218-2225.

914. Church, F. C., Noyes, C. M., and Griffith, M. J. (1985) Inhibition of chymotrypsin by heparin cofactor II, *Proc. Natl. Acad. Sci. U.S.A.* **82**, 6431.

915. Tollefsen, D. M., Sugimori, T., and Maimone, M. M. (1990) Effect of low molecular weight heparin preparations on the inhibition of thrombin by heparin cofactor II, *Sem. Thromb. Hemostas.* **16**, 66-70.

916. Whinna, H. C., Blinder, M. A., Szewczyk, M., Tollefsen, D. M., and Church, F. C. (1991) Role of lysine 173 in heparin binding to heparin cofactor II, *J. Biol. Chem.* **266**, 8129-8135.

917. MacIntosh, S., Jakubowski, H., and Owen, W. G. (1984) Regulation of clearance and inhibition of intra-vascular thrombin, *Fed. Proc.* **43**, 1946.

918. McGuire, E. A., and Tollefsen, D. M. (1987) Activation of heparin cofactor II by by fibroblasts and vascular smooth muscle cells, *J. Biol. Chem.* **262**, 169-175.

919. Whinna, H. C., Chai, H. U., Rosenberg, L. C., and Church, F. C. (1993) Interaction of heparin cofactor II with biglycan and decorin, *J. Biol. Chem.* **268**, 3920-3924.

920. Tollefsen, D. M., and Blank, M. J. (1981) Detection of a new heparin-dependent inhibitor of thrombin in human plasma, *J. Clin. Invest.* **68**, 589-596.

921. Pratt, C. W., Whinna, H. C., Meade, J. M., Treanor, R. E., and Church, F. C. (1989) Physicochemical aspects of heparin cofactor II, in *Heparin and Related Polysaccharides: Structure and Activities, Annals of the New York Academy of Sciences*, eds., Ofosu, F. A., Danishefsky, I. & Hirsh, J., New York Acad. Sci., New York, Vol. 556, pp. 104-111.

922. Ragg, H., Ulshofer, T., and Gerewitz, J. (1990) Glycosaminoglycan-mediated leuserpin-2/thrombin interaction: structure-function relationships, *J. Biol. Chem.* **265**, 22386-22391.

923. Pratt, C. W., and Church, F. C. (1991) Antithrombin: structure and function, *Semin. Hematol.* **28**, 3-9.

924. Tollefsen, D. M., and Pestka, C. A. (1985) Modulation of heparin cofactor II activity by histidine-rich glycoprotein and platelet factor 4, *J. Clin. Invest.* **75**, 496-501.

925. Colombatti, A., and and Bonaldo, P. (1991) The superfamily of proteins with von Willebrand factor type A-like domains: one theme common to components of extracellular matrix, hemostasis, cellular adhesion, and defense mechanisms, *Blood* **77**, 2305-2315.

926. Sadler, J. E. (1991) von Willebrand factor, *J. Biol. Chem.* **266**, 22777-22780.

927. Ruggeri, Z. M., and Ware, J. (1992) The structure and function of von Willebrand factor, *Thromb. Haemostas.* **67**, 594-599.

928. Ginsburg, D., and Bowie, E. J. W. (1992) Molecular genetics of von Willebrand disease, *Blood* **79**, 25-7-2519.

929. Meyer, D., and Girma, J.-P. (1993) von Willebrand factor: structure and function, *Thromb. Haemostas.* **70**, 99-104.

930. Badimon, L., Badimon, J. J., Chesebro, J. H., and Fuster, V. (1993) von Willebrand factor and cardiovascular disease, *Thromb. Haemostas.* **70**, 111-118.

931. Ruggeri, Z. M., and Ware, J. (1993) von Willebrand factor, *FASEB J.* **7**, 308-316.

932. Sobel, M., McNeill, P. M., Carlson, P. L., Jermode, J. C., Adelman, B., Conroy, R., and Marques, D. (1991) Heparin inhibition of von Willebrand factor-dependent platelet function *in vitro* and *in vivo*, *J. Clin. Invest.* **87**, 1787-1793.

933. Madalas, F., Bell, W. R., and Castaldi, P. A. (1978) Isolation and insolubilisation of human FVIII by affinity chromatography, *Haemostasis* **7**, 321-331.

934. Fowler, W. E., Fretto, L. J., Hamilton, K. K., Erickson, H. P., and McKee, P. A. (1985) Structure of human von Willebrand factor, *J. Clin. Invest.* **76**, 1491-1500.

935. Fujima, Y., Titani, K., Holland, L. Z., Roberts, J. R., Kostel, P., Ruggeri, Z. M., and Zimmerman, T. S. (1987) A heparin binding domain of human von Willebrand factor: characterization and localization to a tryptic peptide extending from amino acid residue Val 449 to Lys 728, *J. Biol. Chem.* **262**, 1734-1739.

936. Girma, J. P., Kalafatis, M., Pietu, G., Lavergne, J. M., Chopex, M. W., Edgington, T. S., and Meyer, D. (1986) Mapping of distinct von Willebrand factor domains interacting with platelet GPIb and GPIIb/IIIa and with collagen using monoclonal antibodies, *Blood* **67**, 1356-1366.

937. Andrews, R. K., Gorman, J. J., Booth, W. J., Corino, G. L., Castaldi, P. A., and Berndt, M. C. (1989) Cross linking of a monomeric 39/34 kDa dispase fragment of von Willebrand factor (Leu 480/Val 481-Gly 718) to the N terminal region of the chain of membrane glycoprotein Ib on intact platelet with bis(sulfosuccinimidyl) suberate, *Biochemistry* **28**, 8326-8336.

938. Mohri, H., Yoshioka, A., Zimmerman, T. S., and Ruggeri, Z. M. (1989) Isolation of the von Willeband factor domain interacting with platelet glycoprotein Ib, heparin, and collagen and characterization of its three distinct functional sites, *J. Biol. Chem.* **264**, 17361-17367.

939. Sobel, M., Soler, D. F., Jermode, J. C., and Harris, R. B. (1992) Localization and characterization of a heparin binding domain peptide of human von Willebrand factor, *J. Biol. Chem.* **267**, 8857-8862.

940. Kroner, P. A., and Frey, A. B. (1996) Analysis of the structure and function of the von Willebrand factor A1 domain using targeted deletions and alanine-scanning mutagenesis, *Biochemistry* **35**, 13460-13468.

941. Jahroudi, N., Ardekani, A. M., and Greenberger, J. S. (1996) An NF1-like protein functions as a repressor of the von Willebrand factor promoter, *J. Biol. Chem.* **271**, 21413-21421.

942. Siedlecki, C. A., Lestini, B. J., Kottke-Marchant, K., Eppell, S. J., Wilson, D. L., and Marchant, R. E. (1996) Shear-dependent changes in the three-dimensional structure of human von Willebrand factor, *Blood* **88**, 2939-2950.

943. Sobel, M., Bird, K., Tyler-Cross, R., Toma, N., Conrad, H. E., and Harris, R. B. (1995) Heparins designed to specifically inhibit platelet interactions with von Willebrand factor, *Circulation* **93**, 992-996.

944. Barrow, R. T., Healey, J. F., and Lollar, P. (1994) Inhibition by heparin of thrombin-catalyzed activation of the factor VIII-von Willebrand factor complex, *J. Biol. Chem.* **269**, 593-598.

945. Pannekoek, H., Vries, C. D., and Van Zonneveld, A.-J. (1988) Mutants of human tissue-type plasminogen activator (t-PA): structural aspects and functional properties, *Fibrinolysis* **2**, 123-132.

946. Lijnen, H. R., and Collen, D. (1991) Strategies for the improvement of thrombolytic agents, *Thromb. Haemostas.* **66**, 88-110.

947. Pennica, D., Holmes, W. E., Kohr, W. J., Harkins, R. N., Vehar, G. A., Ward, C. A., Bennett, W., Yelverton, E., Seeburg, P. H., Heyneker, H. L., Goeddel, D. V., and Collen, D. (1983) Cloning and expression of human tissue-type plasminogen activator cDNA in *E. coli, Nature* **301**, 214-221.

948. Bányai, L., Váradi, A., and Patthy, L. (1983) Common evolutionary origin of fibrin-binding structures of fibronectin and tissue-type plasminogen activator, *FEBS Lett.* **163**, 37-41.

949. Tachias, K., and Madison, E. L. (1996) Converting tissue-type plasminogen activator into a zymogen, *J. Biol. Chem.* **271**, 28749-28752.

950. Andrade-Gordon, P., and Strickland, S. (1986) Interaction of heparin with plasminogen activators and plasminogen: effects on the activation of plasminogen, *Biochemistry* **25**, 4033-4040.

951. Andrade-Gordon, P., and Strickland, S. (1990) Fractionation of heparin by chromatography on a tissue plasminogen activator-Sepharose column, *Proc. Natl. Acad. Sci. U.S.A.* **87**, 1865-1869.

952. Bennett, W. F., Paoni, N. F., Keyt, B. A., Botstein, D., Jones, A. J. S., Presta, L., Wurm, F. M., and Zoller, M. J. (1991) High resolution analysis of functional determinants on human tissue-type plasminogen activator, *J. Biol. Chem.* **266**, 5191-5201.

953. Wilhelm, O. G., Jaskinas, S. R., Vlahos, C. J., and Bang, N. U. (1990) Functional properties of the recombinant kringle-2 domain of tissue plasminogen activator produced in *Escherichia coli, J. Biol. Chem.* **265**, 14606-14611.

954. Collen, D., Lijnen, H. R., Bulens, F., and Vandamme, A.-M. (1990) Biochemical and functional characterization of human tissue-type plasminogen activator variants with mutagenized kringle domains, *J. Biol. Chem.* **265**, 12184-12191.

955. Burck, P. J., Berg, D. H., Warrick, M. W., Berg, D. T., Walls, J. D., Jaskunasa, S. R., Crisel, R. M., Weigel, B., Vlahos, C. J., McClure, D. B., and Grinnell, B. W. (1990) Characterization of a modified human tissue plasminogen activator comprising a kringle-2 and a protease domain, *J. Biol. Chem.* **265**, 5170-5177.

956. Langer-Safer, P. R., Ahern, T. J., Angus, L. B., Barone, K. M., Brenner, M. J., Horgan, P. G., Morris, G. E., Stoudemire, J. B., Timony, G. A., and Larsen, G. R. (1991) Replacement of finger and growth factor domains of tissue plasminogen activator with plasminogen kringle-1, *J. Biol. Chem.* **266**, 3715-3723.

957. Larsen, G. R., Timony, G. A., Horgan, P. G., Barone, K. M., Henson, K. S., Angus, L. B., and Stoudemire, J. B. (1991) Protein engineering of novel plasminogen activators with

increased thrombolytic potency in rabbits relative to activase, *J. Biol. Chem.* **266**, 8156-8161.

958. De Serrano, V. S., Menhart, N., and Castellino, F. J. (1992) Expression, purification, and characterization of the recombinant kringle 1 domain from tissue-type plasminogen activator, *Arch. Biochem. Biophys.* **294**, 282-290.

959. De Serrano, V. S., Sehl, L. C., and Castellino, F. J. (1992) Direct identification of lysine-33 as the principal cationic center of the ω-amino acid binding site of the recombinant kringle 2 domain of tissue-type plasminogen activator, *Arch. Biochem. Biophys.* **292**, 206-212.

960. Stack, M. S., and and Pizzo, S. V. (1993) Modulation of tissue plasminogen activator-catalyzed plasminogen activation by synthetic peptides derived from the amino-terminal heparin binding domain of fibronectin, *J. Biol. Chem.* **268**, 18924-18928.

961. De Serrano, V. S., and Castellino, F. J. (1994) Involvement of tyrosine-76 of the kringle 2 domain of tissue-type plasminogen activator in its thermal stability and its ω-amino acid ligand binding site, *Biochemistry* **33**, 3509-3514.

962. Vlahos, C. J., Wilhelm, O. G., Hassell, T., Jaskunas, S. R., and Bang, N. U. (1991) Disulfide pairing of the recombinant kringle-2 domain of tissue plasminogen activator produced in *Escherichia coli*, *J. Biol. Chem.* **266**, 10070-10072.

963. De Serrano, V. S., and Castellino, F. J. (1994) Role of the strictly conserved tryptophan-25 residue in the stabilization of the structure and in the ligand binding properties of the kringle 2 domain of tissue-type plasminogen activator, *Biochemistry* **33**, 1340-1344.

964. Yamada, T., Shimada, Y., and Kikuchi, M. (1996) Integrin-specific tissue-type plasminogen activator engineered by introduction of the Arg-Gly-Asp sequence, *Biochem. Biophys. Res. Commun.* **228**, 306-311.

965. Stein, P. L., van Zonneveld, A.-J., Pannekoek, H., and Strickland, S. (1989) Structural domains of human tissue-type plasminogen activator that confer stimulation by heparin, *J. Biol. Chem.* **264**, 15441-15444.

966. Edelberg, J. M., and Pizzo, S. V. (1990) Kinetic analysis of the effects of heparin and lipoproteins on tissue plasminogen activator mediated plasminogen activation, *Biochemistry* **29**, 5906-5911.

967. Young, T. N., Edelberg, J. M., Stack, S., and Pizzo, S. V. (1992) Ionic modulation of the effects of heparin on plasminogen activation by tissue plasminogen activator: the effects of ionic strength, divalent cations, and chloride, *Arch. Biochem. Biophys.* **296**, 530-538.

968. Edelberg, J. M., Conrad, H. E., and Pizzo, S. V. (1991) Heparin oligosaccharides enhance tissue-type plasminogen activator: a correlation between oligosaccharides length and stimulation of plasminogen activation, *Biochemistry* **30**, 10999-11003.

969. Andrade-Gordon, P., and Strickland, S. (1989) Anticoagulant low molecular weight heparin does not enhance the activation of plasminogen by tissue plasminogen activator, *J. Biol. Chem.* **264**, 15177-15181.

970. Rijken, D. C., de Munk, G. A. W., and Jie, A. F. H. (1993) Interaction of plasminogen activators and plasminogen with heparin: effect of ionic strength, *Thromb. Haemostas.* **70**, 867-872.

971. Weitz, J. I., Kuint, J., Leslie, B., and Hirsh, J. (1991) Standard and low molecular weight heparin have no effect on tissue plasminogen activator induced plasma clot lysis or fibrinolysis, *Thromb. Haemostas.* **65**, 541-544.

972. Agnelli, C., Pascucci, C., Cosmi, B., and Nenci, G. G. (1990) Effects of therapeutic doses of heparin on thrombolysis with tissue-type plasminogen activator in rabbits, *Blood* **76**, 2030-2036.

973. Rapold, H. J., Lu, H. R., Wu, Z., Nijs, H., and Collen, D. (1991) Requirement of heparin for arterial and venous thrombolysis with recombinant tissue-type plasminogen activator, *Blood* **77**, 1020-1024.

974. Nicolini, F. A., Nichols, W. W., Saldeen, T. G. P., Khan, S., and Mehta, J. L. (1992) Adjunctive therapy with low molecular weight heparin with recombinant tissue-type

plasminogen activator causes sustained reflow in canine coronary thrombosis, *Am. Heart J.* **124**, 280-288.

975. Anonick, P. K., Wolf, B. B., and Gonias, S. L. (1990) Regulation of plasmin, miniplasmin, and streptokinase-plasmin complex by α_2-antiplammin, α_2-macroglobulin, and antithrombin III in the presence of heparin, *Thromb. Res.* **59**, 449-462.

976. Liu, J.-N., Tang, W., Sun, Z.-Y., Kung, W., Pannell, R., Sarmientos, P., and Gurewich, V. (1996) A site-directed mutagenesis of pro-urokinase which substantially reduces its intrinsic activity, *Biochemistry* **35**, 14070-14076.

977. Lijnen, H. R., Li, X.-K., Nelles, L., Hu, M.-H., and Collen, D. (1992) Biochemical properties of recombinant single-chain urokinase-type plasminogen activator mutants with deletion of Asn 2 through Phe 157 and/or substitution of Cys 279 with Ala, *Eur. J. Biochem.* **205**, 701-709.

978. Pâques, E. P., Stöhr, H.-A., and Heimburger, N. (1986) Study on the mechanism of action of heparin and related substances on the fibrolytic system: relationship between plasminogen activators and heparin, *Thromb. Res.* **42**, 797-807.

979. Edelberg, J. M., Weissler, M., and Pizzo, S. V. (1991) Kinetic analysis of the effects of glycosaminoglycans and lipoproteins on urokinase-mediated plasminogen activation, *Biochem. J.* **276**, 785-791.

980. Stephens, R. W., Bokman, A. M., Myöhänen, H. T., Reisberg, T., Tapiovaara, H., Pederson, N., Grøndahl-Hansen, J., Llinás, M., and Vaheri, A. (1992) Heparin binding to the urokinase kringle domain, *Biochemistry* **31**, 7572-7579.

981. Barnathan, E. S., Kuo, A., Rosenfeld, L., Karikó, K., Leski, M., Robbiati, F., Nolli, M. L., Henkin, J., and Cines, D. B. (1990) Interaction of single-chain urokinase-type plasminogen activator with human endothelial cells, *J. Biol. Chem.* **265**, 2865-2872.

982. Plough, M., Rønne, E., Behrendt, N., Jensen, A. L., Blasi, F., and Danø, K. (1991) Cellular receptor for urokinase plasminogen activator: carboxyl-terminal processing and membrane anchoring by glycosyl-phosphatidylinositol, *J. Biol. Chem.* **266**, 1926-1933.

983. Behrendt, N., Ploug, M., Patthy, L., Houen, G., Blasi, F., and Danø, K. (1991) The ligand-binding domain of the cell surface receptor for urokinase-type plasminogen activator, *J. Biol. Chem.* **266**, 7842-7847.

984. Plough, M., and Ellis, V. (1994) Structure-function relationships in the receptor for urokinase-type plasminogen activator: comparison to other members of the Ly-6 family and snake venom α-neurotoxins, *FEBS Lett.* **349**, 163-168.

985. Lee, S. W., Ellis, V., and Bichek, B. A. (1994) Characterization of plasminogen activation by glycosylphosphatidylinositol-anchored urokinase, *J. Biol. Chem.* **269**, 2411-2418.

986. Behrendt, N., Rønne, E., and Danø, K. (1996) Domain interplay in the urokinase receptor: requirement for the third domain in high affinity ligand binding and demonstration of ligand contact sites in distinct receptor domains, *J. Biol. Chem.* **271**, 22885-22894.

987. Yahi, N., Sabatier, J.-M., Nickel, P., Mabrouk, K., Gonzalez-Sacarano, F., and Fantini, J. (1994) Suramin inhibits binding of the V3 region of HIV-1 envelope glycoprotein gp120 to galactosylceramide, the receptor for HIV-1 gp120 on human colon epithelial cells, *J. Biol. Chem.* **269**, 24349-24353.

988. Pepper, M. S., Vassalli, J. D., Wilks, J. W., Schweigeren, L., Orci, L., and Mantesano, R. (1994) Modulaton of bovine microvascular endothelial cell proteolytic properties by inhibitors or angiogenesis, *J. Cell.Biochem.* **55**, 419-434.

989. Behrendt, N., Rønne, E., and Danø, K. (1993) Binding of the urokinase-type plasminogen activator to its cell surface receptor is inhibited by low doses of suramin, *J. Biol. Chem.* **268**, 5985-5989.

990. Ellis, V., Behrendt, N., and Danø, K. (1991) Plasminogen activation by receptor-bound urokinase: a kinetic study with both cell-associated and isolated receptor, *J. Biol. Chem.* **266**, 12752-12758.

991. Sordat, B., Reiter, L., and Cajot, J.-F. (1990) Modulation of the malignant phenotype with the urokinase-type plasminogen activator and the type I plasminogen activator inhibitor, *Cell Diff. Dev.* **32**, 277-286.

992. Hollis, W., and Boyd, D. (1991) Regulation of the urokinase receptor by its plasminogen activator, *Thromb. Haemostas.* **66**, 678-683.

993. Pepper, M. S., Sappino, A.-P., Stöcklin, R., Montesano, R., Orci, L., and Vassalli, J.-D. (1993) Upregulation of urokinase receptor expression on migrating endothelial cells, *J. Cell Biol.* **122**, 673-684.

994. Mignatti, P., Mazzieri, R., and Rifkin, D. B. (1991) Expression of the urokinase receptor in vascular endothelial cells is stimulated by basic fibroblast growth factor, *J. Cell Biol.* **113**, 1193-1201.

995. Del Rosso, M., Fibbi, G., Pucci, M., Bini, G., Grappone, C., and Nolli, M. L. (1991) Modulation of surface-associated urokinase: binding, internalization, delivery to lysosomes, and degradation in human keratinocytes, *Exp. Cell Res.* **193**, 346-355.

996. Giambrone, G. J., and McKeown-Longo, P. J. (1992) Vitronectin regulates the synthesis and localization of urokinase-type plasminogen activator in HT-1080 cells, *J. Biol. Chem.* **267**, 13617-13622.

997. Yebra, M., Parry, G. C. N., Strömblad, S., Mackman, N., Rosenberg, S., Mueller, B. M., and Cheresh, D. A. (1996) Requirement of receptor-bound urokinase-type plasminogen activator for integrin αvβ5-directed cell migration, *J. Biol. Chem.* **271**, 29393-29399.

998. Seiffert, D., Mimuro, J., Schleef, R. R., and Loskutoff, D. J. (1990) Interactions between type 1 plasminogen activator inhibitor, extracellular matrix and vitronectin, *Cell Diff. And Dev.* **32**, 287-292.

999. Schneiderman, J., and Loskutoff, D. J. (1991) Plasminogen activator inhibitors, *Trends Cardiovasc. Med.* **1**, 99-102.

1000. Ehrlich, H. J., Gebbink, R. K., Keijer, K., Linders, M., Preissner, K. T., and Pannekoek, H. (1990) Alteration of serpin specificity by a protein cofactor. Vitronectin endows plasminogen activator inhibitor 1 with thrombin inhibitor properties, *J. Biol. Chem.* **265**, 13029-13035.

1001. Aleshkov, S. B., Fa, M., Karolin, J., Strandberg, L., Johansson, L. B.-Å., Wilczynska, M., and Ny, T. (1996) Biochemical and biophysical studies of reactive center cleaved plasminogen activator inhibitor type 1, *J. Biol. Chem.* **271**, 21231-21238.

1002. Sprang, S. R. (1992) The latent tendencies of PAI-1, *Trends Biochem. Sci.* **17**, 49-50.

1003. Ehrlich, H. J., Gebbink, R. K., Keijor, K., and Pannekoek, H. (1992) Elucidation of structural requirements on plasminogen activator inhibitor 1 for binding to heparin, *J. Biol. Chem.* **267**, 11606-11611.

1004. Declerck, P. J., De Mol, M., Alessi, M.-C., Baudner, S., Pâques, E.-P., Preissner, K. T., Müller-Berghaus, G., and Collen, D. (1988) Purification and characterization of a plasminogen activator inhibitor 1 binding protein from human plasma. Identification as a multimeric form of S protein (vitronectin), *J. Biol. Chem.* **263**, 15454-15461.

1005. Westerhausen Jr., D. R., Hopkins, W. E., and Billadello, J. J. (1991) Multiple transforming growth factor-β-inducible elements regulate expression of the plasminogen activator inhibitor type-1 gene in HepG2 cells, *J. Biol. Chem.* **266**, 1092-1100.

1006. Schleef, R. R., Loskutoff, D. J., and Podor, T. J. (1991) Immunoelectron microscopic localization of type 1 plasminogen activator inhibitor on the surface of activated endothelial cells, *J. Cell Biol.* **113**, 1413-1423.

1007. Reilly, C. F., and McFall, R. C. (1991) Platelet-derived growth factor and transforming growth factor-β regulate plasminogen activator inhibitor-1 synthesis in vascular smooth muscle cells, *J. Biol. Chem.* **266**, 9419-9427.

1008. Pepper, M. S., Ferrara, N., Orci, L., and Montesano, R. (1991) Vascular endothelial growth factor (VEGF) induces plasminogen activators and plasminogen activator inhibitor-1 in microvascular endothelial cells, *Biochem. Biophys. Res. Commun.* **181**, 902-906.

1009. Christ, G., Seiffert, D., Hufnagl, P., Gessl, A., Wojta, J., and Binder, B. R. (1993) Type 1 plasminogen activator inhibitor synthesis of endothelial cells is downregulated by smooth muscle cells, *Blood* **81**, 1277-1283.

1010. Keeton, M. R., Curriden, S. A., van Zonneveld, A.-J., and Lostukoff, D. J. (1991) Identification of regulatory sequences in the type 1 plasminogen activator inhibitor gene responsive to transforming growth factor-β, *J. Biol. Chem.* **266**, 23048-23052.

1011. Konkle, B. A., Kollros, P. R., and Kelly, M. D. (1990) Heparin-binding growth factor-1 modulation of plasminogen activator inhibitor-1 expression: interaction with cAMP and protein kinase C-mediated pathways, *J. Biol. Chem.* **265**, 21867-21873.

1012. Fay, W. P., Eitzman, D. T., Shapiro, A. D., Madison, E. L., and Ginsburg, D. (1994) Platelets inhibit fibrinolysis *in vitro* by both plasminogen activator inhibitor-1-dependent and -independent mechanisms, *Blood* **83**, 351-356.

1013. Maurer, F., and Medcalf, R. L. (1996) Plasminogen activator inhibitor type 2 gene induction by tumor necrosis factor and phorbol ester involves transcriptional and post-transcriptional events, *J. Biol. Chem.* **271**, 26074-26080.

1014. Gombau, L., and Schleef, R. R. (1994) Processing of type 1 plasminogen activator inhibitor (PAI-1) into the regulated secretory pathway, *J. Biol. Chem.* **269**, 3875-3880.

1015. Seiffert, D., and Loskutoff, D. J. (1991) Evidence that type 1 plasminogen activator inhibitor binds to the somatomedin domain of vitronectin, *J. Biol. Chem.* **266**, 2824-2830.

1016. Kost, C., Stüber, W., Ehrlich, H. J., Pannekoek, H., and Preissner, K. T. (1992) Mapping of binding sites for heparin, plasminogen activator inhibitor-1, and plasminogen to vitronectin's heparin-binding region reveals a novel vitronectin-dependent feedback mechanism for the control of plasmin formation, *J. Biol. Chem.* **267**, 12098-12105.

1017. Deng, G., Curriden, S. A., Wang, S., Rosenberg, S., and Loskutoff, D. J. (1996) Is plasminogen activator inhibitor-1 the molecular switch that governs urokinase receptor-mediated cell adhesion and release?, *J. Cell Biol.* **134**, 1563-1571.

1018. Owensby, D. A., Morton, P. A., Wun, T.-C., and Schwartz, A. L. (1991) Binding of plasminogen activator inhibitor type-1 to extracellular matrix of HepG2 cells,, *J. Biol. Chem.* **266**, 4334-4340.

1019. Preissner, K. T. (1990) Specific binding of plasminogen to vitronectin: evidence for a modulatory role of vitronectin on fibrin(ogen)-induced plasmin formation by tissue plasminogen activator, *Biochem. Biophys. Res. Commun.* **168**, 966-971.

1020. Keijer, J., Ehrlich, H. J., Linders, M., Preissner, K. T., and Pannekoek, H. (1991) Vitronectin governs the interaction between plasminogen activator inhibitor 1 and tissue-type plasminogen activator, *J. Biol. Chem.* **266**, 10700-10707.

1021. Camani, C., Bachmann, F., and Kruithof, E. K. O. (1994) The role of plasminogen activator inhibitor type 1 in the clearance of tissue-type plasminogen activator by rat hepatoma cells, *J. Biol. Chem.* **269**, 5770-5775.

1022. Moestrup, S. K., Nielson, S., Andreasen, P., Jørgensen, K. E., Nykjær, A., Røgaard, H., Gliemann, J., and Christensen, E. I. (1993) Epithelial glycoprotein-330 mediates endocytosis of plasminogen activator-plasminogen activator inhibitor type-1 complexes, *J. Biol. Chem.* **268**, 16564-16570.

1023. Li, H., Kuo, A., Kochan, J., Strickland, D., Kariko, K., Barnathan, E. S., and Cines, D. B. (1994) Endocytosis of urokinase-plasminogen activator inhibitor type 1 complexes bound to a chimeric transmembrane urokinase receptor, *J. Biol. Chem.* **269**, 8153-8158.

1024. Stefansson, S., and Lawrence, D. A. (1996) The serpin PAI-1 inhibits cell migration by blocking integrin $\alpha_v\beta_3$ binding to vitronectin, *Nature* **383**, 441-443.

1025. Torr-Brown, S. R., and Sobel, B. E. (1993) Attenuation of thrombolysis by release of plasminogen activator inhibitor type-1 from platelets, *Thromb. Res.* **72**, 413-421.

1026. Hopkins, W. E., Westerhausen, D. R., Fujii, S., Billadollo, J. J., and Sobel, B. E. (1991) Mediators of induction of augmented expression of plasminogen activator inhibitor type-1 in HepG2 cells by platelets, *Thromb. Haemostas.* **66**, 239-245.

1027. Patston, P. A., and Schapira, M. (1994) Low-affinity heparin stimulates the inactivation of plasminogen activator inhibitor-1 by thrombin, *Blood* **84**, 1164-1172.

1028. Edelberg, J. M., Reilly, C. F., and Pizzo, S. V. (1991) The inhibition of tissue type plasminogen activator by plasminogen activator inhibitor-1, *J. Biol. Chem.* **266**, 7488-7493.

1029. Preissner, K. T. (1991) Structure and biological activity of vitronectin, *Annu. Rev. Cell Biol.* **7**, 275-310.

1030. Preissner, K. T. (1995) Vitronectin as a link between protease cascades and cell adhesion in hemostasis, in *Vascular Control of Hemostasis*, ed. van Hinsbergh, V. W. M., Harwood Academic Publishers, United Kingdom, pp. 169-186.

1031. Ehrlich, H. J., Richter, B., von der Ahe, D., and Preissner, K. T. (1993) Primary structure of vitronectins and homology with other proteins, in *Biology of Vitronectins and Their Receptors*, eds., Preissner, K. T., Rosenblatt, S., Wegerhoff, J., Kost, C. & Mosher, D. F., Elsevier, Amsterdam, pp. 59-66.

1032. Seiffert, D., Ciambrone, G., Wagner, N. V., Binder, D. R., and Loskutoff, D. J. (1994) The somatomedin B domain of vitronectin: structural requirements for the binding and stabilization of active type 1 plasminogen activator inhibitor, *J. Biol. Chem.* **269**, 2659-2666.

1033. Suzuki, S., Oldberg, Å., Hayman, E. G., Pierschbacher, M. D., and Ruoslahti, E. (1985) Complete amino acid sequence of human vitronectin deduced from cDNA. Similarity of cell attachment sites in vitronectin and fibronectin, *EMBO J.* **4**, 2519-2524.

1034. Jenne, D., and Stanley, K. K. (1985) Molecular cloning of the S protein, a link between complement, coagulation and cell-substrate adhesion, *EMBO J.* **4**, 3153-3557.

1035. Sato, R., Komine, Y., Imanaka, T., and Takano, T. J. (1990) Monoclonal antibody EMR Ia/212D recognizing site of deposition of extracellular lipid in atherosclerosis: isolation and characterization of a cDNA clone for the antigen, *J. Biol. Chem.* **265**, 21232-21236.

1036. Stockmann, S., Hess, P., Declerck, R., Timple, and Preissner, K. T. (1993) Multimeric vitronectin. Identification and characterization of conformation-dependent self-association of the adhesive protein, *J. Biol. Chem.* **268**, 22874-22882.

1037. Tschopp, J., Masson, D., Schäfer, S., Peitsch, M., and Preissner, K. T. (1988) The heparin binding domain of S-protein/vitronectin binds to complement components C7, C8, and C9 and perforin from cytolytic T-cells and inhibits their activities, *Biochemistry* **27**, 4103-4109.

1038. Suzuki, S., Pierschbacher, M. D., Hayman, E. G., Nguyen, K., Öhgren, Y., and Rouslahti, E. (1984) Domain structure of vitronectin. Alignment of active sites, *J. Biol. Chem.* **259**, 1507-15314.

1039. Lane, D. A., Flynn, A. M., Pejler, G. G., Lindahl, U., Choay, J., and Preissner, K. (1987) Structural requirements for the neutralization of heparin-like saccharides by complement S protein/vitronectin, *J. Biol. Chem.* **262**, 16343-16348.

1040. Seiffert, D., and Loskutoff, D. J. (1996) Type 1 plasminogen activator inhibitor induces multimerization of plasma vitronectin. A suggested mechanism for the generation of the tissue form of vitronectin *in vivo*, *J. Biol. Chem.* **271**, 29644-29651.

1041. Høgåsen, J., Mollnes, T. E., and Harboe, M. (1992) Heparin-binding properties of vitronectin are linked to complex formation as illustrated by in vitro polymerization and binding to the terminal complement complex, *J. Biol. Chem.* **267**, 23076-23082.

1042. de Boer, H. C., de Groot, P. G., Bouma, B. N., and Preissner, K. T. (1993) Ternary vitronectin-thrombin/antithrombin III complexes in human plasma: detection and mode of association, *J. Biol. Chem.* **268**, 1279-1283.

1043. de Boer, H. C., Preissner, K. T., Bouma, B. N., and de Groot, P. G. (1992) Binding of vitronectin-thrombin/antithrombin III complex to human endothelial cells is mediated by the heparin binding site of vitronectin, *J. Biol. Chem.* **267**, 2264-2268.

1044. Völker, W., Hess, S., Vischer, P., and Preissner, K. T. (1993) Binding and processing of multimeric vitronectin by vascular endothelial cells, *J. Histochem. Cytochem.* **41**, 1823-1832.

1045. Lim, B.-L., Reid, K. B. M., Ghebrehiwet, B., Peerschke, E. I. B., Leigh, L. A. E., and Preissner, K. T. (1996) The binding protein for globular heads of complement C1q, gC1qR: functional expression and characterization as a novel vitronectin binding factor, *J. Biol. Chem.* **271**, 26739-26744.

1046. Heeb, M. J., España, F., Geiger, M., Collen, D., Stump, D., and Griffin, J. H. (1987) Immunological identity of heparin-dependent plasma and urinary protein C inhibitor and plasminogen activator inhibitor-3, *J. Biol. Chem.* **262**, 15813-15816.

1047. Eldering, E., Verpy, E., Roem, D., Meo, T., and Tosi, M. (1995) COOH-substitutions in the serpin C1 inhibitor that cause loop overinsertion and subsequent multimerization, *J. Biol. Chem.* , 2579-2587.

1048. Pratt, C. W., and Church, F. C. (1992) Heparin binding to protein C inhibitor, *J. Biol. Chem.* **267**, 8789-8794.

1049. St. Charles, R., Walz, D. A., and Edwards, B. F. P. (1989) The three-dimensional structure of bovine platelet factor 4 at 3.0-Å resolution, *J. Biol. Chem.* **264**, 2092-2099.

1050. Griffin, J. H., Evatt, B., Zimmerman, T. S., Kleiss, A. J., and Wideman, C. (1981) Deficiency of protein C in congenital thrombotic disease, *J. Clin. Invest.* **68**, 1370-1373.

1051. Bertina, R. M., Broekmans, A. W., van der Linden, I. K., and Mertens, K. (1982) Protein C deficiency in a Dutch family with thrombotic disease, *Thromb. Haemostas.* **48**, 1-5.

1052. Branson, H., Katz, J., Marble, R., and Griffin, J. H. (1983) Inherited protein C deficiency and a coumarin-responsive chronic relapsing purpura fulminans syndrome in a neonate, *Lancet* **2**, 1165-1168.

1053. Estellés, A., Garcia-Plaza, I., Dasi, A., Aznar, A., Duart, M., Sanz, G., Requejo, J. L., España, F., Jimenez, C., and Abeledo, G. (1984) Severe inherited "homozygous" protein C deficiency in a newborn infant, *Thromb. Haemostas.* **52**, 53-56.

1054. Geiger, M., Priglinger, U. G., J. H., and Binder, B. R. (1991) Urinary protein C inhibitor: glycosaminoglycans synthesized by the epithelial kidney cell line TCL-598 enhance its interaction with urokinase, *J. Biol. Chem.* **266**, 11851-11857.

1055. España, F., Gilabert, J., Estellés, A., Romeu, A., Aznar, J., and Cabo, A. (1991) Functionally active protein C inhibitor/plasminogen activator inhibitor-3 (APC inhibitor/PAI-3) is secreted in seminal vesicles, occurs at high concentrations in human seminal plasma and complexes with prostate-specific antigen, *Thromb. Res.* **64**, 309-320.

1056. Lindahl, A. K., Sandset, P. M., and Abildgaard, U. (1992) The present status of tissue factor pathway inhibitor, *Blood Coag. Fibrinol.* **3**, 439-449.

1057. Pedersen, A. H., Nordfang, O., Norris, F., Wiberg, F. C., Christensen, P. W., Moeller, K. B., Meidahl-Pedersen, J., Beck, T. C., Norris, K., Hedner, U., and Kisiel, W. (1990) Recombinant human extrinsic pathway inhibitor: production, isolation, and characterization of its inhibitory activity on tissue factor-initiated coagulation reactions, *J. Biol. Chem.* **265**, 16786-16793.

1058. Bode, W., and Huber, R. (1992) Natural proteinase inhibitors and their interactions with proteinases, *Eur. J. Biochem.* **204**, 433-451.

1059. Laskowski Jr., M., and Kato, I. (1980) Protein inhibitors of proteinases, *Annu. Rev. Biochem.* **49**, 593-626.

1060. Broze Jr, G. J., Warren, L. A., Novotny, W. F., Higuchi, D. A., Girard, J. J., and Miletich, J. P. (1988) The lipoprotein-associated coagulation inhibitor that inhibits the factor VII/tissue factor complex also inhibits factor Xa: insight into its possible mechanism of action, *Blood* **71**, 335-343.

1061. Wun, T.-Z. (1992) Lipoprotein-associated coagulation inhibitor (LACI) is a cofactor for heparin: synergistic anticoagulant action between LACI and sulfated polysaccharides, *Blood* **79**, 430-438.

1062. Valitin, S., Larnkjoer, A., Østergaard, P., Nielsen, J. I., and Nordfang, O. (1994) Characterization of the binding between tissue factor pathway inhibitor and glycosaminoglycans, *Thromb. Res.* **75**, 173-183.

1063. Kamikubo, Y., Hamuro, T., Matsuda, J., Shinya, N., Miyamoto, S., Funatsu, A., and Kato, H. (1996) Anthithrombotic effect of human recombinant tissue factor pathway inhibitor on endotoxin-induced intravascular coagulation in rats: concerted effect with antithrombin, *Thromb. Haemostas.* **76**, 621-626.

1064. Girard, T. J., Warren, L. A., Novotny, W. F., Likert, T. M., Brown, S. G., Miletich, P., and Broze, G. J. (1989) Functional significance of Kunitz-type inhibitor domains of lipoprotein-associated coagulation inhibitor, *Nature* **338**, 518-520.

1065. Sandset, P. M., Abildgaard, U., and Larsen, M. L. (1988) Heparin induces release of extrinsic pathway inhibitor (EPI), *Thromb. Res.* **50**, 803-813.

1066. Lindahl, A. K., Abildgaard, U., and Stokke, G. (1990) Extrinsic pathway inhibitor after heparin injection: increased response in cancer patients, *Thromb. Res.* **59**, 651-656.

1067. Novotny, W. F., Palmier, M., Wun, T.-C., Broze Jr., G. J., and Miletich, J. P. (1991) Purification and properties of heparin releasable lipoprotein-associated inhibitor, *Blood* **78**, 394-400.

1068. Werling, R. W., Zacharski, L. R., Kisiel, W., Bajaj, S. P., Memoli, V. A., and Rousseau, S. M. (1993) Distribution of tissue factor pathway inhibitor in normal and pathological human tissues, *Thromb. Haemostas.* **69**, 366-369.

1069. Nordfang, O., Bjørn, S. E., Valentn, S., Nielsen, L. S., Wildgoos, P., Beck, T. C., and Hedner, U. (1991) The C-terminus of tissue factor pathway inhibitor is essential to its anticoagulant activity, *Biochemistry* **30**, 10371-10376.

1070. Lindahl, A. K., Abildgaard, U., Larsen, M. L. L., Aamodt, L. M., Nordfang, E., and Beck, T. C. (1991) Extrinsic pathway inhibitor (EPI) and the postheparin anticoagulant effect in tissue thromboplastin induced coagulation, *Thromb. Res. (Suppl. XIV)* **62**, 39-48.

1071. Lindahl, A. K., Abildgaard, U., and Staalesen, R. (1991) The anticoagulant effect in heparinized blood and plasma resulting from interactions with extrinsic pathway inhibitor, *Thromb. Res.* **64**, 155-168.

1072. Jesty, J., Wun, T.-C., and Lorenz, A. (1994) Kinetics of the inhibition of factor Xa and the tissue factor-factor VIIa complex by the tissue factor pathway inhibitor in the presence and absence of heparin, *Biochemistry* **33**, 12686-12694.

1073. Nordfang, O., Kristensen, H. I., Valentin, S., Østergaard, P., and Wadt, J. (1993) The significance of TFPI in clotting assays: comparison and combination with other anticoagulants, *Thromb. Haemostas.* **70**, 448-453.

1074. Warn-Cramer, J., Maki, S. L., and Rapaport, S. I. (1993) Heparin-releasable and platelet pools of tissue factor pathway inhibitor in rabbits, *Thromb. Haemostas.* **69**, 221-226.

1075. Zhang, X., Chen, L., Bancroft, D. P., Lai, C. K., and Maione, T. E. (1994) Crystal structure of recombinant human platelet factor 4, *Biochemistry* **33**, 8361-8366.

1076. Moore, S., Pepper, D. S., and Cash, J. D. (1975) Platelet antiheparin activity: the isolation and characterization of platelet factor 4 released from thrombin aggregated washed human platelets and its dissociation into subunits and the isolation of membrane-bound antiheparin activity, *Biochim. Biophys. Acta* **379**, 379-384.

1077. Bock, P. E., Luscombe, M., Marshall, S. E., Pepper, D. S., and Holbrook, J. J. (1980) The multiple complexes formed by interacton of platelet factor 4 with heparin , *Biochem. J.* **191**, 769-776.

1078. Levine, S. P., and Wohl, H. (1976) Human platelet factor 4: purification and characterization by affinity chromatography: purification of human platelet factor 4, *J. Biol. Chem.* **251**, 324-328.

1079. Zucker, M. B., and and Katz, I. R. (1991) Platelet factor 4: production, structure, and physiologic and immunologic action, *Proc. Soc. Exp. Biol. Med.* **198**, 693-702.

1080. Rucinski, B., Niewiarowski, S., Stryzewski, M., Holt, J. C., and Mayo, K. H. (1990) Human platelet factor 4 and its C-terminal peptides: heparin binding and clearance from the circulation, *Thromb. Haemostas.* **63**, 493-498.

1081. Ibel, K., Poland, G. A., Baldwin, J. P., Pepper, D. S., Luscombe, M., and Holbrook, J. J. (1986) Low resolution structure of the complex of human blood platelet factor 4 with heparin determined by small angle neutron scattering, *Biochim. Biophys. Acta* **870**, 58-63.

1082. Talpas, C. J., Walz, D. A., and Lee, L. (1991) ^{1}H-NMR studies of bovine platelet factor 4: histidine assignments and interactions with heparin, *Biochim. Biophys. Acta* **1078**, 208-218.

1083. Maccarana, M., and Lindahl, U. (1993) Mode of interaction between platelet factor 4 and heparin, *Glycobiology* 3, 271-277.
1084. Barber, A. J., Kaser-Glanzmann, R., Jajabova, M., and Lüscher, E. F. (1972) Characterization of chondroitin 4-sulfate proteoglycan carrier for heparin-neutralizing activity (platelet factor 4) released from human blood platelets, *Biochim. Biophys. Acta* . 286, 312-329.
1085. Levine, S. P., Knieriem, L. K., and Rager, M. A. (1990) Platelet factor 4 and the platelet secreted proteoglycan: immunologic characterization by crossed immunoelectrophoresis, *Blood* 75, 902-910.
1086. Rucinski, B., Knight, L. C., and Niewiarowski, S. (1986) Clearance of human platelet factor 4 by liver and kidney: its alteration by heparin, *Am. J. Physiol.* 251, H800-H807.
1087. Ravanat, C., Gachet, C., Herbert, J.-M., Schuhler, S., Guillemot, J.-C., Uzabiaga, F., Picard, C., Ferrara, P., Freund, M., and Cazenave, J.-P. (1994) Rat platelets contain glycosylated and non-glycosylated forms of platelet factor 4, *Eur. J. Biochem.* 223, 203-210.
1088. Eitzman, D. T., Chi, L., Saggin, L., Schwartz, R. S., Lucchesi, B. R., and Fay, W. P. (1994) Heparin neutralization by platelet-rich thrombi; role of platelet factor 4, *Circulation* 89, 1523-1529.
1089. Dehmer, G. J., Fisher, M., Tate, D. A., Teo, S., and Bonnem, E. M. (1995) Reversal of heparin anticoagulation by recombinant platelet factor 4 in humans, *Circulation* 91, 2188-2194.
1090. Dehmer, G. J., Lange, R. A., Tate, D. A., Pirwitz, M., Daniel, W., Fisher, M., and Bonnem, E. M. (1996) Randomized trial of recombinant platelet factor 4 versus protamine for the reversal of heparin anticoagulation in humans, *Circulation* 94, I-347-II-352.
1091. Jordan, R. E., Favreau, L. V., Braswell, E. H., and Rosenberg, R. D. (1982) Heparin with two binding sites for antithrombin or platelet factor 4, *J. Biol. Chem.* 257, 400-406.
1092. Lijnen, H. R., and Collen, D. (1989) Interaction of heparin with histidine-rich glycoprotein, in *Heparin and Related Polysaccharides: Structure and Activities, Annals of the New York Academy of Sciences*, eds., Ofosu, F. A., Danishefsky, I. & Hirsh, J., New York Acad. Sci., New York, Vol. 556, pp. 181-185.
1093. Lijnen, H. R., Hoylaerts, M., and Collen, D. (1983) Heparin binding properties of human histidine-rich glycoprotein: mechanism and role in the neutralization of heparin in plasma, *J. Biol. Chem.* 258, 3803-3808.
1094. Joide, T., Foster, D., Yoshitake, S., and Davie, E. W. (1986) Amino acid sequence of human histidine-rich glycoprotein derived from the nucleotide sequence of its cDNA, *Biochemistry* 25, 2220-2225.
1095. Koide, T., Odani, S., and Ono, T. (1982) The N-terminal sequence of human plasma histidine-rich proteoglycan homologous to antithrombin with high affinity for heparin, *Fed. Eur. Biochem. Soc. Letts.* 14, 222-224.
1096. Lijnen, R., Rylatt, D. B., and Collen, D. (1983) Physicochemical, immunochemical and functional comparison of human histidine-rich glycoprotein and autorosette inhibition factor, *Biochim. Biophys. Acta* 742, 109-115.
1097. Leung, K. L., Harpel, P. C., Nachman, R. L., and Rabellino, E. M. (1983) Histidine-rich glycoprotein is present in human platelets and is released following thrombin stimulation, *Blood* 62, 1016-1020.
1098. Smith, A., Tatum, A. J., and Morgan, W. T. (1989) Histidine-rich glycoprotein is synthesized by liver and is a negative acute-phase response protein, *J. Cell Biol.* 107, 584 (abst. 3308).
1099. Hennis, C., De Maat, M. P. M., Quax, P. H. A., Le Clercq, E. J., Kuiper, J., and Kluft, C. (1991) Evaluation of sites of synthesis of the histidine-rich glycoprotein, *Thromb. Haemostas.* 65, 884 (abst. 660) .
1100. Lijnen, H. R., Hoylaerts, M., and Collen, D. (1980) Isolation and characterization of a human plasma protein with affinity for the lysine binding sites in plasminogen: role in the

regulation of fibrinolysis and identification as histidine-rich glycoprotein, *J. Biol. Chem.* **255**, 10214-10222.

1101. Leung, L. K. L., Nachman, R. L., and Harpel, P. C. (1984) Complex formation of platelet thrombospondin with histidine-rich glycoprotein, *J. Clin. Invest.* **73**, 5-12.

1102. Leung, L. K. L. (1986) Interaction of histidine-rich glycoprotein with fibrinogen and fibrin, *J. Clin. Invest.* **77**, 1305-1311.

1103. Lerch, G., Nydeggeer, U. E., Kuyas, C., and Haeberli, A. (1988) Histidine-rich glycoprotein binding to activated human platelets, *Br. J. Haematol.* **70**, 219-224.

1104. Hoffman, J. M. L., Hennis, B. C., Kluft, C., and Vijgen, M. (1993) Hereditary increase of plasma histidine-rich glycoprotein associated with abnormal heparin binding (HRG Eindhoven), *Thromb. Haemostas.* **70**, 894-899.

1105. Bornstein, P. (1992) Thrombospondins: structure and regulation of expression, *FASEB J.* **6**, 3290-3299.

1106. Mosher, D. F. (1990) Physiology of thrombospondin, *Annu. Rev. Med.* **41**, 85-97.

1107. Tuszynski, G. P., Srivastava, S., Switalska, H. I., Holt, J. C., Cierniewski, C. S., and Niewiarowski, S. (1985) The interaction of human platelet thrombospondin with fibrinogen; purification and specificity of interaction, *J. Biol. Chem.* **260**, 12240-12245.

1108. Bornstein, P., O'Rourke, K., Wikstrom, K., Wolf, F. W., Katz, R., Li, P., and Dixit, V. M. (1991) A second, expressed thrombospondin gene (Thbs2) exists in the mouse genome, *J. Biol. Chem.* **266**, 12821-12824.

1109. Vos, H. L., Devarayalu, S., de Vries, Y., and Bornstein, P. (1992) Thrombospondin 3 (Thbs3), a new member of the thrombospondin gene family, *J. Biol. Chem.* **267**, 12192-12196.

1110. Lawler, J., and Hynes, R. O. (1986) The structure of human thrombospondin, an adhesive glycoprotein with multiple calcium-binding sites and homologies with several different proteins, *J. Cell Biol.* **103**, 1635-1648.

1111. Dixit, V. M., Hennessy, S. W., Grant, G. A., Rotwoin, P., and Frazier, W. A. (1986) Characterization of a cDNA encoding the heparin and collagen binding domains of human thrombospondin, *Proc. Natl. Acad, Sci. U.S.A.* **83**, 5449-5453.

1112. Kobayashi, S., Eden-McCutchan, F., Framson, P., and Bornstein, P. (1986) Partial amino acid sequence of human thrombospondin as determined by analysis of cDNA clones: homology to malarial circumsporozoite proteins, *Biochemistry* **25**, 8418-8425.

1113. Lawler, J., Ferro, P., and Duquette, M. (1992) Expression and mutagenesis of thrombospondin, *Biochemistry* **31**, 1173-1180.

1114. Dixit, V. M., Grant, G. A., Santoro, S. A., and Frazier, W. A. (1984) Isolation and characterization of a heparin-binding domain from the amino terminus of platelet thrombospondin, *J. Biol. Chem.* **259**, 10100-10105.

1115. Lawler, J., and Hynes, R. O. (1987) The structural organization of the thrombospondin molecule, *Semin. Thromb. Hemostas.* **13**, 245-254.

1116. Lawler, J. (1981) Prediction of secondary structure of platelet factor 4 and β-thromboglobulin from their amino acid sequences, *Thromb. Res.* **21**, 121-127.

1117. Incardona, F., Lawler, J., Cataldo, D., Panet, A., Legrand, Y., Foidart, J. M., and Legrand, C. (1996) Heparin-binding domain, type 1 and type 2 repeats of thrombospondin mediate its interaction human breast cancer cells, *J. Cell. Biochem.* **62**, 431-442.

1118. Guo, N.-H., Krutzsch, H. C., Nègre, E., Vogel, T., Blake, D. A., and Roberts, D. D. (1992) Heparin- and sulfatide-binding peptides from the type I repeats of human thrombospondin promote melanoma cell adhesion, *Proc. Natl. Acad. Sci. U.S.A.* **89**, 3040-3044.

1119. Guo, N.-H., Krutzsch, H. C., Nègre, E., Zabrenestzky, V. S., and Roberts, D. D. (1992) Heparin-binding peptides from the type I repeats of thrombospondin: structural requirements for heparin binding and promotion of melanoma cell adhesion and chemotaxis, *J. Biol. Chem.* **267**, 19349-19355.

1120. Vogel, T., Guo, N.-H., Krutzsch, H. C., Blake, D. A., Hartman, J., Mendelovitz, S., Panet, A., and Roberts, D. D. (1993) Modulation of endothelial cell proliferation, adhesion, and motility by recombinant heparin-binding domain and synthetic peptides from the type 1 repeats of thrombospondin, *J. Cell. Biochem.* **53**, 74-84.

1121. Mumby, S. M., Raugi, G. J., and Bornstein, P. (1984) Interactions of thrombospondin with extracellular matrix proteins: selective binding to type V collagen, *J. Cell Biol.* **98**, 646-652.

1122. Lahav, J., Lawler, J., and Gimbrone, M. A. (1984) Thrombospondin interactions with fibronectin and fibrinogen: mutual inhibition in binding, *Eur. J. Biochem.* **145**, 151-156.

1123. Lawler, J. (1986) The structural and functional properties of thrombospondin, *Blood* **67**, 1197-1209.

1124. Frazier, W. A. (1987) Thrombospondin: a modular adhesive glycoprotein of platelets and nucleated cells, *J. Cell Biol.* **105**, 625-632.

1125. Galvin, N. J., Vance, P. M., Dixit, V. M., Fink, B., and Frazier, W. A. (1987) Interaction of human thrombospondin with types I-IV collagen: direct binding and electron microscopy, *J. Cell Biol.* **104**, 1413-1422.

1126. Roberts, D. D. (1988) Interactions of thrombospondin with sulfated glycolipids and proteoglycans of human melanoma cells, *Cancer Res.* **48**, 6785-6793.

1127. Hogg, P. J., Stenflo, J., and Mosher, D. F. (1992) Thrombospondin is a slow tight-binding inhibitor of plasmin, *Biochemistry* **31**, 265-269.

1128. Dixit, V. M., Galvin, N. J., O'Rourke, K. M., and Frazier, W. A. (1986) Monoclonal antibodies that recognize calcium-dependent structures of human thrombospondin: characterization and mapping of their epitopes, *J. Biol. Chem.* **261**, 1962-1968.

1129. Silverstein, R. I., Leung, L. L. K., and Nachman, R. L. (1986) Thrombospondin: a versatile multifunctional glycoprotein, *Arteriosclerosis* **6**, 245-253.

1130. Roberts, D. D. (1996) Regulation of tumor growth and metastasis by thrombospondin-1, *FASEB J.* **10**, 1183-1191.

1131. Good, D. J., Polverine, P. J., Rastinejad, F., LeBeau, M. M., Lemons, R. S., Frazier, W. A., and Bouck, N. P. (1990) A tumor suppressor-dependent inhibitor of angiogenesis is immunologically and functionally indistinguishable from a fragment of thrombospondin, *Proc. Natl. Acad. Sci. U.S.A.* **87**, 6624-6628 .

1132. Iruela-Arispe, M. L., Bornstein, P., and Sage, H. (1991) Thrombospondin exerts an antiangiogenic effect on cord formation by endothelial cells *in vitro*, *Proc. Natl. Acad. Sci. U.S.A.* **88**, 5026-5030.

1133. Nicosia, R. F., and Tuszynski, G. P. (1994) Matrix-bound thrombospondin promotes angiogenesis *in vitro*, *J. Cell Biol.* **124**, 183-193.

1134. Taraboletti, G., Roberts, D. D., Liotta, L. A., and Giavazzi, R. (1990) Platelet thrombospondin modulates endothelial cell adhesion, motility, and growth: a potential angiogenesis regulatory factor, *J. Cell Biol.* **111**, 765-772.

1135. Murphy-Ullrich, J. E., and Höök, M. (1989) Thrombospondin modulates focal adhesions in endothelial cells, *J. Cell Biol.* **109**, 1309-1319.

1136. Murphy-Ullrich, J. E., Gurusiddappa, S., Frazier, W. A., and Höök, M. (1993) Heparin-binding peptides from thrombospondins 1 and 2 contain focal adhesion-labilizing activity, *J. Biol. Chem.* **268**, 26784-26789.

1137. Kaesberg, P. R., Ershler, W. B., Esko, J. D., and Mosher, D. F. (1989) Chinese hamster ovary cell adhesion to human platelet thrombospondin is dependent on cell surface heparan sulfate proteoglycan, *J. Clin. Invest.* **83**, 994-1001.

1138. Vischer, P., Beeck, H., and Voss, B. (1985) Synthesis, intracellular processing and secretion of thrombospondin in human endothelial cells, *Eur. J. Biochem.* **153**, 435-443.

1139. Mikhailenko, I., Kounnas, M. Z., and Strickland, D. K. (1995) Low density lipoprotein receptor-related protein/α_2-macroglobulin receptor mediates the cellular internalization and degradation of thrombospondin, *J. Biol. Chem.* **270**, 9543-9549.

1140. Völker, W., Shön, P., and Vischer, P. (1991) Binding and endocytosis of thrombospondin and thrombospondin fragments in endothelial cell cultures analyzed by cuprolinic blue staining, colloidal gold labeling, and silver enhancement techniques, *J. Histochem. Cytochem.* **39**, 1385-1394.

1141. Shön, P., Vischer, P., Völker, W., Schmidt, A., and Faber, V. (1992) Cell-associated proteoheparin sulfate mediates binding and uptake of thrombospondin in cultured porcine vascular endothelial cells, *Eur. J. Cell Biol.* **59,** 329-339.

1142. Munjal, I. D., Blake, D. A., Sabet, M. D., and Gordon, S. R. (1990) Thrombospondin: biosynthesis, distribution, and changes associated with wound repair in corneal endothelium, *J. Cell Biol.* **52,** 252-263.

1143. Murphy-Ulrich, J. E., and Mosher, D. F. (1985) Localization of thrombospondin in clots formed *in situ*, *Blood* **66,** 1098-1104.

1144. Raugi, G. J., Olerud, J. E., and Gown, A. M. (1987) Thrombospondin in early human wound tissue, *J. Invest. Dermatol.* **89,** 551-554.

1145. Kehrel, B., Kronenberg, A., Schwippert, B., Niesing-Bresch, D., Niehues, U., Tschöpe, D., van de Loo, J., and Clemetson, K. J. (1991) Thrombospondin binds normally to glycoprotein IIIb deficient platelets, *Biochem. Biophys. Res. Commun.* **179,** 985-991.

1146. Catimel, B., Leung, L., el Ghissasi, H., Mercier, N., and McGregor, J. (1992) Human platelet glycoprotein IIIb binds to thrombospondin fragments bearing the C-terminal region, and/or the type I repeats (CSVTCG motif), but not to the N-terminal heparin-binding region, *Biochem. J.* **284,** 231-236.

1147. Agbanyo, R., Sixma, J. J., de Groot, P. G., Languino, L. R., and Plow, E. F. (1993) Thrombospondin-platelet interactions, *J. Clin. Invest.* **92,** 288-296.

1148. Harpel, P. C., Silverstein, R. L., Pannell, R., Gurewich, V., and Nachman, R. L. (1990) Thrombospondin forms complexes with single-chain and two-chain forms of urokinase, *J. Biol. Chem.* **265,** 11289-11294.

1149. Chang, A. C., and Detwiler, T. C. (1992) Reactions of thrombin-serpin complexes with thrombospondin, *Arch. Biochem. Biophys.* **299,** 100-104.

1150. Tanaka, A., Miyamoto, K., Minamino, N., and Takeda, M. (1992) Cloning and characterization of an androgen-induced growth factor essential for the androgen-dependent growth of mouse mammary carcinoma cells, *Proc. Natl. Acad. Sci. U.S.A.* **89,** 8928-8932.

1151. Ohuchi, H., Yosioka, H., Tanaka, A., Kawakami, Y., Nohno, T., and Soji, S. (1994) Involvement of androgen-induced growth factor (FGF-8) in mouse embryogenesis and morphogenesis, *Biochem. Biophys. Res. Commun.* **204,** 882-888.

1152. Miyamoto, M., Naruo, K. H., Seko, C., Matsumoto, S., Kondo, T., and Kurokawa, T. (1993) Molecular cloning of a novel cytokine cDNA encoding the ninth member of the fibroblast growth factor family, which has unique secretion property, *Mol. Cell Biol.* **13,** 4251-4259.

1153. Santos-Ocampo, S., Colvin, J. S., Chellaiah, A., and Ornitz, D. M. (1996) Expression and biological activity of mouse fibroblast growth factor-9, *J. Biol. Chem.* **271,** 1726-1731.

1154. Lobb, R. R., Harper, J. W., and Fett, J. W. (1986) Purification of heparin-binding growth factors, *Anal. Biochem.* **154,** 1-14.

1155. Ullrich, S., Lagente, O., Lenfant, M., and Courtois, Y. (1986) Effect of heparin on the stimulation of non-vascular cells by human acidic and basic FGF, *Biochem. Biophys. Res. Commun.* **137,** 1205-1213.

1156. Gospodarowicz, D. (1991) Biological activities of fibroblast growth factors, in *The Fibroblast Growth Factor Family, Annals of the New York Academy of Sciences*, eds., Baird, A. & Klagsbrun, M., New York Academy of Sciences, New York, Vol. 638, pp. 1-8.

1157. Hondermarck, H., Deudon, E., and Boilly, B. (1992) Embryonic brain-derived heparan sulfate inhibits cellular membrane binding and biological activity of basic fibroblast growth factor, *Dev. Brain Res.* **68,** 247-253.

1158. Nugent, M. A., Karnovsky, M. J., and Edelman, E. R. (1993) Vascular cell-derived heparan sulfate shows coupled inhibition of basic fibroblast growth factor binding and mitogenesis in vascular smooth muscle cells, *Circ. Res.* **73,** 1051-1060.

1159. Mali, M., Elenius, K., Miettinen, H. M., and Jalkanen, M. (1993) Inhibition of basic fibroblast growth factor-induced growth promotion by overexpression of syndecan-1, *J. Biol. Chem.* **268**, 24215-24222.

1160. Krufka, A., Guimond, O., and Rapraeger, A. (1996) Two heirarchies of FGF-2 signaling in heparin: mitogenic stimulation and high-affinity binding/receptor transphosphorylation, *Biochemistry* **35**, 11131-11141.

1161. Burgess, W. H., and Maciag, T. (1989) The heparin-binding (fibroblast) growth factor family of proteins, *Annu. Rev. Biochem.* **58**, 575-606.

1162. Goldfarb, M. (1990) The fibroblast growth factor family, *Cell Growth Differ.* **1**, 439-445.

1163. Baird, A., and Klagsbrun, M. (1991) The fibroblast growth factor family: an overview, in *The Fibroblast Growth Factor Family, Annals of the New York Academy of Sciences*, eds., Baird, A. & Klagsbrun, M., New York Academy of Sciences, New York, Vol. 638, pp. xiii-xvi.

1164. Friesel, R. E., and Maciag, T. (1995) Molecular mechanisms of angiogenesis: fibroblast growth factor signal transduction, *FASEB J.* **9**, 919-925.

1165. Gospodarowicz, D., and Cheng, J. (1986) Heparin protects basic and acidic FGF from inactivation, *J. Cell. Physiol.* **128**, 475-484.

1166. Copeland, R. A., Ji, H., Halfpenny, A. J., Williams, R. W., Thompson, K. C., Herber, W. K., Thomas, K. A., Bruner, M. W., Ryan, J. A., Marquis-Omer, D., Sanyal, G., Sitrin, R. D., Yamazaki, S., and Middaugh, C. R. (1991) The structure of human acidic fibroblast growth factor and its interaction with heparin, *Arch. Biochem. Biophys.* **289**, 53-61.

1167. Lobb, R. R. (1988) Thrombin inactivates acidic fibroblast growth factor but not basic fibroblast growth factor, *Biochemistry* **27**, 2572-2578.

1168. Damon, D. H., Lobb, R. R., D'Amore, P. A., and Wagner, J. A. (1989) Heparin potentiates the action of acidic fibroblast growth factor by prolonging its biological half-life, *J. Cell. Physiol.* **138**, 221-226.

1169. Shing, Y., Folkman, J., Sullivan, R., Butterfield, C., Murray, J., and Klagsbrun, M. (1984) Heparin affinity: purification of a tumor-derived capillary endothelial cell growth factor, *Science* **223**, 1296-1299.

1170. Jacquot-Dourges, M. A., Zhow, F. L., Muller, D., and Josefonvicz, J. (1991) Affinity chromatography of fibroblast growth factors on coated silica supports grafted with heparin, *J. Chromatog.* **539**, 417-424.

1171. Klagsbrun, M., Smith, S., Sullivan, R., Shing, Y., Davidson, S., Smith, J. A., and Sasse, J. (1987) Multiple forms of basic fibroblast growth factor: amino-terminal cleavages by tumor cell- and brain cell-derived acid proteinases, *Proc. Natl. Acad. Sci. U.S.A.* **84**, 1839-1843.

1172. Klagsbrun, M., Sasse, J., Sullivan, R., and Smith, J. A. (1986) Human tumor cells synthesize an endothelial cell growth factor that is structurally related to basic fibroblast growth factor, *Proc. Natl. Acad. Sci. U.S.A.* **83**, 2448-2452.

1173. Ueno, N., Baird, A., Esch, F., Ling, N., and Guillemin, R. (1986) Isolation of an amino terminal extended form of basic fibroblast growth factor, *Biochem. Biophys. Res. Commun.* **138**, 580-588.

1174. Abraham, J. A., Mergia, A., Whang, J. L., Tumolo, A., Friedman, J., Hjerrild, K. A., Gospodarowicz, D., and Fiddes, J. C. (1986) Nucleotide sequence of a bovine clone encoding the angiogenic protein, basic fibroblast growth factor, *Science* **233**, 545-548.

1175. Moscatelli, D., Joseph-Silverstein, J., Manejias, R., and Rifkin, D. B. (1987) Mr 25,000 heparin-binding protein from guinea pig brain is a high molecular weight form of basic fibroblast growth factor, *Proc. Natl. Acad. Sci. U.S.A.* **84**, 5778-5782.

1176. Presta, M., Rusnati, M., Maier, J. A. M., and Ragnotti, G. (1988) Purification of basic fibroblast growth factor from rat brain: identification of a Mr 22,000 immunoreactive form, *Biochem. Biophys. Res. Commun.* **155**, 1161-1172.

1177. Sommer, A., Moscatelli, D., and Rifkin, D. B. (1989) An amino-terminally extended and post-translationally modified form of a 25 kD form of fibroblast growth factor, *Biochem. Biophys. Res. Commun.* **160**, 1267-1274.

1178. Prats, H., Kaghad, M., Prats, A. C., Klagsbrun, M., Lelias, J. M., Liauzun, P., Chalon, P., Tauber, J. P., Amalric, F., Smith, J. A., and Caput, D. (1989) High molecular mass forms of basic fibroblast growth factor are initiated by alternative GUG codons, *Proc. Natl. Acad. Sci. U.S.A.* **86**, 1836-1840.

1179. Florkiewicz, R., and Sommer, A. (1989) Human basic fibroblast growth factor gene encodes four polypeptides: three initiate translation from non-AUG codons, *Proc. Natl. Acad. Sci. U.S.A.* **86**, 3978-3981.

1180. Baird, A., and Klagsbrun, M. (1991) The fibroblast growth factor family: an overview, in *The Fibroblast Growth Factor Family*, eds., Baird, A. & Klagsbrun, M., New York Academy of Sciences, New York, Vol. 638, pp. xi-xii.

1181. Neufeld, G., and Gospodarowicz, D. (1986) Basic and acidic fibroblast growth factor interact with the same cell surface receptors, *J. Biol. Chem.* **261**, 5631-5637.

1182. Sommer, A., Brewer, M. T., Thompson, R. C., Moscatelli, D., Presta, M., and Rifkin, D. B. (1987) A form of human basic fibroblast growth factor with in extended amino terminus, *Biochem. Biophys. Res. Commun.* **144**, 543-550.

1183. Zhang, J., Cousen, L. S., Bar, P. J., and Sprang, S. R. (1991) Three-dimensional structure of human basic fibroblast growth factor, a structural homolog of interleukin 1β, *Proc. Natl. Acad. Sci. U.S.A.* **88**, 3446-3450.

1184. Eriksson, A. E., Cousens, L. S., Weaver, L. H., and Matthews, B. W. (1991) Three-dimensional structure of human basic fibroblast growth factor, *Proc. Natl. Acad. Sci. U.S.A.* **88**, 3441-3445.

1185. Zhu, X., Komiya, H., Chirino, A., Faham, S., Fox, G. M., Arakawa, T., Hsh, B. T., and Rees, D. C. (1991) Three-dimensional structures of acidic and basic fibroblast growth factors, *Science* **251**, 90-93.

1186. Ago, H., Kitagawa, Y., Fujishima, A., Matsuura, Y., and Katsube, Y. (1991) Crystal structure of basic fibroblast growth factor at 1.6 Å resolution, *J. Biochem. Tokyo* **110**, 360-363.

1187. Marics, I., Adelaide, J., Raybaud, F., Mattei, M. G., Coulier, F., Planche, J., de Lapeyriere, O., and Birnbaum, D. (1989) Characterization of the HST-related FGF-6 gene, a new member of the fibroblast growth factor family, *Oncogene* **4**, 335-340.

1188. Finch, P. W., Rubin, J. S., Miki, T., Ron, D., and Aaronson, S. A. (1989) Human KGF is FGF-related with properties of a paracrine effector of epithelial cell growth, *Science* **245**, 752-755.

1189. Seno, M., Sasada, R., Iwane, M., Sudo, K., Kurokawa, T., Ito, K., and Igarashi, K. (1988) Stabilizing basic fibroblast growth factor using protein engineering, *Biochem. Biophys. Res. Commun.* **151**, 701-708.

1190. Arakawa, T., Hsu, Y.-R., Schiffer, S. G., Tsai, L. B., Curless, C., and Fox, G. M. (1989) Characterization of a cysteine-free analog of recombinant basic fibroblast growth factor, *Biochem. Biophys. Res. Commun.* **161**, 335-341.

1191. Moy, F. J., Seddon, A. P., Böhlen, P., and R., P. (1996) High-resolution solution structure of basic fibroblast growth factor determined by multidimensional heteronuclear magnetic resonance spectroscopy, *Biochemistry* **35**, 13552-13561.

1192. Presta, M., Tiberio, L., Rusnati, M., Dell'Era, P., and Ragnotti, G. (1991) Basic fibroblast growth factor requires a long-lasting activation of protein kinase C to induce cell proliferation in transformed fetal bovine aortic endothelial cells, *Cell Regul.* **2**, 719-726.

1193. Olwin, B. B., and Hauschka, S. D. (1989) Cell type and tissue distribution of the fibroblast growth factor receptor, *J. Cell. Biochem.* **39**, 443-454.

1194. Givol, D., and Yayon, A. (1992) Complexity of FGF receptors: genetic basis for structural diversity and functional specificity, *FASEB J.* **6**, 3362-3369.

1195. Jay, M., and Schlessinger (1991) Structural diversity and binding of FGF receptors, in *The Fibroblast Growth Factor Family, Annals of the New York Academy of Sciences*, eds., Baird, A. & Klagsbrun, M., New York Academy of Sciences, New York, Vol. 638, pp. 161-166.

1196. Johnson, D. E., Lee, P. L., Lu, J., and Williams, L. T. (1990) Diverse forms of a receptor for acidic and basic fibroblast growth factors, *Mol. Cell. Biol.* **10**, 4728-4736.

1197. Johnson, D. E., and Williams, L. T. (1993) Structural and functional diversity in the FGF receptor multigene family, *Adv. Cancer Res.* **251**, 665-668.

1198. Ornitz, D. M., Xu, J., Colvin, J. S., McEwen, D. G., MacArthur, C. A., Coulier, F., Gao, G., and Goldfarb, M. (1996) Receptor specificity of the fibroblast growth factor family, *J. Biol. Chem.* **271**, 15292-15297.

1199. Lee, P. L., Johnson, D. E., Cousens, L. S., Fried, V. A., and Williams, L. T. (1989) Purification and complementary DNA cloning of a receptor for basic fibroblast growth factor, *Science* **245**, 57-60.

1200. Pasquale, E. B., and Singer, S. J. (1989) Identification of a developmentally regulated protein-tyrosine kinase by using anti-phosphotyrosine antibodies to screen a cDNA expression library, *Proc. Natl. Acad, Sci. U.S.A.* **86**, 5449-5453.

1201. Dionne, C. A., Crumley, G., Belot, F., Kaplow, K. M., Searfoss, G., Ruta, M., Burgess, W. H., Jaye, M., and Schlessinger, J. (1990) Cloning and expression of two distinct high-affinity receptors cross-reacting with acidic and basic fibroblast growth factors, *EMBO J.* **9**, 2685-2692.

1202. Musci, T. J., Amaya, E., and Kirschner, M. W. (1990) Regulation of the fibroblast growth factor receptor in early *Xenopus* embryos, *Proc. Natl. Acad. Sci. U.S.A.* **87**, 8365-8369.

1203. Reid, H. H., Wilks, A. F., and Bernard, O. (1990) Two forms of the basic fibroblast growth factor receptor-like mRNA are expressed in the developing mouse brain, *Proc. Natl. Acad. Sci. U.S.A.* **87**, 1596-1600.

1204. Ornitz, D. M., and Leder, P. (1992) Ligand specificity and heparin dependence of fibroblast growth factor receptors 1 and 3, *J. Biol. Chem.* **267**, 16305-16311.

1205. Hoissaint, E., Breathnatch, R., Blanquet, P. R., Champion-Arnaud, P., Gesnel, M. C., Torriglia, A., and Courtois, Y. (1990) Related fibroflast growth factor receptor genes exist in the human genome, *Proc. Natl. Acad. Sci. U.S.A.* **87**, 8180-8184.

1206. Pasquale, E. B. (1990) A distinctive family of embryonic protein-tyrosine kinase receptors, *Proc. Natl. Acad. Sci. U.S.A.* **87**, 5812-5816.

1207. Mansukhani, A., Dell'Era, P., Moscatelli, D., Kornbluth, S., Hanafusa, H., and Basilico, C. (1992) Characterization of the murine BEK fibroblast growth factor (FGF) receptor: activation by three members of the FGF family and requirement for heparin, *Proc. Natl. Acad. Sci. U.S.A.* **89**, 3305-3309.

1208. Keegan, K., Johnson, D. E., Williams, L. T., and Hayman, M. J. (1991) Isolation of an additional member of the fibroblast growth factor receptor family, FGFR-3, *Proc. Natl. Acad. Sci. U.S.A.* **88**, 1095-1099.

1209. Mohammadi, M., Schlessinger, J., and Hubbard, S. R. (1996) Structure of the FGF receptor tyrosine kinase domain reveals a novel autoinhibitory mechanism, *Cell* **86**, 577-587.

1210. Burgess, W. H., Dionne, C. A., Kaplow, J., Mudd, R., Friesel, R., Zilberstein, A., Schlessinger, J., and Jaye, M. (1990) Characterization and cDNA cloning of phospholipase Cα, a major substrate for heparin-binding growth factor 1 (acidic fibroblast growth factor)-activated tyrosine kinase, *Mol. Cell. Biol.* **10**, 4770-4777.

1211. Ullrich, A., and Schlessinger, J. (1990) Signal transduction by receptors with tyrosine kinase activity, *Cell* **61**, 243-254.

1212. Ong, S. H., Goh, K. C., Lim, Y. P., Low, B. C., Klint, P., Claesson-Welsh, L., Cao, X., Tan, Y. H., and Guy, G. R. (1996) SUC1-Associated neurotrophic factor target (SNT) protein is a major FGF-stimulated tyrosine phosphorylated 90-kDa protein which binds to the SH2 domain of BRB2, *Biochem. Biophys. Res. Commun.* **225**, 1021-1026.

1213. Goh, K. C., Lim, Y. P., Ong, S. H., Siak, C. B., Cao, X., Tan, Y. H., and Guy, G. R. (1996) Identification of p90, a prominent tyrosine-phosphorylated protein in fibroblast growth factor-stimulated cells, as 80K-H, *J. Biol. Chem.* **271**, 5832-5838.

1214. Vainikka, S., Joukov, V., Klint, P., and Alitalo, K. (1996) Association of a 85-kDa serine kinase with activated fibroblast growth factor receptor-4, *J. Biol. Chem.* **271**, 1270-1273.

1215. Patstone, G., and Maher, P. (1996) Copper and calcium binding motifs in the extracellular domains of fibroblast growth factor receptors, *J. Biol. Chem.* **271**, 3343-3346.

1216. Wang, L.-Y., Edenson, S. P., Yu, Y.-L., Senderowicz, L., and Turck, C. W. (1996) A natural kinase-deficient variant of fibroblast growth factor receptor 1, *Biochemistry* **35**, 10134-10142.

1217. Neilson, K. M., and Friesel, R. (1996) Ligand-independent activation of fibroblast growth factor receptors by point mutations in the extracellular, transmembrane, and kinase domains, *J. Biol. Chem.* **271**, 25049-25057.

1218. Wennström, S., Sandström, C., and Claesson-Welsh, L. (1991) cDNA cloning and expression of a human FGF receptor which binds acidic and basic FGF, *Growth Factors* **4**, 197-208.

1219. Reich-Slotky, R., Bonneh-Barkay, D., Shaoul, E., Bluma, B., Svahn, C. M., and Ron, D. (1994) Differential effect of cell-associated heparan sulfates on the binding of keratinocyte growth factor (KGF) and acidic fibroblast growth factor to the KGF receptor, *J. Biol. Chem.* **269**, 32279-32285.

1220. Vainikka, S., Partanen, J., Bellosta, P., Coulier, F., Basilico, C., Jaye, M., and Alitalo, K. (1992) Fibroblast growth factor receptor-4 shows novel features in genomic structure, ligand binding and signal transduction, *EMBO J.* **11**, 4273-4280.

1221. Cote, G. J., Huang, E. S.-C., Jin, W., and Morrison, R. S. (1997) Sequence requirements for regulated RNA splicing of the human fibroblast growth factor receptor-1 α exon, *J. Biol. Chem.* **272**, 1054-1060.

1222. Crumbly, G., Bellot, F., Kaplow, J. M., Schlessinger, J., Jaye, M., and Dionne, C. A. (1991) High affinity binding and activation of a truncated FGF receptor by both aFGF and bFGF, *Oncogene* **6**, 2255-2262.

1223. Xu, J., Nakahara, M., Crabb, J. W., Shi, E., Matuo, Y., Fraser, M., Kan, M., Hou, J., and McKeehan, W. L. (1992) Expression and immunochemical analysis of rat and human fibroblast growth factor receptor (flg) isoforms, *J. Biol. Chem.* **267**, 17792-17803.

1224. Hou, J., Kan, M., Wang, F., Xu, J.-M., Nakahara, M., McBride, G., McKeehan, K., and McKeehan, W. L. (1992) Substitution of putative half-cystine residues in heparin-binding fibroblast growth factor receptors: loss of binding activity in both two and three loop isoforms, *J. Biol. Chem.* **267**, 17804-17808.

1225. Zimmer, Y., Givol, D., and Yayon, A. (1993) Multiple structural elements determine ligand binding of fibroblast growth factors. Evidence that both Ig domain 2 and 3 define receptor specificity, *J. Biol. Chem.* **268**, 7899-7903.

1226. Cheon, H.-G., LaRochell., W. J., Bottaro, D. P., Burgess, W. H., and Aaronson, S. A. (1994) High-affinity binding sites for related fibroblast growth factor ligands reside within different receptor immunoglobulin-like domains, *Proc. Natl. Acad. Sci. U.S.A.* **91**, 989-999.

1227. Mathieu, M., Chatelain, E., Ornitz, D., Bresnick, J., Mason, I., Kiefer, P., and Dickson, C. (1995) Receptor binding and mitogenic properties of mouse fibroblast growth factor 3, *J. Biol. Chem.* **270**, 24197-24203.

1228. Meisinger, C., Zeschnigk, C., and Grothe, C. (1996) *In vivo* and *in vitro* effect of glucocorticoids on fibroblast growth factor (FGF)-2 and FGF receptor 1 expression, *J. Biol. Chem.* **271**, 16520-16525.

1229. Chedid, M., Rubin, J. S., Csaky, K. G., and Aaronson, S. A. (1994) Regulation of keratinocyte growth factor gene expression by interleukin 1, *J. Biol. Chem.* **269**, 10753-10757.

1230. Schmidt, A., Skaletz-Rorowski, A., Breithardt, G., and Buddecke, E. (1995) Growth status-dependent changes of bFGF compartmentalization and heparan sulfate structure in arterial smooth muscle cells, *Eur. J. Cell Biol.* **67**, 130-134.

1231. Steinfeld, R., Van Den Berghe, H., and David, G. (1996) Stimulation of fibroblast growth factor receptor-1 occupancy and signaling by cell surface-associated syndecans and glypican, *J. Cell Biol.* **133**, 405-416.

1232. Whitlock, J. M., Murdoch, A. D., Iozzo, R. V., and Underwood, P. A. (1996) The degradation of human endothelial cell-derived perlecan and release of bound basic fibroblast growth factor by stromelysin, collagenase, plasmin, and hepananases, *J. Biol. Chem.* **271**, 10079-10086.

1233. Guillonneau, X., Tassin, J., Berrou, E., Bryckaert, M., Courtois, Y., and Mascarelli, F. (1996) *In vitro* changes in plasma membrane heparan sulfate proteoglycans and in perlecan expression participate in the regulation of fibroblast growth factor 2 mitogenic activity, *J. Cell. Physiol.* **166**, 170-187.

1234. Berrou, E., Quarck, R., Fontenay-Roupie, M., Lévy-Toledano, S., Tobelem, G., and Bryckaert, M. (1995) Transforming growth factor-β1 increases internalization of basic fibroblast growth factor by smooth muscle cells: implication of cell-surface heparan sulphate proteoglycan endocytosis, *Biochem. J.* **331**, 393-399.

1235. Mignatti, P., Morimoto, T., and B., R. D. (1992) Basic fibroblast growth factor, a protein devoid of secretory signal sequence, is released by cells via a pathway independent of the endoplasmic reticulum-Golgi complex, *J. Cell. Physiol.* **151**, 81-93.

1236. Rogelj., S., Weinberg, R. A., Fanning, P., and Klagsbrun, M. (1988) Basic fibroblast growth factor fused to a signal peptide transforms cells, *Nature* **331**, 173-175.

1237. Brooks, R. A., Burrin, J. M., and Kohner, E. M. (1991) Characterization of release of basic fibroblast growth factor from bovine retinal endothelial cells in monolayer cultures, *Biochem. J.* **276**, 113-120.

1238. Vlodavsky, I., Folkman, J., Sullivan, R., Fridman, R., Ishai-Michaeli, R., Sasse, J., and Klagsbrun, M. (1987) Endothhelial cell-derived basic fibroblast growth factor: synthesis and deposition into subendothelial extracellular matrix, *Proc. Natl. Acad. Sci. U.S.A.* **84**, 2292-2296.

1239. Muthukrishnan, L., Warder, E., and McNeil, P. L. (1991) Basic fibroblast growth factor is efficiently released from a cytosolic storage site through plasma membrane disruptions of endothelial cells, *J. Cell. Physiol.* **148**, 1-16.

1240. Folkman, J., Klagsbrun, M., Sasse, J., Wadzinski, M., Ingber, D., and Vlodavsky, I. (1988) A heparin-binding angiogenic protein-basic fibroblast growth factor-is stored within basement membranes, *Am. J. Pathol.* **130**, 393-400.

1241. Salmivirta, M., Heino, J., and Jalkanen, M. (1992) Basic fibroblast growth factor-syndecan complex at cell surface or immobilized to matrix promotes cell growth , *J. Biol. Chem.* **267**, 17606-17610.

1242. Bashkin, P., Neufeld, G., Gitay-Goren, H., and Vlodavsky, I. (1992) Release of cell surface-associated basic fibroblast growth factor by glycosylphosphatidylinositol-specific phospholipase, *J. Cell. Physiol.* **151**, 126-137.

1243. Ishai-Michaeli, R., Svahn, C. M., Weber, M., Chajek-Shaul, T., Korner, G., Ekre, H.-P., and Vlodavsky, I. (1992) Importance of size and sulfation of heparin in release of basic fibroblast growth factor from the vascular endothelium and extracellular matrix, *Biochemistry* **31**, 2080-2088.

1244. Bashkin, P., Doctrow, S., Klagsbrun, M., Svahn, C. M., Folkman, J., and Vlodavsky, I. (1989) Basic fibroblast growth factor binds to subendothelial extracellular matrix and is released by heparitinase and heparin-like molecules, *Biochemistry* **26**, 1737-1743.

1245. Gleizes, P.-E., Noaillac-Depeyre, J., Dupont, M.-A., and Gas, N. (1996) Basic fibroblast growth factor (FGF-2) is addressed to caveolae after binding to the plasma membrane of BHK cells, *Eur. J. Cell Biol.* **71**, 144-153.

1246. Houssaint, E., Blanquet, P. R., Champion-Arnaud, P., Gesnel, M. C., Torriglia, A., Courtois, Y., and Breathnach, R. (1990) Related fibroblast growth factor genes exist in the human genome, *Proc. Natl. Acad. Sci. U.S.A.* **87**, 8180-8184.

1247. Partanen, J., Mäkelä, T. P., Eerola, E., Korhonen, J., Hirvonen, H., Claesson-Welsh, L., and Alitalo, K. (1990) FGFR-4, a novel acidic fibroblast growth factor receptor with a distinct expression pattern, *EMBO J.* **10**, 1347-1354.

1248. Mansukhani, A., Moscatelli, D., Talarico, D., Levytska, V., and Basilico, C. (1990) A murine fibroblast growth factor (FGF) receptor expressed in CHO cells is activated by basic FGF and Kaposi FGF, *Proc. Natl. Acad. Sci. U.S.A.* **87**, 4378-4382.

1249. Gannoun-Zaki, L., Pieri, I., Badet, J., Moenner, M., and Barritault, D. (1991) Internalization of basic fibroblast growth factor by Chinese hamster lung fibroblast cells: involvement of several pathways, *Exp. Cell Res.* **197**, 272-279.

1250. Moscatelli, D. (1987) High and low affinity binding sites for basic fibroblast growth factor on cultured cells: absence of a role for low affinity binding in the stimulation of plasminogen activator production by bovine capillary endothelial cells, *J. Cell. Physiol.* **131**, 123-130.

1251. Roghani, M., and Moscatelli, D. (1992) Basic fibroblast growth factor is internalized through both receptor-mediated and heparan sulfate-mediated mechanisms, *J. Biol. Chem.* **267**, 22156-22162.

1252. Quarto, N., and Amalric, F. (1994) Heparan sulfate proteoglycans as transducers of FGF-2 signalling, *J. Cell Sci.* **107**, 3201-3212.

1253. Fannon, M., and Nugent, M. A. (1996) Basic fibroblast growth factor binds its receptors, is internalized, and stimulates DNA synthesis in Balb/c3T3 cells in the absence of heparan sulfate, *J. Biol. Chem.* **271**, 17949-17956.

1254. Richard, C., Liuzzo, J. P., and Moscatelli, D. (1995) Fibroblast growth factor-2 can mediate cell attachment by linking receptor and heparan sulfate proteoglycans on neighboring cells, *J. Biol. Chem.* **270**, 24188-24196.

1255. Moscatelli, D. (1988) Metabolism of receptor-bound and matrix-bound basic fibroblast growth factor by bovine capillary endothelial cells, *J. Cell Biol.* **107**, 753-759.

1256. Moenner, M., Gannoun-Zaki, L., Badet, J., and Barritault, D. (1989) Internalization and limited processing of basic fibroblast growth factor in Chinese hamster lung fibroblasts, *Growth Factors* **1**, 115-123.

1257. Saksela, O., Moscatelli, D., Sommer, A., and Rifkin, D. B. (1988) Endothelial cell-derived heparan sulfate binds basic fibroblast growth factor and protects it from proteolytic degradation, *J. Cell. Physiol.* **107**, 743-751.

1258. Hawker Jr., J. R., and Granger, H. J. (1994) Nuclear accumulation of exogenous basic fibroblast growth factor in endotheilal, fibroblast, and myoblast cell lines results in diverse biological responses, *In Vitro Cell. Dev. Biol.* **30A**, 653-663.

1259. Reiland, J., and Rapraeger, A. C. (1993) Heparan sulfate proteoglycan and FGF receptor target basic FGF to different intracellular destinations, *J. Cell Sci.* **105**, 1085-1093.

1260. Rusnati, M., Urbinati, C., and Presta, M. (1993) Internalization of basic fibroblast growth factor (bFGF) in cultured endothelial cells: role of the low affinity heparin-like bFGF receptors, *J. Cell. Physiol.* **154**, 152-161.

1261. Burgess, W. H., Shaheen, A. M., Ravera, M., Jaye, M., Donohue, P. J., and Winkles, J. A. (1990) Possible dissociation of the heparin-binding and mitogenic activities of heparin-binding (acidic fibroblast) growth factor-1 from its receptor-binding activities by site-directed mutagenesis of a single lysine residue, *J. Cell Biol.* **111**, 2129-2138.

1262. Isacchi, A., Satuto, M., Chiesa, R., Bergonzoni, L., Rusnati, M., Sarmientos, P., Ragnotti, G., and Presta, M. (1991) A six-amino acid deletion in basic fibroblast growth factor dissociates its mitogenic activity from its plasminogen activator-inducing capacity, *Proc. Natl. Acad. Sci. U.S.A.* **88**, 2628-2632.

1263. Wiedlocha, Falnes, P. Ø., Madshus, I. H., Sandvig, K., and Olsnes, S. (1994) Dual mode of signal transduction by externally added acidic fibroblast growth factor, *Cell* **76**, 1039-1051.

1264. Imamura, T., Oka, S., Tanahashi, T., and Okita, Y. (1994) Cell cycle-dependent nuclear localization of exogenously added fibroblast growth factor-1 in BALB/c 3T3 and human vascular endothelial cells, *Exp. Cell Res.* **215**, 363-372.

1265. Olwin, B. B., and Rapraeger, A. (1992) Repression of myogenic differentiation by aFGF, bFGF, and K-FGF is dependent on cellular heparan sulfate, *J. Cell Biol.* **118**, 631-639.

1266. Zhan, X., Hu, X. G., Friedman, S., and T., M. (1992) Analysis of endogenous and exogenous nuclear translocation of fibroblast growth factor-1 in NIH 3T3 cells, *Biochem. Biophys. Res. Commun.* **188**, 982-991.

1267. Zhan, X., Hu, X., Friesel, R., and T., M. (1993) Long term growth factor exposure and differential tyrosine phosphorylation are required for DNA synthesis in BALB/c 3T3 cells, *J. Biol. Chem.* **268**, 9611-9620.

1268. Cao, Y., Exström, M., and Pettersson, R. F. (1993) Characterization of the nuclear translocation of acidic fibroblast growth factor, *J. Cell Sci.* **104**, 77-87.

1269. Dang, C. V., and Lee, W. M. F. (1989) Nuclear and nucleolar targeting sequences of c-erb-A, c-myb, N-myc, p53, HSP70, and HIV tat proteins, *J. Biol. Chem.* **264**, 18019-18023.

1270. Florkiewicz, R. Z., Baird, A., and Gonzalez, A.-M. (1991) Multiple forms of bFGF: differential nuclear and cell surface localization, *Growth Factors* **4**, 265-275.

1271. Bugler, B., Amalric, F., and Prats, H. (1991) Alternative initiation of translation determines cytoplasmic or nuclear location of basic fibroblast growth factor, *Mol. Cell Biol.* **11**, 573-577.

1272. Monzat, V., Ratovo, G., Estival, A., Fanjul, M., Bertrand, C., Clément, B., Vaysse, N., Hollande, E., and Clemente, F. (1996) Expression of two FGF-2 isoforms in pancreatic acinar cells (AR4-2J): intracellular localization and role in the regulation of the extracellular matrix biosynthesis, *Eur. J. Cell Biol.* **69**, 316-326.

1273. Quarto, N., Finger, F. P., and Rifkin, D. B. (1991) The NH_2-terminal extension of high molecular weight bFGF is a nuclear targeting signal, *J. Cell. Physiol.* **147**, 311-318.

1274. Jaye, M., Howk, R., Burgess, W., Ricca, G. A., Chiu, I.-M., Ravera, M. W., O'Brien, S. J., Modi, W. S., Maciag, T., and Drohan, W. N. (1986) Human endothelial growth factor: cloning, nucleotide sequence, and chromosome localization, *Science* **233**, 531-545.

1275. Imamura, T., Engleka, K., Zhan, X., Tokita, T., Forough, R., Roeder, D., Jackson, A., Maier, J. A. M., Hla, T., and Maciag, T. (1990) Recovery of mitogenic activity of a growth factor mutant with a nuclear translocation sequence, *Science* **249**, 1567-1570.

1276. Imamura, T., Tokita, Y., and Mitsui, Y. (1992) Identification of a heparin-binding growth factor-1 nuclear translocation sequence by deletion mutation analysis, *J. Biol. Chem.* **267**, 5676-5679.

1277. Lin, Y.-Z., Yao, S., and Hawiger, J. (1996) Role of the nuclear localization sequence in fibroblast growth factor-1-stimulated mitogenic pathways, *J. Biol. Chem.* **271**, 5305-5308.

1278. Kiefer, P., and Dickson, C. (1995) Nucleolar association of fibroblast growth factor 3 via specific sequence motifs has inhibitor effects on cell growth, *Mol. Cell. Biol.* **15**, 4364-4374.

1279. Harper, J. W., and Lobb, R. R. (1988) Reductive methylation of lysine residues in acidic fibroblast growth factor: effect on mitogenic activity and heparin affinity, *Biochemistry* **27**, 671-678.

1280. Feige, J.-J., and Baird, A. (1989) Basic fibroblast growth factor is a substrate for protein phosphorylation and is phosphorylated by capillary endothelial cells in culture, *Proc. Natl. Acad. Sci. U.S.A.* **86**, 3174-3178.

1281. Seno, M., Sasada, R., Kurokawwa, T., and Igarashi, K. (1990) Carboxyl-terminal structure of basic fibroblast growth factor significantly contributes to its affinity for heparin , *Eur. J. Biochem.* **188**, 239-245.

1282. Springer, B. A., Pantoliano, M. W., Barbera, F. A., Gunyuzlu, P. L., Thompson, L. D., Herblin, W. F., Rosenfeld, S. A., and Book, G. W. (1994) Identification and concerted function of two receptor binding surfaces on basic fibroblast growth factor required for mitogenesis, *J. Biol. Chem.* **269**, 26879-26884.

1283. Fan, M., Wang, F., Xu, J., Crabb, J. W., Hou, J., and McKeenan (1993) An essential heparin-binding domain in the fibroblast growth factor receptor kinase, *Science* **2592**, 1918-1921.

1284. Brickman, Y. G., Ford, M. D., Small, D. H., Bartlett, P. F., and Nurcombe, V. (1995) Heparan sulfates mediate the binding of basic fibroblast growth factor to a receptor on neural precursor cells, *J. Biol. Chem.* **270**, 24941-24948.

1285. Deisenhofer, J. (1981) Crystallographic refinement and atomic models of the human Fc fragment and its complex with fragment B of protein A from *Staphylococcus aureus*, *Biochemistry* **20**, 2361-2370.

1286. McCoon, P. E., Angerer, R. C., and Angerer, L. M. (1996) SpFGFR, a new member of the fibroblast growth factor receptor family, is developmentally regulated during early sea urchin development, *J. Biol. Chem.* **271**, 20119-20125.

1287. Hou, J., Kan, M., McKeehan, K., McBride, G., Adams, P., and McKeenan, W. L. (1991) Fibroblast growth factor receptors from liver vary in three structural domains, *Science* **251**, 665-668.

1288. Kiefer, M. C., Baird, A., Nguyen, T., George-Nascimento, C., Mason, O. B., Boley, L. J., Valenzuela, P., and Barr, P. J. (1991) Molecular cloning of human basic fibroblast growth factor receptor cDNA and expression of a biologically active extracullular domain in a baculovirus system, *Growth Factors* **5**, 115-127.

1289. Kouhara, H., Kasayama, S., Saito, H., Matsumoto, K., and Sato, B. (1991) Expression cDNA cloning of fibroblast growth factor (FGF) receptor in mouse breast cancer cells: a variant form in FGF-responsive transformed cells, *Biochem. Biophys. Res. Commun.* **176**, 31-37.

1290. Yan, G., Wang, F., Fukabori, Y., Sussman, D., Hou, J., and McKeehan (1992) Expression and transforming activity of a variant of the heparin-binding fibroblast growth factor receptor (flg) gene resulting from splicing of the alpha exon at an alternate 3'-acceptor site, *Biochem. Biophys. Res. Commun.* **183**, 423-430.

1291. Dell, K. R., and Williams, L. T. (1992) A novel form of fibroblast growth factor receptor 2: alternative splicing of the third immunoglobulin-like domain confers ligand binding specificity, *J. Biol. Chem.* **267**, 21125-21129.

1292. Tagagi, Y., Shrivastav, S., Miki, T., and Sakaguchi, K. (1994) Molecular cloning and expression of the acid fibroblast growth factor receptors in a rat parathyroid cell line (PT-r): parathyroid cell-specific calcium-dependent change of ligand accessibility and covalent attachment of heparan sulfate glycosaminoglycan to the receptors, *J. Biol. Chem.* **269**, 23743-23749.

1293. Arakawa, T., Wen, J., and Philo, J. S. (1994) Stoichiometry of heparin binding to basic fibroblast growth factor, *Arch. Biochem. Biophys.* **308**, 267-273.

1294. Bârzu, T., Lormeau, J.-C., Petitou, M., Michelson, S., and Choay, J. (1989) Heparin-derived oligosaccharides: affinity for acidic fibroblast growth factor and effect on its growth-promoting activity for human endothelial cells, *J. Cell. Physiol.* **14**, 538-548.

1295. Mach, H., Volkin, D. B., Burke, C. J., Middaugh, C. R., Linhardt, R. J., Fromm, J. R., Loganathan, D., and Mattsson, L. (1993) Nature of the interaction of heparin with acidic fibroblast growth factor, *Biochemistry* **32**, 5480-5489.

1296. Volkin, D. B., Tsai, P. K., Dabora, J. M., Gress, J. O., Burkee, C. J., Linhardt, R. J., and Middaugh, C. R. (1993) Physical stabilization of acidic fibroblast growth factor by polyanions, *Arch. Biochem. Biophys.* **300**, 30-41.

1297. Aviezer, D., Levy, E., Safran, M., Svahn, C., Buddecke, E., Schmidt, A., David, G., Vlodavsky, I., and Yayon, A. (1994) Differential structural requirements of heparin and heparan sulfate proteoglycans that promote binding of basic fibroblast growth factor to its receptor, *J. Biol. Chem.* **269**, 114-121.

1298. Ornitz, D. M., Herr, A. B., Nilsson, M., Westman, J., Svahn, C.-M., and Waksman, G. (1995) FGF binding and FGF receptor activation by synthetic heparan-derived di- and trisaccharides, *Science* **268**, 432-436.

1299. Kaplow, J. M., Bellot, F., Crumley, G., Dionne, C. A., and Jaye, M. (1990) Effect of heparin on the binding affinity of acidic FGF for the cloned human FGF receptors, flg and bek, *Biochem. Biophys. Res. Commun.* **172**, 107-112.

1300. Nugent, M. A., and Edelman, E. R. (1992) Kinetics of basic fibroblast growth factor binding to its receptor and heparan sulfate proteoglycan: a mechanism for cooperativity, *Biochemistry* **31**, 8876-8883.

1301. Moscatelli, D. (1992) Basic fibroblast growth factor (bFGF) dissociates rapidly from heparan sulfates but slowly from receptors: implications for mechanisms of bFGF release from pericellular matrix, *J. Biol. Chem.* **267**, 25803-25809.

1302. Roghani, M., Mansukhan, A., Dell'Era, P., P., B., Bosilico, C., Rifkin, D. B., and Moscatelli, D. (1994) Heparin increases the affinity of basic fibroblast growth factor for its receptor but is not required for binding, *J. Biol. Chem.* **269**, 3976-3984.

1303. Prestrelski, S. J., Fox, G. M., and Arakawa, T. (1992) Binding of heparin to basic fibroblast growth factor induces a conformational change, *Arch. Biochem. Biophys.* **293**, 314-319.

1304. Reich-Slotky, R., Shaoul, E., B., B., Graziani, G., and Ron, D. (1995) Chimeric molecules between keratinocyte growth factor and basic fibroblast growth factor define domains that confer receptor binding specificities, *J. Biol. Chem.* **270**, 29813-29818.

1305. Zhu, X., Hsu, B. T., and Rees, D. C. (1993) Structural studies of the binding of the anti-ulcer drug sucrose octasulfate to acidic fibroblast growth factor, *Structure* **1**, 27-34.

1306. Folkman, J., Szabo, S., Stovroff, M., McNeil, P., Li, W., and Shing, Y. (1991) Duodenal ulcer: discovery of a new mechanism and development of angiogenic therapy that accelerates healing, *Ann. Surg.* **214**, 414-425.

1307. Venkataraman, G., Sasisekharan, V., Herr, A. B., Ornitz, D. M., Waksman, G., Cooney, C. L., Langer, R., and Sasisekharan, R. (1996) Preferential self-association of basic fibroblast growth factor is stabilized by heparin during receptor dimerization and activation, *Proc. Natl. Acad. Sci. USA* **93**, 845-850.

1308. Nurcombe, V., Ford, M. D., Wildschut, J. A., and Bartlett, P. F. (1993) Developmental regulation of neural response to FGF-1 and FGF-2 by heparan sulfate proteoglycan, *Science* **260**, 103-106.

1309. Curnutte, J. T., Orkin, S. H., and Dinauer, M. C. (1994) Genetic disorders of phagocyte function, in *Molecular Basis of Blood Diseases, 2nd edition*, eds., Stamatoyannopoulos, G., Nienhuis, A. W., Majerus, P. & Varmus, H., W. B. Saunders Company, Philadelphia, pp. pp. 493-540.

1310. Wahl, L. M., and Wahl, S. M. (1992) Inflammation, in *Wound Healing. Biochemical and Clinical Aspects*, eds., Cohen, I. K., Diegelmann, R. F. & Lindblad, W. J., W. B. Saunders Company, Philadelphia, pp. 40-62.

1311. Ushio-Fukai, M., Zafari, A. M., Fukui, T., Ishizaka, N., and Griendling, K. K. (1996) p22^{phox} is a critical component of the superoxide-generating NADH/NADPH oxidase system and regulates angiotensin II-induced hypertrophy in vascular smooth muscle cells, *J. Biol. Chem.* **271**, 23317-23321.

1312. Fridovich, I. (1995) Superoxide radical and superoxide dismutases, *Annu. Rev. Biochem.* **64**, 97-112.

1313. Fridovich, I. (1979) Superoxide and superoxide dismutase, in *Adv. In Inorg. Chem.*, eds., Eichhorn, G. & Marzilli, L., Elsevier/North-Holland, Amsterdam, Vol. 1, pp. 67-90.

1314. Benov, L., and Fridovich, I. (1996) Functional significance of Cu,ZnSOD in *Escherichia coli*, *Arch. Biochem. Biophys.* **327**, 249-253.

1315. McCord, J. M., and Fridovich, I. (1969) Superoxide dismutase: an enzymic function for erythrocuprein (hemocuprein), *J. Biol. Chem.* **244**, 6049-6055.

1316. Weisinger, R. A., and Fridovich, I. (1973) Mitochondrial superoxde dismutase: site of synthesis and intramitochondrial localization, *J. Biol. Chem.* **248**, 4793-4796.

1317. Steinman, H. M., and Hill, R. L. (1973) Sequence homologies among bacterial and mitochondrial superoxide dismutases, *Proc. Natl. Acad. Sci. U.S.A.* **70**, 3725-3729.

1318. Marklund, S. L. (1982) Human copper-containing superoxide dismutase of high molecular weight, *Proc. Natl. Acad. Sci. U.S.A.* **79**, 7634-7638.

1319. Marklund, S. L. (1984) Extracellular superoxide dismutase in human tissues and human cell lines, *J. Clin. Invest.* **74**, 1398-1403.

1320. Karlsson, K., and Marklund, S. L. (1989) Binding of human extracellular-superoxide dismutase C to cultured cell lines and to blood cells, *Lab. Invest.* **60**, 659-666.
1321. Karlsson, K., Sandström, J., Edlund, A., and Marklund, S. L. (1994) Turnover of extracellular superoxide dismutase in tissues, *Lab. Invest.* **70**, 705-710.
1322. Marklund, S. L., Holme, E., and Hellner, L. (1982) Superoxide dismutase in extracellular fluids, *Clin. Chim. Acta* **126**, 41-51.
1323. Marklund, S. L., Bjelle, A., and Elmqvist, L.-G. (1986) Superoxide dismutase isoenzymes of the synovial fluid in rheumatoid arthritis and reactive arthritides, *Ann. Rheumatol. Dis.* **45**, 847-851.
1324. Karlsson, K., and Marklund, S. L. (1988) Extracellular superoxide dismutase in the vascular system of mammals, *Biochem. J.* **255**, 223-228.
1325. Ohta, H., Adachi, T., and Hirano, K. (1993) The nature of heterogeneous components of extracellular-superoxide dismutase purified from human umbilical cords, *Free Rad. Biol. Med.* **15**, 151-158.
1326. Marklund, S. L. (1984) Extracellular superoxide dismutase and other superoxide dismutase isoenzymes in tissues from nine mammalian species, *Biochem. J.* **222**, 649-655.
1327. Marklund, S. L. (1990) Expression of extracellular superoxide dismutase by human cell lines, *Biochem. J.* **266**, 213-219.
1328. Karlsson, K., and Marklund, S. L. (1987) Heparin-induced release of extracellular superoxide dismutase to human blood plasma, *Biochem. J.* **242**, 55-59.
1329. Marklund, S. L. (1984) Properties of extracellular superoxide dismutase from human lung, *Biochem. J.* **220**, 269-272.
1330. Hjalmarsson, K., Marklund, S. L., Engström, Å., and Edlund, T. (1987) Isolation and sequence of complementary DNA encoding human extracellular superoxide dismutase, *Proc. Natl. Acad. Sci. U.S.A.* **84**, 6340-6344.
1331. Tainer, J. A., Getzoff, E. D., Beem, K. M., Richardson, J. S., and Richardson, D. C. (1982) Determination and analysis of the 2 Å structure of copper, zinc superoxide dismutase, *J. Mol. Biol.* **160**, 181-217.
1332. Willems, J., Zwijsen, A., Slegers, H., Micolaï, S., Bettadapura, J., Raymackers, J., and Scarcez, T. (1993) Purification and sequence of rat extracellular superoxide dismutase B secreted by C6 glioma, *J. Biol. Chem.* **268**, 24614-24621.
1333. Adachi, T., and Marklund, S. L. (1989) Interactions between human extracellular superoxide dismutase C and sulfated polysaccharides, *J. Biol. Chem.* **264**, 8537-8541.
1334. Sandström, J., Carlsson, L., Marklund, S. L., and Edlund, T. (1992) The heparin-binding domain of extracellular superoxide dismutase and formation of variants with reduced heparin affinity, *J. Biol. Chem.* **267**, 18205-18209.
1335. Adachi, T., Kodera, T., Ohta, H., Hayashi, K., and Hirano, K. (1992) The heparin binding site of human extracellular-superoxide dismutase, *Arch. Biochem. Biophys.* **297**, 155-161.
1336. Adachi, T., Ohta, H., Hirano, K., Hayashi, K., and Marklund, T. L. (1991) Non-enzymic glycation of human extracellular superoxide dismutase, *Biochem. J.* **279**, 263-267.
1337. Sandström, J., Nilsson, P., Karlsson, K., and Marklund, S. L. (1994) 10-Fold increase in human plasma extracellular superoxide dismutase content caused by a mutation in heparin-binding domain, *J. Biol. Chem.* **269**, 19163-19166.
1338. Adachi, T., Yamada, H., Yamada, Y., Morihara, N., Yamazaki, N., Murakami, T., Futenma, A., Kato, K., and Hirana, K. (1996) Substitution of glycine for arginine-213 in extracellular-superoxide dismutase impairs affinity for heparin and endothelial cell surface, *Biochem. J.* **313**, 235-239.
1339. Adachi, T., Morihara, N., Yamazaki, N., Yamada, H., Futenma, A., Kato, K., and Hirano, K. (1996) An arginine-213 to glycine mutation in human extracellular-superoxide dismutase reduces susceptibility to trypsin-like proteinases, *J.Biochem.* **120**, 184-188.
1340. Karlsson, K., Edlund, A., Sandström, J., and Marklund, S. L. (1993) Proteolytic modification of the heparin-binding affinity of extracellular superoxide dismutase, *Biochem. J.* **290**, 623-626.

1341. Adachi, T., Yamada, H., Futenma, A., Kato, K., and Hirano, K. (1995) Heparin-induced release of extracellular-superoxide dismutase form (V) to plasma, *J. Biochem. Tokyo* **117**, 586-590.

1342. Carlsson, L. M., Marklund, S. L., and Edlund, T. (1996) The rat extracellular superoxide dismutase dimer is converted to a tetramer by the exchange of a single amino acid, *Proc. Natl. Acad. Sci. USA* **93**, 5219-5222.

1343. Sandström, J., Karlsson, K., Edlund, T., and Marklund, S. L. (1993) Heparin-affinity patterns and composition of extracellular superoxide dismutase in human plasma and tissues, *Biochem. J.* **294**, 853-857.

1344. Karlsson, K., and Marklund, S. L. (1988) Heparin-, dextran sulfate-, and protamine-induced release of extracellular-superoxide dismutase to plasma in pigs, *Biochim. Biophys. Acta* **967**, 110-114.

1345. Becker, M., Henger, M. D., and Lehr, H.-A. (1994) Heparin-released superoxide dismutase inhibits postischemic leukocyte adhesion to venacular endothelium, *Am. J. Physiol. (Heart Circ. Physiol. 36)* **267**, H925-H930.

1346. Hansson, L., Edlund, M., Edlund, A., Johansson, T., Marklund, S. L., Fromm, S., Strömqvist, and Törnell, J. (1994) Expression and characterization of biologically active human extracellular superoxide dismutase in milk of transgenic mice, *J. Biol. Chem.* **269**, 5358-5363.

1347. Karlsson, K., and Marklund, S. L. (1988) Plasma clearance of human extracellular-superoxide dismutase C in rabbits, *J. Biol. Chem.* **82**, 762-766.

1348. Karlsson, K., Sandström, J., Edlund, J., Edlund, E., and Marklund, S. L. (1993) Pharmacokinetics of extracellular-superoxide dismutase in the vascular system, *Free Rad. Biol. Med.* **14**, 185-190.

1349. Adachi, T., Ohta, H., Yamada, H., Futenma, A., Kato, K., and Hirano, K. (1992) Quantitative analysis of extracellular-superoxide dismutase in serum and urine by ELISA with monoclonal antibody, *Clin. Chim. Acta* **213,**, 89-102.

1350. Adachi, T., Nakamura, M., Yamada, H., Futenma, A., Kato, K., and Hirano, K. (1994) Quantitative and qualitative changes of extracellular-superoxide dismutase in patients with various diseases, *Clin. Chim. Acta* **229**, 123-131.

1351. Tibell, L., Hjalmarsson, K., Edlund, T., Skogman, G., Engström, Å., and Marklund, S. L. (1987) Expression of human extracellular superoxide dismutase in Chinese hamster ovary cells and characterization of the product, *Proc. Natl. Acad. Sci. U.S.A.* **84**, 6634-6638.

1352. Marklund, S. L. (1992) Regulation by cytokines of extracellular superoxide dismutase and other superoxide dismutase isoenzymes in fibroblasts, *J. Biol. Chem.* **267**, 6696-6701.

1353. Miles, A. M., Bohle, D. S., Glassbrenner, P. A., Hansert, B., Wink, D. A., and Grisham, M. B. (1996) Modulation of superoxide-dependent oxidation and hydroxylation reactions by nitric oxide, *J. Biol. Chem.* **271**, 40-47.

1354. Carlsson, L. M., Jonsson, J., Edlund, T., and Marklund, S. L. (1995) Mice lacking extracellular superoxide dismutase are more sensitive to hyperoxia, *Proc. Natl. Acad. Sci. U.S.A.* **92**, 6264-6268.

1355. Tsan, M. F., White, J. E., Santana, T. A., and Lee, C. Y. (1990) Tracheal insufflation of tumor necrosis factor protects rats against oxygen toxicity, *J. Appl. Physiol.* **68**, 1211-1219.

1356. Tsan, M. F., White, J. E., Treanor, C., and Shaffner, J. B. (1990) Molecular basis for tumor necrosis factor-induced increase in pulmonary superoxide dismutase activities, *Am. J. Physiol.* **259**, L506-L512.

1357. White, C. W., Avraham, K. B., Shanley, P. F., and Groner, Y. (1991) Transgenic mice with expression of elevated levels of copper-zinc superoxide dismutase in the lungs are resistant to pulmonary oxygen toxicity, *J. Clin. Invest.* **87**, 2162-2168.

1358. Wispé, J. R., Warner, B. B., Clark, J. C., Dey, C. R., Neumann, J., Glasser, S. W., Crapo, J. D., Chang, L.-Y., and Whitsett, J. A. (1992) Human Mn-superoxide dismutase in pulmonary epithelial cells of transgenic mice confers protection form oxygen injury, *J. Biol. Chem.* **267**, 23937-23941.

1359. Ferrari, R., Ceconi, C., Curello, S., Ghielmi, S., and Albertini, A. (1989) Superoxide dismutase: possible therapeutic use in cardiovascular disease, *Pharmacol. Res.* **21**, 57-66.
1360. Omar, B. A., Flores, S. C., and McCord, J. M. (1992) Superoxide dismutase: pharmacological developments and applications, *Adv Pharmacol.* **23**, 109-161.
1361. Sjöquist, P. O., and Marklund, S. L. (1992) Endothelium bound extracellular superoxide dismutase type C reduces damage in reperfused ischaemic rat, *Cardiovasc. Res.* **26**, 34-350.
1362. Boissinot, M., Kuhn, L. A., Lee, P. F., C. L., Wang, Y., Hallewell, R. A., and Tainer, J. A. (1993) Rational design and expression of a heparin-targeted human superoxide dismutase, *Biochem. Biophys. Res. Commun.* **190**, 250-256.
1363. Salin, M. L., and McCord, J. M. (1975) Free radicals and inflammation:studies on superoxide-mediated NBT reduction by leokocytes, in *Erythrocyte Structure and Function*, ed. Brewer, G. J., Liss, New York, pp. 731-752.
1364. Riesenberg, K., Sclaeffer, F., Katz, A., and Levy, R. (1995) Inhibition of superoxide production in human neutrophils by combinations of heparin and thrombolytic agents, *Br. Heart J.* **73**, 14-19.
1365. Ohta, H., Adachi, T., and Hirano, K. (1994) Internalization of human extracellular superoxide dismutase by bovine aortic endothelial cells, *Free Rad. Biol. Med.* **16**, 501-507.
1366. Havel, R. J. (1986) Lipid transport function of lipoproteins in blood plasma, *Am. J. Physiol.* **253**, E1-E5.
1367. Dammerman, M., and Breslow, J. L. (1995) Genetic basis for lipoprotein disorders, *Circulation* **91**, 505-512.
1368. Grundy, S. M. (1995) Role of low-density lipoproteins in atherogenesis and development of coronary heart disease, *Clin. Chem.* **41**, 139-146.
1369. Nilsson-Ehle, P., Garfinkel, A. S., and Schotz, M. C. (1980) Lipolytic enzymes and plasma lipoprotein metabolism, *Annu. Rev. Biochem.* **49**, 667-693.
1370. Olivecrona, T., and Bengtsson-Olivecrona, G. (1989) Heparin and lipases, in *Heparin: Chemical and Biological Properties, Clinical Applications*, eds., Lane, D. A. & Lindahl, U., CRC Press, Boca Raton, FL, pp. 335-361.
1371. Flower, D. R. (1994) The lipocalin protein family, *FEBS Lett.* **354**, 7-11.
1372. Tall, A., Swenson, T., Hesler, C., and Granot, E. (1987) Mechanisms of facilitated lipid transfer mediated by plasma lipid transfer proteins, in *Plasma Lipoproteins*, ed. Gotto Jr., A. M., Elsevier, Amsterdam, pp. 277-297.
1373. Goldberg, I. J. (1996) Lipoprotein lipase and lipolysis: central roles in lipoprotein metabolism and atherogenesis, *J. Lipid Res.* **37**, 693-707.
1374. Chappell, D. A., Fry, G. L., Waknitz, M. A., Iverius, P.-H., Williams, S. E., and Strickland, D. K. (1992) The low density lipoprotein receptor-related protein/α_2-macroglobulin receptor binds and mediates catabolism of bovine milk lipoprotein lipase, *J. Biol. Chem.* **267**, 25764-25767.
1375. Goldberg, I. J., Kandel, J. J., Blum, C. B., and Ginsberg, H. N. (1986) Association of plasma lipoproteins with postheparin lipase activities, *J. Clin. Invest.* **78**, 1523-1528.
1376. Kern, P. A., Martin, R. A., Carty, J., Goldberg, I. J., and Ong, J. M. (1990) Identification of lipoprotein lipase immunoreactive protein in pre- and postheparin plasma from normal subjects and patients with type I hyperlipoproteinemia, *J. Lipid Res.* **31**, 17-26.
1377. Peterson, J., Olivecrona, T., and Bengtsson-Olivecrona, G. (1985) Distribution of lipoprotein lipase and hepatic lipase between plasma and tissues: effect of hypertriglyceridemia, *Biochim. Biophys. Acta* **837**, 262-270.
1378. Vilaró, S., Reina, M., Ramirez, I., and Llobera, M. (1986) Intralipid administration induces a lipoprotein lipase-like activity in the liver of adult rats, *Biochem. J.* **236**, 273-278.
1379. Saxena, U., Witte, L. D., and Goldberg, I. J. (1989) Release of endothelial cell lipoprotein lipase by plasma lipoproteins and free fatty acids, *J. Biol. Chem.* **264**, 4349-4355.

1380. Hultin, M., Bengtsson-Olivecrona, G., and Olivecrona, T. (1992) Release of lipoprotein lipase to plasma by triacylglycerol emulsions: comparison to the effect of heparin, *Biochim. Biophys. Acta* **1125**, 97-103.

1381. Vilella, E., Joven, J., Fernández, M., Vilaró, S., Brunzell, J. D., Olivecrona, T., and Bengtsson-Olivecrona, G. (1993) Lipoprotein lipase in human plasma is mainly inactive and associated with cholesterol-rich lipoproteins, *J. Lipid Res.* **34**, 1555-1564.

1382. Wallinder, L., Peterson, J., Olivecrona, T., and Bengtsson-Olivecrona, G. (1984) Hepatic and extrahepatic uptake of intravenously injected lipoprotein lipase, *Biochim. Biophys. Acta* **795**, 513-524.

1383. Vilaró, S., Llobera, M., Bengtsson-Olivecrona, G., and Olivecrona, T. (1988) Lipoprotein lipase uptake by the liver: localization, turnover, and metabolic role, *Am. J. Physiol.* **254**, G711-G722.

1384. Eisenberg, S., Sehayek, E., Olivecrona, T., and Vlodavsky, I. (1992) Lipoprotein lipase enhances binding of lipoproteins to heparan sulfate on cell surfaces and extracellular matrix, *J. Clin. Invest.* **90**, 2013-2021.

1385. Sato, T., Aoki, J., Nagai, Y., Dohmae, N., Takio, K., Doi, T., Arai, H., and Inoue, K. (1997) Serine phospholipid-specific phospholipase A that is secreted from activated platelets: a new member of the lipase family, *J. Biol. Chem.* **272**, 2192-2198.

1386. Hide, W. A., Chan, L., and Li, W.-H. (1992) Structure and evolution of the lipase family, *J. Lipid Res.* **33**, 167-178.

1387. Kirchgessner, T. G., Svenson, K. L., Lusis, A. J., and Schotz, M. C. (1987) The sequence of cDNA encoding lipoprotein lipase: a member of a lipase gene family, *J. Biol. Chem.* **262**, 8463-8466.

1388. Derewenda, Z. S., and Cambillau, C. (1991) Effects of gene mutations in lipoprotein and hepatic lipases as interpreted by a molecular model of the pancreatic triglyceride lipase, *J. Biol. Chem.* **266**, 23112-23119.

1389. Bruin, T., Groot, N. B., Jansen, J., and Kastelein, J. J. P. (1994) The C-terminus of lipoprotein lipase is essential for biological function but contains no domain for glycosylphosphatidylinositol anchoring, *Eur. J. Biochem.* **221**, 1019-1025.

1390. van Tilbeurgh, H., Roussel, A., Lalouel, J.-M., and Cambillau, C. (1994) Lipoprotein lipase. Molecular model based on the pancreatic lipase X-ray structure: consequences for heparin binding and catalysis, *J. Biol. Chem.* **269**, 4626-4633.

1391. Winkler, F. K., D'Arcy, A., and Hunziker, W. (1990) Structure of human pancreatic lipase, *Nature* **343**, 771-774.

1392. Faustinella, F., Smith, L. C., Semenkovich, C. F., and Chan, L. (1991) Structural and functional roles of highly conserved serines in human lipoprotein lipase: evidence that serine 132 is essential for enzyme catalysis, *J. Biol. Chem.* **266**, 9481-94385.

1393. Oka, K., Wang-Iverson, P., Parterniti Jr., J. R., and Brown, W. V. (1989) Interaction of lipoprotein lipase with heparin, in *Heparin and Related Polysaccharides: Structure and Activities*, eds., Ofosu, F. A., Danishefsky, I. & Hirsh, J., Ann. New York Acad. Sci., Vol. 556, pp. 173-180.

1394. Deeb, S. S., and Peng, R. (1989) Structure of the human lipoprotein lipase gene, *Biochemistry* **28**, 4131-4135.

1395. Kirchgessner, T. G., Chuat, J.-C., Heinzmann, C., Etienne, J., Guilhot, S., Svenson, K., Ameis, D., Pilon, C., D'Auriol, L., Andalibi, A., Schotz, M. C., Galibert, F., and Lusis, A. J. (1989) Organization of the human lipoprotein lipase gene and evolution of the lipase gene family, *Proc. Natl. Acad. Sci. U.S.A.* **86**, 9647-9651.

1396. Osbourne Jr., J. C., Bengtsson-Olivecrona, G., Lee, N. S., and Olivecrona, T. (1985) Studies on inactivation of lipoprotein lipase: role of the dimer to monomer dissociation, *Biochemistry* **24**, 5606-5611.

1397. Hill, J. S., Davis, R. C., Yang, D., Wen, J., Philo, J. S., Poon, P. H., Phillips, M. L., Kempner, E. S., and Wong, H. (1996) Human hepatic lipase subunit structure determination, *J. Biol. Chem.* **271**, 22931-22936.

1398. Schoonderwoerd, K., Hom, M. L., Luthjens, L. H., Vieira van Bruggen, D., and Jansen, H. (1996) Functional molecular mass of rat hepatic lipase in liver, adrenal gland and ovary is different, *Biochem. J.* **318**, 463-467.

1399. Shirai, K., and Jackson, R. L. (1982) Lipoprotein lipase-catalyzed hydrolysis of p-nitrophenyl butyrate. Interfacial activation by phospholipid vesicles, *J. Biol. Chem.* **257**, 1253-1258.

1400. Dugi, K. A., Dichek, H. L., Talley, G. D., Brewer Jr., H. B., and Santamarina-Fojo, S. (1992) Human lipoprotein lipase: the loop covering the catalytic site is essential for interaction with lipid substrates, *J. Biol. Chem.* **267**, 25086-25091.

1401. Kabayashi, J., Nishida, T., Ameis, D., Stahnke, G., Schots, M. C., Haoshimoto, H., Fgukamachi, I., Shirai, K., Saito, Y., and Yoshida, S. (1992) A heterozygous mutation (the codon for $Ser^{447} \rightarrow$ a stop codon) in lipoprotein lipase contributes to a defect in lipid interface recognition in a case with type I hyperlipidemia, *Biochem. Biophys. Res. Commun.* **182**, 70-77.

1402. Shirai, K., Saito, Y., and Yoshida, S. (1984) Post-heparin plasma hepatic triacylglycerol lipase-catalyzed tributyrin hydrolysis: effect of trypsin treatment, *Biochim. Biophys. Acta* **795**, 9-14.

1403. Uusi-Oukari, M., Ehnholm, C., and Jauhiainen, M. (1996) Inhibition of hepatic lipase by m-aminophenylboronate: application of phenylboronate affinity chromatography for purification of human postheparin plasma lipases, *J. Chromatog. B* **682**, 233-242.

1404. Deckelbaum, R. J., Ramakrishnan, R., Eisenberg, S., Olivecrona, T., and Bengtsson-Olivecrona, G. (1992) Triacylglycerol and phospholipid hydrolysis in human plasma lipoproteins: role of lipoprotein and hepatic lipase, *Biochemistry* **31**, 8544-8551.

1405. Davis, R. C., Wong, H., Nikazy, J., Wang, K., Han, Q., and Schotz, M. C. (1992) Chimeras of hepatic lipase and lipoprotein lipase: domain localization of enzyme-specific properties, *J. Biol. Chem.* **267**, 21499-21504.

1406. Assmann, G., Krauss, R. M., Fredrickson, D. S., and Levy, R. J. (1973) Positional specificity of triglyceride lipases in post-heparin plasma, *J. Biol. Chem.* **248**, 7184-7190.

1407. Henderson, A. D., Richmond, W., and Elkeles, R. S. (1993) Hepatic and lipoprotein lipases selectively assayed in postheparin plasma, *Clin. Chem.* **39**, 218-223.

1408. Bensadoun, A., Ehnholm, C., Steinberg, D., and Brown, W. V. (1974) Purification and characterization of lipoprotein lipase from pig adipose tissue, *J. Biol. Chem.* **249**, 2220-2227.

1409. Iverius, P.-H., and Lindahl, U. (1972) Effects of heparin on lipoprotein lipase from bovine milk, *J. Biol. Chem.* **247**, 6610-6616.

1410. Semenkovich, C. F., Luo, C.-C., Nakanishi, M. K., Chen, S.-H., Smith, L. C., and Chan, L. (1990) *In vitro* expression and site-specific mutagenesis of the cloned human lipoprotein lipase gene: potential N-linked glycosylation site asparagine 43 is important for both enzyme activity and secretion, *J. Biol. Chem.* **265**, 5429-5433.

1411. Clarke, A. R., and Holbrook, J. J. (1985) The mechanism of activation of lipoprotein lipase by apolipoprotein C-II: the formation of a protein-protein complex in free solution and at a triacylglycerol/water interface, *Biochim. Biophys. Acta* **827**, 358-368.

1412. LaRosa, J. C., Levy, R. I., Herbert, P., Lux, S. E., and Fredrickson, D. S. (1970) A specific apoprotein activator for lipoprotein lipase, *Biochem. Biophys. Res. Commun.* **41**, 57-62.

1413. Heng, Q., Blackett, P., Jackson, K. W., McConathy, W. J., and Wang, C.-S. (1990) C-Terminal domain of apolipoprotein CII as both activator and competitive inhibitor of lipoprotein lipase, *Biochem. J.* **269**, 403-407.

1414. Hoffmann, M. M., and Stoffel, W. (1996) Construction and functional characterization of recombinant fusion proteins of human lipoprotein lipase and apolipoprotein CII, *Eur. J. Biochem.* **237**, 545-552.

1415. Thuren, T., Wilcox, R. W., Sisson, P., and Waite, M. (1991) Hepatic lipase hydrolysis of lipid monolayers: regulation by apolipoproteins, *J. Biol. Chem.* **266**, 4853-4861.

1416. Parkin, S. M., Speake, B. K., and Robinson, D. S. (1982) Purification and characterization of rat adipose tissue lipoprotein lipase, *Biochem. J.* **207**, 485-495.
1417. Kinnunen, P. K. J., Huttunen, J. K., and Ehnholm, C. (1976) Properties of purified bovine milk lipoprotein lipase, *Biochim. Biophys. Acta* **450**, 342-351.
1418. Semb, H., and Olivecrona, T. (1986) Lipoprotein lipase in guinea pig tissues: molecular size and rates of synthesis, *Biochim. Biophys. Acta* **878**, 330-337.
1419. Garfinkel, A. S., and Schotz, M. C. (1987) Lipoprotein lipase, in *Plasma Lipoproteins*, ed. Gotto, A. M., Elsevier, Amsterdam, pp. 335-356.
1420. Bensadoun, A. (1991) Lipoprotein lipase, *Annu. Rev. Nutr.* **11**, 217-237.
1421. Hahn, P. F. (1943) Abolishment of alimentary lipemia following injection of heparin, *Science* **98**, 19-25.
1422. Olivecrona, T., and Bengtsson, G. (1981) Heparin and lipoprotein lipase, in *Chemistry and Biology of Heparin*, eds., Lundblad, R. L., Brown, W. V., Mann, K. G. & Roberts, H. R., Elsevier/North Holland, Amsterdam, pp. 187-194.
1423. Beisiegel, U., Weber, W., and Bengtsson-Olivecrona, G. (1991) Lipoprotein lipase enhances the binding of chylomicrons to low density lipoprotein receptor-related protein, *Proc. Natl. Acad. Sci. U.S.A.* **88**, 8342-8346.
1424. Nykjær, A., Bengtsson-Olivecrona, G., Lookene, A., Moestrup, S. K., Petersen, C. M., Weber, W., Beisiegel, U., and Gliemann, J. (1993) The α_2-macroglobulin receptor/low density lipoprotein receptor-related protein binds lipoprotein lipase and β-migrating very low density lipoprotein associated with the lipase, *J. Biol. Chem.* **268**, 15048-15055.
1425. Edwards, I. J., Goldberg, I. J., Parks, J. S., Ku, H., and Wagner, W. D. (1993) Lipoprotein lipase enhances the interaction of low density lipoproteins with artery-derived extracellular matrix proteoglycans, *J. Lipid Res.* **34**, 1155-1163.
1426. Chappell, D. A., Fry, G. L., Waknitz, M. A., Muhonen, L. E., Pladet, M. W., Iverius, P.-H., and Strickland, D. K. (1993) Lipoprotein lipase induces catabolism of normal triglyceride-rich lipoproteins via the low density lipoprotein receptor-related protein/α_2-macroglobulin receptor *in vitro*, *J. Biol. Chem.* **268**, 14168-14175.
1427. Mulder, M., Lombardi, P., Jansen, H., van Berkel, T. J. C., Frants, R. R., and Hovekes, L. M. (1992) Heparan sulphate proteoglycans are involved in the lipoprotein-mediated enhancement of the cellular binding of very low density and low density lipoproteins, *Biochem. Biophys. Res. Commun.* **185**, 582-587.
1428. Hoogewerf, A. J., and Bensadoun, A. (1991) Occurrence of sulfate in an asparagine-linked complex oligosaccharide of chicken adipose lipoprotein lipase, *J. Biol. Chem.* **266**, 1048-1057.
1429. Liu, G., Hultin, M., Østergaard, P., and Olivecrona, T. (1992) Interaction of size-fractionated heparins with lipoprotein lipase and hepatic lipase in the rat, *Biochem. J.* **285**, 731-736.
1430. Kirchgessner, T. G., LeBoeu, R. C., Langner, C. A., Zollman, S., Chang, C. H., Taylor, B. A., Schotz, M. C., Gordon, J. I., and Lusis, A. J. (1989) Genetic and developmental regulation of the lipoprotein lipase gene: loci both distal and proximal to the lipoprotein lipase structural gene control enzyme expression, *J. Biol. Chem.* **264**, 1473-1482.
1431. Semb, H., and Olivecrona, T. (1986) Nutritional regulation of lipoprotein lipase in guinea pig tissues, *Biochim. Biophys. Acta* **876**, 249-255.
1432. Doolittle, M. H., Ben-Zeev, O., Olovson, J., Martin, D., and Kirchgessner, T. G. (1990) The response of lipoprotein lipase to feeding and fasting: evidence of posttranslational regulation, *J. Biol. Chem.* **265**, 4570-4577.
1433. Faustinella, F., Chang, A., Van Biervliet, J. P., Rossene, M., Vinaimant, N., Smith, L. C., Chen, S.-H., and Chan, L. (1991) Catalytic triad residue mutation (Asp^{156}→Gly) causing familial lipoprotein lipase deficiency: co-inheritance with a nonsense mutation (Ser^{447}→Term) in a Turkish family, *J. Biol. Chem.* **266**, 14418-14424.
1434. Kobayashi, J., Nishiba, T., Ameiss, D., Stahnke, G., Schotz, M. C., Hashimoto, H., Fukamachi, I., Shirai, K., Saito, Y., and Yoshida, S. (1992) A heterozygous mutation (the

codon for Ser[447]→a stop codon) in lipoprotein lipase contributes to a defect in lipid interface recognition in a case with type I hyperlipidemia, *Biochem. Biophys. Res. Commun.* **182**, 70-77.

1435. Olivecrona, T., Chernick, S. S., Bengtsson-Olivecrona, G., Paterniti Jr., J. R., Brown, W. V., and Scow, R. O. (1985) Combined lipase deficiency (cld/cld) in mice. demonstration that an inactive form of lipoprotein lipase is synthesized, *J. Biol. Chem.* **260**, 2552-2557.

1436. Ginzinger, D. G., Lewis, M. E. S., Ma, Y., Jones, B. R., Liu, G., Jones, S. D., and Hayden, M. R. (1996) A mutation in the lipoprotein lipase gene is the molecular basis of chylomicronemia in a colony of domestic cats, *J. Clin. Invest.* **97**, 1257-1266.

1437. Buscà, R., Martínez, M., Vilella, E., Pognonec, P., Deeb, S., Auwerx, J., Reina, M., and Vilaró, S. (1996) The mutation Gly[142]→Glu in human lipoprotein lipase produces a missorted protein that is diverted to lysosomes, *J. Biol. Chem.* **271**, 2139-2146.

1438. Zieve, F. J., and Zieve, L. (1972) Post-heparin phospholipase and post-heparin lipase have different tissue origins, *Biochem. Biophys. Res. Commun.* **47**, 1480-1485.

1439. Iverius, P.-H. (1972) The interaction between human plasma lipoproteins and connective tissue glycosaminoglycans, *J. Biol. Chem.* **247**, 2607-2613.

1440. Chevreuil, O., HUltin, M., Østergaard, P., and Olivecrona, T. (1996) Heparin-decasaccharides impair the catabolism of chylomicrons, *Biochem. J.* **320**, 437-444.

1441. Merchant, Z. M., Erbe, E. E., Eddy, W. P., Patel, D., and Linhardt, R. J. (1986) Effect of very low molecular weight heparin-derived oligosaccharides on lipoprotein lipase release in rabbits, *Atherosclerosis* **62**, 151-158.

1442. Eckel, R. H., Fujimoto, W. Y., and Brunzell, J. D. (1977) Development of lipoprotein lipase in cultured 3T3-L1 cells, *Biochem. Biophys. Res. Commun.* **78**, 288-293.

1443. Chapman, A. B., Knight, D. M., Dieckmann, B. S., and Ringold, G. M. (1984) Analysis of gene expression during differentiation of adipogenic cells in culture and hormonal control of the developmental program, *J. Biol. Chem.* **259**, 15548-15555.

1444. Spooner, P. M., Chernick, S. S., Garrison, M. M., and Scow, R. O. (1979) Development of lipoprotein lipase activity and accumulation of triacylglycerol in differentiating 3T3-L1 adipocytes, *J. Biol. Chem.* **254**, 1305-1311.

1445. Olivercrona, T., Chernick, S. S., Bengtsson-Olivecrona, G., Garrison, M., and Crow, R. O. (1987) Synthesis and secretion of lipoprotein lipase in 3T3-L1 adipocytes: demonstration of inactive forms of lipase in cells, *J. Biol. Chem.* **262**, 10748-10759.

1446. Vannier, C., Amri, E.-Z., Etienne, J., Négrel, R., and Ailhaud, G. (1985) Maturation and secretion of lipoprotein lipase in cultured adipose cells. I. Intracellular activation of the enzyme, *J. Biol. Chem.* **260**, 4424-4431.

1447. Vannier, C., and Ailhaud, G. (1989) Biosynthesis of lipoprotein lipase in cultured mouse adipocytes. II. Processing, subunit assembly, and intracellular transport, *J. Biol. Chem.* **264**, 13206-13216.

1448. Iverius, P.-H., and Östlund-Lindqvist (1976) Lipoprotein lipase from bovine milk: isolation procedure, chemical characterization, and molecular weight analysis, *J. Biol. Chem.* **251**, 7791-7795.

1449. Ben-Zeev, O., Doolittle, M. H., Davis, R. C., Elovson, J., and Schotz, M. C. (1992) Maturation of lipoprotein lipase: expression of full catalytic activity requires glucose trimming but not translocation to the cis-Golgi compartment, *J. Biol. Chem.* **267**, 6219-6227.

1450. Ong, J. M., and Kern, P. A. (1989) The role of glucose and glycosylation in the regulation of lipoprotein lipase synthesis and secretion in rat adipocytes, *J. Biol. Chem.* **264**, 3177-3182.

1451. Semb, H., and Olivecrona, T. (1989) The relation between glycosylation and activity of guinea pig lipoprotein lipase, *J. Biol. Chem.* **264**, 4195-4200.

1452. Rothblat, G. H., and DeMartinis, F. D. (1977) Release of lipoprotein lipase from rat adipose tissue cells grown in culture, *Biochem. Biophys. Res. Commun.* **78**, 45-50.

1453. Braun, J. E. A., and Severson, D. L. (1993) Release of lipoprotein lipase from cardiac myocytes by low-molecular weight heparin, *Lipid* **28**, 59-61.

1454. Shimada, K., Gill, P. J., Silbert, J. E., Douglas, W. H. J., and Fanburg, B. L. (1981) Involvement of cell surface heparan sulfate in the binding of lipoprotein lipase to cultured bovine endothelial cells, *J. Clin. Invest.* **68**, 995-1002.

1455. Cheng, C.-F., Oosta, G. M., Bensadoun, A., and Rosenberg, R. D. (1981) Binding of lipoprotein lipase to endothelial cells in culture, *J. Biol. Chem.* **256**, 12893-12898.

1456. Saxena, U., Klein, M. G., and Goldberg, I. J. (1991) Transport of lipoprotein lipase across endothelial cells, *Proc. Natl. Acad. Sci. U.S.A.* **88**, 2254-2258.

1457. Björntorp, P., Karlsson, M., Pettersson, P., and Sypniewska, G. (1980) Differentiation and function of rat adipocyte precursor cells in primary culture, *J. Lipid Res.* **21**, 714-723.

1458. Edwards, I. J., Xu, H., Obunike, J. C., Goldberg, I. J., and Wagner, W. D. (1995) Differentiated macrophages synthesize a heparan sulfate proteoglycan and an oversulfated chondroitin sulfate proteoglycan that bind lipoprotein lipase, *Arterioscler. Thromb. Vasc. Biol.* **15**, 400-409.

1459. Spooner, P. M., Chernick, S. S., Garrison, M. M., and Scow, R. O. (1979) Insulin regulation of lipoprotein lipase activity and release in 3T3-L1 adipocytes: separation and dependence of hormonal effects on hexose metabolism and synthesis of RNA and protein, *J. Biol. Chem.* **254**, 10021-10029.

1460. Sakayama, K., Masuno, H., Okumura, H., Shibata, T., and Okuda, H. (1996) Recombinant human tumor necrosis factor-α suppresses synthesis, activity and secretion of lipoprotein lipase in cultures of human osteosacroma cell line, *Biochem. J.* **316**, 813-817.

1461. Sakayama, K., Masuno, H., Okumura, H., Shibata, T., and Okuda, H. (1996) Recombinant human tumour necrosis factor-α suppresses synthesis, activity and secretion of lipoprotein lipase in cultures of a human osteosarcoma cell line, *Biochem. J.* **316**, 813-817.

1462. Olivecrona, G., Vilaró, S., Esko, J. D., and Olivecrona, T. (1996) Synthesis and secretion of lipoprotein lipase in heparan sulfate-deficient Chinese hamster ovary cells, *Isr. J. Med. Sci.* **32**, 430-444.

1463. Saxena, U., Klein, M. G., and Goldberg, I. J. (1991) Identification and characterization of the endothelial cell surface lipoprotein lipase receptor, *J. Biol. Chem.* **266**, 17516-17521.

1464. Khoo, J. C., Mahoney, E. M., and Witztum, J. L. (1981) Secretion of lipoprotein lipase by macrophages in culture, *J. Biol. Chem.* **256**, 7105-7108.

1465. Vannier, C., and Ailhaud, G. (1986) A continuous flow method for the study of lipoprotein lipase secretion in adipose cells, *Biochim. Biophys. Acta* **875**, 324-333.

1466. Semb, H., and Olivecrona, T. (1987) Mechanisms for turnover of lipoprotein lipase in guinea pig adipocytes, *Biochim. Biophys. Acta* **921**, 104-115.

1467. Cupp, M., Bensadoun, A., and Melford, K. (1987) Heparin decreases the degradation rate of lipoprotein lipase in adipocytes, *J. Biol. Chem.* **262**, 6383-6388.

1468. Berryman, D. E., and Bensadoun, A. (1995) Heparan sulfate proteoglycans are primarily responsible for the maintenance of enzyme activity, binding, and degradation of lipoprotein lipase in Chinese hamster ovary cells, *J. Biol. Chem.* **270**, 24525-24531.

1469. Liu, G., Bengtsson-Olivecrona, G., and Olivecrona, T. (1993) Assembly of lipoprotein lipase in perfused guinea pig hearts, *Biochem. J.* **292**, 277-282.

1470. Friedman, G., Chajek-Shaul, T., Olivecrona, T., Stein, O., and Stein, Y. (1982) Fate of milk ^{125}I-labeled lipoprotein lipase in cells in culture: comparison of lipoprotein lipase- and n-lipoprotein lipase-synthesizing cells, *Biochim. Biophys. Acta.* **711**, 114-122.

1471. Yang, C.-Y., Gu, Z.-V., Yang, H.-K., Rohde, M. F., Gotto Jr., A. M., and Pownall, H. J. (1989) Structure of bovine milk lipoprotein lipase, *J. Biol. Chem.* **2674**, 16822-16827.

1472. Chajek-Shaul, T., Friedman, G., Knobler, H., Stein, O., Etienne, J., and Stein, Y. (1985) Importance of the different steps of glycosylation for the activity and secretion of lipoprotein lipase in rat preadipocytes studied with monensin and tunicamycin, *Biochim. Biophys. Acta.* **837**, 123-134.

1473. Buscà, R., Pujana, M. A., Pognonec, P., Auwerx, J., Deeb, S. S., Reina, M., and Vilaró, S. (1995) Absence of N-glycosylation at asparagine 43 in human LPL induces its accumulation in the rough endoplasmic reticulum and alters this cellular compartment, *J. Lipid Res.* **36**, 939-951.

1474. Masuno, H., Schultz, C. J., Park, K.-W., Blanchette-Mackie, E. J., Mateo, C., and Scow, R. O. (1991) Glycosylation, activity and secretion of lipoprotein lipase in cultured brown adipocytes of newborn mice, *Biochem. J.* **277**, 801-809.

1475. Masuno, H., and Okuda, H. (1994) Role of processing of the oligosaccharide chains in the affinity of lipoprotein lipase for heparin, *Biochim. Biophys. Acta* **1212**, 125-128.

1476. Park, J.-W., Oh, M.-S., Yang, J.-Y., Park, B.-H., Rho, H.-W., Lim, S.-N., Jhee, E.-C., and Kim, H.-R. (1995) Glycosylation, dimerization, and heparin affinity of lipoprotein lipase in 3T3-L1 adipocytes, *Biochim. Biophys. Acta* **1254**, 45-50.

1477. Masuno, H., Blanchette-Mackie, E. J., Schultz, C. J., Spaeth, A. E., Scow, R. O., and Okuda, H. (1992) Retention of glucose by N-linked oligosaccharide chains impedes expression of lipoprotein lipase activity: effect of castanospermine, *J. Lipid Res.* **33**, 1343-1349.

1478. Park, J.-W., Blanchette-Mackie, E. J., and Scow, R. O. (1996) Brefeldin A enables synthesis of active lipoprotein lipase in cld/cld and castanospermine-treated mouse brown aditpcytes via translocation of Golgi components to endoplasmic reciculum, *Biochem. J.* **317**, 125-134.

1479. Sivaram, P., Klein, M. G., and Goldberg, I. J. (1992) Identification of a heparin-releasable lipoprotein binding protein from endothelial cells, *J. Biol. Chem.* **267**, 16517-16522.

1480. Wölle, J., Ferguson, E., Devall, L. J., Neuton, R. S., and Saxena, U. (1995) Identification of a novel 85-kDa lipoprotein lipase binding protein on human aortic endothelial cell surface, *Biochem. Biophys. Res. Commun.* **216**, 906-912.

1481. Sivaram, P., Vanni-Reyes, T., and Goldberg, I. J. (1996) Endothelial cells synthesize and process apolipoprotein B, *J. Biol. Chem.* **271**, 15261-15266.

1482. Garfinkel, A. S., Kempner, E. S., Ben-Zeev, O., Nikazy, J., James, S. J., and Schotz, M. C. (1983) Lipoprotein lipase: size of the functional unit determined by radiation inactivation, *J. Lipid Res.* **24**, 775-780.

1483. Kerfélec, B., LaForge, K. S., Pulgserver, A., and Scheele, G. (1986) Primary structures of canine pancreatic lipase and phospholipase A2 messenger RNA, *Pancreas* **1**, 430-437.

1484. DeCaro, J., Boubouard, M., Bonicel, J., Guidoni, B. A., Desnuelle, P., and Rovery, M. (1981) Porcine pancreatic lipase: completion of the primary structure, *Biochim. Biophys. Acta.* **671**, 129-138.

1485. Hermoso, J., Pignol, D., Kerfelec, B., Crenon, I., Chapus, C., and Fontecilla-Camps, J. C. (1996) Lipase activation by nonionic detergents: the crystal structure of the porcine lipase-colipase-tetraethylene glycol monooctyl ether complex, *J. Biol. Chem.* **271**, 18007-18016.

1486. van Tilbeurgh, H., Sarda, L., Verger, R., and Cambillau, C. (1992) Structure of the pancreatic lipase-procolipase complex, *Nature* **359**, 159-162.

1487. van Tilbeurgh, H., Egloff, M.-P., Martinez, C., Rugani, N., Verger, R., and Cambillau, C. (1993) Interfacial activation of the lipase-procolipase complex by mixed micelles revealed by X-ray crystallography, *Nature* **362**, 814-820.

1488. Schmit, G. D., Momsen, M. M., Owen, W. G., Naylor, S., Tomlinson, A., Wu, G., and Stark, R. E. (1996) The affinities of procolipase and colipase for interfaces are regulated by lipids, *Biophys. J.* **71**, 3421-3429.

1489. Clarke, A. R., Luscombe, M., and Holbrook, J. J. (1983) The effect of chain length of heparin on its interaction with lipoprotein lipase, *Biochim. Biophys. Acta* **747**, 130-137.

1490. Dugi, K. A., Dichek, H. L., and Santamarina-Fojo, S. (1995) Human hepatic and lipoprotein lipase: the loop covering the catalytic site mediates lipase substrate specificity, *J. Biol. Chem.* **270**, 25396-35401.

1491. Kobayashi, J., Applebaum-Bowden, D., Dugi, K. A., Brown, D. R., Kashyap, V. S., Parrott, C., Duarte, C., Maeda, N., and Santamarina-Fojo, S. (1996) Analysis of protein structure-function *in vivo*. Adenovirus-mediated transfer of lipase lid mutants in hapatic lipase-deficient mice, *J. Biol. Chem.* **271**, 26296-26301.

1492. Carrière, F., Thirstrup, K., Hjorth, S., Ferrato, F., Nielsen, P. F., Withers-Martinez, C., Cambillau, C., Boel, E., L., T., and Verger, R. (1997) Pancreatic lipase structure-function relationships by domain exchange, *Biochemistry* **36**, 239-248.

1493. Wong, H., Davis, R. C., Nikazy, J., Seebart, K. E., and Schotz, M. C. (1991) Domain exchange: characterization of a chimeric lipase of hepatic lipase and lipoprotein lipase, *Proc. Natl. Acad. Sci. U.S.A.* **88,** 11290-11294.

1494. Senda, M., Oka, K., Brown, W. V., Qasba, P. K., and Furiuchi, Y. (1987) Molecular cloning and sequence of a cDNA coding for bovine lipoprotein lipase, *Proc. Natl. Acad. Sci. U.S.A.* **84,** 4369-4373.

1495. Olivecrona, T., Egelrud, T., Iverius, P.-H., and Lindahl, U. (1971) Evidence for an ionic binding of lipoprotein lipase to heparin, *Biochem. Biophys. Res. Commun.* **43,** 524-529.

1496. Hata, A., Ridinger, D. N., Sutherland, S. D., Emi, M., Kwong, L. K., Shuhua, J., Lubbers, A., Guy-Grand, B., Basdevant, A., Iverius, P.-H., Wilson, D. E., and Lalouel, J.-M. (1992) Missense mutations in exon 5 of the human lipoprotein lipase gene. Inactivation correlates with loss of dimerization, *J. Biol. Chem.* **267,** 21032-20139.

1497. Bengtsson-Olivecrona, G., and Olivecrona, T. (1985) Binding of active and inactive forms of lipoprotein lipase to heparin: effects of pH, *Biochem. J.* **226,** 409-413.

1498. Bengtsson, G., and Olivecrona, T. (1977) Interaction of lipoprotein lipase with heparin-Sepharose, *Biochem. J.* **167,** 109-119.

1499. Bengtsson, G., and Olivecrona, T. (1980) Interaction of lipoprotein lipase with native and modified heparin-like polysaccharides, *Biochem. J.* **189,** 625-633.

1500. Bengtsson, G., and Olivecrona, T. (1977) Interaction of heparin with proteins: demonstration of different binding sites for antithrombin and lipoprotein lipase, *FEBS Lett.* **79,** 59-63.

1501. Jackson, R. L., Socorro, L., Fletcher, G. M., and Cardin, A. D. (1985) Heparin binding to lipoprotein lipase and low density lipoproteins, *FEBS Lett.* **190,** 297-300.

1502. Parthasarathy, N., Goldberg, I. J., Sivaram, P., Mulloy, B., Flory, D. M., and Wagner, W. D. (1994) Oligosaccharide sequences of endothelial cell surface heparan sulfate proteoglycan with affinity for lipoprotein lipase, *J. Biol. Chem.* **269,** 22391-22396.

1503. Lookene, A., Chevreuil, O., Østergaard, P., and Olivecrona, G. (1996) Interaction of lipoprotein lipase with heparin fragments and with heparan sulfate: stoichiometry, stabilization, and kinetics, *Biochemistry* **35,** 12155-12163.

1504. Persson, B., Bengtsson-Olivecrona, G., Enerbäck, S., Olivecrona, T., and Jörnvall, H. (1989) Structural features of lipoprotein lipase. Lipase family relationships, binding interactions, non-equivalence of lipase cofactors, vitellogenin similarities and functional subdivision of lipoprotein lipase , *Eur. J. Biochem.* **179,** 39-45.

1505. Komaromy, M. C., and Schotz, M. C. (1987) Cloning of rat hepatic lipase cDNA: evidence for a lipase gene family, *Proc. Natl. Acad. Sci. U.S.A.* **84,** 1526-1530.

1506. Stahnke, G., Sprengel, R., Augustin, J., and Will, H. (1987) Human hepatic triglyceride lipase: cDNA cloning, amino acid sequence and expression in a cultured cell line, *Differentiation* **35,** 45-52.

1507. Datta, S., Luo, C.-C., Li., W.-H., VanTuinen, P., Ledbetter, D. H., Brown, M. A., Chen, S.-H., Liu., S.-W., and Chan, L. (1988) Human hepatic lipase. Cloned cDNA sequence, restriction fragments, full length polymorphisms, chromosomal localization, and evolutionary relationships with lipoprotein lipase and pancreatic lipase, *J. Biol. Chem.* **263,** 1107-1110.

1508. Martin, G. A., Busch, S. J., Meredith, G. D., Cardin, A. D., Blankenship, D. T., Mao, S. J. T., Rechtin, A. E., Woods, C. W., Racke, M. M., Schafer, M. P., Fitzgerald, M. C., Burke, D. M., Flanagan, M. A., and Jackson, R. L. (1988) Isolation and cDNA sequence of human postheparin plasma hepatic triglyceride lipase, *J. Biol. Chem.* **263,** 10907-10914.

1509. Semenkovich, C. F., Chen, S.-H., Wims, M., Luo, C.-C., Li, W.-H., and Chan, L. (1989) Lipoprotein lipase and hepatic lipase mRNA tissue specific expression, developmental regulation, and evolution, *J. Lipid Res.* **30,** 423-431.

1510. Cai, S.-J., Wong, D. M., Chen, S.-H., and Chan, L. (1989) Structure of the human hepatic triglyceride lipase gene, *Biochemistry* **28,** 8966-8971.

1511. Ameis, D., Stahnke, G., Kabayashi, J., McLean, J., Lee, G., Büscher, M., Schotz, M. C., and Will, H. (1990) Isolation and characterization of the human hepatic lipase gene, *J. Biol. Chem.* **265**, 6552-6555.

1512. Lookene, A., and Bengtsson-Olivecrona, G. (1993) Chymotryptic cleavage of lipoprotein lipase: identification of cleavage sites and functional studies of the truncated molecule, *Eur. J. Biochem.* **213**, 185-194.

1513. Lookene, A., Groot, N. B., Kastelein, J. J. P., Olivecrona, G., and Bruin, T. (1997) Mutation of tryptophan residues in lipoprotein lipase: effects on stability, immunoreactivity, and catalytic properties, *J. Biol. Chem.* **272**, 766-772.

1514. Pepe, G., Chimienti, G., Resta, F., Di Perna, V., Tarricone, C., Lovecchio, M., Colacicco, A. M., and Capurso, A. (1994) A new Italian case of lipoprotein lipase deficiency: a leu^{365}-val change resulting is loss of enzyme activity, *Biochem. Biophys. Res. Commun.* **199**, 570-576.

1515. Berryman, D. E., and Bensadoun, A. (1993) Site-directed mutagenesis of a putative heparin binding domain of avian lipoprotein lipase, *J. Biol. Chem.* **268**, 3272-3276.

1516. Ma, Y., Henderson, H. E., Siu, M.-S., Zhang, H., Forsythe, I. J., Clarke-Lewis, I., Hayden, M. R., and Brunzell, J. D. (1994) Mutagenesis in four candidate heparin binding regions (residues 279-282, 291-304, 390-393, and 439-448) and identification of residues affecting heparin binding of human lipoprotein lipase, *J. Lipid Res.* **35**, 2049-2059.

1517. Chajek-Shaul, T., Halimi, O., Ben-Naim, M., Stein, O., and Stein, Y. (1989) Phosphatidylinositol-specific phospholipase C releases lipoprotein lipase from the heparin releasable pool in rat heart cell cultures, *Biochim. Biophys. Acta* **1014**, 178-183.

1518. Hata, A., Ridinger, D. N., Sutherland, S. D., Emi, M., Shuhua, J., Myers, R. L., Ren, K., Cheng, T., Inoue, I., Wilson, D. E., Iverius, P.-H., and Lalouel, J.-M. (1993) Binding of lipoprotein lipase to heparin: identification of five critical residues in two distinct segments of the amino-terminal domain, *J. Biol. Chem.* **268**, 8447-8457.

1519. Emi, M., Wilson, D. E., Iverius, P.-H., Wu, L., Hata, A., Hegele, R., Williams, R. R., and Lalouel, J.-M. (1990) Missense mutation (Gly → Glu188) of human lipoprotein lipase imparting functional deficiency, *J. Biol. Chem.* **265**, 5910-5916.

1520. Beg, O. U., Meng, M. S., Skarlatos, S. I., Previato, L., Brunzell, J. D., Brewer Jr., H. B., and Fojo, S. S. (1990) Lipoprotein lipase Bethesda: a single amino acid substitution (Ala-176 ⑧ Thr) leads to abnormal heparin binding and loss of enzymic activity, *Proc. Natl. Acad. Sci. U.S.A.* **87**, 3474-3478.

1521. Yamamoto, T., Davis, C. G., Brown, M. S., Schneider, W. J., Casey, M. L., Goldstein, J. L., and Russell, D. W. (1984) The human LDL receptor: a cysteine-rich protein with multiple Alu sequences in its mRNA, *Cell* **39**, 27-38.

1522. Goldstein, J. L., Brown, M. S., Anderson, R. G. W., Russell, D. W., and Schneider, W. J. (1985) Receptor-mediated endocytosis: concepts emerging from the LDL system, *Annu. Rev. Cell Biol.* **1**, 1-39.

1523. Brown, M. S., and Goldstein, J. L. (1986) A receptor-mediated pathway for cholesterol homeostasis, *Science* **232**, 34-47.

1524. Gianturco, S. H., and Bradley, W. A. (1987) Lipoprotein receptors, in *Plasma Lipoproteins*, ed. Gotto Jr., A. M., Elsevier, Amsterdam, pp. 183-220.

1525. Brown, M. S., Herz, J., Kowal, R. C., and Goldstein, J. L. (1991) LRP target or decoy?, *Curr. Opin. Lipidol.* **2**, 65-72.

1526. Willnow, T. E., Goldstein, J. L., Orth, K., Brown, M. S., and Herz, J. (1992) Low density lipoprotein receptor-related protein and gp330 bind similar ligands, including plasminogen activator-inhibitor complexes and lactoferrin, an inhibitor of chylomicron remnant clearance, *J. Biol. Chem.* **267**, 26172-26180.

1527. Moestrup, S. K. (1994) The α_2-macroglobulin receptor and epithelial glycoprotein-330: two giant receptors mediating endocytosis of multiple ligands, *Biochim. Biophys. Acta* **1197**, 197-213.

1528. Krieger, M., and Herz, J. (1994) Structures and functions of multiligand lipoprotein receptors: macrophage scavenger receptors and LDL receptor-related protein (LRP), *Annu. Rev. Biochem.* **63**, 601-637.

1529. Strickland, D. K., Kounnas, M. Z., and Argraves, W. S. (1995) LDL receptor-related protein: a multiligand receptor for lipoprotein and proteinase catabolism, *FASEB J.* **9**, 890-898.

1530. Kowal, R. C., Herz, K., Goldstein, J. L., Esser, V., and Brown, M. S. (1989) Low density lipoprotein receptor-related protein mediates uptake of cholesteryl esters derived from apoprotein E-enriched lipoproteins, *Proc. Natl. Acad. Sci. U.S.A.* **86**, 5810-5814.

1531. Williams, K. J., Fless, G. M., Petrie, K. A., Snyder, M. L., Brocia, R. W., and Swenson, T. L. (1992) Mechanisms by which lipoprotein lipase alters cellular metabolism of lipoprotein(a), low density lipoprotein, and nascent lipoproteins: roles for low density lipoprotein receptors and heparan sulfate proteoglycans, *J. Biol. Chem.* **267**, 13284-13292.

1532. Windler, E., and Havel, R. J. (1985) Inhibitory effects of C apolipoproteins from rats and humans on the uptake of triglyceride-rich lipoproteins and their remnants by perfused rat liver, *J. Lipid Res.* **26**, 556-565.

1533. Kowal, R. C., Herz, J., Weisgraber, K. H., Mahley, R. W., Brown, M. S., and Goldstein, J. L. (1990) Opposing effects of apolipoproteins E and C on lipoprotein binding to low density lipoprotein receptor-related protein, *J. Biol. Chem.* **265**, 10771-10779.

1534. Weisgraber, K. H., Mahley, R. W., Kowal, R. C., Herz, J., Goldstein, J. L., and Brown, M. S. (1990) Apolipoprotein C-I modulates the interaction of apolipoprotein E with β-migrating very low density lipoproteins (β-VLDL) and inhibits binding of β-VLDL to low density lipoprotein receptor-related protein, *J. Biol. Chem.* **265**, 22453-22459.

1535. Mahley, R. W. (1988) Apolipoprotein E: cholesterol transport protein with expanding role in cell biology, *Science* **240**, 622-630.

1536. Mahley, R. W., and Rall Jr., S. C. (1989) Type III hyperlipoproteinemia (dysbetalipoproteinemia): the role of apolipoprotein E in normal and abnormal lipoprotein metabolism, in *The Metabolic Basis of Inherited Disease, Sixth Edition*, eds., Scriver, C. R., Beaubet, A. L., Sly, W. S. & Valle, D., McGraw-Hill, New York, pp. 1195-1213.

1537. Rall Jr., S. C., and Mahley, R. W. (1992) The role of apolipoprotein E genetic variants in lipoprotein disorders, *J. Intern. Med.* **231**, 653-659.

1538. Herz, J., Hamann, U., Rodne, S., Myklebost, O., Gausepohl, H., and Stanley, K. K. (1988) Surface location and high affinity for calcium of a 500-kd liver membrane protein closely related to the LDL-receptor suggest a physiological role as lipoprotein receptor, *EMBO J.* **7**, 4119-4127.

1539. Mahley, R. W., Ji, Z.-S., Brecht, W. J., Miranda, R. D., and He, D. (1994) Role of heparan sulfate proteoglycans and the LDL receptor-related protein in remnant lipoprotein metabolism, *Ann. N. Y. Acad. Sci.* **737**, 39-52.

1540. Mamo, J. C. L., Elsegood, C. L., Gennat, H. C., and Yu, K. (1996) Degradation of chylomicron remnants by macrophages occurs via phagocytosis, *Biochemistry* **35**, 10210-10214.

1541. Fredrickson, D. S., Levy, R. I., and Lindgren, F. T. (1969) A comparison of heritable abnormal lipoprotein patterns as defined by two techniques, *J. Clin. Invest.* **47**, 2446-2457.

1542. Mahley, R. W., and Holcombe, K. S. (1977) Alteration of the plasma lipoproteins and apoproteins following cholesteral feeding in the rat, *J. Lipid Res.* **18**, 314-324.

1543. van Dijk, M. C. M., Ziere, G. J., Boers, W., Linthorst, C., Bijsterbosch, M. K., and van Berkle, T. J. C. (1991) Recognition of chylomicron remnants and β-migrating very-low-density lipoproteins by the remnant receptor of parenchymal liver cells is distinct from the liver α_2-macroglobulin-recognition site, *Biochem. J.* **279**, 863-870.

1544. Herz, J., Kowal, R. C., Ho, Y. K., Brown, M. S., and Goldstein, J. L. (1990) Low density lipoprotein receptor-related protein mediates endocytosis of monoclonal antibodies in cultured cells and rabbit liver, *J. Biol. Chem.* **265**, 21355-21362.

1545. Strickland, D. K., Ashcom, J. D., Williams, S., Burgess, W. H., Migliorini, M., and Argraves, W. S. (1990) Sequence identity between the α_2-macroglobulin receptor and low density lipoprotein receptor-like protein suggests that this molecule is a multifunctional receptor, *J. Biol. Chem.* **265**, 17401-17404.

1546. Kristensen, T., Moestrup, S. K., Gliemann, J., Bendtsen, L., Sand, O., and Sottrup-Jensen, L. (1990) Evidence that newly cloned low-density-lipoprotein receptor related protein (LRP) is the α_2-macroglobulin receptor, *FEBS Lett.* **276**, 151-155.

1547. Sottrup-Jensen, L. (1989) α_2-Macroglobulins: structure, shape, and mechanism of protease complex formation, *J. Biol. Chem.* **264**, 11539-11542.

1548. Willingham, M. C., Maxfield, F. R., and Pastan, I. H. (1979) α_2-Macroglobulin binds to the plasma membranes of cultured fibroblasts: diffuse binding followed by clustering in the coated regions, *J. Cell Biol.* **82**, 614-625.

1549. Kaplan, J., and Nielsen, M. L. (1979) Analysis of macrophage surface receptors. I. Binding of α-macroglobulin-protease complexes in rabbit alveolar macrophages, *J. Biol. Chem.* **254**, 7323-7328.

1550. Moestrup, S. K., Holtet, T. L., Etzerodt, M., Thøgersen, H. C., Nykjær, A., Andreasen, P. A., Rasmussen, H. H., Sottrup-Jensen, L., and Gliemann, J. (1993) α_2-Macroglobulin-proteinase complexes, plasminogen activator inhibitor type-1-plasminogen activator complexes, and receptor-associated protein bind to a region of the α_2-macroglobulin receptor containing a cluster of 8 complement-type repeats, *J. Biol. Chem.* **268**, 13691-13696.

1551. Nielsen, K. L., Holtet, T. L., Etzerodt, M., Moestrup, S. K., Gliemann, J., Suttrup-Jensen, L., and Thøgersen, H. C. (1996) Identification of residues in α-macroglobulins important for binding to the α_2-macroglobulin receptor/low density lipoprotein receptor-related protein, *J. Biol. Chem.* **271**, 12909-12912.

1552. Kounnas, M. Z., Chappell, D. A., Wong, H., Argraves, W. S., and Strickland, D. K. (1995) The cellular internalization and degradation of hepatic lipase is mediated by low density lipoprotein receptor-related protein and requires cell surface proteoglycans, *J. Biol. Chem.* **270**, 9307-9312.

1553. Nykjær, A., Petersen, C. M., Møller, B., Jensen, P. H., Moestrup, S. K., Holtet, T. L., Etzerodt, M., Thøgersen, H. C., Munch, M., Andreasen, P. A., and Gliemann, J. (1992) Purified α_2-macroglobulin receptor/LDL receptor-related protein binds urokinase.plasminogen activator inhibitor type I complex: evidence that the α_2-macroglobulin receptor mediates cellular degradation of urokinase receptor-bound complexes, *J. Biol. Chem.* **267**, 14543-14546.

1554. Bu, G., Williams, S., Strickland, D. K., and Schwartz, A. L. (1992) Low density lipoprotein receptor-related protein/α_2-macroglobulin receptor is an hepatic receptor for tissue-type plasminogen activator, *Proc. Natl. Acad. Sci. U.S.A.* **89**, 7427-7431.

1555. Orth, K., Madison, E. L., Gething, M.-J., Sambrook, J. F., and Herz, J. (1992) Complexes of tissue-type plasminogen activator and its serpin inhibitor plasminogen-activator inhibitor type 1 are internalized by means of the low density lipoprotein receptor-related protein/α_2-macroglobulin receptor, *Proc. Natl. Acad. Sci. U.S.A.* **89**, 7422-7426.

1556. Huettinger, M., Retzek, H., Hermann, M., and Goldenberg, H. (1992) Lactoferrin specifically inhibits endocytosis of chylomicron remnants but not α_2-macroglobulin, *J. Biol. Chem.* **267**, 18551-18557.

1557. Godyna, S., Liau, G., Popa, I., Stefansson, S., and Argraves, W. S. (1995) Identification of the low density lipoprotein receptor-related protein (LRP) as an endocytic receptor for thrombospondin-1, *J. Cell Biol.* **129**, 1403-1410.

1558. Strickland, D. K., Ashcom, J. D., Williams, S., Battey, F., Behre, E., McTigue, K., Battey, J. F., and Argraves, W. S. (1991) Primary structure of α_2-macroglobulin receptor-associated protein: human homologue of a Heymann nephritis antigen, *J. Biol. Chem.* **266**, 13364-13369.

1559. Moestrup, S. J., and Gliemann, J. (1991) Analysis of ligand recognition by the purified α_2-macroglobulin receptor (low density lipoprotein receptor-related protein): evidence that high affinity of α_2-macroglobulin-proteinase complex is achieved by binding to adjacent receptors, *J. Biol. Chem.* **266**, 14011-14017.

1560. Kounnas, M. Z., Argraves, W. S., and Strickland, D. K. (1992) The 39-kDa receptor-associated protein interacts with two members of the low density lipoprotein receptor family, α_2-macroglobulin receptor and glycoprotein 330, *J. Biol. Chem.* **267**, 21162-21166.

1561. Van Leuven, F., Hilliker, C., Serneels, L., Umans, L., Overbergh, L., De Strooper, B., Fryns, J. P., and Van den Berghe, H. (1995) Cloning, characterization, and chromosomal localization to 4p16 of the human gene (LRPAP1) coding for the α_2-macroglobulin receptor-associated protein and structural comparison with the murine gene coding for the 44-kDa heparin-binding protein, *Genomics* **25**, 492-500.

1562. Hiesberger, T., Hodits, R., Ullrich, R., Exner, M., Kerjaschki, D., Schneider, W. J., and Nimpf, J. (1996) Receptor-associated protein and members of the low density lipoprotein family share a common epitope, *J. Biol. Chem.* **271**, 28792-28797.

1563. Howard, G. C., Roberts, B. C., Epstein, D. L., and Pizzo, S. V. (1996) Characterization of α_2-macroglobulin binding to human trabecular meshwork cells: presence of the α_2-macroglobulin signaling receptor, *Arch. Biochem. Biophys.* **333**, 19-26.

1564. Petersen, C. M., Ellgaard, L., Nykjær, A., Vilhardt, F., Vorum, H., Thørgersen, H. C., Nielsen, M. S., Jacobsen, C., Moestrup, S. K., and Gliemann, J. (1996) The receptor-associated protein (RAP) binds calmodulin and is phosphorylated by calmodulin-dependent kinase II, *EMBO J.* **15**, 4165-4173.

1565. Shakibaei, M., and Frevert, U. (1996) Dual interaction of the malaria circumsporozoite protein with the low density lipoprotein receptor-related protein (LRP) and heparan sulfate proteoglycans, *J. Exp. Med.* **184**, 1699-1711.

1566. Herz, J., Goldstein, J. L., Strickland, D. K., Ho, Y. K., and Brown, M. S. (1991) 39-kDa protein modulates binding of ligands to low density lipoprotein receptor-related protein/α_2-macroglobulin receptor, *J. Biol. Chem.* **266**, 21232-21238.

1567. Williams, S. E., Ashcom, J. D., Argraves, W. S., and Strickland, D. K. (1992) A novel mechanism for controlling the activity of α_2-macroglobulin receptor/low density lipoprotein receptor-related protein: multiple regulatory sites for 39-kDa receptor-associated protein, *J. Biol. Chem.* **267**, 9035-9040.

1568. Battey, F. D., Gåfvels, M. E., FitzGerald, D. J., Argraves, W. S., Chappell, D. A., Strauss, J. F. I., and K., S. D. (1994) The 39-kDa receptor-associated protein regulates ligand binding by the very low density lipoprotein receptor, *J. Biol. Chem.* **269**, 23268-23273.

1569. Bu, G., and Rennke, S. (1996) Receptor-associated protein is a folding chaperone for low density lipoprotein recetpor-related protein, *J. Biol. Chem.* **271**, 22218-22224.

1570. Kerjaschki, D., and Farquhar, M. G. (1983) The pathogenic antigen of Heymann nephritis is a membrane glycoprotein of the renal proximal tubule brush border, *Proc. Natl. Acad. Sci. U.S.A.* **79**, 5557-5561.

1571. Chatelet, F., Brianti, E., Ronco, P., Roland, J., and Verroust, P. (1986) Ultrastructural localization by monoclonal antibodies of brush-border antigens expressed by glomeruli. II. Extrarenal distribution, *Am. J. Pathol.* **122**, 512-519.

1572. Saito, A., Pietromonaco, S., Loo, A. K.-C., and Farquhar, M. G. (1994) Complete cloning and sequencing of rat gp330/'megalin', a distinct member of the low-density lipoprotein receptor gene family, *Proc. Natl. Acad. Sci. U.S.A.* **91**, 9725-9729.

1573. Buc-Caron, M. H., Condamine, H., and Kerjaschki, D. (1987) Rat Heymann nephritis antigen is closely related to brushin, a glycoprotein present in early mouse embryo epithelia, *Ann. Inst. Pasteur Immunol.* **138**, 707-722.

1574. Raychowdhury, R., Niles, J. H., McCluskey, R. T., and Smith, J. A. (1989) Autoimmune target in Heymann nephritis is a glycoprotein with homology to the LDL receptor, *Science* **244**, 1163-1165.

1575. Hjälm, G., Murray, E., Crumley, G., W., H., Lundgren, S., Onyango, I., Ek, B., Larsson, M., Juhlin, C., Hellman, P., Davis, H., Åkerstöm, G., Rask, L., and Morse, B. (1996) Cloning and sequencing of human gp330, a Ca^{2+}-binding receptor with potential intracellular signaling properties, *Eur. J. Biochem.* **239**, 132-137.

1576. Orlando, R. A., Kerjaschki, D., Kurihara, H., Biemesderfer, D., and Farquhar, M. G. (1992) gp330 associates with a 44-kDa protein in the rat kidney to form the Heymann nephritis antigen complex, *Proc. Natl. Acad. Sci. U.S.A.* **89**, 6698-6702.

1577. Kanalas, J. J., and Makker, S. P. (1991) Identification of the rat Heymann nephritis autoantigen (GP330) as a receptor site for plasminogen, *J. Biol. Chem.* **266**, 10825-10829.

1578. Kounnas, M. Z., Chappell, D. A., Strickland, D. K., and Argraves, W. S. (1993) Glycoprotein 330, a member of the low density lipoprotein receptor family, binds lipoprotein lipase *in vitro*, *J. Biol. Chem.* **268**, 14176-14181.

1579. Takahashi, S., Kawarabayasi, Y., Nakai, T., Sakai, K., and Yamamoto, T. (1992) Rabbit very low density lipoprotein receptor: a low density lipoprotein receptor-like protein with distinct ligand specificity, *Proc. Natl. Acad. Sci. U.S.A.* **89**, 9252-9256.

1580. Argraves, K. M., Battey, F. D., MacCalman, C. D., McCrae, K. R., Gåfvels, M., Kozarsky, K. F., Chappell, D. A., Strauss, J. F. I., and Strickland, D. K. (1995) The very low density lipoprotein receptor mediates the cellular catabolism of lipoprotein lipase and urokinase-plasminogen activator inhibitor type I complexes, *J. Biol. Chem.* **270**, 26550-26557.

1581. Kim, D.-H., Iijima, H., Goto, K., Sakai, J., Ishii, H., Kim, H.-J., Suzuki, H., Kondo, H., Saeki, S., and Yamamoto, T. (1996) Human apolipoprotein E receptor 2: a novel lipoprotein receptor of the low density lipoprotein receptor family predominantly expressed in brain, *J. Biol. Chem.* **271**, 8373-8380.

1582. Novak, S., Hiesberger, T., Schneider, W. J., and Nimpf, J. (1996) A new low density lipoprotein receptor homologue with 8 ligand binding repeats in brain of chicken and mouse, *J. Biol. Chem.* **271**, 11732-11736.

1583. Jacobsen, L., Madsen, P., Moestrup, S., Lund, A. H., Tommerup, N., Nykjær, A., Sottrup-Jensen, L., Gliemann, J., and Petersen, C. M. (1996) Molecular characterization of a novel human hydrid-type receptor that binds the α_2-macroglobulin receptor-associated protein, *J. Biol. Chem.* **271**, 31379-31383.

1584. Mahley, R. W., Innerarity, T. L., Rall, J., S. C., and Weisgraber, K. H. (1984) Plasma lipoproteins: apolipoprotein structure and function, *J. Lipid Res.* **25**, 1277-1294.

1585. Breslow, J. L. (1987) Lipoprotein genetics and molecular biology, in *Plasma Lipoproteins*, ed. Gotto, A. M., Elsevier, Amsterdam, pp. 359-397.

1586. Chen, S.-H., Yang, C.-Y., Chen, P.-F., Setzer, D., Tanimura, M., Li, W.-H., Gotto, J., A. M., and Chan, L. (1986) The complete cDNA and amino acid sequence of human apolipoprotein B-100, *J. Biol. Chem.* **261**, 12918-12921.

1587. Scanu, A. M., and Fless, G. M. (1990) Lipoprotein (a): heterogeneity and biological relevance, *J. Clin. Invest.* **85**, 1709-1715.

1588. Young, S. G., Bertics, S. J., Scott, T. M., Bubois, B. W., Curtiss, L. K., and Witztum, J. L. (1986) Parallel expression of the M19 genetic polymorphism in apoprotein B-100 and apoprotein B-48, *J. Biol. Chem.* **261**, 2995-2998.

1589. Greeve, J., Altkemper, I., Dieterich, J.-H., Greten, H., and Windler, E. (1993) Apolipoprotein B mRNA editing in 12 different mammalian species: hepatic expression is reflected in low concentrations of apoB-containing plasma lipoproteins, *J. Lipid Res.* **34**, 1367-1383.

1590. Milne, R., Theolis, R. J., Maurice, R., Pease, R. J., Weech, P. J., Rassart, E., Fruchart, J. C., Scott, J., and Marcel, Y. L. (1989) The use of monoclonal antibodies to localize the low density lipoprotein receptor-binding domain of apolipoprotein B, *J. Biol. Chem.* **264,** 19754-19760.

1591. Ebert, D. L., Maeda, N., Lowe, S. W., Hasler-Rapacz, J., Rapacz, J., and Attie, A. D. (1989) Primary structure comparison of the proposed low density lipoprotein (LDL) receptor binding domains of human and pig apolipoprotein B: implications for LDL-receptor interactions, *J. Lipid Res.* **29,** 1501-1509.

1592. Sivaram, P., Choi, S. Y., Curtiss, L. K., and Goldberg, I. J. (1994) An amino-terminal fragment of apolipoprotein B binds to lipoprotein lipase and may facilitate its binding to endothelial cells, *J. Biol. Chem.* **269,** 9409-9412.

1593. Choi, S. Y., Sivaram, P., Walker, D., Curtiss, L. K., Gretch, D. G., Sturley, S. L., Atti., D., A., Deckelbaum, R. J., and Goldberg, I. J. (1995) Lipoprotein lipase association with lipoproteins involves protein-protein interactions with apolipoprotein B, *J. Biol. Chem.* **270,** 8081-8086.

1594. Weisgraber, K. H., and Rall Jr., S. C. (1987) Human apolipoprotein B-100 heparin-binding sites, *J. Biol. Chem.* **262,** 11097-11103.

1595. Hirose, N., Blankenship, D. T., Krivanek, M. A., Jackson, R. L., and Cardin, A. D. (1987) Isolation and characterization of four heparin-binding cyanogen bromide peptides of human plasma apolipoprotein B, *Biochemistry* **26,** 5505-5512.

1596. Cardin, A. D., Witt, K. R., and Jackson, R. L. (1984) Visualization of heparin-binding proteins by ligand blotting with ^{125}I-heparin, *Anal. Biochem.* **137,** 368-373.

1597. Goldstein, J. L., Basu, S. K., Brunschede, G. Y., and Brown, M. S. (1976) Release of low density lipoprotein from its cell surface receptor by sulfated glycosaminoglycans, *Cell* **7,** 85-95.

1598. Mahley, R. W., Weisgraber, K. H., and Innerarity, T. L. (1979) Interaction of plasma lipoproteins containing apolipoproteins B and E with heparin and cell surface receptors, *Biochim. Biophys. Acta.* **575,** 81-91.

1599. Weisgraber, K. H., Rall Jr., S. C., and Mahley, R. W. (1981) Human E apolipoprotein heterogeneity: cysteine-arginine interchanges in the amino acid sequence of the apo-E isoforms, *J. Biol. Chem.* **256,** 9077-9083.

1600. Rall Jr., S. C., Weisgraber, K. H., and Mahley, R. W. (1982) Human apolipoprotein E: the complete amino acid sequence, *J. Biol. Chem.* **257,** 4171-4178.

1601. Zannis, V. I., McPherson, J., Goldberger, G., Karathanasis, S. K., and Breslow, J. L. (1984) Synthesis, intracellular processing, and signal peptide of human apolipoprotein E, *J. Biol. Chem.* **259,** 5495-5499.

1602. McLean, J. W., Elshourbagy, N. A., Chang, D. J., Mahley, R. W., and Taylor, J. M. (1984) Human apolipoprotein E mRNA. cDNA cloning and nucleotide sequencing of a new variant, *J. Biol. Chem.* **259,** 6498-6504.

1603. Roth, R. I., Jackson, R. L., Pownal, H. J., and Gotto Jr., A. M. (1977) Interaction of plasma "arginine-rich" apoliprotein with dimyristoylphosphatidylcholine, *Biochemistry* **16,** 5030-5036.

1604. Lucas, M., and Mazzone, T. (1996) Cell surface proteoglycans modulate net synthesis and secretion of macrophage apolipoprotein E, , *J. Biol. Chem.* **271,** 13454-13460.

1605. Rall Jr., S. C., Weisgraber, K. H., and Mahley, R. W. (1986) Isolation and characterization of apolipoprotein E, *Methods Enzymol.* **128,** 273-287.

1606. Quarfordt, S. H., and Shelburne, F. A. (1981) Human serum apoliprotein binding to heparin, in *Chemistry and Biology of Heparin,* eds., Lundblad, R. L., Brown, W. V., Mann, K. G. & Roberts, H. R., Elsevier North Holland, Amsterdam, pp. 207-215.

1607. Weisgraber, K. H., and Mahley, R. W. (1986) Characterization of apoliprotein E-containing lipoproteins, *Methods Enzymol.* **129,** 145-166.

1608. Weisgraber, K. H., Rall Jr., S. C., Mahley, R. W., Milne, R. W., Marcel, Y. L., and Sparrow, J. T. (1986) Human apolipoprotein E: determination of the heparin binding sites of apolipoprotein E3, *J. Biol. Chem.* **261,** 2068-2076.

1609. Cardin, A. D., Hirose, N., Blankenship, D. T., Jackson, R. L., and Harmony, J. A. K. (1986) Binding of a high reactive heparin to human apolipoprotein E: identification of two heparin-binding domains, *Biochem. Biophys. Res. Commun.* **134**, 783-789.

1610. Cardin, A. D., Jackson, R. L., Sparrow, D. A., and Sparrow, J. T. (1989) Interaction of glycosaminoglycans with lipoproteins, in *Heparin and Related Polysaccharides: Structure and Activities, Annals of the New York Academy of Sciences*, eds., Ofosu, F. A., Danishefsky, I. & Hirsh, J., Ann. N. Y. Acad. Sci., New York, Vol. 556, pp. 186-193.

1611. Innerarity, T. L., Friedlander, E., Rall Jr., S. C., H., W. K., and Mahley, R. W. (1983) The receptor-binding domain of human apolipoprotein E: binding of apolipoprotein E fragments, *J. Biol. Chem.* **258**, 12341-12347.

1612. Weisgraber, K. H., Innerarity, T. L., Harder, K. J., Mahley, R. W., Milne, R. W., Marcel, Y. L., and Sparrow, J. T. (1983) The receptor binding domain of human apolipoprotein E: monoclonal antibody inhibition of binding, *J. Biol. Chem.* **258**, 12348-12354.

1613. Lalazar, A., Weisgraber, K. H., Rall Jr., S. C., Giladi, H., Innerarity, T. L., Levanon, A. Z., Boyles, J. K., Amit, B., Gorecki, M., Mahley, R. W., and Vogel, T. (1988) Site-specific mutagenesis of human apolipoprotein E: receptor binding activity of variants with single amino acid substitutions, *J. Biol. Chem.* **263**, 3542-3545.

1614. Lalazar, A., Ignatius, S.-H., and Mahley, R. W. (1989) Human apolipoprotein E: receptor binding activity of truncated variants with carboxyl-terminal deletions, *J. Biol. Chem.* **264**, 8447-8450.

1615. Innerarity, T. L., Pitas, R. E., and Mahley, R. W. (1986) Lipoprotein-receptor interactions, *Methods Enzymol.* **129**, 542-566.

1616. Mahley, R. (1996) Heparan sulfate proteoglycan/low density lipoprotein receptor-related protein pathway involved in type III hyperlipoproteinemia and Alzheimer's disease, *Isr. J. Med. Sci.* **32**, 414-429.

1617. Ji, Z.-S., Fazio, S., Lee, Y.-L., and Mahley, R. W. (1994) Secretion-capture role for apolipoprotein E in remnant lipoprotein metabolism involving cell surface heparan sulfate proteoglycans, *J. Biol. Chem.* **269**, 2764-2772.

1618. van Vlijmen, B. J. M., van Dijk, K. M., van't Hof, H. B., van Gorp, P. J. J., van der Zee, A., van der Boom, H., Breuer, M. L., Hofker, M. H., and Havekes, L. M. (1996) In the absence of endogenous mouse apolipoprotein E, apolipoprotein E*2(Arg-158-Cys) transgenic mice develop more severe hyperlipoproteinemia than apolipoprotein E*3-Leiden transgenic mice, *J. Biol. Chem.* **271**, 30595-30602.

1619. Takahashi, S., Suzuki, J., Kohno, M., Oida, K., Tamai, T., Miyabo, S., Yamamoto, T., and Nakai, T. (1995) Enhancement of the binding of triglyceride-rich lipoproteins to the very low density lipoprotein receptor by apolipoprotein E and lipoprotein lipase, *J. Biol. Chem.* **270**, 15747-15754.

1620. Beisiegel, U., Weber, W., Ihrke, G., Herz, J., and Stanley, K. K. (1989) The LDL-receptor-related protein, LRP, is an apolipoprotein E-binding protein, *Nature* **341**, 162-164.

1621. Hussain, M. M., Maxfield, F. R., Más-Oliva, J., Tabas, I., Ji, Z.-S., Innerarity, T. L., and Mahley, R. W. (1991) Clearance of chylomicron remnants by the low density lipoprotein receptor-related protein/α_2-macroglobulin receptor, *J. Biol. Chem.* **268**, 139367-13940.

1622. Ji, Z.-S., Brecht, W. J., Miranda, R. D., Hussain, M. M., Innerarity, T. L., and Mahley, R. W. (1993) Role of heparan sulfate proteoglycans in the binding and uptake of apolipoprotein E-enriched remnant lipoproteins by cultured cells, *J. Biol. Chem.* **268**, 10160-10167.

1623. Shamano, H., Fgukazawa, C., Shibasaki, Y., Mori, N., Gotoda, T., Harada, K., Shimada, K., Yamada, M., Yazaki, N., and Takaku, F. (1991) The effect of apo E secretion on lipoprotein uptake in transfected cells, *Biochim. Biophys. Acta* **1086**, 245-254.

1624. Ji, Z.-S., Sanan, D. A., and Mahley, R. W. (1995) Intravenous heparinase inhibits remnant lipoprotein clearance from the plasma and uptake by the liver: *in vivo* role of heparan sulfate proteoglycans, *J. Lipid Res.* **36**, , 583-592.

1625. Felts, J. M., Itakura, H., and Crane, R. T. (1975) The mechanism of assimilation of constituents of chylomicrons, very low density lipoproteins and remnants: a new theory, *Biochem. Biophys. Res. Commun.* **66**, 1467-1475.

1626. Saxena, U., Ferguson, E., and Bisgaier, C. L. (1993) Apolipoprotein E modulates low density lipoprotein retention by lipoprotein lipase anchored to the subendothelial matrix, *J. Biol. Chem.* **268**, 14812-14829.

1627. Mulder, M., Lombardi, P., Jansen, H., van Berkel, T. J. C., Frants, R. R., and Hovekes, L. M. (1993) Low density lipoprotein receptor internalizes low density and very low density lipoproteins that are bound to heparan sulfate proteoglycans via lipoprotein lipase, *J. Biol. Chem.* **268**, 9369-9375.

1628. Obunike, K. C., Edwards, I. J., Rumsey, S. C., Curtiss, L. K., Wagner, W. D., Deckelbaum, R. J., and Goldberg, I. J. (1994) Cellular differences in lipoprotein lipase-mediated uptake of low density lipoproteins, *J. Biol. Chem.* **269**, 13129-13135.

1629. Fernández-Borja, M., Bellido, D., Vilella, E., Olivecrona, G., and Vilaró, S. (1996) Lipoportein lipase-mediated uptake of lipoprotein in human fibroblasts: evidence for and LDL receptor-independent internalization pathway, *J. Lipid Res.* **37**, 464-481.

1630. van Barlingen, H. H. J. J., de Jong, H., Erkelens, D. W., and de Bruin, T. W. A. (1996) Lipoprotein lipase-enhanced binding of human triglyceride-rich lipoproteins to heparan sulfate: modulation by apolipoprotein E and apolipoprotein C, *J. Lipid Res.* **37**, 754-763.

1631. Chang, S., Maeda, N., and Borensztajn, J. (1996) The role of lipoprotein lipase and apoprotein E in the recognition of chylomicrons and chylomicron remnants by cultured isolated mouse hepatocytes, *Biochem. J.* **318**, 29-34.

1632. Komaromy, M., Azhar, S., and Cooper, A. D. (1996) Chinese hamster ovary cells expressing a cell surface-anchored form of hepatic lipase. Characterization of low density lipoprotein and chylomicron remnant uptake and selective uptake of high density lipoprotein-cholesteryl ester, *J. Biol. Chem.* **271**, 16906-16914.

1633. Shafi, S., Brady, S. E., Bensadoun, A., and Havel, R. J. (1994) Role of hepatic lipase in the uptake and processing of chylomicron remnants in rat liver, *J. Lipid Res.* **35**, 709-720.

1634. Ji, Z.-S., Lauer, S. J., Fazio, S., Bensadoun, A., Taylor, J. M., and Mahley, R. W. (1994) Enhanced binding and uptake of remnant lipoproteins by hepatic lipase-secreting hepatoma cells in culture, *J. Biol. Chem.* **269**, 13429-13436.

1635. Chappell, D. A., Inoue, I., Fry, G. L., Pladet, M. W., Bowen, S. L., Iverius, P.-H., Lalouel, J.-M., and Strickland, D. K. (1994) Cellular catabolism of normal very low density lipoproteins via the low density lipoprotein receptor-related protein/α_2-macroglobulin receptor is induced by the C-terminal domain of lipoprotein lipase, *J. Biol. Chem.* **269**, 18001-18006.

1636. Williams, S. E., Inoue, I., Tran, H., Fry, G. L., Pladet, M. W., Iverius, P.-H., Lalouel, J.-M., Chappell, D. A., and Strickland, D. K. (1994) The carboxyl-terminal domain of lipoprotein lipase binds to the low density lipoprotein receptor-related protein/α_2-macroglobulin receptor (LRP) and mediates binding of normal very low density lipoproteins to LRP, *J. Biol. Chem.* **269**, 8653-8658.

1637. Nykjær, A., Nielsen, M., Lookene, A., Meyer, N., Røgaard, H., Etzerodt, M., Beisiegel, U., Bengtsson-Olivecrona, G., and Gliemann, J. (1994) A carboxy-terminal fragment of lipoprotein lipase binds to the low density lipoprotein receptor-related protein and inhibits lipase-mediated uptake of lipoprotein in cells, *J. Biol. Chem.* **268**, 31747-31755.

1638. Rensen, P. C. N., and van Berkel, T. J. C. (1996) Apolipoprotein E effectively inhibits lipoprotein lipase-mediated lipolysis of chylomicron-like triglyceride-rich lipid emulsions *in vitro* and *in vivo*, *J. Biol. Chem.* **271**, 14791-14799.

1639. Ji, Z.-S., Lauer, S. J., Fazio, S., Bensadoun, A., Taylor, J. M., and Mahley, R. W. (1994) Enhanced binding and uptake of remnant lipoproteins by HL-secreting hepatoma cells in culture, *J. Biol. Chem*. **269**, 13429-13436.

1640. Beisegel, U. (1995) Receptors for triglyceride-rich lipoproteins and their role in lipoprotein metabolism, *Curr. Opin. Lipidol.* **6**, 117-122.

1641. Dasso, L. T., and Taylor, C. W. (1991) Heparin and other polyanions uncouple α_1-adrenoceptors from G-proteins, *Biochem. J.* **280**, 791-795.

1642. Merenmies, J., Pihlaskari, R., Laitinen, J., Wartiovaara, J., and Rauvala, H. (1991) 30-kDa heparin-binding protein of brain (amphoterin) involved in neurite outgrowth: amino acid sequence and localization in the filopodia of the advancing plasma membrane, *J. Biol. Chem.* **266**, 16722-13729.

1643. Rupert, R., Hoffmann, E., and Sebald, W. (1996) Human bone morphogenetic protein 2 contains a heparin-binding site which modifies its biological activity, *Eur. J. Biochem.* **237**, 295—302.

1644. Calvete, J. J., Mann, K., Sanz, L., Raida, M., and Töpfer-Petersen, E. (1996) The primary structure of BSP-30K, a major lipid-, gelatin-, and heparin-binding glycoprotein of bovine seminal plasma, *FEBS Lett.* **399**, 147-152.

1645. Hessing, M., Vlooswijk, R. A. A., Hackeng, T. M., Kanters, D., and Bouma, B. N. (1990) the localization of the heparin-binding fragments on human C4b-binding protein, *J. Immunol.* **144**, 204-208.

1646. Ferrer-Lopez, P., Renesto, P., Prevost, M.-C., Goudon, P., and Chignard, M. (1991) Heparin inhibits neutrophil-induced platelet activation via chathepsin G, *J. Lab. Clin. Med.* **119**, 231-239.

1647. San Antonio, J. D., Lander, A. D., Karnovsky, M. J., and Slayter, H. S. (1994) Mapping the heparin-binding sites on type I collagen monomers and fibrils, *J. Cell Biol.* **125**, 1179-1188.

1648. Specks, U., Mayer, U., Nischt, R., Spissinger, T., Mann, K., Timpl, R., Engel, J., and Chu, M.-L. (1992) Structure of recombinant N-terminal gobule of type VI collagen $\alpha3$ chain and its binding to heparin and hyaluronan, *EMBO J.* , 4821-4890.

1649. Blackmore, T. K., Sadlon, T. A., Ward, H. M., Lublin, D. M., and Gordon, D. L. (1996) Identification of a heparin binding domain in the seventh short consensus repeat of complement factor H[1], *J. Immunol.* **157**, 5422-.

1650. Pangburn, M. K., Atkinson, M. A. L., and Meri, S. (1991) Localization of the heparin-binding site on complement factor H*, *J. Biol. Chem.* **266**, 16847-16853.

1651. Bágel'ová, J., Antalík, M., and Bona, M. (1994) Studies on cytochrome c-heparin interactions by differential scanning calorimetry, *Biochem. J.* **297**, 99-101.

1652. Hansson, R., Holmberg, C. G., Tibbling, G., Tryding, N., Westling, H., and Wetterquist, H. (1996) *Acta Med. Scand.* **180**, 533-536.

1653. Robinson-White, A., Baylin, S. B., Olivecrona, T., and Beaven, M. A. (1980) J. Clin. Invest. 76, 93- (1985) Binding of diamine oxidase active to rat and guinea pig microvascular endothelial cells: comparisons with lipoprotein lipase binding, *J. Clin. Invest.* **76**, 93-100.

1654. Frommherz, K. J., Faller, B., and Bieth, J. G. (1991) Heparin strongly decreases the rate of inhibition of neutrophil elastase by α_1-proteinase inhibitor, *J. Biol. Chem.* **266**, 15356-15362.

1655. Li, Y. H., Liang, X. P., Little-van den Hurk, S. V., Attah-Poku, S., and Babiuk, L. A. (1996) Glycoprotein Bb, the N-terminal subunit of bovine herpesvirus 1 gB, can bind to heparan sulfate on the surfacet of Madin-Darby bovine kidney cells, *J. Virol.* **70**, 2032-2037.

1656. Sawitzky, D., Voigt, A., Zeichhardt, H., and Habermehl, K. O. (1996) Glycoprotein b (gB) of pseudorabies virus interacts specifically with the glycosaminoglycan heparin, *Virus Res.* **41**, 101-.

1657. Roberts, R., Gallagher, J., Spooncer, E., Allen, T. D., Bloomfield, F., and Dexter, T. M. (1988) Heparan sulphate bound growth factors: a mechanism for stromal cell mediated haemopoiesis, *Nature* **332**, 376-378.

1658. Thompson, S. A., Higashiyama, S., Wood, K., Pollitt, N. S., Damm, D., McEnroe, G., Garrick, B., Ashton, N., Lau, K., Hancock, N., Klagsbrun, M., and Abraham, J. A. (1994)

Characterization of sequences within heparin-binding EGF-like growth factor that mediate interaction with heparin, *J. Biol. Chem.* **269**, 2541-2549.

1659. Furukawa, T., Ozawa, M., Huang, R.-P., and Muramatsu, T. (1990) A heparin binding protein whose expression increases during differentiation of embryonal carcinoma cells to parietal endoderm cells: cDNA cloning and sequence anasysis, *J. Biochem. (Tokyo)* **108**, 297-302.

1660. Lyon, M., Deakin, J. A., Mizuno, K., Nakamura, T., and Gallagher, J. T. (1994) Interaction of hepatocyte growth factor with heparan sulfate: elucidation of the major heparan sulfate structural determinants, *J. Biol. Chem.* **269**, 11216-11223.

1661. Zarnegar, R., Muga, S., Rahija, R., and Michaloupoulos, G. (1990) Tissue distribution of hepatopoietin-A: a heparin-binding polypeptide growth factor for hepatocytes, *Proc. Natl. Acad. Sci. U.S.A.* **87**, 1252-1256.

1662. Liu, S., Smith, S. E., Julian, J., Rohde, L. H., Karin, N. J., and Carson, D. D. (1996) cDNA cloning and expression of HIP, a novel cell surface heparan sulfate/heparin-binding protein of human uterine epithelial cells and cell lines, *J. Biol. Chem.* **271**, 11817-11823.

1663. WuDunn, D., and Spear, P. G. (1989) Initial interaction of herpes simplex virus with cells is binding to heparan sulfate, *J. Virol.* **63**, 52-58.

1664. Herold, B. C., Visalli, R. J., Susmarski, N., Brandt, C. R., and G., S. P. (1994) Glycoprotein C-independent binding of herpes symplex virus to cells requires cell surface heparan sulphate and glycoprotein b, *J. Gen. Virol.* **75**, 1211-1222;.

1665. Jones, C. P., and Sawyer, R. T. (1989) Heparin inhibits mmmalian, but not leech, hyaluronidase, *Thromb. Res.* **55**, 791-796.

1666. Arai, T., Clarke, J., Parker, A., Busby Jr., W., Nam, T., and Clemmons, D. R. (1996) Substitution of specific amino acids in insulin-like growth factor (IGF) binding protein 5 alters heparin binding and its change in affinity for IGF-I in response to heparin, *J. Biol. Chem.* **271**, 6099-6106.

1667. Arai, T., Parker, A., Busby Jr., W., and Clemmons, D. R. (1994) Heparin, heparan sulfate, and dermatan sulfate regulate formation of the insulin-like gorwth factor-I and insulin-like growth factor-binding protein complexes, *J. Biol. Chem.* **269**, 20388-20393.

1668. Lortat-Jacob, H., Turnbull, J. E., and Grimaud, J.-A. (1995) Molecular organization of the interferon-γ-binding domain in heparan sulphate, *Biochem. J.* **310**, 497-550.

1669. Björk, I., Olson, S. T., Sheffer, R. G., and Shore, J. D. (1989) Binding of heparin to human high molecular weight kininogen, *Biochemistry* **28**, 1231-1221.

1670. Mahley, R. W., Ji, S.-S., Brecht, W. J., Miranda, R. D., and He, D. (1994) Biology of α_2-macroglobulin, its receptor, and related proteins, Ann. New York Acad. Sci., Vol. 737, pp. 39-52.

1671. Beck, K., Hunter, I., and Engel, J. (1990) Structure and function of laminin: anatomy of a multidomain glycoprotein, *FASEB J.* **4**, 148-160.

1672. Yurchenco, P. D., Sung, U., Ward, M. D., Yamada, Y., and O'Rear, J. J. (1993) Recombinant laminin G domain mediates myoblast adhesion and heparin binding, *J. Biol. Chem.* **268**, 8356-8365.

1673. Skinner, M. P., Fournier, D. J., Andrews, R. K., Gorman, J. J., Chesterman, C. N., and Berndt, M. C. (1989) Characterization of human platelet GMP-140 as a heparin-binding protein, *Biochem. Biophys. Res. Commun.* **164**, 1373-1379.

1674. Knaus, H.-G., Moshammer, T., Friedrich, K., Kang, H. C., Haugland, R. P., and Glossmann, H. (1992) *In vivo* labeling of L-type Ca^{2+} channels by fluorescent dihydropyridines: evidence for a functional, extracellular heparin-binding site, *Proc. Natl. Acad. Sci. U.S.A.* **89**, 3586-3590.

1675. Shakibaei, M., and Frevert, U. (1996) Dual interaction of malaria circumsporozoite protein with low density lipoprotein receptor-related proteir (LRP) and heparan sulfate proteoglycans, *J. Exp. Med.* **184**, 1699-1711.

1676. Pejler, G. (1996) Mast cell chymase in complex with heparin proteoglycan is regulated by protamine, *FEBS Lett.* **383**, 170-174.

1677. Pejler, G. (1994) Interaction of heparin with rat mast cell protease 1, *J. Biol. Chem.* **269**, 14451-14456.

1678. Kinnuenen, T., Raulo, E., Nolo, R., Maccarana, M., Lindahl, U., and Rauvala, H. (1996) Neurite outgrowth in brain neurons induced by heparin-binding growth-associated molecule (HB-GAM) depends on the specific interaction of HB-GAM with heparan sulfate at the cell surface, *J. Biol. Chem.* **271**, 2243-2248.

1679. Volpi, N., Bianchini, P., and Bolognani, L. (1991) Competitive inhibition of myosin ATPase activity by different molecular weight heparins, *Biochem. Int.* **24**, 243-253.

1680. Kallapur, S. G., and Akeson, R. A. (1992) The neural cell adhesion molecule (NCAM) heparin binding domain binds to cell surface heparan sulfate proteoglycans, *J. Neurosci. Res.* **33**, 538-548.

1681. Block, B., Normand, E., Kovesdi, I., and Böhlen, P. (1992) Expression of the HBNF (heparin-binding neurite-promoting factor gene in the brain of fetal, neonatal and adult rat: an in situ hybridization study, *Dev. Brain Res.* **70**, 267-278.

1682. Leculier, C., Benzerara, O., Couprie, N., Francina, A., Lasne, Y., Archimbaud, E., and Fiere, D. (1992) Specific binding between human neutrophils and heparin, *Brit. J. Haematol.* **81**, 81-85.

1683. Paralkar, V. M., Nandedkar, A. K. N., Pointer, R. H., Kleinman, H. K., and Reddi, A. H. (1990) Interaction of onteogenenin, a heparin binding bone morphogetetic protein, with collagen, *J. Biol. Chem.* **265**, 17281-17284.

1684. Smith, C. D., Wen, D., Mooberry, S. L., and Chang, K.-J. (1992) Inhibition of phosphatidylinositol 4-phosphate kinase by heparin: a possible mechanism for the antiproliferative effects of heparin, *Biochem. J.* **281**, 803-808.

1685. Diccianni, M. B., McLean, L. R., Stuart, W. D., Mistry, M. J., Gil, C. M., Diccianni, M. B., McLean, L. R., Stuart, W. D., Mistry, M. J., Gil, C. M., and Harmony, J. A. K. (1991) Porcine pancreatic phospholipase A2 isoforms; differential regulation by heparin, *Biochim. Biophys. Acta* **1082**, 85-93.

1686. Dua, R., and Cho, W. (1994) Inhibition of human secretory class II phospholipase A2 by heparin, *Eur. J. Biochem.* **221**, 481-490.

1687. Kohnke-Godt, B., and Gabius, H.-J. (1991) Heparin-binding lectin from human placenta: further characterization of ligand binding and structural properties and its relationship to histones and heparin-binding growth factors, *Biochemistry* **30**, 55-65.

1688. Debbage, P. L., Lange, W., Hellmann, T., and Gabius, H.-J. (1988) Detection of receptors for sulfated polysaccharides in human placenta by biotinylated probes, *J. Histochem. Cytochem.* **36**, 1097-1102.

1689. Lustig, F., Hoebeke, J., Östergren-Lundèn, G., Velge-Roussel, F., Bondjers, G., Olsson, U., Rüetschi, U., and Fager, G. (1996) Alternative splicing determines the binding of platelet-derived growth factor (PDGF-AA) the glycosaminoglycans, *Biochemistry* **35**, 12077-12085.

1690. Matuo, Y., Adams, P. S., Nishi, N., Yasumitsu, H., Crabb, J. W., Matusik, R. J., and McKeehan, W. L. (1989) The androgen-dependent rat prostate protein, probasin, is a heparin-binding protein that co-purifies with heparin-binding growth factor-1, *In Vitro Cell. & Dev. Biol.* **25**, 581-584.

1691. Raulais, D., Lagente-Chevallier, O., Guettet, C., Duprez, D., Courtois, Y., and Vigny, M. (1991) A new heparin binding protein regulated by retinoic acid from chick embryo, *Biochem. Biophys Res. Commun.* **174**, 708-715.

1692. Urios, P., Duprez, D., Le Caer, J.-P., Cortois, Y., Vigny, M., and Laurent, M. (1991) Molecular cloning of RI-HB, a heparin binding protein regulated by retinoic acid, *Biochem. Biophys. Res. Commun.* **175**, 617-624.

1693. Calvete, K. K., Dostàlovà, S., L., Adermann, K., Thole, H. H., and Töpfer-Petersen, E. (1996) Mapping the heparin-binding domain of boar spermadhesins, *FEBS Lett.* **379**, 207-211.

1694. Aukhil, I., Joshi, P., Yan, Y., and Erickson, H. P. (1993) Cell- and heparin-binding domains of the hexabrachion arm identified by tenascin expression proteins, *J. Biol. Chem.* **268,** 2542-2553.

1695. Erickson, H. P. (1993) Tenascin-C, tenascin-R and tenascin-X: a family of talented proteins in search of functions, *Curr. Opin. Cell Biol.* **5,** 869-876.

1696. McCaffrey, T. A., Falcone, D. J., Vicente, D., Du, B., Consigli, S., and Borth, W. (1994) Protection of transforming growth factor-b1 activity by heparin and fucoidan, *J. Cell. Physiol.* **159,** 51-59.

1697. Finotti, P. (1996) Separation by heparin-affinity chromatography of catalytically active and inactive forms of trypsin which retain the (NA-K)ATPase stitumating property, *Clin. Chim. Acta* **256,** 37-51.

1698. Lantz, M., Thysell, H., Nilsson, E., and Olsson, I. (1991) On the binding of tumor necrosis factor (TNF) to heparin and the release in vivo of the TNF-binding protein I by heparin, *J. Clin. Invest.* **88,** 2026-2031.

1699. Cohen, T., Gitay-Goren, H., Sharon, R., Shibuya, M., Halaban, R., Levi, B.-Z., and Neufeld, G. (1995) VEGF121, a vascular endothelial growth factor (VEGF) isoform lacking heparin bindinng ability, requires cell-surface heparan sulfates for efficient binding to the VEGF receptors of human melanoma cells, *J. Biol. Chem.* **270,** 11322-11326.

1700. Reichsman, F., Smith, L., and Cumberledge, S. (1996) Glycosaminoglycans can modulate extracellular localization of the wingless protein and promote signal transduction, , *J. Cell Biol.* **135,** 819-827.

1701. Adachi, T., Fukushima, T., Usami, Y., and Hirano, K. (1993) Binding of human xanthine oxidase to sulphated glycosaminoglycans on the endothelial-cell surface, *Biochem. J.* **289,** 523-527.

1702. Radi, R., Rubbo, H., Bush, K., and Freeman, B. A. (1997) Xanthine oxidase binding to glycosaminoglycans: kinetics and superoxide dismutase interactions of immobilized xanthine oxidase-heparin complexes, *Arch. Biochem. Biophys.* **339,** 125-135.

Index

A

G